NONLINEAR AND DISTRIBUTED CIRCUITS

NONLINEAR AND DISTRIBUTED CIRCUITS

Edited by
Wai-Kai Chen
University of Illinois
Chicago, U.S.A.

CRC Taylor & Francis
Taylor & Francis Group
Boca Raton London New York

A CRC title, part of the Taylor & Francis imprint, a member of the Taylor & Francis Group, the academic division of T&F Informa plc.

This material was previously published in *The Circuits and Filters Handbook, Second Edition.* © CRC Press LLC 2002.

Published in 2006 by
CRC Press
Taylor & Francis Group
6000 Broken Sound Parkway NW, Suite 300
Boca Raton, FL 33487-2742

Printed in the United States of America on acid-free paper
10 9 8 7 6 5 4 3 2 1

International Standard Book Number-10: 0-8493-7276-3 (Hardcover)
International Standard Book Number-13: 978-0-8493-7276-6 (Hardcover)
Library of Congress Card Number 2005050565

Library of Congress Cataloging-in-Publication Data

Nonlinear and distributed circuits / Wai-Kai Chen, editor-in-chief.
p. cm.
Includes bibliographical references and index.
ISBN 0-8493-7276-3 (alk. paper)
1. Electronic circuits. 2. Electric circuits, Nonlinear. I. Chen, Wai-Kai, 1936-

TK7867.N627 2005
621.3815--dc22 2005050565

T&F informa

Taylor & Francis Group
is the Academic Division of T&F Informa plc.

Visit the Taylor & Francis Web site at
http://www.taylorandfrancis.com

and the CRC Press Web site at
http://www.crcpress.com

Preface

The purpose of *Nonlinear and Distributed Circuits* is to provide in a single volume a comprehensive reference work covering the broad spectrum of analysis, synthesis, and design of nonlinear circuits; their representation, approximation, identification, and simulation; cellular neural networks; multiconductor transmission lines; and analysis and synthesis of distributed circuits. The book is written and developed for the practicing electrical engineers and computer scientists in industry, government, and academia. The goal is to provide the most up-to-date information in the field.

Over the years, the fundamentals of the field have evolved to include a wide range of topics and a broad range of practice. To encompass such a wide range of knowledge, the book focuses on the key concepts, models, and equations that enable the design engineer to analyze, design, and predict the behavior of nonlinear and distributed systems. While design formulas and tables are listed, emphasis is placed on the key concepts and theories underlying the processes.

The book stresses fundamental theory behind professional applications. In order to do so, it is reinforced with frequent examples. Extensive development of theory and details of proofs have been omitted. The reader is assumed to have a certain degree of sophistication and experience. However, brief reviews of theories, principles, and mathematics of some subject areas are given. These reviews have been done concisely with perception.

The compilation of this book would not have been possible without the dedication and efforts of Professors Leon O. Chua and Thomas Koryu Ishii, and, most of all, the contributing authors. I wish to thank them all.

Wai-Kai Chen
Editor-in-Chief

Editor-in-Chief

Wai-Kai Chen, Professor and Head Emeritus of the Department of Electrical Engineering and Computer Science at the University of Illinois at Chicago, is now serving as a member of the Board of Trustees at International Technological University. He received his B.S. and M.S. degrees in electrical engineering at Ohio University, where he was later recognized as a Distinguished Professor. He earned his Ph.D. in electrical engineering at the University of Illinois at Urbana/Champaign.

Professor Chen has extensive experience in education and industry and is very active professionally in the fields of circuits and systems. He has served as visiting professor at Purdue University, University of Hawaii at Manoa, and Chuo University in Tokyo, Japan. He was Editor of the *IEEE Transactions on Circuits and Systems, Series I and II*, President of the IEEE Circuits and Systems Society, and is the Founding Editor and Editor-in-Chief of the *Journal of Circuits, Systems and Computers*. He received the Lester R. Ford Award from the Mathematical Association of America, the Alexander von Humboldt Award from Germany, the JSPS Fellowship Award from Japan Society for the Promotion of Science, the Ohio University Alumni Medal of Merit for Distinguished Achievement in Engineering Education, the Senior University Scholar Award and the 2000 Faculty Research Award from the University of Illinois at Chicago, and the Distinguished Alumnus Award from the University of Illinois at Urbana/Champaign. He is the recipient of the Golden Jubilee Medal, the Education Award, the Meritorious Service Award from IEEE Circuits and Systems Society, and the Third Millennium Medal from the IEEE. He has also received more than a dozen honorary professorship awards from major institutions in China and Taiwan.

A fellow of the Institute of Electrical and Electronics Engineers and the American Association for the Advancement of Science, Professor Chen is widely known in the profession for his *Applied Graph Theory* (North-Holland), *Theory and Design of Broadband Matching Networks* (Pergamon Press), *Active Network and Feedback Amplifier Theory* (McGraw-Hill), *Linear Networks and Systems* (Brooks/Cole), *Passive and Active Filters: Theory and Implements* (John Wiley), *Theory of Nets: Flows in Networks* (Wiley-Interscience), and *The VLSI Handbook* (CRC Press) and *The Electrical Engineering Handbook* (Academic Press).

Advisory Board

Leon O. Chua
University of California
Berkeley, California

John Choma, Jr.
University of Southern California
Los Angeles, California

Lawrence P. Huelsman
University of Arizona
Tucson, Arizona

Contributors

Guanrong Chen
City University of Hong Kong
Kowloon, Hong Kong

Daniël De Zutter
Gent University
Gent, Belgium

Manuel Delgado-Restituto
Universidad de Sevilla
Sevilla, Spain

Martin Hasler
Swiss Federal Institute of Technology
Lausanne, Switzerland

Jose L. Huertas
Universidad de Sevilla
Sevilla, Spain

Thomas Koryu Ishii
Marquette University
Milwaukee, Wisconsin

Michael Peter Kennedy
University College
Dublin, Ireland

Erik Lindberg
Technical University of Denmark
Lyngby, Denmark

Luc Martens
Gent University
Gent, Belgium

Wolfgang Mathis
University of Hannover
Hannover, Germany

Csaba Rekeczky
Hungarian Academy of Sciences
Budapest, Hungary

Angel Rodríguez-Vázquez
Universidad de Sevilla
Sevilla, Spain

Tamás Roska
Hungarian Academy of Science
Budapest, Hungary

Vladimír Székely
Budapest University of Technology
and Economics
Budapest, Hungary

Lieven Vandenberghe
University of California
Los Angeles, California

F. Vidal
Universidad de Malaga
Malaga, Spain

Joos Vandewalle
Katholieke Universiteit Leuven
Leuven Heverlee, Belgium

Ákos Zarándy
Hungarian Academy of Science
Budapest, Hungary

Table of Contents

1

Qualitative Analysis

Martin Hasler
Swiss Federal Institute of Technology

1.1 Introduction

The main goal of circuit analysis is to determine the solution of the circuit, i.e., the voltages and the currents in the circuit, usually as functions of time. The advent of powerful computers and circuit analysis software has greatly simplified this task. Basically, the circuit to be analyzed is fed to the computer through some circuit description language, or it is analyzed graphically, and the software will produce the desired voltage or current waveforms. Progress has rendered the traditional paper-and-pencil methods obsolete, in which the engineer's skill and intution led the way through series of clever approximations, until the circuits equations can be solved analytically.

A closer comparison of the numerical and the approximate analytical solution reveals, however, that the two are not quite equivalent. Although the former is precise, it only provides the solution of the circuit with given parameters, whereas the latter is an approximation, but the approximate solutions most often is given explicity as a function of some circuit parameters. Therefore, it allows us to assess the influence of these parameters on the solution.

If we rely entirely on the numerical solution of a circuit, we never get a global picture of its behavior, unless we carry out a huge number of analyses. Thus, the numerical analysis should be complemented by a qualitative analysis, one that concentrates on general properties of the circuit, properties that do not depend on the particular set of circuit parameters.

1.2 Resistive Circuits

The term *resistive circuits* is not used, as one would imagine, for circuits that are composed solely of resistors. It admits all circuit elements that are not dynamic, i.e., whose constitutive relations do not involve time derivatives, integrals over time, or time delays, etc. Expressed positively, resistive circuit elements are described by constitutive relations that involve only currents and voltages at the same time instants.

Physical circuits can never be modeled in a satisfactory way by resistive circuits, but resistive circuits appear in many contexts as auxiliary constructs. The most important problem that leads to a resistive circuit is the determination of the equilibrium points, or, as is current use in electronics, the **DC-operating points**, of a dynamic circuit. The DC-operating points of a circuit correspond in a one-to-one fashion

FIGURE 1.1 Symbols of the *V*- and the *I*-resistor.

to the solutions of the resistive circuit obtained by removing the capacitors and by short circuiting the inductors. The resistive circuit associated with the state equations of a dynamic circuit is discussed in [1].

Among the resistive circuit elements we find, of course, the resistors. For the purposes of this introduction, we distinguish between, linear resistors, *V*-resistors and *I*-resistors. ***V*-resistors** are **voltage controlled**, i.e., defined by constitutive relations of the form

$$i = g(v) \tag{1.1}$$

In addition, we require that g is a continuous, increasing function of v, defined for all real v. Dually, an **I-resistor** is **current controlled**, i.e., defined by a constitutive relation of the form

$$v = h(i) \tag{1.2}$$

In addition, we require that h is a continuous, increasing function of i, defined for all real i. We use the symbols of Figure 1.1 for *V*- and *I*-resistor. Linear resistors are examples of both *I*- and *V*-resistors. An example of a *V*-resistor that is not an *I*-resistor is the junction diode, modeled by its usual exponential constitutive relation

$$i = I_s\left(e^{v/nV_T} - 1\right) \tag{1.3}$$

Although (1.3) could be solved for v and thus the constitutive relation could be written in the form (1.2), the resulting function h would be defined only for currents between $-I_s$ and $+\infty$, which is not enough to qualify for an *I*-resistor. For the same reason, the static model for a Zener diode would be an *I*-resistor, but not a *V*-resistor. Indeed, the very nature of the Zener diode limits its voltages on the negative side.

A somewhat strange by-product of our definition of *V*- and *I*-resistors is that **independent voltage sources** are *I*-resistors and **independent current sources** are *V*-resistors. Indeed, a voltage source of value E has the constitutive relation

$$v = E \tag{1.4}$$

which clearly is of the form (1.2), with a constant function h, and a current source of value I has the form

$$i = I \tag{1.5}$$

which is of the form (1.1) with a constant function g. Despite this, we shall treat the independent sources as a different type of element.

Another class of resistive elements is the **controlled sources**. We consider them to be two-ports, e.g., a **voltage-controlled voltage source** (VCVS). A VCVS is the two-port of Figure 1.2, where the constitutive relations are

$$v_1 = \alpha v_2 \tag{1.6}$$

$$i_1 = 0 \tag{1.7}$$

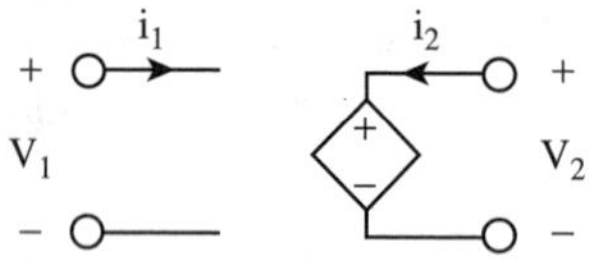

FIGURE 1.2 VCVS as a two-port.

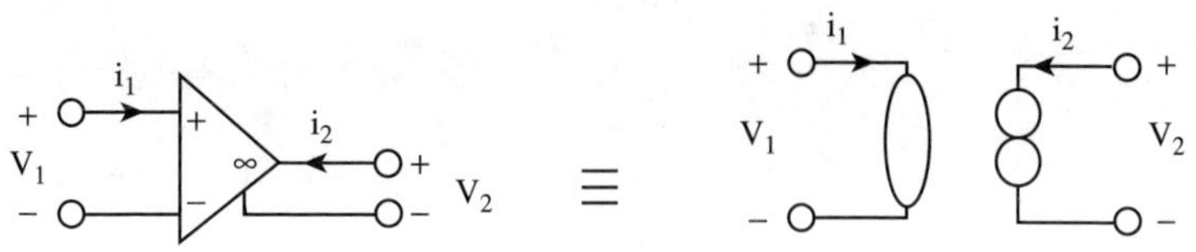

FIGURE 1.3 Operational amplifier as a juxtaposition of a nullator and a norator.

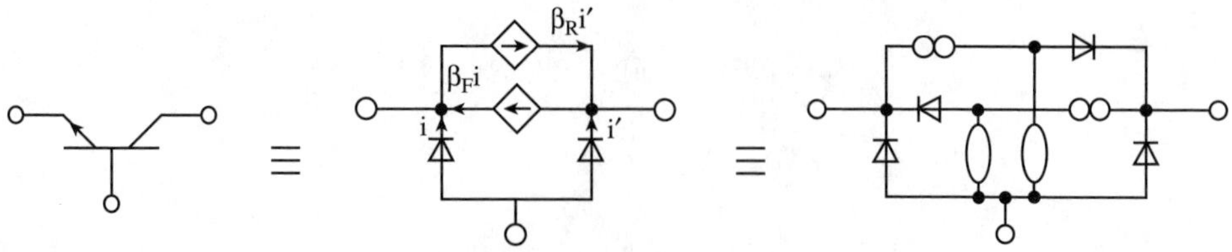

FIGURE 1.4 Equivalent circuit of a bipolar npn transistor.

The other controlled sources have similar forms. Another useful resistive circuit element is the **ideal operational amplifier**. It is a two-port defined by the two constitutive relations

$$v_1 = 0 \tag{1.8}$$

$$i_1 = 0 \tag{1.9}$$

This two-port can be decomposed into the juxtaposition of two singular one-ports, the nullator and the norator, as shown in Figure 1.3. The nullator has two constitutive relations:

$$v = 0 \quad i = 0 \tag{1.10}$$

whereas the norator has no constitutive relation.

For all practical purposes, the resistive circuit elements mentioned thus far are sufficient. By this we mean that all nonlinear resistive circuits encountered in practice possess an equivalent circuit composed of nonlinear resistors, independent and controlled sources, and nullator–norator pairs. Figure 1.4 illustrates this fact. Here, the equivalent circuit of the bipolar transistor is modeled by the Ebers–Moll equations:

$$\begin{pmatrix} i_1 \\ i_2 \end{pmatrix} = \begin{pmatrix} 1+\dfrac{1}{\beta_F} & -1 \\ -1 & 1+\dfrac{1}{\beta_R} \end{pmatrix} \begin{pmatrix} g(v_1) \\ g(v_2) \end{pmatrix} \tag{1.11}$$

The function g is given by the right-hand side of (1.3).

Actually, the list of basic resistive circuit elements given so far is redundant, and the nullator–norator pair renders the controlled sources superfluous. An example of a substitution of controlled sources by nullator–norator pairs is given in Figure 1.4. Equivalent circuits exist for all four types of controlled sources with nullator–norator pairs. Figure 1.5 gives an equivalent circuit for a **voltage-controlled current source** (VCCS), where the input port is floating with respect to the output port.

The system of equations that describes a resistive circuit is the collection of Kirchhoff equations and the constitutive relations of the circuit elements. It has the following form (if we limit ourselves to resistors, independent sources, nullators, and norators):

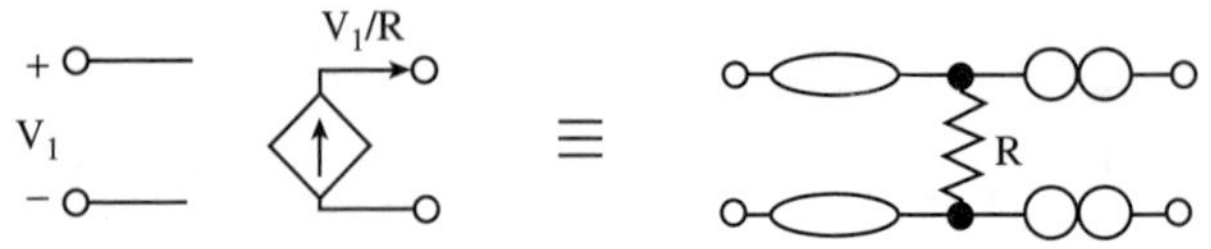

FIGURE 1.5 Equivalent circuit for a floating voltage-controlled current source.

$$\mathbf{Ai} = \mathbf{0} \quad \text{(Kirchhoff's voltage law)} \tag{1.12}$$

$$\mathbf{Bv} = \mathbf{0} \quad \text{(Kirchhoff's voltage law)} \tag{1.13}$$

$$i_k = g(v_k) \quad (V - \text{resistor}) \tag{1.14}$$

$$v_k = h(i_k) \quad (I - \text{resistor}) \tag{1.15}$$

$$v_k = E_k \quad \text{(independent voltage source)} \tag{1.16}$$

$$i_k = I_k \quad \text{(independent current source)} \tag{1.17}$$

$$\left.\begin{aligned} v_k &= 0 \\ i_k &= 0 \end{aligned}\right\} \quad \text{(nullators)} \tag{1.18}$$

In this system of equations, the unknowns are the branch voltages and the branch currents

$$\mathbf{v} = \begin{pmatrix} v_1 \\ v_2 \\ \vdots \\ v_b \end{pmatrix} \qquad \mathbf{i} = \begin{pmatrix} i_1 \\ i_2 \\ \vdots \\ i_b \end{pmatrix} \tag{1.19}$$

where the b is the number of branches. Because we have b linearly independent Kirchhoff equations [2], the system contains $2b$ equations and $2b$ unknowns. A solution $\xi = \binom{\mathbf{v}}{\mathbf{i}}$ of the system is called a solution of the circuit. It is a collection of branch voltages and currents that satisfy (1.12) to (1.19).

Number of Solutions of a Resistive Circuit

As we found earlier, the number of equations of a resistive circuit equals the number of unknowns. One may therefore expect a unique solution. This may be the norm, but it is far from being generally true. It is not even true for linear resistive circuits. In fact, the equations for a linear resistive circuit are of the form

$$\mathbf{H}\boldsymbol{\xi} = \mathbf{e} \tag{1.20}$$

where the $2b \times 2b$ matrix $\mathbf{H}$ contains the resistances and elements of value 0, ± 1, whereas the vector $\mathbf{e}$ contains the source values and zeroes. The solution of (1.20) is unique iff the determinant of $\mathbf{H}$ differs from zero. If it is zero, then the circuit has either infinitely many solutions or no solution at all. Is such a case realistic? The answer is yes and no. Consider two voltages sources connected as shown in Figure 1.6.

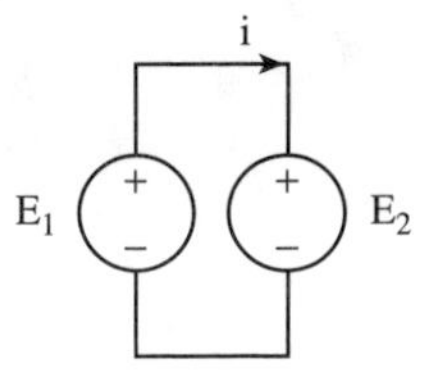

FIGURE 1.6 Circuit with zero or infinite solutions.

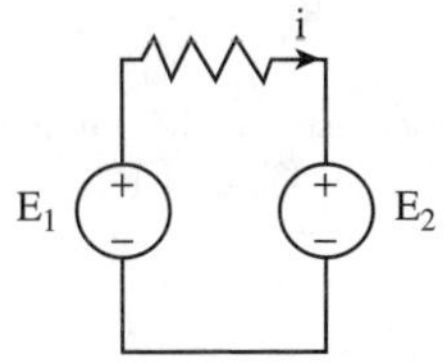

FIGURE 1.7 Circuit with exactly one solution.

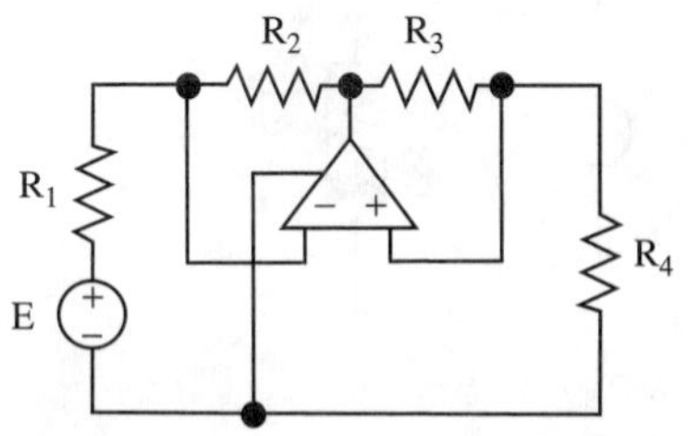

FIGURE 1.8 Circuit with one, zero, or infinite solutions.

If $E_1 \neq E_2$, the constitutive relations of the sources are in contradiction with Kirchhoff's voltage law (KVL), and thus the circuit has no solution, whereas when $E_1 = E_2$, the current i in Figure 1.6 is not determined by the circuit equations, and thus the circuit has infinitely many solutions. One may object that the problem is purely academic, because in practice wires as connections have a small, but positive, resistance, and therefore one should instead consider the circuit of Figure 1.7, which has exactly one solution.

Examples of singular linear resistive circuits exist that are much more complicated. However, the introduction of parasitic elements always permits us to obtain a circuit with a single solution, and thus the special case in which the matrix **H** in (1.9) is singular can be disregarded. Within the framework of linear circuits, this attitude is perfectly justified. When a nonlinear circuit model is chosen, however, the situation changes. An example clarifies this point.

Consider the linear circuit of Figure 1.8. It is not difficult to see that it has exactly one solution, except when

$$R_1R_3 = R_2R_4 \tag{1.21}$$

In this case, the matrix **H** in (1.29) is singular and the circuit of Figure 1.8 has zero or infinitely many solutions, depending on whether E differs from zero. From the point of view of linear circuits, we can disregard this singular case because it arises only when (1.21) is exactly satisfied with infinite precision.

Now, replace resistor R_4 by a nonlinear resistor, where the characteristic is represented by the bold line in Figure 1.9. The resulting circuit is equivalent to the connection of a voltage source, a linear resistor, and the nonlinear resistor, as shown in Figure 1.10. Its solutions correspond to the intersections of the nonlinear resistor characteristic and the load line (Figure 1.9). Depending on the value of E, either one, two, or three solutions are available. Although we still need infinite precision to obtain two solutions, this is not the case for one or three solutions. Thus, more than one DC-operating point may be observed in electronic circuits. Indeed, for static memories, and multivibrators in general, multiple DC-operating points are an essential feature.

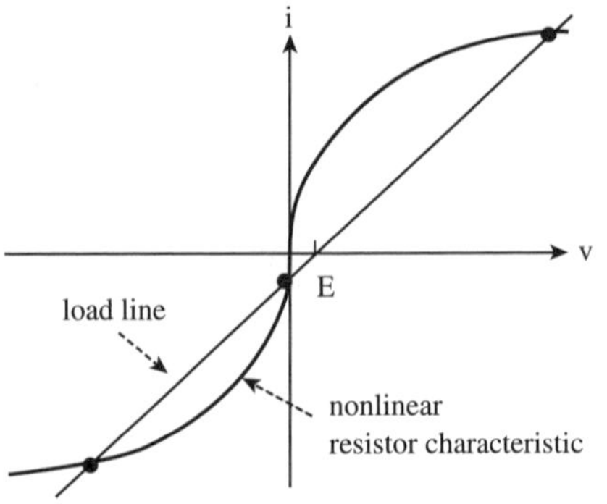

FIGURE 1.9 Characteristic of the nonlinear resistor and solutions of the circuit of Figure 1.10.

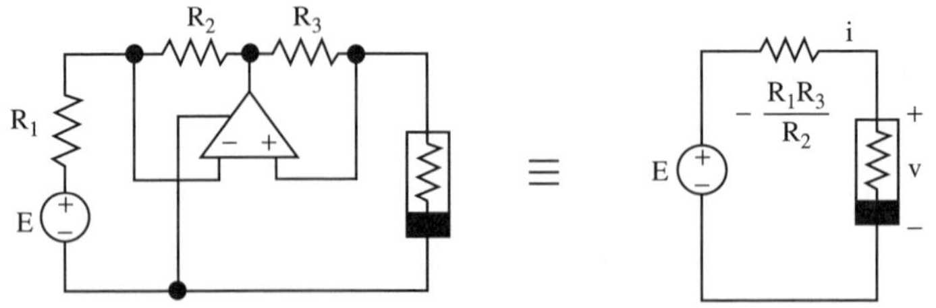

FIGURE 1.10 Circuit with one, two or three solutions.

The example of Figure 1.10 shows an important aspect of the problem. The number of solutions depends on the parameter values of the circuit. In the example the value of E determines whether one, two, or three solutions are available. This is not always the case. An important class of nonlinear resistive circuits always has exactly one solutions, irrespective of circuit parameters. In fact, for many applications, e.g., amplification, signal shaping, logic operations, etc., it is necessary that a circuit has exactly one DC-operating point. Circuits that are designed for these functionalities should thus have a unique DC-operating point for any choice of element values.

If a resistive circuit contains only two-terminal resistors with increasing characteristics and sources, but no nonreciprocal element such as controlled sources, operational amplifiers, or transistors, the solution is usually unique. The following theorem gives a precise statement.

Theorem 1.1: *A circuit composed of independent voltage and current sources and strictly increasing resistors without loop of voltage sources and without cutset of current sources has at most one solution.*

The interconnection condition concerning the sources is necessary. The circuit of Figure 1.6 is an illustration of this statement. Its solution is not unique because of the loop of voltage sources. The loop is no longer present in the circuit of Figure 1.7, which satisfies the conditions of Theorem 1.1, and which indeed has a unique solution.

If the resistor characteristics are not *strictly* increasing but only increasing (i.e., if the v-i curves have horizontal or vertical portions), the theorem still holds, if we exclude loops of voltage sources and I – resistors, and cutsets of current sources and V – resistors.

Theorem 1.1 guarantees the uniqueness of the solution, but it cannot assure its existence. On the other hand, we do not need increasing resistor characteristics for the existence.

Theorem 1.2: *Let a circuit be composed of independent voltage and current sources and resistors whose characteristics are continuous and satisfy the following passivity condition at infinity:*

$$v \to +\infty \Leftrightarrow i \to +\infty \quad \text{and} \quad v \to -\infty \Leftrightarrow i \to -\infty \tag{1.22}$$

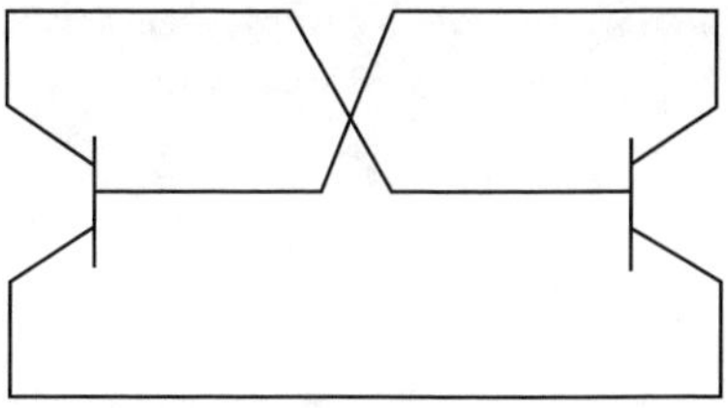

FIGURE 1.11 Feedback structure.

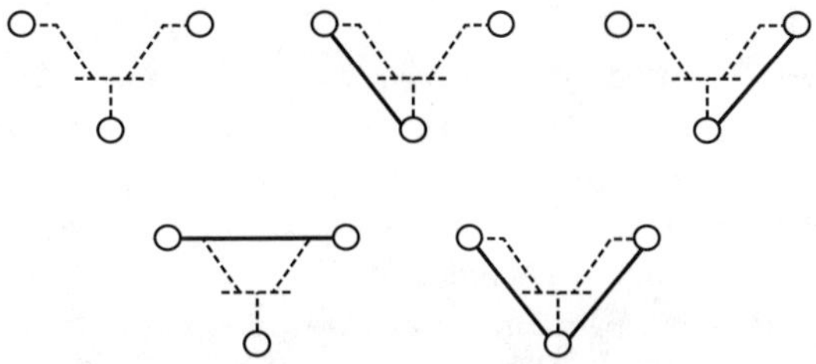

FIGURE 1.12 Short-open-circuit combinations for replacing the transistors.

If no loop of voltage sources and no cutset of current sources exist, then we have at least one solution of the circuit.

For refinements of this theorem, refer to [1] and [3].

If we admit nonreciprocal elements, neither Theorem 1.1 nor 1.2 remain valid. Indeed, the solution of the circuit of Figure 1.10 may be nonunique, even though the nonlinear resistor has a strictly increasing characteristic. In order to ensure the existence and uniqueness of a nonreciprocal nonlinear resistive circuit, nontrivial constraints on the interconnection of the elements must be observed. The theorems below give different, but basically equivalent, ways to formulate these constraints.

The first results is the culminating point of a series of papers by Sandberg and Wilson [3]. It is based on the following notion.

Definition 1.1.

- The connection of the two bipolar transistors shown in Figure 1.11 is called a **feedback structure**. The type of the transistors and the location of the collectors and emitters is arbitrary.
- A circuit composed of bipolar transistors, resistors, and independent sources contains a feedback structure, if it can be reduced to the circuit of Figure 1.11 by replacing each voltage source by a short circuit, each current source by an open circuit, each resistor and diode by an open or a short circuit, and each transistor by one of the five short-open-circuit combinations represented in Figure 1.12.

Theorem 1.3: *Let a circuit be composed of bipolar transistors, described by the Ebers–Moll model, positive linear resistors, and independent sources. Suppose we have no loop of voltage sources and no cutset of current sources. If the circuit contains no feedback structure, it has exactly one solution.*

This theorem [4] is extended in [5] to MOS transistors.

The second approach was developed by Nishi and Chua. Instead of transistors, it admits controlled sources. In order to formulate the theorem, two notions must be introduced.

Definition 1.2. A circuit composed of controlled sources, resistors, and independent sources satisfies the **interconnection condition**, if the following conditions are satisfied:

- No loop is composed of voltage sources, output ports of (voltage or current) controlled voltage sources, and input ports of current controlled (voltage or current) sources.
- No cutset is composed of current sources, outputs ports of (voltage or current) controlled current sources, and input ports of voltage controlled (voltage or current) sources.

Definition 1.3. A circuit composed exclusively of controlled sources has a **complementary tree** structure if both the input and output ports each form a tree. The fundamental loop matrix of the input port tree has the form

$$B = \left[B_{\mathrm{T}} \middle| 1 \right] \tag{1.23}$$

The circuit is said to have a **positive (negative) complementary tree structure**, if the determinant of $\boldsymbol{B}_{\mathrm{T}}$ is positive (negative).

Theorem 1.4: *Suppose a circuit composed of controlled sources, strictly increasing resistors satisfying* (1.22), *and independent sources satisfies the interconnection condition. If, by replacing each resistor either by a short circuit or an open circuit, all independent and some dependent voltage sources by short circuits, and all independent and some dependent current sources by open circuits, one never obtains a negative complementary tree structure, the circuit has exactly one solution* [6].

A similar theorem for circuits with operational amplifiers instead of controlled sources is proved in [7].

The third approach is that of Hasler [1, 8]. The nonreciprocal elements here are nullator–norator pairs. Instead of reducing the circuit by some operations in order to obtain a certain structure, we must orient the resistors in certain way. Again, we must first introduce a new concept.

Definition 1.4. Let a circuit be composed of nullator-norator pairs, resistors, and independent voltage and current sources. A **partial orientation** of the resistors is **uniform**, if the following two conditions are satisfied:

- Every oriented resistor is part of an evenly directed loop composed only of oriented registors and voltage sources
- Every oriented resistor is part of an evenly directed cutset composed only of norators, oriented resistors, and voltage sources

Theorem 1.5: *Let a circuit be composed of nullator-norator pairs, V- and I-resistors, and independent voltage and current sources. If the following conditions are satisfied, the circuit has exactly one solution:*

- *The norators, I-resistors, and the voltage sources together form a tree.*
- *The nullators, I-resistors, and the voltage sources together form a tree.*
- *The resistors have no uniform partial orientation, except for the trivial case, in which no resistor is oriented.*

We illustrate the conditions of this theorem with the example of Figure 1.10. In Figure 1.13 the resistors are specified as *V*- and *I*-resistors and a uniform orientation of the resistors is indicated. Note that the nonlinear resistor is a *V*-resistor, but not an *I*-resistor, because its current saturates. The linear resistors, however, are both *V*- and *I*-resistors. The choice in Figure 1.13 is made in order to satisfy the first two conditions of Theorem 1.5. Correspondingly, in Figure 1.14 and 1.15 the norator–*I*-resistor–voltage source tree and the nullator-*I*-resistor voltage source tree are represented. Because the third condition is not satisfied, Theorem 1.5 cannot guarantee a unique solution. Indeed, as explained earlier, this circuit may have three solutions.

Theorem 1.5 has been generalized to controlled sources, to resistors that are increasing but neither voltage nor current controlled (e.g., the ideal diode), and to resistors that are decreasing instead of increasing [9].

Theorems 1.3, 1.4, and 1.5 have common features. Their conditions concern the **circuit structure** — the circuit graph that expresses the interconnection of the elements and the type of elements that occupy the branches of the graph, but not the element values. Therefore, the theorems guarantee the existence and uniqueness of the solution for whole classes of circuits, in which the individual circuits differ by their element values and parameters. In this sense, the conditions are not only sufficient, but also

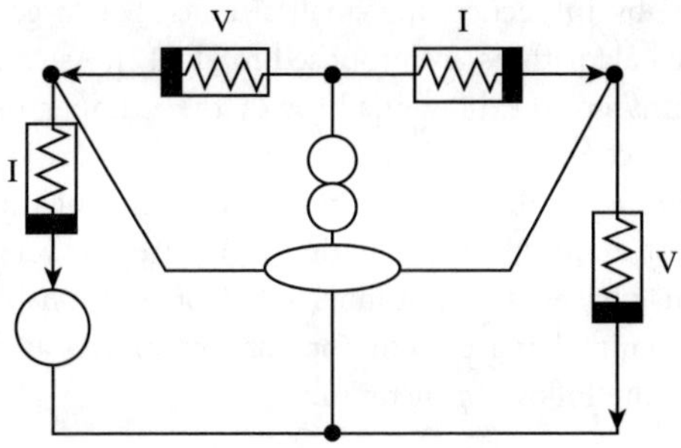

FIGURE 1.13 Circuit of Figure 1.10 with nullator and norator.

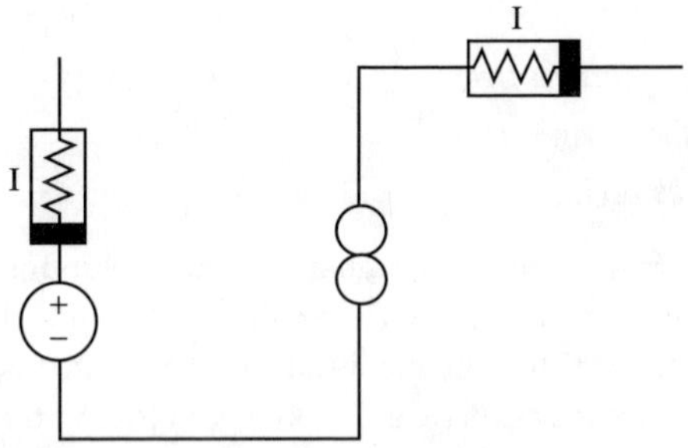

FIGURE 1.14 Norator–*I*-resistor–voltage source tree.

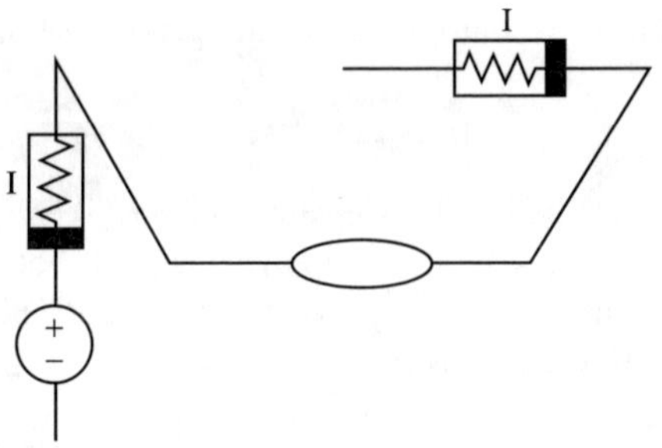

FIGURE 1.15 Nullator–*I*-resistor–voltage source tree.

necessary. This means, for example, in the case of Theorem 1.5 that if all circuits with the same structure have exactly one solution, then the three conditions must be satisfied. However, by logical contraposition, if one of the three conditions is not satisfied for a given circuit structure, a circuit with this structure exists which has either no solution or more than one solutions.

On the other hand, if we consider a specific circuit, the conditions are only sufficient. They permit us to prove that the solution exists and is unique, but some circuits do not satisfy the conditions and still have exactly one solution. However, if the parameters of such a circuit are varied, one eventually falls onto a circuit with no solution or more than one solution.

The main conditions of Theorems 1.3 and 1.4 have an evident intuitive meaning. The orientations to look for in Theorems 1.5 are linked to the sign of the currents and the voltages of the difference of two solutions. Because the resistors are increasing, these signs are the same for the voltage and current differences. If we extend the analysis of the signs of solutions or solution differences to other elements, we must differentiate between voltages and currents. This approach, in which two orientations for all branches are considered, one corresponding to the currents and one corresponding to the voltages, is pursued in [10].

The conditions of Theorems 1.3 to 1.5 can be verified by inspection for small circuits. For larger circuits, one must resort to combinatorial algorithms. Such algorithms are proposed in [11,12]. As can be expected from the nature of conditions, the algorithms grow exponentially with the number of resistors. It is not known whether algorithms of polynomial complexity exist.

Some circuits always have either no solution or an infinite number of solutions, irrespective of the element and parameter values. Figure 1.6 gives the simplest example. Such circuits clearly are not very useful in practice. The remaining circuits are those that may have a finite number $n > 1$ of solutions if the circuit parameters are chosen suitably. These are the circuits that are useful for static memories and for multivibrators in general. This class is characterized by the following theorem.

Theorem 1.6: *Let circuit be composed of nullator-norator pairs, V- and I-resistors, and independent voltage and current sources. If the following three conditions are satisfied, the circuit has more than one, but a finite number of solutions for a suitable choice of circuit parameters:*

- *The norators, I-resistors, and the voltage sources together form a tree.*
- *The nullators, I-resistors, and the voltage sources together form a tree.*
- *A nontrivial, uniform partial orientation of the resistors occurs.*

Can we be more precise and formulate conditions on the circuit structure that guarantee four solutions, for example? This is not possible because changing the parameters of the circuit will lead to another number of solutions. Particularly with a circuit structure that satisfies the conditions of Theorem 1.6, there is a linear circuit that always has an infinite number of solutions. If we are more restrictive on the resistor characteristics, e.g., imposing convex or concave characteristics for certain resistors, it is possible to determine the **maximum number of solutions.** A method to determine an upper bound is given in [14], whereas the results of [15] allow us to determine the actual maximum number under certain conditions. Despite these results, however, the maximum number of solutions is still an open problem.

Bounds on Voltages and Currents

It is common sense for electrical engineers that in an electronic circuit all node voltages lie between 0 and the power supply voltage, or between the positive and the negative power supply voltages, if both are present. Actually, this is only true for the DC-operating point, but can we prove it in this case? The following theorems give the answer. They are based on the notion of passivity.

Definition 1.5. A resistor is **passive** if it can only absorb, but never produce power. This means that for any point (v, i) on its characteristic we have

$$v \cdot i \geq 0 \tag{1.24}$$

A resistor is **strictly passive**, if in addition to (1.24) it satisfies the condition

$$v \cdot i = 0 \rightarrow v = i = 0 \tag{1.25}$$

Theorem 1.7: *Let a circuit be composed of strictly passive resistors and independent voltage and current sources. Then, for every branch k of the circuit the following bounds can be given:*

$$|v_k| \leq \sum_{\text{source branches } j} |v_j| \tag{1.26}$$

$$|i_k| \leq \sum_{\text{source branches } j} |i_j| \tag{1.27}$$

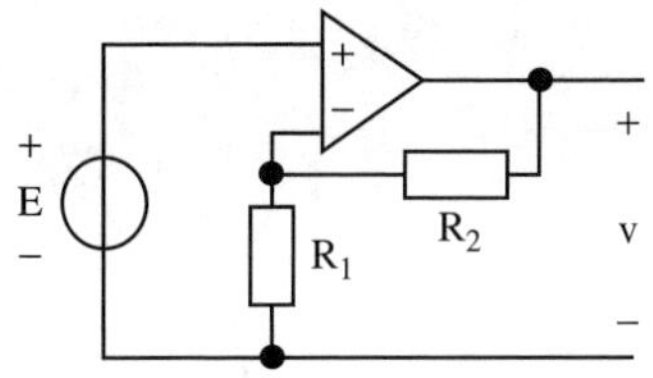

FIGURE 1.16 Voltage amplifier.

If, in addition, the circuit is connected and all sources have a common node, the ground node, then the maximum and the minimum node voltage are at a source terminal.

The theorem implies in particular that in a circuit with a single voltage source, all branch voltages are bounded by the source voltage in magnitude, and all node voltages lie between zero and the source voltage. Similarly, if a circuit has a single current source, all branch currents are bounded by the source current in magnitude. Finally, if several voltage sources are present that are all connected to ground and have positive value, then the node voltages lie between zero and the maximum source voltage. If some sources have positive values and others have negative values, then all node voltages lie between the maximum and the minimum source values.

This theorem and various generalizations can be found in [1]. The main drawback is that it does not admit nonreciprocal elements. A simple counterexample is the voltage amplifier of Figure 1.16. The voltage of the output node of the operational amplifier is

$$v = \frac{R_1 + R_2}{R_1} E \tag{1.28}$$

Thus, the output node voltage is higher than the source voltage. Of course, the reason is that the operational amplifier is an active element. It is realized by transistors and needs a positive and a negative voltage source as the power supply. The output voltage of the operational amplifier cannot exceed these supply voltages. This fact is not contained in the model of the ideal operational amplifier, but follows from the extension of Theorem 1.7 to bipolar transistors [1, 16].

Theorem 1.8: *Let a circuit be composed of bipolar transistors modeled by the Ebers–Moll equations, of strictly passive resistors, and of independent voltage and current sources. Then, the conclusion of Theorem 1.7 hold.*

At first glance, Theorem 1.8 appears to imply that it is impossible to build an amplifier with bipolar transistors. Indeed, it is impossible to build such an amplifier with a single source, the input signal. We need at least one power supply source that sets the limits of dynamic range of the voltages according to Theorem 1.8. The signal source necessarily has a smaller amplitude and the signal can be amplified roughly up to the limit set by the power supply source.

Theorem 1.8 can be extended to MOS transistors. The difficulty is that the nonlinear characteristics of the simplest model is not strictly increasing, and therefore some interconnection condition must be added to avoid parts with undetermined node voltages.

Monotonic Dependence

Instead of looking at single solutions of resistive circuits, as done earlier in the chapter, we consider here a solution as a function of a parameter. The simplest and at the same time the most important case is the dependence of a solution on the value of a voltage or current source. To have a well-defined situation, we suppose that the circuit satisfies the hypotheses of Theorem 1.5. In this case [1, 8], the solution is a continuous function of the source values.

As an example, let us consider the circuit of Figure 1.17. We are interested in the dependence of the various currents on the source voltage E. Because the circuit contains only strictly increasing resistors,

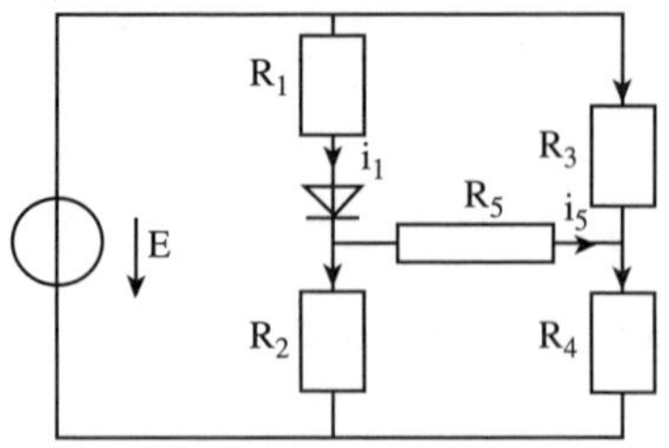

FIGURE 1.17 Circuit example for source dependence.

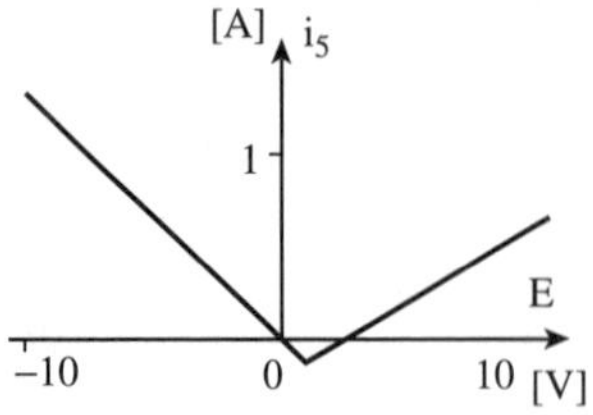

FIGURE 1.18 Nonmonotonic dependence.

we expect all currents to be strictly monotonic functions of E. This is not true. In Figure 1.18, the current $i_5(E)$ is represented for $R_1 = R_2 = R_3 = 2R_4 = R_5 = 1\ \Omega$ and for standard diode model parameters. Clearly, it is nonmonotonic.

1.3 Autonomous Dynamic Circuits

Introduction

This section adds to the resistive elements of the previous section — the capacitors and the inductors. A nonlinear capacitor is defined by the constitutive relation

$$v = h(q) \tag{1.29}$$

where the auxiliary variable q is the charge of the capacitor, which is linked to the current by

$$i = \frac{dq}{dt} \tag{1.30}$$

The dual element, the nonlinear inductor, is defined by

$$i = g(\varphi) \tag{1.31}$$

where the auxiliary variable φ, the flux, is linked to the voltage by

$$v = \frac{d\varphi}{dt} \tag{1.32}$$

The symbols of these two elements are represented in Figure 1.19.

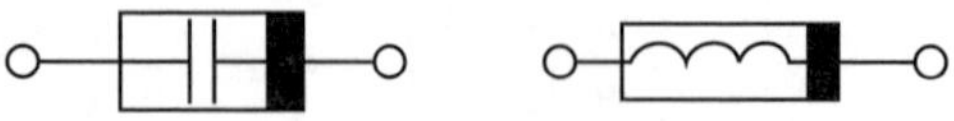

FIGURE 1.19 Symbols of the nonlinear capacitor and the nonlinear inductor.

The system of equations that describes an autonomous dynamic circuit is composed of (1.12) to (1.17), completed with (1.29) and (1.30) for capacitor branches and (1.31) and (1.32) for inductor branches. Hence, it becomes a mixed differential–nondifferential system of equations. Its solutions are the voltages, currents, charges, and fluxes as functions of time. Because it contains differential equations, we have infinitely many solutions, each one determined by some set of initial conditions.

If all variables except the charges and fluxes are eliminated from the system of equations, one obtains a reduced, purely differential system of equations

$$\frac{d\mathbf{q}}{dt} = \mathbf{f}(\mathbf{q}, \varphi) \tag{1.33}$$

$$\frac{d\varphi}{dt} = \mathbf{g}(\mathbf{q}, \varphi) \tag{1.34}$$

where $\mathbf{q}$ and φ are the vectors composed of, respectively, the capacitor charges and the inductor fluxes. These are the **state equations** of the circuit. Under mild assumptions on the characteristics of the nonlinear elements (local Lipschitz continuity and eventual passivity), it can be shown that the solutions are uniquely determined by the initial values of the charges and fluxes at some time t_0, $\mathbf{q}(t_0)$, and $\varphi(t_0)$, and that they exist for all times $t_0 \leq t < \infty$ [1, 17].

It cannot be taken for granted, however, that the circuit equations actually can be reduced to that state, Eqs. (1.33) and (1.34). On the one hand, the charges and fluxes may be dependent and thus their initial values cannot be chosen freely. However, the state equations may still exist, in terms of a subset of charges and fluxes. This means that only these charges and fluxes can be chosen independently as initial conditions. On the other hand, the reduction, even to some alternative set of state variables, may be simply impossible. This situation is likely to lead to **impasse points**, i.e., nonexistence of the solution at a finite time. We refer the reader to the discussion in [1]. In the sequel, we suppose that the solutions exist from the initial time t_0 to $+\infty$ and that they are determined by the charges and fluxes at t_0.

We are interested in the asymptotic behavior, i.e., the behavior of the solutions when the time t goes to infinity. If the dynamic circuit is linear and strictly stable, i.e., if all its natural frequencies are in the open left half of the complex plane, then all solutions converge to 1 and the same DC-operating (equilibrium) point. This property still holds for many nonlinear circuits, but not for all by far. In particular, the solutions may converge to different DC-operating points, depending on the initial conditions (static memories), they may converge to periodic solutions (free-running oscillators), or they may even show chaotic behavior (e.g., Chua's circuit). Here, we give conditions that guarantee the solutions converge to a unique solution or one among several DC-operating points.

Convergence to DC-Operating Points

The methods to prove convergence to one or more DC-operating points is based on **Lyapunov functions**. A Lyapunov function is a continuously differentiable function $W(\xi)$, where ξ is the vector composed of the circuit variables (the voltages, currents, charges, and fluxes). In the case of autonomous circuits, a Lyapunov function must have the following properties:

1. W is bounded below, i.e., there exists a constant W_0 such that

$$W(\xi) \geq W_0 \quad \text{for all } \xi \tag{1.35}$$

2. The set of voltages, currents, charges, and fluxes of the circuit such that $W(\xi) \leq E$ is bounded for any real E.
3. For any solution $\xi(t)$ of the circuit

$$\frac{d}{dt} W(\xi(t)) \leq 0 \tag{1.36}$$

4. If

$$\frac{d}{dt}W(\xi(t)) = 0 \tag{1.37}$$

then $\xi(t)$ is a DC-operating point.

If an autonomous circuit has a Lyapunov function and if it has at least one, but a finite number of DC-operating points, then every solution converges to a DC-operating point. The reason is that the Lyapunov function must decrease along each solution, and thus must result in a local minimum, a stable DC-operating point. If more than one DC-operating point exists, it may, as a mathematical exception that cannot occur in practice, end up in a saddle point, i.e., an unstable DC-operating point.

The problem with the Lyapunov function method is that it gives no indication as to how to find such a function. Basically, three methods are available to deal with this problem:

1. Use some standard candidates for Lyapunov functions, e.g., the stored energy.
2. Use a certain kind of function and adjust the parameters in order to satisfy 2 and 3 in the previous list. Often, quadratic functions are used.
3. Use an algorithm to generate Lyapunov functions [18-20].

The following theorems were obtained via approach 1, and we indicate which Lyapunov was used to prove them. At first glance, this may seem irrelevant from an engineering point of view. However, if we are interested in designing circuits to solve optimization problems, we are likely to be interested in Lyapunov functions. Indeed, as mentioned previously, along any solution of the circuit, the Lyapunov function decreases and approaches a minimum of the function. Thus, the dynamics of the circuit solve a minimization problem. In this case, we look for a circuit with a given Lyapunov function, however, usually we look for a Lyapunov function for a given circuit.

Theorem 1.9: *Let a circuit be composed of capacitors and inductors with a strictly increasing characteristic, resistors with a strictly increasing characteristic, and independent voltage and current sources. Suppose the circuit has a DC-operating point* $\bar{\xi}$. *By Theorem 1.1, this DC-operating point is unique. Finally, suppose the circuit has no loop composed of capacitors, inductors, and voltage sources and no cutset composed of capacitors, inductors, and current sources. Then, all solutions of the circuit converge to* $\bar{\xi}$.

The Lyapunov function of this circuit is given by a variant of the stored energy in the capacitors and the resistors, the stored energy with respect to $\bar{\xi}$ [1, 17]. If the constitutive relations of the capacitors and the inductors are given by $v_k = h_k(q_k)$ and $i_k = g_k(v_k)$, respectively, then this Lyapunov function becomes

$$\begin{aligned} W(\xi) = &\sum_{\substack{\text{capacitor}\\ \text{branches } k}} \int_{\bar{q}k}^{qk} \left(h_k(q) - h_k(\bar{q}_k)\right) dq \\ &+ \sum_{\substack{\text{inductor}\\ \text{branches } k}} \int_{\bar{\varphi}_k}^{\varphi_k} \left(g_k(\varphi) - g_k(\bar{\varphi}_k)\right) d\varphi \end{aligned} \tag{1.38}$$

The main condition (1.36) for a Lyapunov function follows from the fact that the derivative of the stored energy is the absorbed power, here in incremental form:

$$\frac{d}{dt}W(\xi) = \sum_{\substack{\text{capacitor}\\ \text{and inductor}\\ \text{branches } k}} \Delta v_k \, \Delta i_k = -\sum_{\substack{\text{resistor}\\ \text{branches } k}} \Delta v_k \, \Delta i_k \le 0 \tag{1.39}$$

Various generalizations of Theorem 1.9 have been given. The condition "strictly increasing resistor characteristic" has been relaxed to a condition that depends on ξ in [1, 17] and mutual inductances and capacitances have been admitted in [17].

The next theorem admits resistors with nonmonotonic characteristics. However, it does not allow for both inductors and capacitors.

Theorem 1.10: *Let a circuit be composed of capacitors with a strictly increasing characteristic, voltage-controlled resistors such that*

$$v \to +\infty \Rightarrow i > I_+ > 0 \text{ and } v \to -\infty \Rightarrow i < I_- < 0 \tag{1.40}$$

and independent voltage sources. Furthermore, suppose that the circuit has a finite number of DC-operating points. Then every solution of the circuit converges toward a DC-operating point.

This theorem is based on then following a Lyapunov function, called **cocotent**:

$$W(\xi(t)) = \sum_{\substack{\text{resistor}\\ \text{branches } k}} \int_o^{v_k} g_k(v)dv \tag{1.41}$$

where $i_k = g_k(v_k)$ is the constitutive relation of the resistor on branch k. The function W is decreasing along a solution of the circuit because

$$\begin{aligned} \frac{d}{dt}W(\xi(t)) &= \sum_{\substack{\text{resistor}\\ \text{branches } k}} \frac{dv_k}{dt} i_k = -\sum_{\substack{\text{capacitor}\\ \text{branches } k}} \frac{dv_k}{dt} i_k \\ &= -\sum_{\substack{\text{capacitor}\\ \text{branches } k}} \frac{dh_k}{dq} i_k^2 \le 0 \end{aligned} \tag{1.42}$$

where $h_k\ (q_k)$ is the constitutive relation of the capacitor on branch k.

Theorem 1.10 has a dual version. It admits inductors instead of capacitors, current-controlled resistors, and current sources. The corresponding Lyapunov function is the content:

$$W(\xi) = \sum_{\substack{\text{resistor}\\ \text{branches } k}} \int_o^{i_k} h_k(i)di \tag{1.43}$$

where $v_k = h_k(i_k)$ is the constitutive relation of the resistor on branch k.

The main drawback of the two preceding theorems is that they do not admit nonreciprocal elements such as controlled sources, operational amplifiers, etc. In other words, no statement about the analog neural network of Figure 1.20 can be made. In this network the nonreciprocal element is the VCVS with the nonlinear characteristics $v_2 = \sigma(v_1)$. However, Theorem 1.10 can be generalized to a reciprocal voltage controlled N-port resistor closed on capacitors and voltage sources. Such an N-port (Figure 1.21) is described by a constitutive relation of the form

$$i_k = g_k(v_1, \ldots, v_N) \tag{1.44}$$

and it is **reciprocal**, if for all $\mathbf{v}$, and all k, j we have

$$\frac{\partial g_k}{\partial v_j}(\mathbf{v}) = \frac{\partial g_j}{\partial v_k}(\mathbf{v}) \tag{1.45}$$

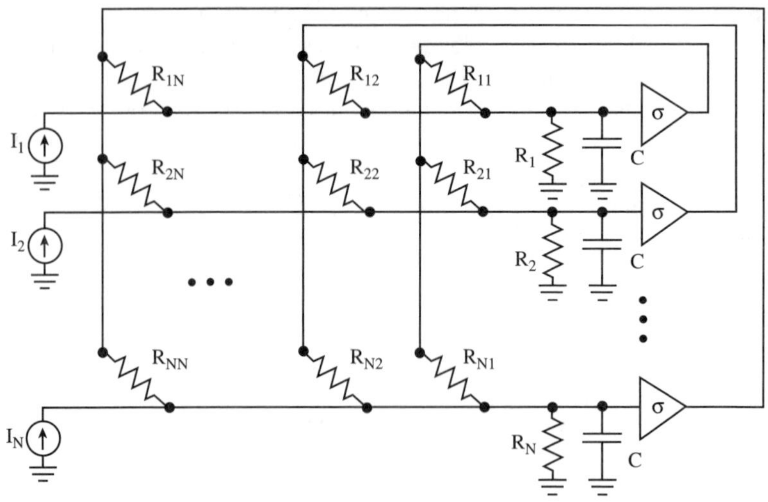

FIGURE 1.20 Analog neural network.

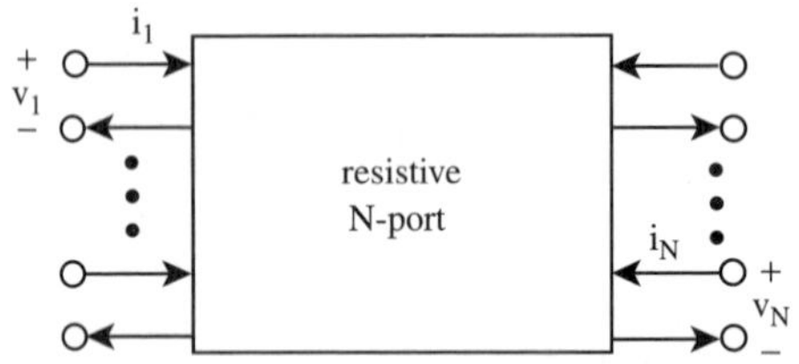

FIGURE 1.21 Resistive *N*-port.

Theorem 1.11: *Let a circuit be composed of charge-controlled capacitors with a strictly increasing characteristic and independent voltage sources that terminate a reciprocal voltage-controlled N-port with constitutive relation (1.42) so that we find constants V and P > 0 such that*

$$\|\mathbf{v}\| \geq V \Rightarrow \mathbf{g} \cdot \mathbf{v} = \sum_{k=1}^{N} g_k(\mathbf{v}) v_k \geq P \tag{1.46}$$

If the number of DC-operating points is finite, then all solutions converge toward a DC-operating point.

The proof of this theorem is based on the Lyapunov function $W(\mathbf{v})$ that satisfies

$$\frac{\partial W}{\partial v_k}(\mathbf{v}) = g_k(\mathbf{v}) \tag{1.47}$$

Thanks to condition (1.45), function W exists. The first two conditions for a Lyapunov function are a consequence of (1.46). Finally,

$$\begin{aligned} \frac{d}{dt} W(\xi(t)) &= \sum_{\substack{\text{resistor} \\ \text{branches}\, k}} g_k(\mathbf{v}) \frac{dv_k}{dt} \\ &= \sum_{\substack{\text{resistor} \\ \text{branches}\, k}} i_k \frac{dv_k}{dt} \end{aligned} \tag{1.48}$$

$$= -\sum_{\substack{\text{capacitor}\\ \text{branches } k}} \frac{dh_k}{dq} i_k^2 \leq 0$$

where $h_k(q_k)$ is the constitutive relation of the capacitor on branch k.

To illustrate how Theorem 1.11 can be applied when Theorem 1.10 fails, consider the analog neural network of Figure 1.20. If the capacitor voltages are denoted by u_i and the voltages at the output of the voltage sources by v_i, the state equations for the network of Figure 1.1 become

$$-C_i \frac{du_i}{dt} = \frac{u_i}{R_i} + \sum_{j=1}^{N} \frac{u_i - v_j}{R_{ij}} + I_i \tag{1.49}$$

Suppose that the nonlinear characteristic $\sigma(u)$ is invertible. The state equations can be written in terms of the voltages v_i:

$$-C \frac{d\sigma^{-1}}{dv}(v_i) \frac{dv_i}{dt} = G_i \sigma^{-1}(v_i) - \sum_{j=1}^{N} \frac{v_j}{R_{ij}} + I_i \tag{1.50}$$

where

$$G_i = \frac{1}{R_i} + \sum_{j=1}^{N} \frac{1}{R_{ij}} \tag{1.51}$$

Equations (1.40) can be reinterpreted as the equations of a resistive N-port with the constitutive relations

$$g_i(\mathbf{v}) = G_i \sigma^{-1}(v_i) - \sum_{j=1}^{N} \frac{v_j}{R_{ij}} + I_i \tag{1.52}$$

closed on nonlinear capacitors with the constitutive relation

$$v = \sigma\left(\frac{q}{C}\right) \tag{1.53}$$

If σ is a sigmoidal function, as is most often supposed in this context (i.e., a strictly increasing function with $s(u) \to \pm 1$ for $u \to \pm\infty$), then the capacitors have a strictly increasing characteristic, as required by Theorem 1.11. Furthermore, the resistive N-port is reciprocal if for $i \neq j$

$$\frac{\partial g_i}{\partial v_j} = -\frac{1}{R_{ij}} = \frac{\partial g_j}{\partial v_i} = -\frac{1}{R_{ji}} \tag{1.54}$$

In other words, if for all i, j

$$R_{ij} = R_{ji} \tag{1.55}$$

On the other hand, inequality (1.46) must be modified because the sigmoids have values only in the interval $[-1, +1]$ and thus (1.50) are defined only on the invariant bounded set $S = \{\mathbf{v} \,|\, -1 < v_i < +1\}$. Therefore, inequality (1.50) must be satisfied for vectors $\mathbf{v}$ sufficiently close to the boundary of S. This is indeed the case, because $\sigma^{-1}(v) \to \pm\infty$ as $v \to \pm 1$, whereas the other terms of the right-hand side of (1.52) remain bounded.

It follows that all solutions of the analog neural network of Figure 1.20 converge to a DC-operating point as $t \to \infty$, provided σ is a sigmoid function and the connection matrix R_{ij} (synaptic matrix) is symmetrical. The Lyapunov function can be given explicitly:

$$W(\mathbf{v}) = \sum_{i=1}^{N} G_i \int_0^{v_i} \sigma^{-1}(v)dv - \frac{1}{2}\sum_{i,j=1}^{N} \frac{v_i v_j}{R_{ij}} + \sum_{i=1}^{N} v_i I_i \tag{1.56}$$

1.4 Nonautonomous Dynamic Circuits

Introduction

This section is a consideration of circuits that contain elements where constitutive relations depend explicitly on time. However, we limit time dependence to the independent sources. For most practical purposes, this is sufficient. A time-dependent voltage source has a constitutive relation

$$v = e(t) \tag{1.57}$$

and a time-dependent current source

$$i = e(t) \tag{1.58}$$

where $e(t)$ is a given function of time which we suppose here to be continuous. In information processing circuits, $e(t)$ represents a signal that is injected into the circuit, whereas in energy transmission circuits $e(t)$ usually is a sinusoidal or nearly sinusoidal function related to a generator.

The time-dependent sources may drive the voltages and the currents to infinity, even if they only inject bounded signals into the circuit. Therefore, the discussion begins with the conditions that guarantee the boundedness of the solutions.

Boundedness of the Solutions

In electronic circuits, even active elements become passive when the voltages and currents grow large. This is the reason that solutions remain bounded.

Definition 1.6. A resistor is **eventually passive** if, for sufficiently large voltages and/or currents, it can only absorb power. More precisely, eventual passivity means that constants V and I exist such that, for all points (v, i) on the resistor characteristic with $|v| > V$ or $|i| > I$, we have

$$v \cdot i \geq 0 \tag{1.59}$$

Note that sources are not eventually passive, but as soon as an internal resistance of a source is taken into account, the source becomes eventually passive. The notion of eventual passivity can be extended to time-varying resistors.

Definition 1.7. A time-varying resistor is eventually passive if constants V and I are independent of time and are such that all points (v, i), with $|v| > V$ or $|i| > I$ that at some time lie on the characteristic of the resistor, satisfy the passivity condition (1.59). According to this definition, time-dependent sources with internal resistance are eventually passive if the source signal remains bounded.

Eventual passivity allows us to deduce bounds for the solutions. These bounds are uniform in the sense that they do not depend on the particular solution. To be precise, this is true only asymptotically, as $t \to \infty$.

Definition 1.8. The solutions of a circuit are **eventually uniformly bounded** if there exist constants V, I, Q, and Φ are such that, for any solution there exists a time T such that for any $t > T$, the voltages $v_k(t)$ are bounded by V, the currents $i_k(t)$ are bounded by I, the charges $q_k(t)$ are bounded by Q, and the fluxes $\varphi_k(t)$ are bounded by Φ.

Another manner of expressing the same property is to say that an attracting domain exists in state space [1].

Theorem 1.12: *A circuit composed of eventually passive resistors with* $v \cdot i \to +\infty$ *as* $|v| \to \infty$ *or* $|i| \to \infty$, *capacitors with* $v \to \pm\infty$ *as* $q \to \neq\infty$, *and inductors with* $i \to \pm\infty$ *as* $\varphi \to \infty$ *has eventually uniformly bounded solutions if no loop or cutset exists without a resistor* [1,17].

Again, this theorem is proved by using a Lyapunov function, namely the stored energy

$$W(\xi) = \sum_{\substack{\text{capacitor} \\ \text{branches } k}} \int_0^{q_k} h_k(q)\,dq + \sum_{\substack{\text{capacitor} \\ \text{branches } k}} \int_0^{\varphi_k} g_k(\varphi)\,d\varphi \tag{1.60}$$

Inequality (1.36) holds only outside of a bounded domain.

Unique Asymptotic Behavior

In the presence of signals with complicated waveforms that are injected into a circuit, we cannot expect simple waveforms for the voltages and the currents, not even asymptotically, as $t \to \infty$. However, we can hope that two solutions, starting from different initial conditions, but subject to the same source, have the same steady-state behavior. The latter term needs a more formal definition.

Definition 1.9. A circuit has unique asymptotic behavior if the following two conditions are satisfied:

1. All solutions are bounded.
2. For any two solutions $\xi_1(t)$ and $\xi_2(t)$

$$\left\| \xi_1(t) - \xi_2(t) \right\| \to_{t \to \infty} 0 \tag{1.61}$$

In order to prove unique asymptotic behavior, it is necessary to extend the notion of the Lyapunov function [1]. This does not lead very far, but at least it permits us to prove the following therorem.

Therorem 1.13: *Suppose a circuit is composed of resistors with a strictly increasing characteristic such that* $v \cdot i \to \infty$ *as* $|v| \to \infty$ *or* $|i| \to \infty$, *positive linear capacitors, positive linear inductors, time-depending voltage (current) sources with bounded voltage (current) and a positive resistor in series (parallel). If no loop or cutset is composed exclusively of capacitors and inductors, the circuit has unique asymptotic behavior* [1, 17].

This theorem is unsatisfactory because linear reactances are required and real devices are never exactly linear. It has been shown that slight nonlinearities can be tolerated without losing the unique asymptotic behavior [21]. On the other hand, we cannot expect to get much stronger general results because nonautonomous nonlinear circuits may easily have multiple steady-state regimes and even more complicated dynamics, such as chaos, even if the characteristics of the nonlinear elements are all strictly increasing.

Another variant of Theorem 1.13 considers linear resistors and nonlinear reactances [17].

References

[1] M. Hasler and J. Neirynck, *Nonlinear Circuits,* Boston: Artech House, 1986.
[2] L. O. Chua, C. A. Desoer, and E. S. Kuh, *Linear and Nonlinear Circuits,* Electrical & Electronic Engineering Series, Singapore: McGraw-Hill International Editors, 1987.
[3] A. N. Willson, Ed., *Nonlinear Networks: Theory and Analysis,* New York: IEEE Press, 1974.
[4] R. O. Nielsen and A. N. Willson, "A fundamental result concerning the topology of transistor circuits with multiple equilibria," *Proc. IEEE,* vol. 68, pp. 196–208, 1980.
[5] A. N. Willson, "On the topology of FET circuits and the uniqueness of their dc operating points," *IEEE Trans. Circuits Syst.,* vol. 27, pp. 1045–1051, 1980.
[6] T. Nishi and L. O. Chua, "Topological criteria for nonlinear resistive circuits containing controlled sources to have a unique solution," *IEEE Trans. Circuits Syst.,* vol. 31, pp. 722–741, Aug. 1984.
[7] T. Nishi and L. O. Chua, "Nonlinear op-amp circuits: Existence and uniqueness of solution by inspection," *Int. J. Circuit Theory Appl.,* vol 12, pp. 145–173, 1984.
[8] M. Hasler, "Nonlinear nonreciprocal resistive circuits with a unique solution," *Int. J. Circuit Theory Appl.,* vol. 14, pp. 237–262, 1986.
[9] M. Fosséprez, *Topologie et Comportement des Circuits non Linéaires non Réciproques,* Lausanne: Presses Polytechnique Romands, 1989.
[10] M. Hasler, "On the solution of nonlinear resistive networks," *J. Commun.* (*Budapest, Hungary*), special issue on nonlinear circuits, July 1991.
[11] T. Parker, M. P. Kennedy, Y. Lioa, and L. O. Chua, "Qualitative analysis of nonlinear circuits using computers," *IEEE Trans. Circuits Syst.,* vol. 33, pp. 794–804, 1986.
[12] M. Fosséprez and M. Hasler, "Algorithms for the qualitative analysis of nonlinear resistive circuits," *IEEE ISCAS Proc.,* pp. 2165–2168, May 1989.
[13] M. Fosséprez and M. Hasler, "Resistive circuit topologies that admit several solutions," *Int. J. Circuit Theory Appl.,* vol. 18, pp. 625–638, Nov. 1990.
[14] M. Fosséprez, M. Hasler, and C. Schnetzler, "On the number of solutions of piecewise linear circuits," *IEEE Trans. Circuits Syst.,* vol. CAS-36, pp. 393–402, March 1989.
[15] T. Nishi and Y. Kawane, "On the number of solutions of nonlinear resistive circuits," *IEEE Trans.,* vol. E74, pp. 479–487, 1991.
[16] A. N. Willson, "The no-gain property for networks containing three-terminal elements," *IEEE Trans. Circuits Syst.,* vol. 22, pp. 678–687, 1975.
[17] L. O. Chua, "Dynamic nonlinear networks: state of the art, "*IEEE Trans. Circuits Syst.,* vol. 27, pp. 1059–1087, 1980.
[18] R. K. Brayton and C.H. Tong, "Stability of dynamical systems," *IEEE Trans. Circuits Syst.,* vol. 26, pp. 224–234, 1979.
[19] R. K. Brayton and C. H. Tong, "Constructive stability and asymptotic stability of dynamical systems," *IEEE Trans. Circuits Syst.,* vol. 27, pp. 1121–1130, 1980.
[20] L. Vandenberghe and S. Boyd, "A polynomial-time algorithm for determining quadratic Lyapunov functions for nonlinear systems," *Proc. ECCTD*-93, pp. 1065–1068, 1993.
[21] M. Hasler and Ph. Verburgh, "Uniqueness of the steady state for small source amplitudes in nonlinear nonautonomous circuits," *Int. J. Circuit Theory Appl.,* vol. 13, pp. 3–17, 1985.

2

Synthesis and Design of Nonlinear Circuits

A. Rodríguez-Vázquez
Universidad de Sevilla, Spain

M. Delgado-Restituto
Universidad de Sevilla, Spain

J. L. Huertas
Universidad de Sevilla, Spain

F. Vidal
Universidad de Malaga, Spain

2.1 Introduction

Nonlinear synthesis and design can be informally defined as a constructive procedure to interconnect components from a catalog of available primitives, and to assign values to their constitutive parameters to meet a specific nonlinear relationship among electrical variables. This relationship is represented as an implicit integro-differential operator, although we primarily focus on the synthesis of *explicit algebraic* functions,

$$y = f(\mathbf{x}) \tag{2.1}$$

where y is a voltage or current, $f(\cdot)$ is a nonlinear real-valued function, and $\mathbf{x}$ is a vector with components that include voltages and currents. This synthesis problem is found in two different circuit-related areas: device **modeling** [8, 76] and analog **computation** [26]. The former uses ideal circuit elements as primitives to build computer models of real circuits and devices (see Chapter 1). The latter uses real circuit components, available either off the shelf or integrable in a given fabrication technology, to realize hardware for nonlinear signal processing tasks. We focus on this second area, and intend to outline

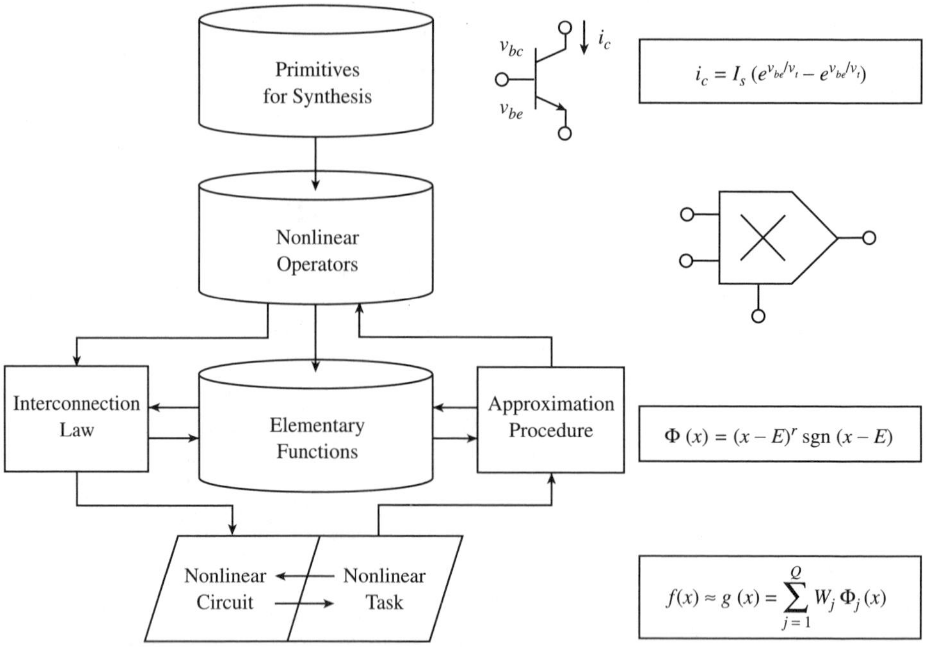

FIGURE 2.1 Hierarchical decomposition of the synthesis problem.

systematic approaches to devise electronic function generators. Synthesis relies upon hierarchical decomposition, conceptually shown in Figure 2.1, which encompasses several subproblems listed from top to bottom:

- Realization of nonlinear **operators** (multiplication, division, squaring, square rooting, logarithms, exponentials, sign, absolute value, etc.) through the interconnection of primitive components (transistors, diodes, operational amplifiers, etc.).
- Realization of **elementary functions** (polynomials, truncated polynomials, Gaussian functions, etc.) as the interconnection of the circuit blocks devised to build nonlinear operators.
- **Approximation** of the target as a combination of elementary functions and its realization as the interconnection of the circuit blocks associated with these functions.

Figure 2.1 illustrates this hierarchical decomposition of the synthesis problem through an example in which the function is approximated as a linear combination of truncated polynomials [30], where realization involves analog multipliers, built by exploiting the nonlinearities of bipolar junction transistors (BJTs) [63]. Also note that the subproblems cited above are closely interrelated and, depending on the availability of primitives and the nature of the nonlinear function, some of these phases can be bypassed. For instance, a logarithmic function can be realized exactly using BJTs [63], but requires approximation if our catalog includes only field-effect transistors whose nonlinearities are polynomic [44].

The technical literature contains excellent contributions to the solution of all these problems. These contributions can hardly be summarized or even quoted in just one section. Many authors follow a block-based approach which relies on the pervasive voltage operational amplifier (or op amp), the rectification properties of junction diodes, and the availability of voltage multipliers, in the tradition of classical analog computation. Examples are [7], [59], and [80]. Remarkable contributions have been made which focus on qualitative features such as negative resistance or hysteresis, rather than the realization of well-defined approximating functions [9, 20, 67]. Other contributions focus on the realization of nonlinear operators in the form of IC units. **Translinear** circuits, BJTs [23, 62] and MOSFETs [79] are particularly well suited to realize algebraic functions in IC form. This IC orientation is shared by recent developments in analog VLSI computational and signal processing systems for neural networks [75], fuzzy logic [81], and other nonlinear signal processing paradigms [56, 57, 71].

This chapter is organized to fit the hierarchical approach in Figure 2.1. We review a wide range of approximation techniques and circuit design styles, for both discrete and monolithic circuits. It is based on the catalog of primitives shown in Appendix A. In addition to the classical op-amp-based continuous-time circuits, we include current-mode circuitry because nonlinear operators are realized simply and accurately by circuits that operate in *current domain* [23, 57, 62, 79]. We also cover discrete-time circuits realized using analog dynamic techniques based on charge transfer, which is very significant for mixed-signal processing and computational microelectronic systems [27, 72]. Section 2.2 is devoted to approximation issues and outlines different techniques for uni- and multidimensional functions, emphasizing hardware-oriented approaches. These techniques involve several nonlinear operators and the linear operations of scaling and aggregation (covered in Section 2.3, which also presents circuits to perform transformations among different kinds of characteristics). Sections 2.4 and 2.5 present circuits for piecewise-linear (PWL) and piecewise-polynomial (PWP) functions, 2.6 covers neural and fuzzy approximation techniques, and 2.7 outlines an extension to dynamic circuits.

2.2 Approximation Issues

Unidimensional Functions

Consider a target function, $f(x)$, given analytically or as a collection of measured data at discrete values of the independent variable. The approximation problem consists of finding a multiparameter function, $g(x, \mathbf{w})$, which yields proper fitting to the target, and implies solving two different subproblems: (1) which approximating functions to use, and (2) how to adjust the parameter vector, $\mathbf{w}$, to render optimum fitting. We only outline some issues related to this first point. Detailed coverage of both problems can be found in mathematics and optimization textbooks [73, 78]. Other interesting views are found in circuit-related works [6, 11, 30], and the literature on neural and fuzzy networks [12, 21, 33, 43, 51].

An extended technique to design nonlinear electronic hardware for both discrete [63, 80] and monolithic [35, 62, 79] design styles uses **polynomial approximating functions,**

$$g(x) = \sum_{j=0}^{Q} \alpha_j x^j \tag{2.2}$$

obtained through expansion by either Taylor series or orthogonal polynomials (Chebyshev, Legendre, or Laguerre) [26]. Other related approaches use **rational functions,**

$$g(x) = \frac{\sum_{j=0,Q} \alpha_j x^j}{\sum_{j0,R} \beta_j x^j} \tag{2.3}$$

to improve accuracy in the approximation of certain classes of functions [14]. These can be realized by polynomial building blocks connected in feedback configuration [63]. In addition, [39] presents an elegant synthesis technique relying on linearly controlled resistors and conductors to take advantage of linear circuits synthesis methods (further extended in [28]).

From a more general point of view, hardware-oriented approximating functions can be classified into two major groups:

1. Those involving the linear combination of basis functions

$$g(x) = \sum_{j=1}^{Q} w_j \Phi_j(x) \tag{2.4}$$

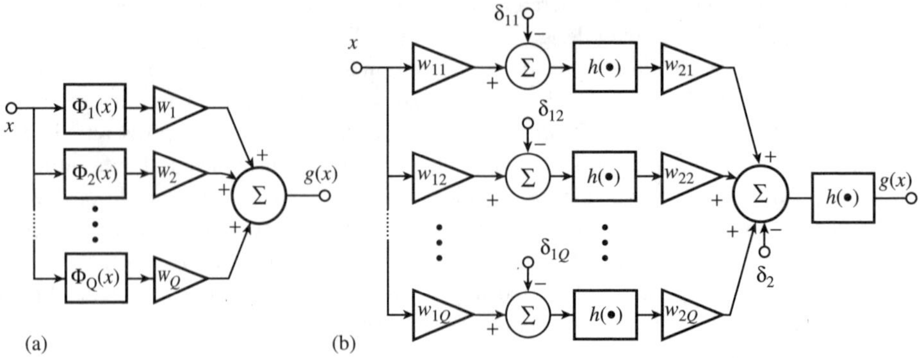

FIGURE 2.2 Block diagram for approximating function hardware. (a) Using linear combination of basis functions; (b) using two layers of nested sigmoids.

which include polynomial expansions. PWL and PWP interpolation and **radial basis functions** (RBF). The hardware for these functions consists of two layers, as shown in Figure 2.2 (a). The first layer contains Q nonlinear processing nodes to evaluate the basis functions; the second layer scales the output of these nodes and aggregates these scaled signals in a summing node.

2. Those involving a multilayer of nested **sigmoids** [51]; for instance, in the case of two layers [82],

$$g(x) = h\left[\left\{\sum_{j=1,Q} w_{2j} h\left(w_{1j}x - \delta_{1j}\right)\right\} - \delta_2\right] \tag{2.5}$$

with the sigmoid function given by

$$h(x) = \frac{2}{1 + \exp(-\lambda x)} - 1 \tag{2.6}$$

where $\lambda > 0$ determines the steepness of the sigmoid. Figure 2.2(b) shows a hardware concept for this approximating function, also consisting of two layers.

Piecewise-Linear and Piecewise-Polynomial Approximants

A drawback of polynomial and rational approximants is that their behavior in a small region determines their behavior in the whole region of interest [78]. Consequently, they are not appropriate to fit functions that are uniform throughout the whole region [see Figure 2.3 (a)]. Another drawback is their lack of modularity, a consequence of the complicated dependence of each fitting parameter on multiple target data, which complicates the calculation of optimum parameter values. These drawbacks can be overcome by splitting the target definition interval into Q subintervals, and then expressing approximating function as a linear combination of basis functions, each having **compact** support over only one subinterval, i.e.,

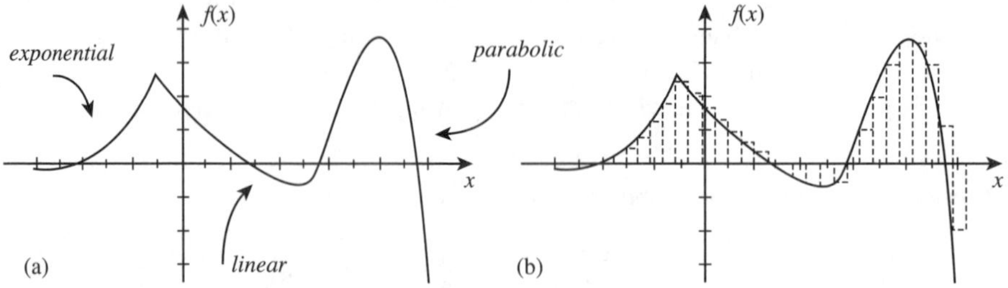

FIGURE 2.3 Example of nonuniform function.

zero value outside this subinterval. For the limiting case in which $Q \to \infty$, this corresponds to interpolating the function by its samples associated to infinitely small subintervals [Figure 2.3(b)]. Such action is functionally equivalent to expressing a signal as its convolution with a delta of Dirac [10].

This splitting and subsequent approximation can be performed ad hoc, by using different functional dependences to fit each subregion. However, to support the systematic design of electronic hardware it is more convenient to rely on well-defined classes of approximating functions. In particular, Hermite PWPs provide large modularity by focusing on the interpolation of measured data taken from the target function. Any lack of flexibility as compared to the ad hoc approach may be absorbed in the splitting of the region.

Consider the more general case in which the function, $y = f(x)$, is defined inside a real interval $[\delta_0, \delta_{N+1}]$ and described as a collection of data measured at *knots* of a given interval partition, $\Delta = \{\delta_0, \delta_1, \delta_2, \ldots, \delta_N, \delta_{N+1}\}$. These data may include the function values at these points, as well as their derivatives, up to the $(M-1)$th order,

$$f^{(k)}(\delta_i) = \left.\frac{d^k}{dx^k} f(x)\right|_{x=\delta_i} \qquad i = 0,1,2,\ldots,N,N+1 \tag{2.7}$$

where k denotes the order of the derivative and is zero for the function itself. These data can be interpolated by a linear combination of basis polynomials of degree $2M-1$,

$$g(x) = \sum_{i=0}^{N+1}\sum_{k=0}^{M-1} f^{(k)}(\delta_i)\Phi_{ik}(x) \tag{2.8}$$

where the expressions for these polynomials are derived from the interpolation data and continuity conditions [78]. Note that for a given basis function set and a given partition of the interval, each coefficient in (2.8) corresponds to a single interpolation kust.

The simplest case uses *linear* basis functions to interpolate only the function values,

$$g(x) = \sum_{i=0}^{N+1} f(\delta_i) l_i(x) \tag{2.9}$$

with no function derivatives interpolated. Figure 2.4 shows the shape of the inner jth linear basis function, which equals 1 at δ_i and decreases to 0 at δ_{i-1} and δ_{i+1}. Figure 2.5(a) illustrates the representation in (2.9). By increasing the degree of the polynomials, the function derivatives also can be interpolated. In particular, two sets of third-degree basis functions are needed to retain modularity in the interpolation of the function and its first derivative at the knots

$$g(x) = \sum_{i=0}^{N+1} f(\delta_i) v_i(x) + \sum_{i=0}^{N+1} f^{(1)}(\delta_i) s_i(x) \tag{2.10}$$

where Appendix B shows the shapes and expressions of the value, $v_i(x)$, and slope, $s_i(x)$, basis functions.

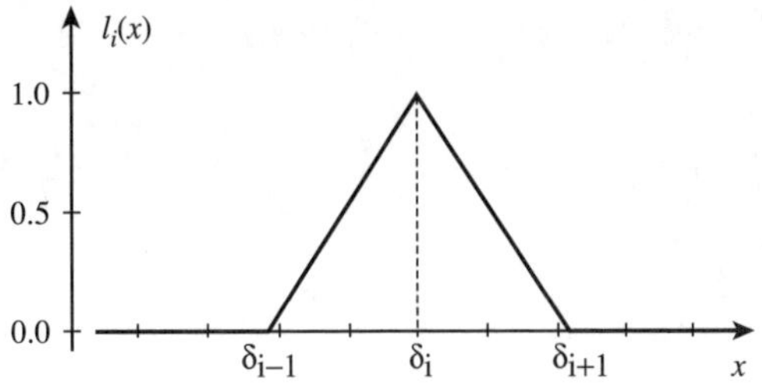

FIGURE 2.4 Hermite linear basis function.

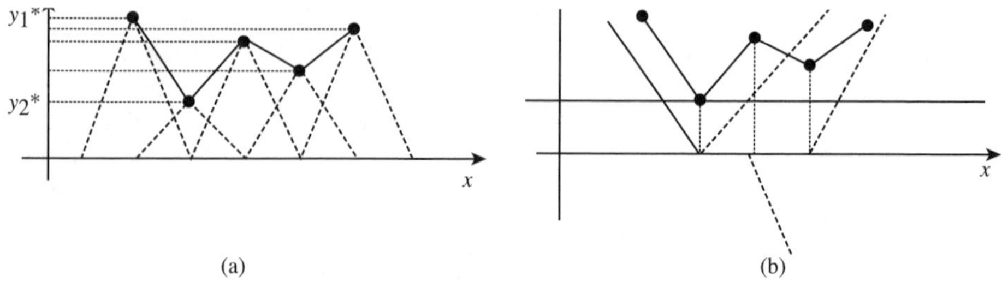

FIGURE 2.5 Decomposition of a PWL function using the extension operator.

The modularity of Hermite polynomials is not free; their implementation is not cheapest in terms of components and, consequently, may not be optimal for an application in which the target function is fixed. These applications are more conveniently handled by the so-called **canonical representation** of PWP functions. A key concept is the **extension** operator introduced in [6]; the basic idea behind this concept is to build the approximating function following an iterative procedure. At each iteration, the procedure starts from a function that fits the data on a subinterval, enclosing several pieces of the partition interval, and then adds new terms to also fit the data associated to the next piece. Generally, some pieces are fit from left to right and others from right to left, to yield

$$g(x) = g^0(x) + \sum_{i=1}^{N+} \Delta^+ g_i(x) + \sum_{i=-N_-}^{-1} \Delta^- g_i(x) \tag{2.11}$$

It is illustrated in Figure 2.5(b). The functions in (2.11) have the following general expressions

$$\begin{aligned} \Delta^+ g(x) &= wu_+(x-\delta) \equiv w(x-\delta)\,\text{sgn}(x-\delta) \\ \Delta^- g(x) &= wu\_(x-\delta) \equiv w(x-\delta)\,\text{sgn}(\delta - x) \\ g^0(x) &= ax + b \end{aligned} \tag{2.12}$$

where sgn $(\cdot)$ denotes the *sign* function, defined as an application of the real axis onto the discrete set $\{0,1\}$.

This representation, based on the extension operator, is elaborated in [6] to obtain the following canonical representation for unidimensional PWL functions:

$$g(x) = ax + b + \sum_{i=1}^{N} w_i |x - \delta_i| \tag{2.13}$$

which has the remarkable feature of involving only one nonlinearity: the **absolute value** function.

The extension operator concept was applied in [30] to obtain canonical representations for cubic Hermite polynomials and B-splines. Consequently, it demonstrates that a PWP function admits a global expression consisting of a linear combination of powers of the input variable, plus truncated powers of shifted versions of this variable. For instance, the following expression is found for a cubic B-spline:

$$g(x) = \sum_{r=0}^{3} \alpha_r x^r + \sum_{i=1}^{N} \beta_i (x - \delta_i)^3 \,\text{sgn}(x - \delta_i) \tag{2.14}$$

with α_r and β_i obtainable through involved operations using the interpolation data. Other canonical PWP representations devised by these authors use

$$(x-\delta_i)^r \operatorname{sgn}(x-\delta_i) = \frac{1}{2}\{|x-\delta_i| + (x-\delta_i)\}(x-\delta_i)^{r-1} \tag{2.15}$$

to involve the absolute value, instead of the sign function, in the expression of the function.

Gaussian and Bell-Shaped Basis Functions

The Gaussian basis function belongs to the general class of radial basis functions [51, 52], and has the following expression:

$$\Phi(x) = \exp\left(-\frac{(x-\delta)^2}{2\sigma^2}\right) \tag{2.16}$$

plotted in Figure 2.6. The function value is significant only for a small region of the real axis centered around its *center*, δ, and its shape is controlled by the *variance* parameter, σ^2. Thus, even through the support of Gaussian functions is not exactly compact, they are negligible except for well-defined *local* domains of the input values.

By linear combination of a proper number of Gaussians, and a proper choice of their centers and variances, as well as the weighting coefficients, it is possible to approximate nonlinear functions to any degree of accuracy [51]. Also, the local feature of these functions renders this adjustment process simpler than for multilayer networks composed of nested sigmoids, whose components are global [43, 50].

A similar interpolation strategy arises in the framework of fuzzy reasoning, which is based on local *membership* functions whose shape resembles a Gaussian. For instance, in the ANFIS system proposed by Jang [33]

$$\Phi(x) = \frac{1}{1+\left[\left(\frac{x-\delta}{\sigma}\right)^2\right]^\beta} \tag{2.17}$$

as plotted in Figure 2.7(a) where the shape is controlled by β and σ, and the position is controlled by δ. Other authors, for instance, Yamakawa [81], use the PWL membership function shape of Figure 2.7(b),

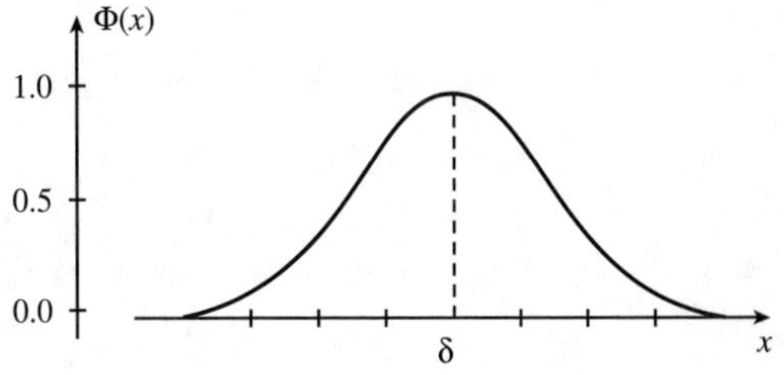

FIGURE 2.6 Guassian basis function.

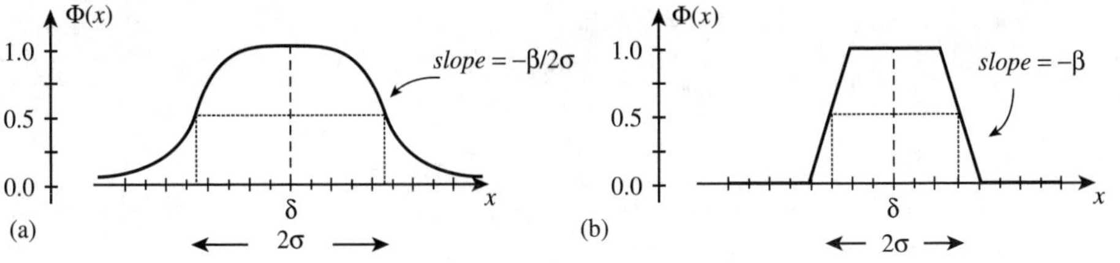

FIGURE 2.7 Fuzzy membership functions: (a) polynomial; (b) piecewise-linear.

which is similar to the Hermite linear basis function of Figure 2.4. From a more general point of view, cubic B-splines [78] used to build hardware [59] and for device modeling [76] also can be considered to be members of this class of functions.

Multidimensional Functions

Approximation techniques for multidimensional functions are informally classified into five groups:

1. Sectionwise piecewise polynominal functions [6, 30]
2. Canonical piecewise linear representations [11]
3. Neuro-fuzzy interpolation [33, 81]
4. Radial basis functions [51, 52]
5. Multilayers of nested sigmoids [82]

Sectionwise Piecewise-Polynomial Functions

This technique reduces the multidimensional function to a sum of products of functions of only one variable:

$$g(\mathbf{x}) = \sum_{k_1=1}^{M_1} \sum_{k_2=1}^{M_2} \cdots \sum_{k_P=1}^{M_P} \alpha(k_1, k_2, \cdots, k_P) \prod_{j=1}^{P} \Phi_{k_j}(x_j) \tag{2.18}$$

where $\alpha(k_1, k_2, \ldots, k_P)$ denotes a constant coefficient. These function representations were originally proposed by Chua and Kang for the PWL case [6] where

$$\begin{aligned} &\Phi_1(x_j) = 1 \quad \Phi_2(x_j) = x_j \quad \Phi_3(x_j) = |x_j - \delta_{j1}| \\ &\cdots \Phi_{M_P}(x_j) = |x_j - \delta_{jM_P - 2}| \end{aligned} \tag{2.19}$$

Similar to the unidimensional case, the only nonlinearity involved in these basis functions is the absolute value. However, multidimensional functions not only require weighted summations, but also multiplications. The extension of (2.18) to PWP functions was covered in [30], and involves the same kind of nonlinearities as (2.14) and (2.15).

Canonical Piecewise Linear Representations

The canonical PWL representation of (2.13) can be extended to the multidimensional case, based on the following representation:

$$g(\mathbf{x}) = \mathbf{a}^T\mathbf{x} + b + \sum_{i=1}^{Q} c_i |\mathbf{w}_i^T \mathbf{x} - \delta_i| \tag{2.20}$$

where $\mathbf{a}$ and $\mathbf{w}_i$ are P-vectors; b, c_i, and δ_i are scalars; and Q represents the number of hyperplanes that divide the whole space R^P into a finite number of polyhedral regions where $g(\cdot)$ can be expressed as an affine representation. Note that (2.20) avoids the use multipliers. Thus, $g(\cdot)$ in (2.20) can be realized through the block diagram of Figure 2.8, consisting of Q absolute value nonlinearities and weighted summers.

Radial Basis Functions

The idea behind radial basis function expansion is to represent the function at each point of the input space as a linear combination of kernel functions whose arguments are the radial distance of the input point to a selected number of centers

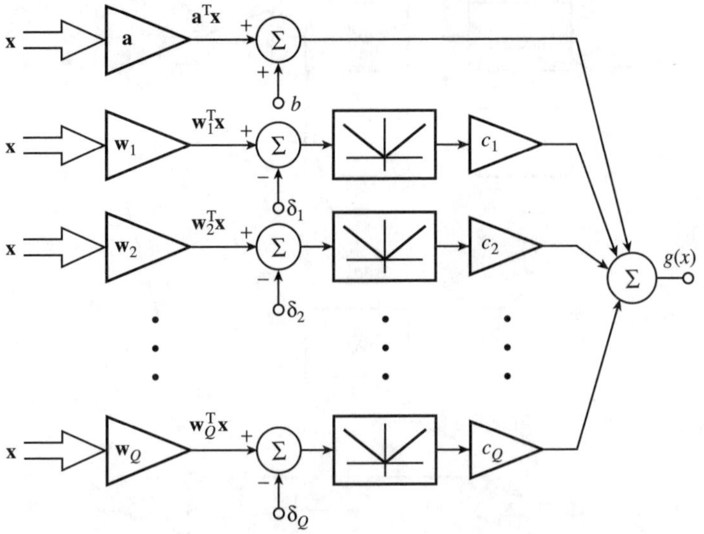

FIGURE 2.8 Canonical block diagram for a canonical PWL function.

$$g(\mathbf{x}) = \sum_{j=1}^{Q} w_j \Phi_j\left(\left\|\mathbf{x} - \boldsymbol{\delta}_j\right\|\right) \tag{2.21}$$

where $\|\cdot\|$ denotes a norm imposed on R^P, usually assumed Euclidean. The most common basis function is a Gaussian kernel similar to (2.16),

$$\Phi(\mathbf{x}) = \exp\left(-\frac{\|\mathbf{x} - \boldsymbol{\delta}\|^2}{2\sigma^2}\right) \tag{2.22}$$

although many other alternatives are available [51], for instance,

$$\Phi(r) = \left(\sigma^2 + r^2\right)^{-\alpha} \quad \Phi(r) = r \quad \alpha \geq -1 \tag{2.23}$$

where r is the radial distance to the center of the basis function, $r \equiv \|\mathbf{x} - \delta\|$. Micchelli [42] demonstrated that any function where the first derivative is monotonic qualifies as a radial basis function. As an example, as (2.23) displays, the identity function $\Phi(r) = r$ falls into this category, which enables connecting the representation by radial basis functions to the canonical PWL representation [40]. Figure 2.9 is a block diagram for the hardware realization of the radial basis function model.

Neuro-Fuzzy Interpolation

This technique exploits the interpolation capabilities of fuzzy inference, and can be viewed as the multidimensional extension of the use of linear combination of bell-shaped basis functions to approximate nonlinear functions of a single variable [see (2.4) and (2.17)]. Apart from its connection to approximate reasoning and artificial intelligence, this extension exhibits features similar to the sectionwise PWP representation, namely, it relies on a well-defined class of unidimensional functions. However, neuro-fuzzy interpolation may be advantageous for hardware implementation because it requires easy-to-build collective computation operators instead of multiplications.

Figure 2.10 depicts the block diagram of a neuro-fuzzy interpolator for the simplest case in which inference is performed using the singleton algorithm [33] to obtain

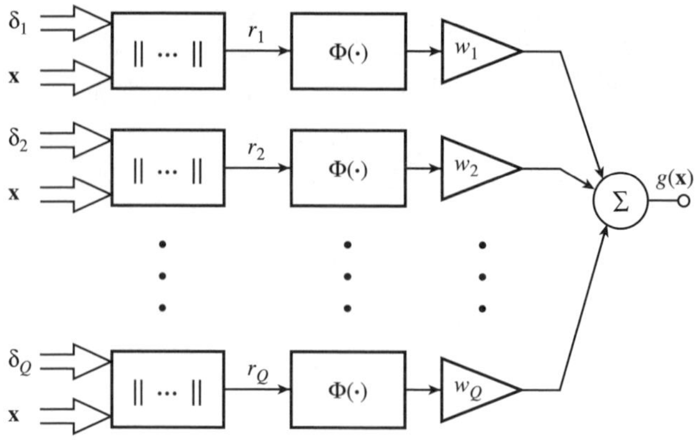

FIGURE 2.9 Concept of radial basis function hardware.

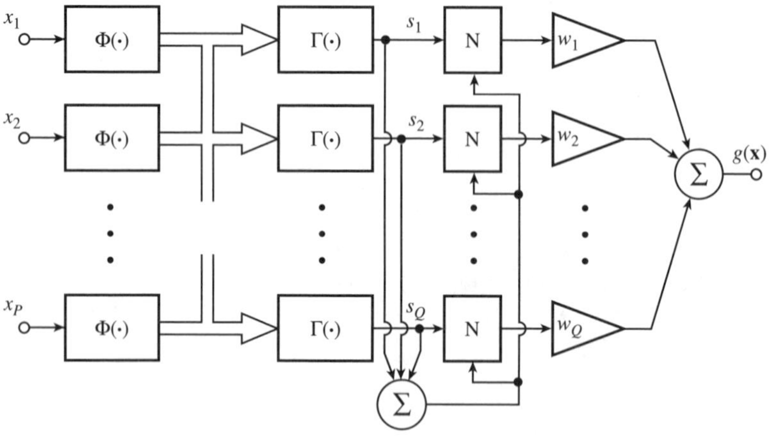

FIGURE 2.10 Conceptual architecture of a neuro-fuzzy interpolator.

$$g(\mathbf{x}) = \sum_{j=1}^{Q} w_j \frac{s_j(\mathbf{x})}{\sum_{i=1,Q} s_i(\mathbf{x})} \tag{2.24}$$

where the functions $s_i(\mathbf{x})$, called activities of the fuzzy *rules*, are given as

$$s_j(\mathbf{x}) = \Gamma\{\Phi_{j1}(x_1), \Phi_{j2}(x_2), \ldots, \Phi_{jP}(x_P)\} \tag{2.25}$$

where $\Gamma(\cdot)$ is any T-norm operator, for instance, the *minimum*, and $\Phi(\cdot)$ has a bell-like shape (see Figure 2.7).

Multilayer Perceptron

Similar to (2.5), but multilayer perceptron consists of the more general case of several layers, with the input to each nonlinear block given as a linear combination of the multidimensional input vector [82].

2.3 Aggregation, Scaling, and Transformation Circuits

The mathematical techniques presented in Section 2.2 require several nonlinear operators and the linear operators of scaling and aggregation (covered for completeness in this section). This section also covers

FIGURE 2.11 First-order models for voltage op amps and CCIIs using nullators and norators.

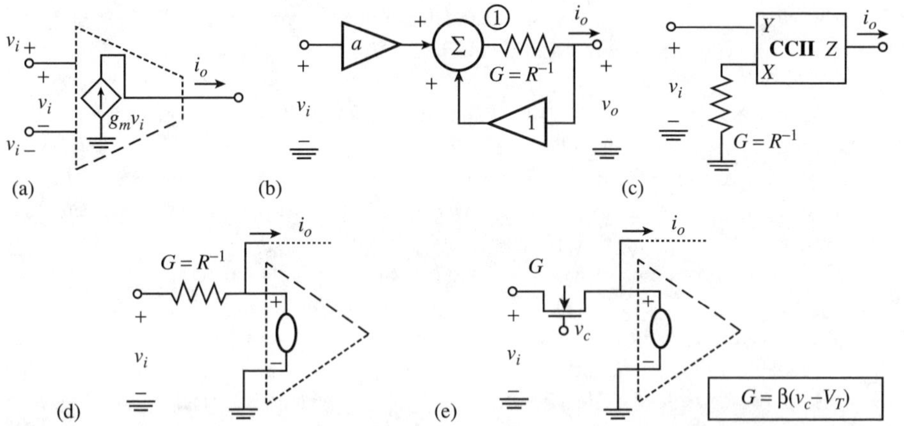

FIGURE 2.12 Voltage-to-current transformation: (a) using an OTA; (b) using voltage feedback; (c) using a current conveyor; (d) using virtual ground of an op amp; (e) same as d, but with active resistors.

transformation circuits. This is because in many practical situations we aim to exploit some nonlinear mechanism which intrinsically involves a particular kind of characteristics. For instance, a MOS transistor has inherent square-law transconductance, while a diode exhibits an exponential driving-point. Similarly, many nonlinear operators are naturally realized in current-mode domain and involve currents at both the input and the output. Thus, transformation circuits are needed to exploit these mechanisms for other types of characteristics.

Transformation Circuits

Two basic problems encountered in the design of transformation circuits are how to convert a voltage node into a current node and vice versa. We know no unique way to realize these functions. Instead, there are many alternatives which depend on which active component from Appendix A is used. The OTA can be represented to a first-order model as a voltage-controlled current source (VCCS) with linear transconductance parameter g_m. Regarding the op amp and CCII, it is convenient to represent them by the first-order models of Figure 2.11, which contain **nullators** and **norators**.[1] A common appealing feature of both models is the virtual ground created by the input nullator. It enables us to sense the current drawn by nodes with fixed voltage — fully exploitable to design transformation circuits.

Voltage-to-Current Transformation

A straightforward technique for voltage-to-current conversion exploits the operation of the OTA as a VCCS [see Figure 2.12(a)] to obtain $i_0 = g_m v_i$, where g_m is the OTA transconductance parameter [22]. A

[1]A nullator simultaneously yields a short circuit and an open circuit, while the voltage and the current at a norator are determined by the external circuitry. The use of a nullator to model the input port of an op amp is valid only if the component is embedded in a negative feedback configuration. With regard to the CCII, the required feedback is created by the internal circuitry.

drawback is that its operation is linear only over a limited range of the input voltage. Also, the scaling factor is inaccurate and strongly dependent on temperature and technology. Consequently, voltage-to-current conversion using this approach requires circuit strategies to increase the OTA linear operation range [17, 70], and tuning circuits to render the scaling parameter accurate and stable [70]. As counterparts, the value of the scaling factor is continuously adjustable through a bias voltage or current. Also, because the OTA operates in open loop, its operation speed is not restricted by feedback-induced pole displacements.

The use of feedback attenuates the linearity problem of Figure 2.12(a) by making the conversion rely on the constitutive equation of a passive resistor. Figure 2.12(b) illustrates a concept commonly found in op-amp-based voltage-mode circuits [29, 59]. The idea is to make the voltage at node ① of the resistor change linearly with v_o, $v_1 = v_o + av_i$, and thus render the output current independent of v_o, to obtain $i_o = G(v_o + av_i - v_o) = aGv_i$. The summing node in Figure 2.12 (b) is customarily realized using op amps and resistors, which is very costly in the more general case in which the summing inputs have high impedance. The circuits of Figure 2.12(c) and (d) reduce this cost by direct exploitation of the virtual ground at the input of current conveyors [Figure 2.12(c)] and op amps [Figure 2.12 (d)]. For both circuits, the virtual ground forces the input voltage v_i across the resistor. The resulting current is then sensed at the virtual ground node and routed to the output node of the conveyor, or made to circulate through the feedback circuitry of the op amp, to obtain $i_o = Gv_i$.

Those implementations of Figure 2.12(b), (c), and (d) that use off-the-shelf passive resistors overcome the accuracy problems of Figure 2.12(a). However, the values of monolithic components are poorly controlled. Also, resistors may be problematic for standard VLSI technologies, where high-resistivity layers are not available and consequently, passive resistors occupy a large area. A common IC-oriented alternative uses the ohmic region of the MOS transistor to realize an active resistor [69] [Figure 2.12(e)]. Tuning and linearity problems are similar to those for the OTA. Circuit strategies to overcome the latter are ground in [13, 32, 66, 69].

Current-to-Voltage Transformation

The most straightforward strategy consists of a single resistor to draw the input current. It may be passive [Figure 2.13(a)] or active [Figure 2.13(b)]. Its drawback is that the node impedance coincides with the resistor value, and thus makes difficult impedance matching to driving and loading stages. These matching problems are overcome by Figure 2.13(c), which obtains low impedances at both the input and the output ports. On the other hand, Figure 2.13(d) obtains low impedance at only the input terminal, but maintains the output impedance equal to the resistor value. All circuits in Figure 2.13 obtain $v_o = Ri_i$, where $R = g_m^{-1}$ for the OTA.

Voltage/Charge Domain Transformations for Sampled-Data Circuits

The linearity and tuning problems of previous IC-related transformation approaches are overcome through the use of *dynamic* circuit design techniques based on **switched-capacitors** [72]. The price is that the operation is no longer asynchronous: relationships among variables are only valid for a discrete set of time instants. Variables involved are voltage and charge, instead of current, and the circuits use capacitors, switches, and op amps.

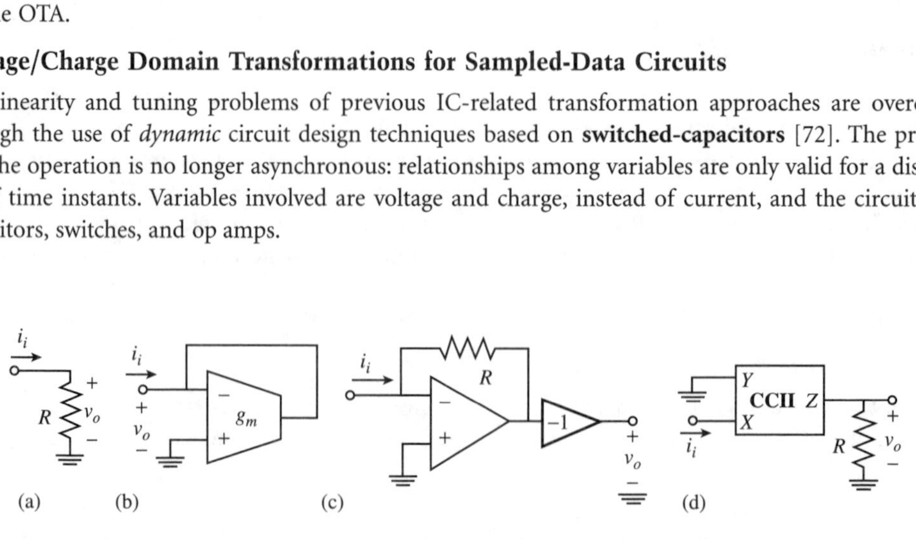

FIGURE 2.13 Current-to-voltage transformation: (a) using a resistor; (b) using a feedback OTA; (c) using op amps; (d) using current conveyors.

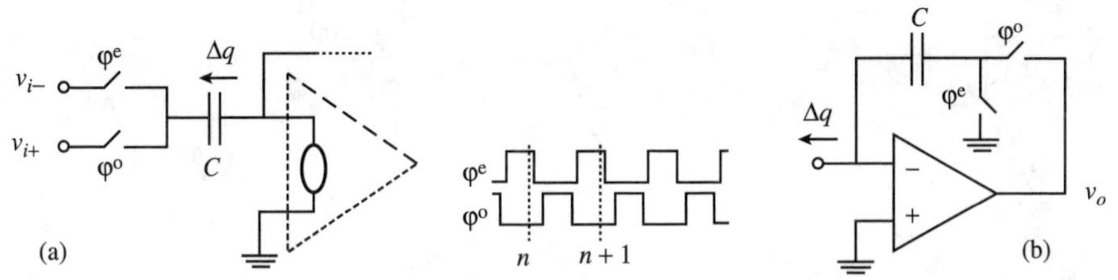

FIGURE 2.14 Transformations for sampled-data circuits: (a) *V*-to-*q*; (b) *q*-to-*V*.

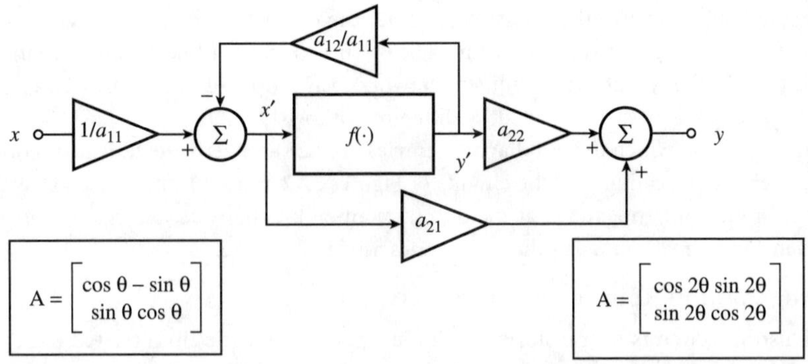

FIGURE 2.15 Concept of linear transformation converter for transfer characteristics: general architecture, and transformation matrices for rotation (left) and reflection (right).

Figure 2.14(a) is for voltage-to-charge transformation, while Figure 2.14(b) is for charge-to-voltage transformation. The switches in Figure 2.14(a) are controlled by nonoverlapping clock signals, so that the structure delivers the following incremental charge to the op amp virtual ground node:

$$\Delta q^e = C\left(v_{i^+} - v_{i^-}\right) = -\Delta q^o \tag{2.26}$$

where the superscript denotes the clock phase during which the charge is delivered. Complementarily, the structure of Figure 2.14(b) initializes the capacitor during the even clock phase, and senses the incremental charge that circulates through the virtual ground of the op amp during the odd clock phase. Thus, it obtains

$$v_0^o = C\left(\Delta q^o\right) \tag{2.27}$$

References [45, 46] and [68] contain alternative circuits for the realization of the scaling function. Such circuits have superior performance in the presence of parasitics of actual monolithic op amps and capacitors.

Transformation among Transfer Characteristics

Figure 2.15 depicts the general architecture needed to convert one kind of transfer characteristics, e.g., voltage transfer, into another, e.g., current transfer. Variables x' and y' of the original characteristics can be either voltage or current, and the same occurs for x and y of the converted characteristic. The figure depicts the more general case, which also involves a linear transformation of the characteristics themselves:

$$\begin{bmatrix} x \\ y \end{bmatrix} = \mathbf{A}\begin{bmatrix} x' \\ y' \end{bmatrix} = \begin{bmatrix} a_{11} & a_{12} \\ a_{21} & a_{22} \end{bmatrix}\begin{bmatrix} x' \\ y' \end{bmatrix} \tag{2.28}$$

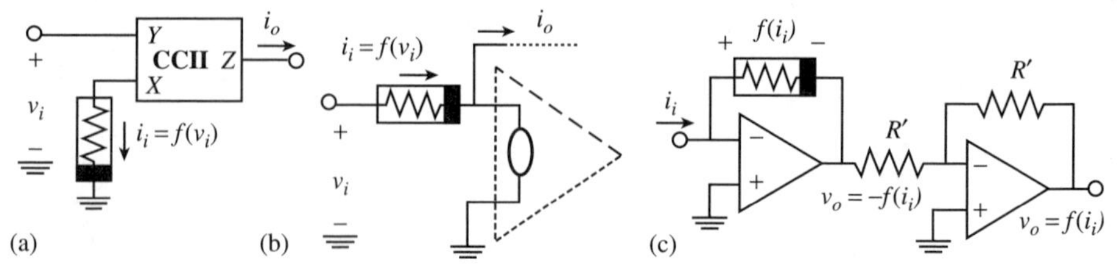

FIGURE 2.16 From driving-point to transfer characteristics: (a) and (b) transconductance from voltage-controlled driving-point; (c) transimpedance from current-controlled driving-point.

For example, the figure encloses the matrices to rotate the characteristics by an angle θ, and to reflect the characteristics with respect to an edge with angle θ. This concept of linear transformation converters and its applications in the synthesis of nonlinear networks was proposed initially by Chua [5] for driving-point characteristics, and further extended by different authors [24, 29].

In the simplest case, in which the nondiagonal entries in (2.28) are zero, the transformation performed over the characteristics is scaling, and the circuits of Figures 2.12 and 2.13 can be used directly to convert x into x' at the input front end, and y' at the output front end. Otherwise, aggregation operation is also required, which can be realized using the circuits described elsewhere.

From Driving-Point to Transfer and Vice Versa

Figure 2.16 illustrates circuits to transform driving-point characteristics into related transfer characteristics. Figure 2.16(a) and (b) use the same principle as Figure 2.12(c) and (d) to transform a voltage-controlled *driving-point* characteristic, $i_i = f(v_i)$, into a *transconductance* characteristics. On the other hand Figure 2.16(c) operates similarly to Figure 2.13(c) to transform a current-controlled *driving-point* characteristic, $v_i = f(i_i)$, into a transimpedance characteristic. If the resistance characteristics of the resistor in Figure 2.16(a) and (b), or the conductance characteristic of the resistor in Figure 2.16(c), is invertible, these circuits serve to invert nonlinear functions [63]. For instance, using a common base BJT in Figure 2.16(c) obtains a logarithmic function from the BJT exponential transconductance. Also, the use of a MOST operating in the ohmic region serves to realize a division operation.

Lastly, let us consider how to obtain driving-point characteristics from related transfer characteristics. Figure 2.17(a) and (b) correspond to the common situation found in op amp-based circuits, where the transfer is between voltages. Figure 2.17(a) is for the voltage-controlled case and Figure 2.17(b) is for the current-controlled case. They use feedback strategies similar to Figure 2.17(b) to render either the input

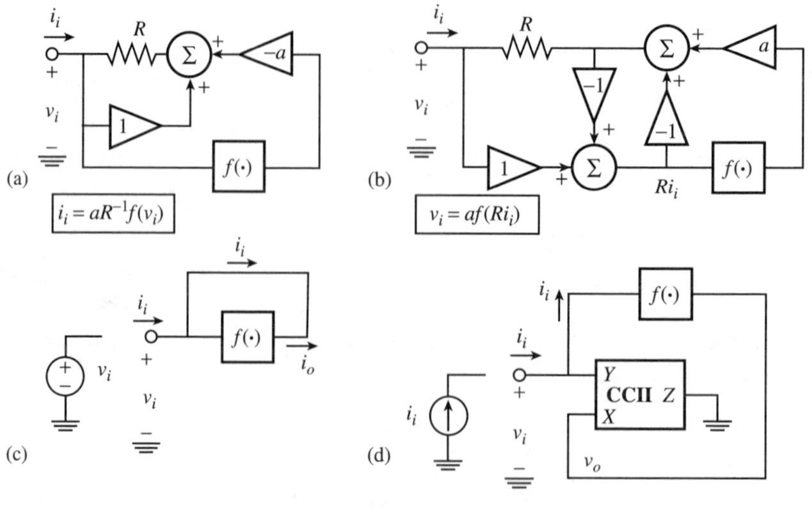

FIGURE 2.17 From transfer to driving-point characteristics.

voltage or the input current independent of the linear contributions of the other port variable. A general theory for this kind of transformation converter can be found in [29].

Note that these figures rely on a Thévenin representation. Similar concepts based on Norton representations allow us to transform current transfer characteristics into driving-point characteristics. However, careful design is needed to preserve the input current while sensing it.

Other interesting transformation circuits are depicted in Figure 2.17(c) and (d). The block in Figure 2.17(c) is a transconductor that obtains $i_o = -f(v_i)$ with very large input impedance. Then, application of feedback around it obtains a voltage-controlled resistor, $i_o = f(v_i)$. Figure 2.17(d) obtains a current-controlled resistor, $v_i = f(i_i)$, using a current conveyor to sense the input current and feedback the output voltage of a transimpedance device with $v_o = f(i_i)$.

Scaling and Aggregation Circuitry

Scaling Operation

Whenever the weights are larger than unity, or are negatives, the operation of scaling requires active devices. Also, because any active device acts basically as a transconductor, the scaling of voltages is performed usually through the transformation of the input voltage into an intermediate current and the subsequent transformation of this current into the output voltage. Figure 2.18 illustrates this for an op-amp-based amplifier and an OTA-based amplifier. The input voltage is first scaled and transformed in i_o, and then this current is scaled again and transformed into the output voltage. Thus, the scaling factor depends on two design parameters. Extra control is achieved by also scaling the intermediate current.

Let us now consider how to scale currents. The most convenient strategy uses a **current mirror**, whose simplest structure consists of two matched transistors connected as shown in Figure 2.19(a) [25]. Its operating principle relies on functional cancellation of the transistor nonlinearities to yield a linear relationship

$$i_o = p_2 f(v_i) = p_2 f\left[f^{-1}\left(\frac{i_i}{p_1}\right)\right] = \frac{p_2}{p_1} i_i \tag{2.29}$$

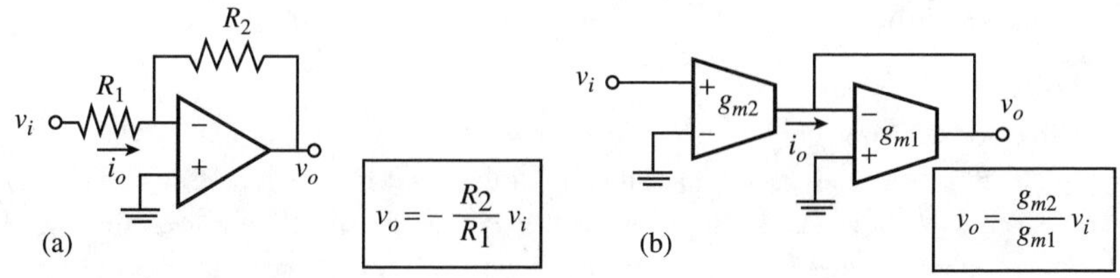

FIGURE 2.18 Mechanisms for voltage scaling.

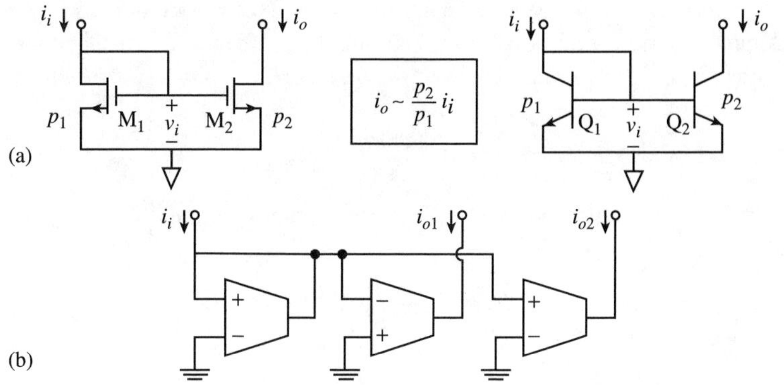

FIGURE 2.19 Current scaling using current mirrors.

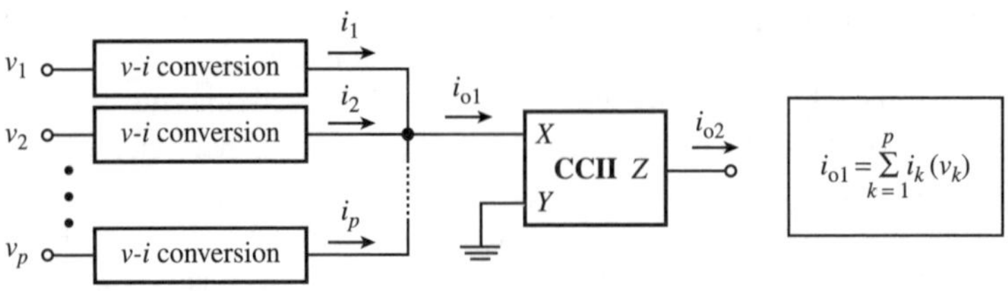

FIGURE 2.20 Aggregation of voltages through intermediate currents and current conveyor.

where p_1 and p_2 are parameters with value that can be designer controlled; for instance, β of the MOST or I_s of the BJT (see Appendix A and [44]). The input and output currents in Figure 2.19(a) must be positive. Driving the input and output nodes with bias currents I_B and $(p_2/p_1)I_B$, respectively, one obtains $i_i = i_i' + I_B$ and $i_o = I_o' + (p_2/p_1)I_B$, and this enables bilateral operation on i_i' and i_o'.

In practical circuits, this simple design concept must be combined with circuit strategies to reduce errors due to nonnegligible input current of BJTs, DC voltage mismatch between input and output terminals, finite input resistance, and finite output resistance. Examples of these strategies can be found in [25, 56, 77]. On the other hand, sizing and layout strategies for other problems related to random mismatches between input and output devices are found in [41] and [48], which are applicable to most matching problems in MOS IC design.

The current mirror concept is extensible to any pair of matched transconductors, provided their transconductance characteristics are invertible and parameterized by a designer-controlled scale factor p, and that the dependence of the output current with the output voltage is negligible. In particular, the use of differential transconductors enables us to obtain bilateral operation simply, requiring no current-shifted biasing at the input and output nodes. It also simplifies achieving noninverting amplification (that is, positive scale factors), as Figure 2.19(b) illustrates. This figure also serves to illustrate the extension of the mirror concept to multiple current outputs. Note that except for loading considerations, no other limitations exist on the number of output transconductors that can share the input voltage. Also, because fan-out of a current source is strictly one, this replication capability is needed to enable several nodes to be excited by a common current. On the other hand, the fact that the different current output replicas can be scaled independently provides additional adjusting capability for circuit design.

Signal Aggregation

As for the scaling operation, aggregation circuitry operates in current domain, based on Kirchhoff's current law (KCL). Thus, the aggregation of voltages requires that first they be transformed into currents (equivalently, charge packets in the switched-capacitor circuitry) and then added through KCL, while currents and incremental charges are added by routing all the components to a common node. If the number of components is large, the output impedance of the driving nodes is not large enough, and/or the input importance of the load is not small enough, this operation will encompass significant loading errors due to variations of the voltage at the summing node. This is overcome by clamping the voltage of this node using a virtual ground, which in practical circuits is realized by using either the input port of an op amp, or terminals X and Y of a current conveyor. Figure 2.20 illustrates the current conveyor case.

2.4 Piecewise-Linear Circuitry

Consider the elementary PWL function that arise in connection with the different methods of representation covered in Section 2.2:

- Two-piece concave and convex characteristics [see (2.12)]
- Hermite linear basis function (see Figure 2.4 and Appendix B)
- Absolute value [see (2.13)]

where **rectification** is the only nonlinear operator involved. The circuit primitives in Appendix A exhibit several mechanisms which are exploitable in order to realize rectification:

- Cut-off of diodes and transistors — specifically, current through a diode negligible for negative voltage, output current of BJTs, and MOSTs negligible under proper biasing
- Very large resistance and zero offset voltage of an analog switch for negative biasing of the control terminal
- Digital encoding of the sign of a differential voltage signal using a comparator

Similar to scaling and aggregation operations, rectification is performed in current domain, using the mechanisms listed previously to make the current through a branch negligible under certain conditions. Three techniques are presented, which use current transfer in a transistor-based circuit, current-to-voltage transfer using diodes and op amp, and charge transfer using switches and comparators, respectively.

Current Transfer Piecewise-Linear Circuitry

Figure 2.21(a) and (b) presents the simplest technique to rectify the current transferred from node ① to node ②. They exploit the feature of diodes and diode-connected transistors to support only positive currents. Figure 2.21(a) operates by precluding negative currents to circulate from node ① to node ②, while Figure 2.21(b) also involves the nonlinear transconductance of the output transistor M_o; negative currents driving the node ① force v_i to become smaller than the cut-in voltage and, consequently, the output current becomes negligible. A drawback to both circuits is that they do not provide a path for negative input currents, which accumulates spurious charge at the input node and forces the driving stage to operate outside its linear operating regime. Solutions to these problems can be found in [57] and [61]. Also, Figure 2.21(a) produces a voltage displacement equal to the cut-in voltage of the rectifying device, which may be problematic for applications in which the voltage at node ① bears information. A common strategy to reduce the voltage displacements uses feedback to create superdiodes (shown in Figure 2.21(c) for the grounded case and Figure 2.21(d) for the floating case), and where the reduction of the voltage displacement is proportional to the DC gain of the amplifier.

Figure 2.22(a), called a current switch, provides paths for positive and negative currents entering node ①, and obtains both kinds of elementary PWL characteristics exploiting cut-off of either BJTs or MOSTs. It consists of two complementary devices: npn (top) and pnp BJTs, or *n*-channel (top) and *p*-channel MOSTs. Its operation is very simple: any positive input current increases the input voltage, turning the bottom device ON. Because both devices share the input voltage, the top device becomes OFF. Similarly, the input voltage decreases for negative input currents, so that the top device becomes

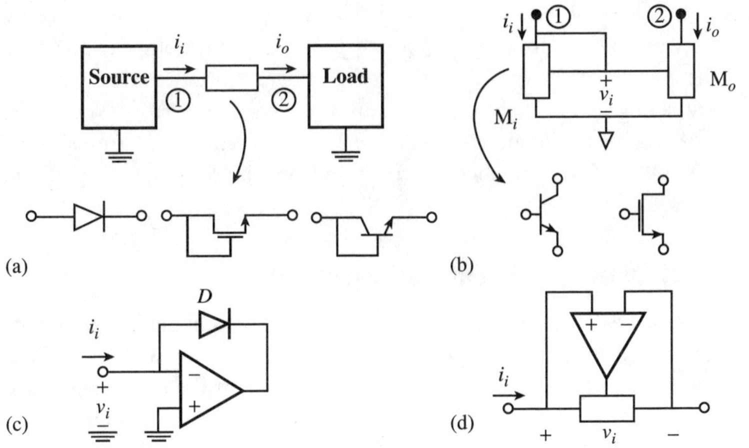

FIGURE 2.21 (a) and (b) circuit techniques for current rectification; (c) and (d) superdiodes.

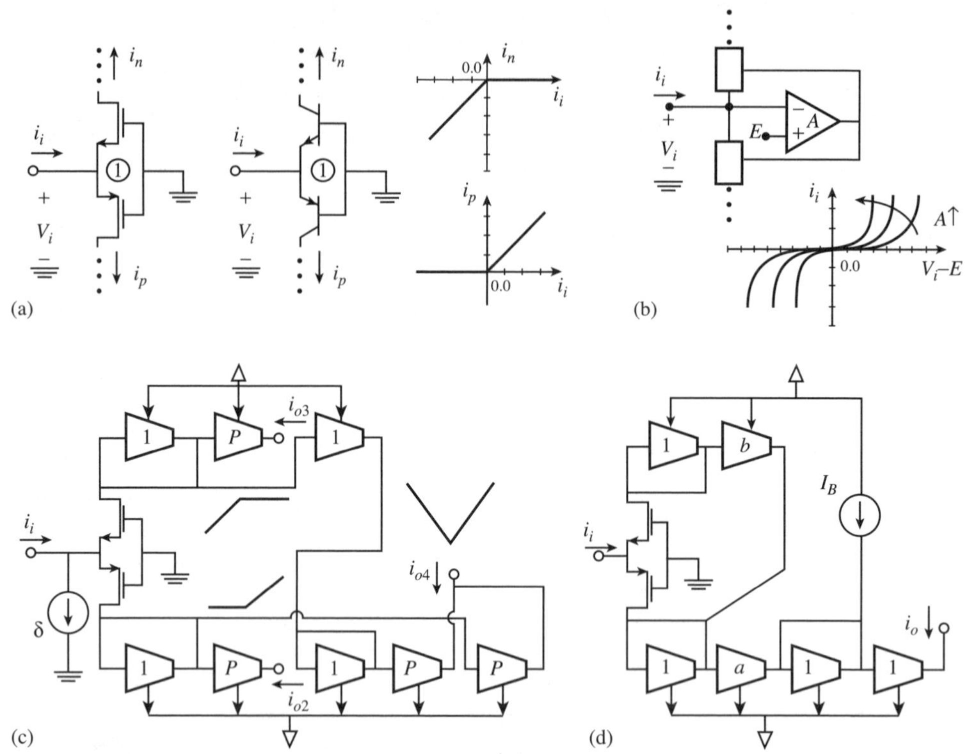

FIGURE 2.22 Current switch and its application for different basic PWL curves.

ON and the bottom OFF. In sum, positive input currents are drawn to the bottom device, while negative currents are drawn to the top device.

An inconvenience of Figure 2.22(a) is the dead zone exhibited by its input driving-point characteristics, which is very wide for MOSTs. It may produce errors due to nonlinear loading of the circuitry that drives the input node. Figure 2.22(b) overcomes this by using a circuit strategy similar to that of the superdiodes. The virtual ground at the op amp input renders the dead-zone centered around the voltage level E, and its amplitude is reduced by a factor proportional to the amplifier DC gain. Some considerations related to the realization of this amplifier are found in [58].

Proper routing and scaling of the currents i_p and i_n in Figure 2.22(a) gives us the concave and convex basic characteristics with full control of the knot and position and the slope in the conducting region. Figure 2.22(c) is the associated circuit, in which the input bias current controls the knot position, and the slope in the conducting region is given by the gain of the current mirrors. Note that this circuit also obtains the absolute value characteristics, while Figure 2.22(d) obtains the Hermite linear basis function. The way to obtain the PWL fuzzy membership function from this latter circuit is straightforward, and can be found in [58].

Transresistance Piecewise-Linear Circuitry

The circuit strategies involved in PWL current transfer can be combined in different ways with the transformation circuits discussed previously to obtain transconductance and voltage-transfer PWL circuits. In many cases, design ingenuity enables optimum merging of the components and consequently, simpler circuits. Figure 2.23(a) depicts what constitutes the most extended strategy to realize the elementary PWL functions using off-the-shelf components [63, 80]. The input current is split by the feedback circuitry around the op amp to make negative currents circulate across D_n and positive currents circulate

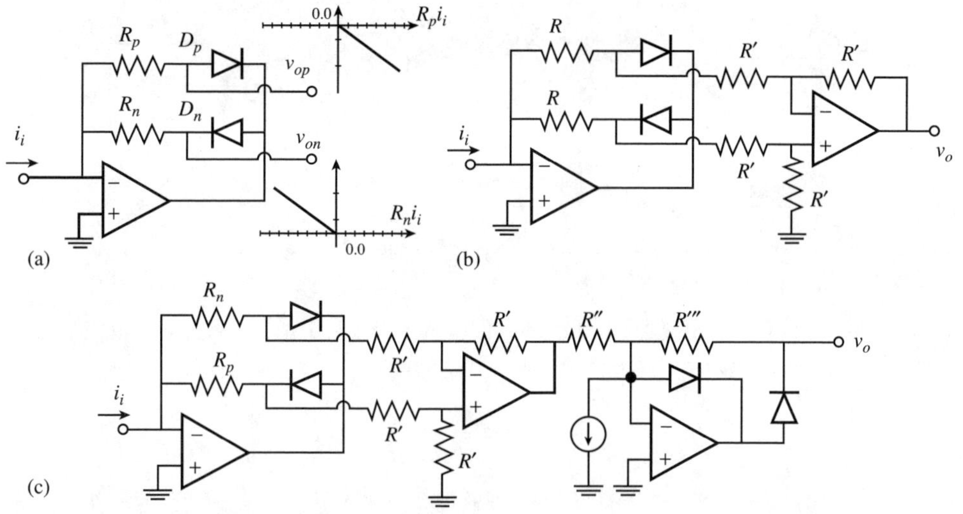

FIGURE 2.23 PWL transimpedance circuits.

across D_p. Consequently, this feedback renders the input node of the op amp a virtual ground and thus reduces errors due to finite diode cut-in voltage in the transresistance characteristics. Similar to Figure 2.22, the position of the knot in these elementary characteristics is directly controlled by an input bias current. Also note that the virtual ground can be exploited to achieve voltage-to-current transformation using the strategy of Figure 2.12(d) and thus, voltage-transfer operation.

Algebraic combination of the elementary curves provided by Figure 2.23(a) requires transforming the voltages v_{on} and v_{op} into currents and then aggregating these currents by KCL. For example, Figure 2.23(b) is the circuit for the absolute value and Figure 2.23(c) presents a possible implementation of the Hermite basis function.

Other related contributions found in the literature focus on the systematic realization of PWL driving-point resistors, and can be found in [7] and [10].

Piecewise-Linear Shaping of Voltage-to-Charge Transfer Characteristics

The realization of PWL relationships among sampled-data signals is based on nonlinear voltage-to-charge transfer and uses analog switches and comparators. Figure 2.24(a) is a circuit structure, where one of the capacitor terminals is connected to virtual ground and the other to a switching block. Assume that nodes ① and ② are both grounded. Note that for $(v - \delta) > 0$ the switch arrangement set node ④ to δ, while node ⑤ is set to v. For $(v - \delta) < 0$, nodes ④ and ⑤ are both grounded. Consequently, voltage at node ③ in this latter situation does not change from one clock phase to the next, and consequently, the incremental

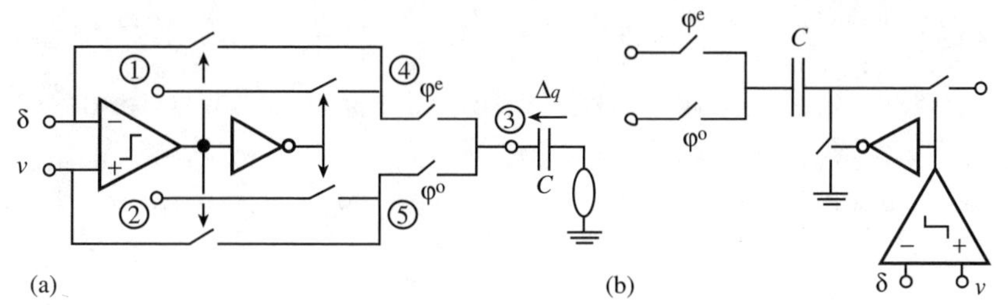

FIGURE 2.24 Circuits for rectification in voltage-to-charge domain.

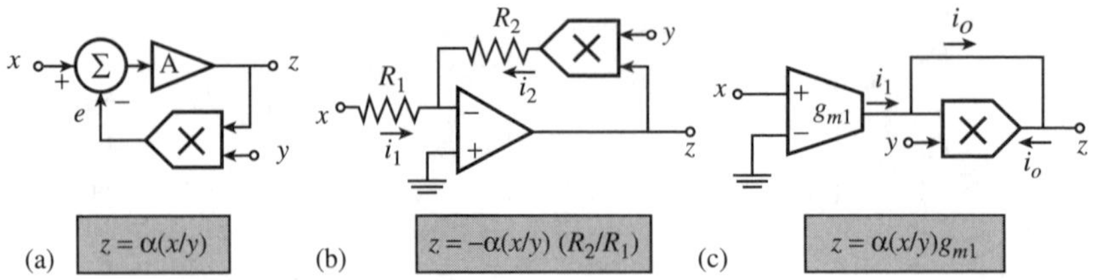

FIGURE 2.25 Division operator using a feedback multiplier: (a) concept; (b) with voltage multiplier and op amp; (c) with transconductance multiplier and OTA.

charge becomes null for $(v - \delta) < 0$. On the other hand, for $(v - \delta) > 0$, the voltage at node ③ changes from one clock phase to the next, and generates an incremental charge

$$\Delta q^e = C(v - \delta) = -\Delta q^o \tag{2.30}$$

which enables us to obtain negative and positive slopes using the same circuit, as shown in Figure 2.24(a). To make the characteristics null for $(v - \delta) > 0$, it suffices to interchange the comparator inputs. Also, the technique is easily extended to the absolute value operation by connecting terminal ① to v, and terminal ② to δ. The realization of the Hermite linear basis function is straightforward and can be found in [55].

Other approaches to the realization of PWL switched-capacitor circuitry use series rectification of the circulating charge through a comparator-controlled switch [Figure 2.24(b)], and can be found in [16] and [31]. The latter also discusses exploitation of these switched-capacitor circuits to realize continuous time driving-point characteristics, the associated transformation circuits, and the dynamic problematics.

2.5 Polynomials, Rational, and Piecewise-Polynomial Functions

These functions use rectification (required for truncation operation in the PWP case) and analog *multiplication*,

$$z = \frac{xy}{\alpha} \tag{2.31}$$

as basic nonlinear operators.[2] Joining the two inputs of the multiplier realizes the *square* function. Analog *division* is realized by applying feedback around a multiplier, illustrated at the conceptual level in Figure 2.25(a); the multiplier obtains $e = (zy)/\alpha$, and, for $A \to \infty$, the feedback forces $x = e$. Thus, if $y \neq 0$, the circuit obtains $z = \alpha(x/y)$. Joining y and z terminals, the circuit realizes the *square root*, $z = (\alpha x)^{1/2}$. This concept of division is applicable regardless of the physical nature of the variables involved. In the special case in which e and x are current and z is a voltage, the division can be accomplished using KCL to yield $x = e$. Figure 2.25(b) shows a circuit for the case in which the multiplication is in voltage domain, and Figure 2.25(c) is for the case in which multiplication is performed in transconductance domain. The transconductance gain for input z in the latter case must be negative to guarantee stability.

Concepts and Techniques for Polynomic and Rational Functions

Figure 2.26 illustrates conceptual hardware for several polynomials up to the fifth degree. Any larger degree is realized similarly. Figure 2.27 uses polynomials and analog division to realize rational functions

[2] Scale factor α in (2.31) must be chosen to guarantee linear operation in the full variation range of inputs and outputs.

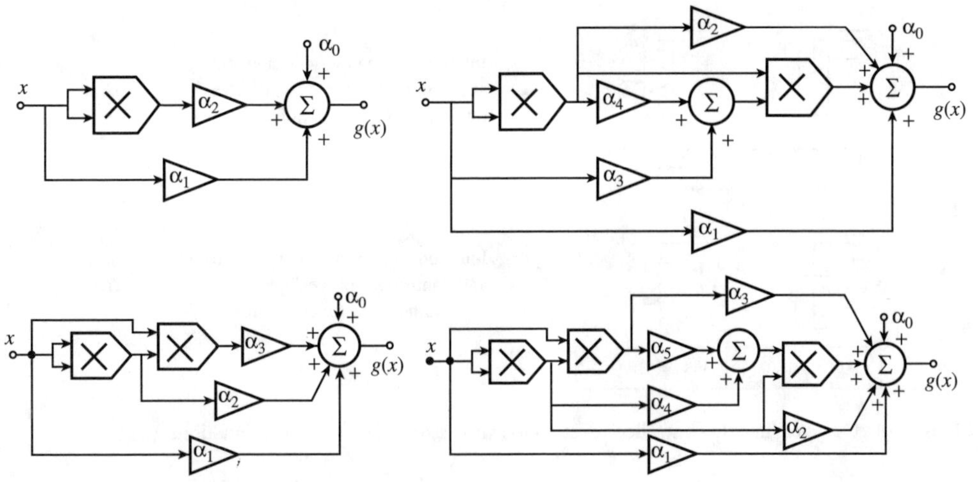

FIGURE 2.26 Conceptual hardware for polynomial functions.

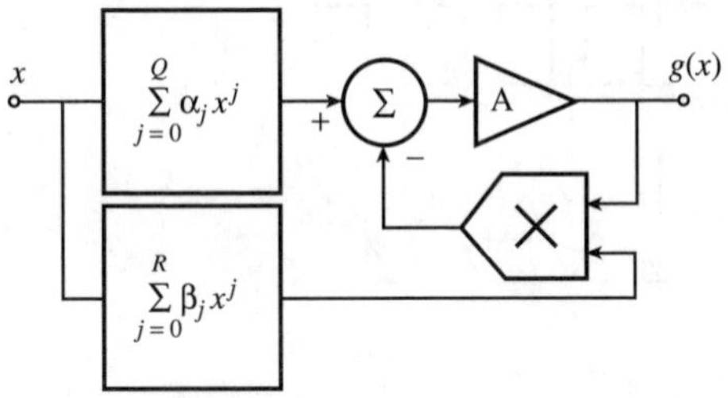

FIGURE 2.27 Rational function generation through feedback division.

$$g(x) = \frac{\sum_{i=0,Q} \alpha_j x^j}{\sum_{j=0,R} \beta_j x^j} \tag{2.32}$$

For simplicity, we have assumed that the internal scaling factors of the multipliers in Figure 2.26 and Figure 2.27 equal one.

An alternative technique to realize rational functions is based on **linearly controlled resistors**, described as $v = (Lx)i$, and **linearly controlled conductors**, $i = (Cx)v$, where L and C are real parameters. This technique exploits the similarity between these characteristics and those which describe inductors and capacitors in the frequency domain, to take advantage of the synthesis techniques for rational transfer function in the s-plane through interconnection of these linear components [28] [39] (Figure 2.28). As for the previous cases, realization of linearly controlled resistors and conductors require only multipliers and, depending upon the nature of the variables involved in the multipliers, voltage-to-current and current-to-voltage transformation circuits.

Multiplication Circuitry

Two basic strategies realize multiplication circuitry: using signal processing and exploiting some nonlinear mechanism of the primitive components. Signal processing multipliers rely on the generation of a pulsed signal whose amplitude is determined by one of the multiplicands and its duty cycle by the other, so that the area is proportional to the result of the multiplication operation. Figure 2.29(a) presents an implementation concept based on averaging. This is performed by a low-pass filter where the input is a pulse train

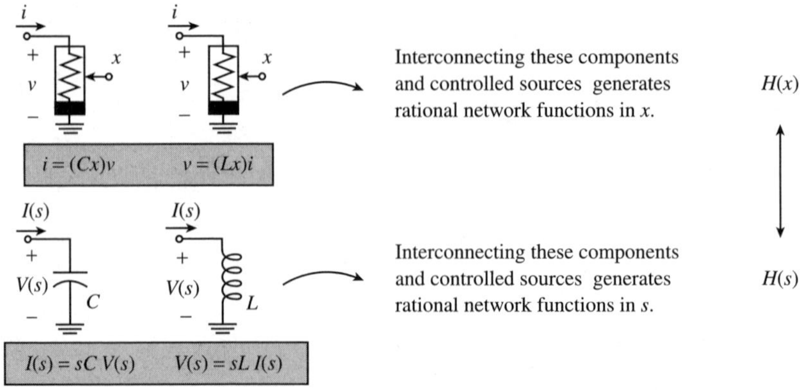

FIGURE 2.28 Usage of linearly controlled resistors to synthesize rational network functions.

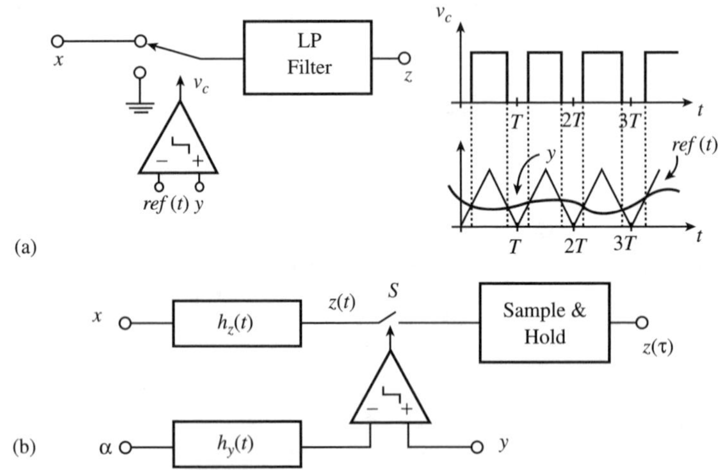

FIGURE 2.29 Signal processing multipliers: (a) by averaging; (b) by shaping in time domain.

with amplitude proportional to x and duty cycle proportional to y. The latter proportionality is achieved through nonlinear sampling by comparing y with a time reference sawtooth signal. Thus, the area under each pulse in the train is the product of $x \times y$, extracted by the low-pass filter. This implementation concept is discussed in further detail in classical texts on analog computation [63], and applied more recently to analog VLSI signal processing [72].

Figure 2.29(b) is an alternative implementation concept based on signal shaping in the time domain. It uses two linear blocks with normalized unit step response given as $h_z(t)$ and $h_y(t)$. The first is driven by level x to obtain

$$z(t) = xh_z(t) \quad 0 \leq t < \tau \tag{2.33}$$

where τ denotes the duration of the time interval during which the switch S remains closed. The other is driven by a references level α, to render τ given by

$$\tau = h_y^{-1}\left(\frac{y}{\alpha}\right) \tag{2.34}$$

Assuming both linear blocks are identical and the time function invertible, one obtains the steady-state value of z, $z(\tau)$, as the product of levels x and y.

The simplest implementation of Figure 2.29 uses integrators, i.e., $h(t) = t$, as linear blocks [see Figure 2.41(b)]. Also note that the principle can be extended to the generation of powers of an input signal by higher order shaping in time domain. In this case, both linear blocks are driven by reference levels. The block $h_y(t)$ consists of a single integrator, $\tau = y/\alpha$. The other consists of the cascade of P integrators, and obtains $z(t) = \beta t^P$. Thus, $z(t) = \beta(y/\alpha)^P$. Realizations suitable for integrated circuits are found in [34] and [55].

Multipliers Based on Nonlinear Devices

The primitives in Appendix A display several mechanisms that are exploitable to realize analog multipliers:

- Exponential functionals associated to the large-signal transconductance of BJTS, and the possibility of obtaining logarithmic dependencies using feedback inversion
- Square-law functionals associated to the large-signal transconductance of the MOS transistor operating in saturation region
- Small-signal transconductance of a BJT in active region as a linear function of collector current
- Small-signal transconductance of a MOST in saturation as a linear function of gate voltage
- Small-signal self-conductance of a MOS transistor in ohmic region as a linear function of gate voltage

These and related mechanisms have been explained in different ways and have resulted in a huge catalog of practical circuits. To quote all the related published material is beyond the scope of this section. The references listed at the end were selected because of their significance, and their cross-references contain a complete view of the state of the art. Also, many of the reported structures can be grouped according to the theory of *translinear circuits,* which provides a unified framework to realize nonlinear algebraic functions through circuits [23, 62, 79].

Log-Antilog Multipliers

Based on the exponential large-signal transconductance of the BJT, and the following relationships,

$$\begin{aligned} z' &= \ln(x) + \ln(y) = \ln(xy) \\ z &= e^{z'} = e^{\ln(xy)} = xy \end{aligned} \tag{2.35}$$

which can be realized as illustrated in Figure 2.30(a) [65]. This circuit operates on positive terminal currents to obtain $i_0 = (i_1 i_2)/i_3$, which can be understood from translinear circuit principles by noting that the four base-to-emitter voltages define a **translinear loop**,

$$\begin{aligned} 0 &= v_{be1} + v_{be2} - v_{be3} - v_{be4} \\ &= \ln\left(\frac{i_1}{I_s}\right) + \ln\left(\frac{i_2}{I_s}\right) - \ln\left(\frac{i_3}{I_s}\right) - \ln\left(\frac{i_o}{I_s}\right) \end{aligned} \tag{2.36}$$

The circuit can be made to operate in four-quadrant mode, though restricted to currents larger than $-I_B$, by driving each terminal with a bias current source of value I_B. Also, because all input terminals are virtual ground the circuit can be made to operate on voltages by using the voltage-to-current transformation concept of Figure 2.12(d). Similarly, the output current can be transformed into a voltage by using an extra op amp and the current-to-voltage transformation concept of Figure 2.13(c). Extension

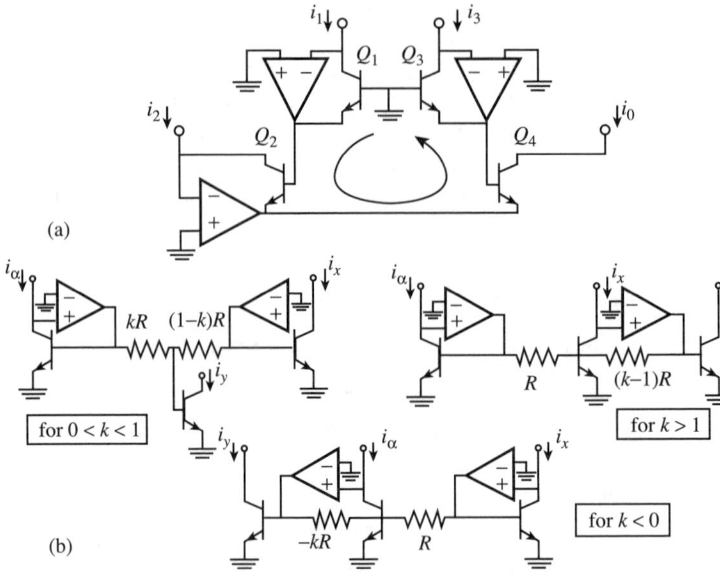

FIGURE 2.30 (a) Core block of a log-antilog multiplier; (b) circuits to elevate to a power.

of this circuit structure to generate arbitrary powers is discussed in [23]. Figure 2.30(b) [1] uses similar techniques, based on introducing scaling factors in the translinear loop, to obtain

$$i_y = i_\alpha^{1-k} i_x^k \tag{2.37}$$

Square-Law Multipliers

These are based on the algebraic properties of the square function, most typically

$$z = \frac{1}{4}\left[(x+y)^2 - (x-y)^2\right] = xy \tag{2.38}$$

shown conceptually in Figure 2.31(a), and the possibility of obtaining the square of a signal using circuits, typically consisting of a few MOS transistors operating in saturation region. Figure 2.31(b) through (f) depict some squarer circuits reported in the literature.

The completeness of square-law operators for the realization of nonlinear circuits was demonstrated from a more general point of view in [47], and their exploitation has evolved into systematic circuit design methodologies to perform both linear and nonlinear functions [3].

Transconductance Multipliers

A direct, straightforward technique to realize the multiplication function exploits the possibility of controlling the transconductance of transistors through an electrical variable (current or voltage). Although this feature is exhibited also by unilateral amplifiers, most practical realizations use differential amplifiers to reduce offset problems and enhance linearity [25]. Figure 2.32 presents a generic schematic for a differential amplifier, consisting of two identical three-terminal active devices with common bias current. The expressions on the right display its associated transconductance characteristics for npn-BJTs and n-channel MOSTs, respectively [25]. These characteristics are approximated to a first-order model as

$$i_{zBJT} \approx \frac{i_y}{4U_t} v_x \qquad i_{zMOST} \approx \left(\sqrt{\beta i_y}\right) v_x \tag{2.39}$$

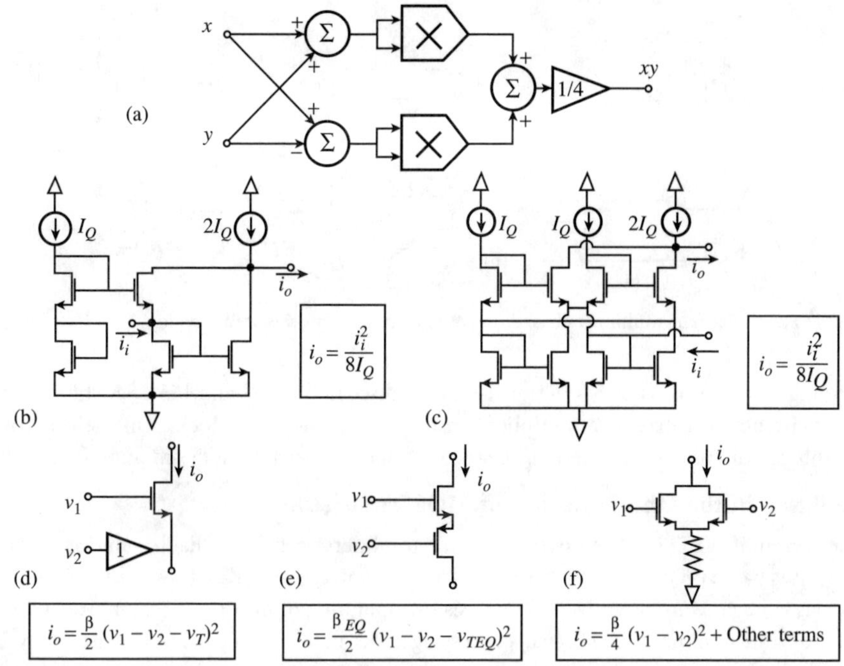

FIGURE 2.31 (a) Block diagram of the quarter-square multiplier; (b) current-mode squarer circuit in [3]; (c) current-mode squarer circuit in [79]; (d) voltage-mode squarer circuit in [36]; (e) voltage-mode squarer circuit in [60]; (f) voltage-mode squarer circuit in [49].

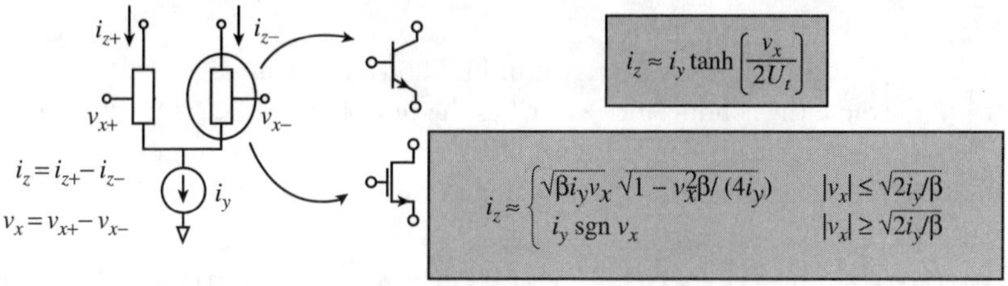

FIGURE 2.32 Differential amplifiers and their associated large-signal transconductances.

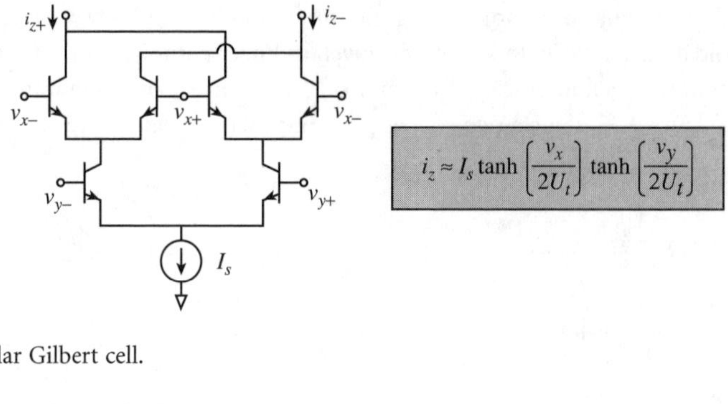

FIGURE 2.33 Bipolar Gilbert cell.

which clearly displays the multiplication operation, although restricted to a rather small linearity range. Practical circuits based on this idea focus mainly on increasing this range of linearity, and follow different design strategies. Figure 2.33 gives an example, known as the Gilbert cell, or Gilbert multiplier [23].

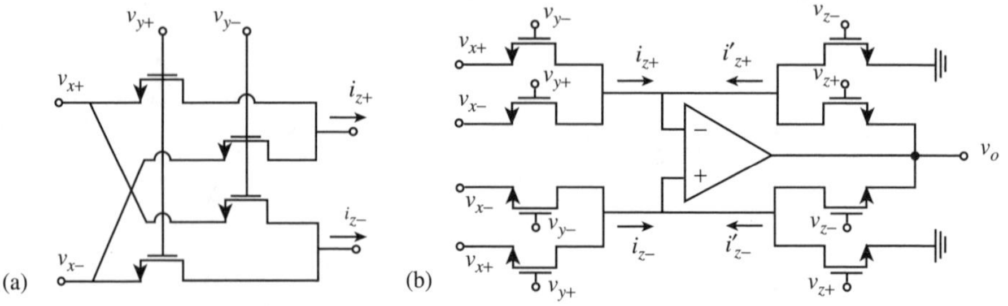

FIGURE 2.34 Four-quadrant multipliers based on MOS transistors in the ohmic region.

Corresponding realizations using MOS transistors are discussed in [2] and [53]. Sánchez-Sinencio et al. [61] present circuits to realize this multiplication function using OTA blocks. On the other hand, [17] presents a tutorial discussion of different linearization techniques for MOS differential amplifiers.

Multiple Based in the Ohmic Region of MOS Transistors

The ohmic region of JFETs has been used to realize amplifiers with controllable gain for automatic gain control [54]. It is based on controlling the equivalent resistance of the JFET transistor in its ohmic region through a bias voltage. More recently, MOS transistors operating in the ohmic region were used to realize linear [69, 70] and nonlinear [35] signal processing tasks in VLSI chips. There exist many ingenious circuits to eliminate second- and higher-order nonlinearities in the equivalent resistance characteristics. The circuit in Figure 2.34(a) achieves very good nonlinearity cancellation through cross-coupling and fully differential operation, obtaining

$$i_{z+} - i_{z-} = 2\beta\left(v_{x+} - v_{x-}\right)\left(v_{y+} - v_{y-}\right) \tag{2.40}$$

and its use in multiplication circuits is discussed in [35] and [66]. A more general view is presented in Figure 2.34(b) [35], where the conductance as well as the resistance of the MOS ohmic region are used to obtain a versatile amplifier-divider building block. Enomoto and Yasumoto [18] report another interesting multiplier that combines the ohmic region of the MOS transistor and sampled-data circuits.

2.6 Sigmoids, Bells, and Collective Computation Circuits

Sigmoidal Characteristics

As (2.5) illustrates, approximating a nonlinear function through a multilayer perceptron requires the realization of sigmoidal functions, with arguments given as linear combinations of several variables. The exact shape of the sigmodial is not critical for the approximation itself, although it may play an important role in fitting [82]. Figure 2.35 depicts two shapes used in practice. Figure 2.35(a), the hard limiter, has

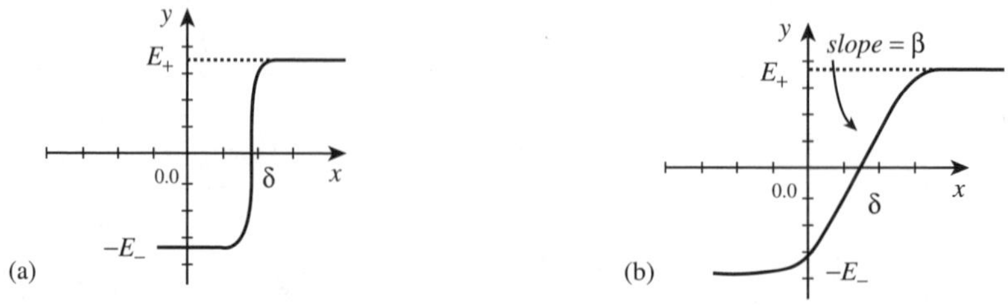

FIGURE 2.35 Typical sigmoidal shapes: (a) hard limiter; (b) soft limiter.

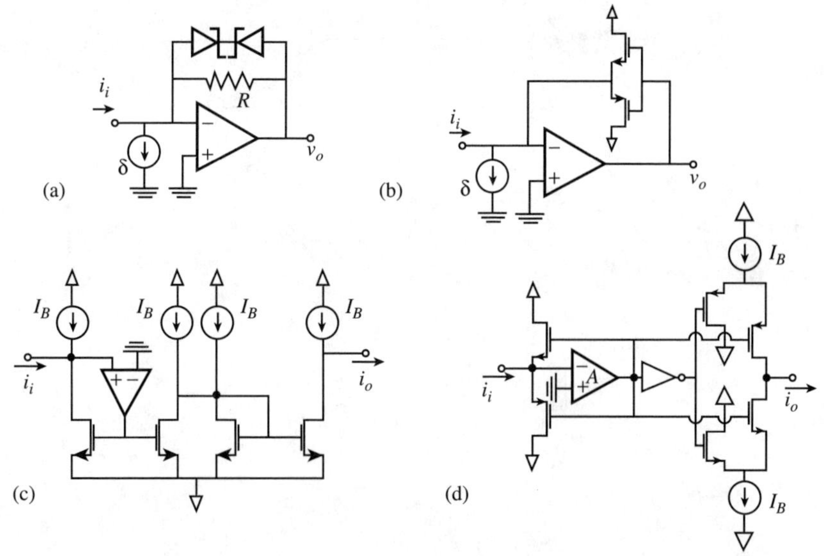

FIGURE 2.36 Realization of sigmoidal characteristics with input current: (a) transimpedance soft limiter; (b) transimpedance hard limiter; (c) and (d) soft and hard limiters in current transfer domain.

an inner piece of large (ideally infinite) slope, while for Figure 2.35(b), the soft limiter, this slope is smaller and can be used as a fitting parameter.

Most amplifiers have large-signal transfer characteristics whose shape is a sigmoid or an inverted sigmoid. We present only those circuits whose inputs are currents because this simplifies the circuitry needed to obtain these inputs as linear combinations of other variables. The op amp circuit of Figure 2.36(a) realizes the soft limiter characteristics in transimpedance form. The center is set by the input bias current and the slope through the resistor ($\beta = R$). If the branch composed of the two Zener diodes is eliminated, the saturation levels E_+ and E_- are determined through the internal op amp circuitry, inappropriate for accurate control. (Otherwise, they are determined through the Zener breakdown voltages.) On the other hand, Figure 2.36(b) also realizes the hard sigmoid in transimpedance domain [58]. The output saturation levels for this structure are $E_+ = V_{Tn}$ and $E_- = |V_{Tp}|$, where V_{Tn} and V_{Tp} are the threshold voltages of the NMOS transistor and the PMOS transistor, respectively. To obtain the output represented by a current, one can use voltage-to-current transformation circuits. References [15, 57, 58] discuss simpler alternatives operating directly in current domain. For instance, Figure 2.36(c) and (d) depict circuits for the soft limiter characteristics and the hard limiter characteristics.

With regard to the calculation of the input to the sigmoid as a linear combination of variables, note that the input node of all circuits in Figure 2.36 is virtual ground. Consequently, the input current can be obtained as a linear combination of voltages or currents using the techniques for signal scaling and aggregation presented in Section 2.3.

Bell-Like Shapes

The exact shapes of (2.16) and (2.17) involve the interconnection of squarers, together with blocks to elevate to power, and exponential blocks — all realizable using techniques previously discussed in this chapter. However, these exact shapes are not required in many applications, and can be approximated using simpler circuits. Thus, let us consider the differential amplifier of Figure 2.32, and define $v_i = v_x$, $I_B = i_y$, and $i_o = i_z$ for convenience. The expressions for the large-signal transconductance displayed along with the figures show that they are sigmoids with saturation levels at I_B and $-I_B$. They are centered at $v_i = 0$, with the slope at this center point given by (2.39). The center can be shifted by making $v_i = v_{x+}$ and $\delta = v_{x-}$.

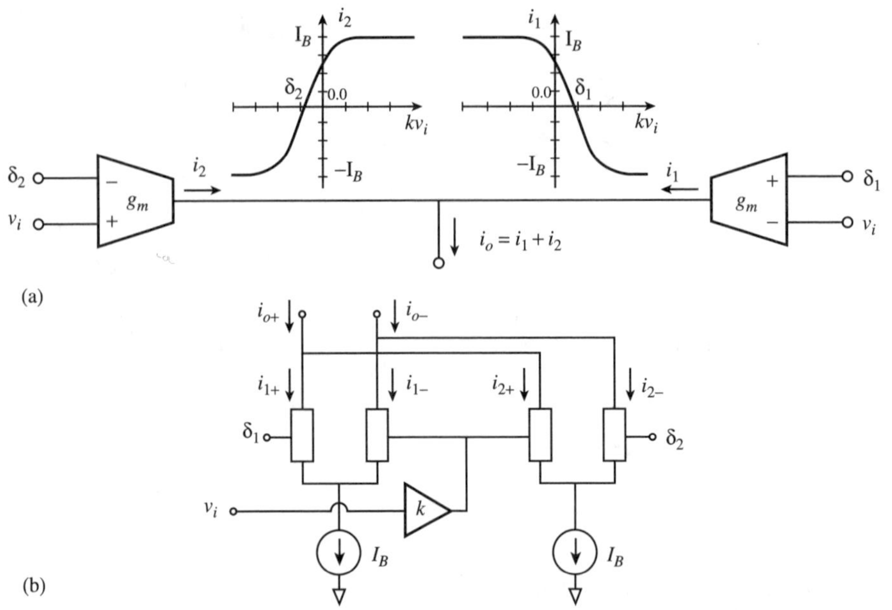

FIGURE 2.37 Transconductance circuits for bell-shaped function: (a) using OTAs; (b) using differential amplifiers.

Similar to the differential amplifier, most OTAs exhibit sigmoid-like characteristics under large-signal operation, exploitable to realize nonlinear functions [19, 37, 56, 61, 71]. This may rely on the mathematical techniques behind multilayer perceptrons, or on those behind radial basis functions and fuzzy interpolation.

Figure 2.37(a) obtains a bell-shaped transconductance through a linear, KCL combination of the two sigmoidal characteristics, one of negative slope and the other of positive slope. The width and center of the bell (see Figure 2.7) are given respectively by

$$2\sigma = \delta_2 - \delta_1 \qquad \delta = \frac{\delta_2 + \delta_1}{2} \tag{2.41}$$

controlled by the designer. The slope of the bell at the crossover points is also controlled through the transconductance of the OTAs.

For simpler circuit realizations, this technique can be used directly with differential amplifiers, as shown in Figure 2.37(b). The differential output current provided by the circuit can be transformed into a unilateral one using a *p*-channel current mirror. Equation (2.41) also applies for this circuit, and the slope at the crossovers is

$$\text{slope}_{\text{MOST}} = k\sqrt{\beta I_B} \qquad \text{slope}_{\text{BJT}} = \frac{kI_B}{4U_t} \tag{2.42}$$

Note that the control of this slope through the bias current changes the height of the bell. It motivates the use of a voltage gain block in Figure 2.37. Thus, the slope can be changed through its gain parameter *k*. The slope can also be changed through β for the MOSTs. Practical realizations of this concept are found in [4], [71], and [74]. The voltage amplifier block can be realized using the techniques presented in this chapter. Simpler circuits based on MOS transistors are found in [53].

Collective Computation Circuitry

Radial basis functions and fuzzy inference require multidimensional operators to calculate radial distances in the case of radial basis functions, and to normalize vectors and calculate T-norms in the case of fuzzy

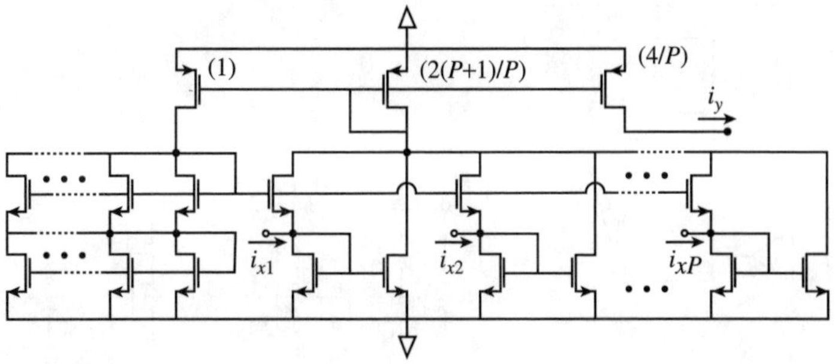

FIGURE 2.38 CMOS self-biased Euclidean distance circuit [38].

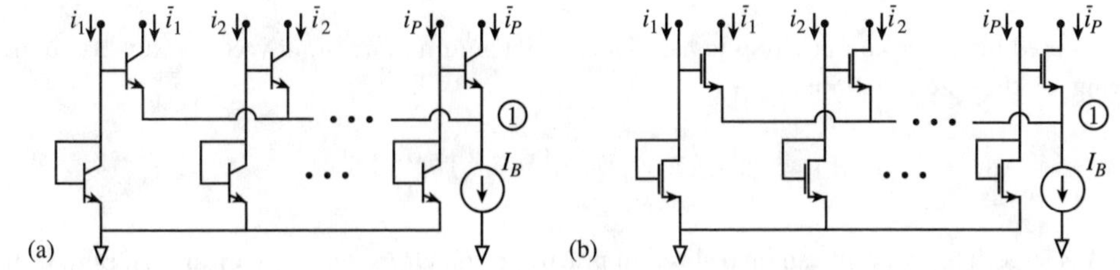

FIGURE 2.39 Current-mode normalization circuits: (a) BJT; (b) CMOS.

inference. These operators can be expressed as the interconnection of the nonlinear blocks discussed previously, or realized in a simpler manner through dedicated collective computation circuitry. Most of these circuits operate intrinsically in current domain and are worth mentioning because of this simplicity and relevance for parallel information processing systems.

Euclidean Distance

Figure 2.38 [38] presents a current-mode circuit to compute

$$i_y = \sqrt{\sum_{k=1,P} i_{xk}^2} \tag{2.43}$$

based on the square-law of MOS transistors in the saturation region. If the current i_k at each terminal is shifted through a bias current of value δ_k, the circuit serves to compute the Euclidean distance between the vector of input currents and the vector $\boldsymbol{\delta}$.

Normalization Operation

Figure 2.39 depicts circuits to normalize an input current vector, for the BJT [23] and the CMOS [74] cases, respectively. Their operation is based on KCL and the current mirror principle. Kirchhoff's circuit law forces the sum of the output currents at node ① to be constant. On the other hand, the current mirror operation forces a functional dependency between each pair of input and output currents. Thus, they obtain

$$\bar{i}_k \approx \frac{i_k}{\sum_{j=1,P} i_j} \tag{2.44}$$

for each current component.

T-Norm Operator

The calculation of the minimum of an input vector $\mathbf{x}$ is functionally equivalent to obtaining the complement of the maximum of the complements of its components. Figure 2.40(a) illustrates a classical

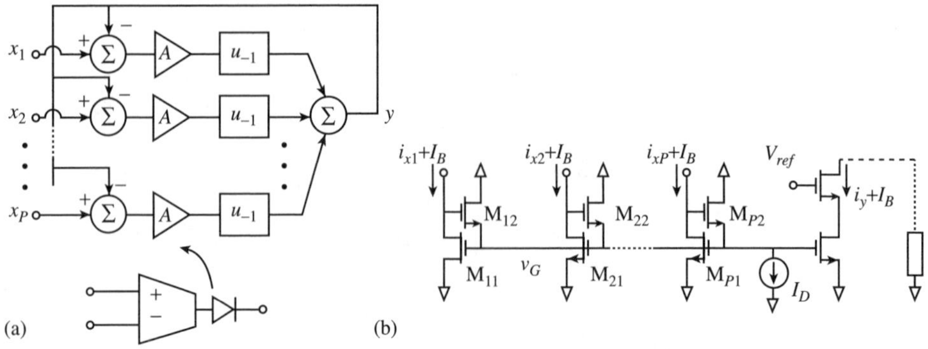

FIGURE 2.40 Concept for maximum operator and current-mode realization.

approach used in analog computation to calculate the maximum of an input vector **x**. It is based on the following steady-state equation:

$$-y + \sum_{k=1,P} u_{-1}\left(A\left(x_k - y\right)\right) = 0 \tag{2.45}$$

where A is large. This concept can be realized in practice using OTAs, op amps, or diodes. Both of these have voltage input and output. Alternatively, Figure 2.40(b) shows a CMOS current-mode realization [74]. In this circuit the maximum current determines the value of the common gate voltage, v_G. The only input transistor operating in the saturation region is that which is driven by maximum input current; the rest operate in the ohmic region.

2.7 Extension to Dynamic Systems

A dynamic system with state vector **x** and dynamics represented as

$$T_k \frac{dx_k}{dt} = f_k\left(\mathbf{X}\right),\ 1 \le k \le P \tag{2.46}$$

can be mapped on the block diagram of Figure 2.41(a), and realized by the interconnection of non-linear resistive blocks and integrators. This approach is similar to that followed in classical *analog computation* [26]

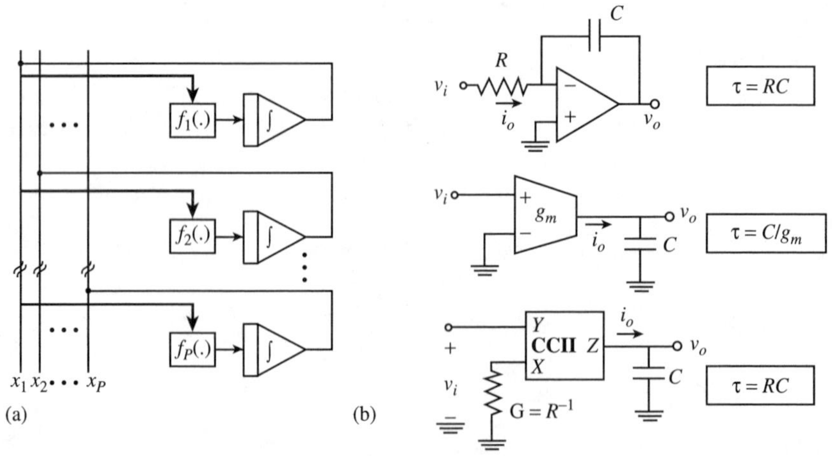

FIGURE 2.41 Conceptual state-variable block diagram of dynamic systems integrator circuits.

and has integrators as key components. Figure 2.41(b) illustrates several integrator circuits. Combining these circuits with the circuitry for nonlinear functions provides systematic approaches to synthesize nonlinear dynamic systems based on the approximations presented in this chapter [56]. On the other hand, Rodríguez-Vázquez and Delgado-Restituto [57] discuss related techniques to synthesize nonlinear systems described by finite-difference equations.

2.8 Appendix A: Catalog of Primitives

Figure 2.42 outlines our catalog of primitive components, all of which are available off-the-shelf, and, depending on the fabrication technology, can be realized on a common semiconductor substrate [44]. Generally, the catalog differs between individual technologies; for instance, no npn BJTs are available in a CMOS n-well technology. The use of linear capacitors may appear surprising because we constrain ourselves to cover only static characteristics. However, we will not exploit their dynamic i–v relationship, but instead their constitutive equation in the charge-voltage plane, which is algebraic.

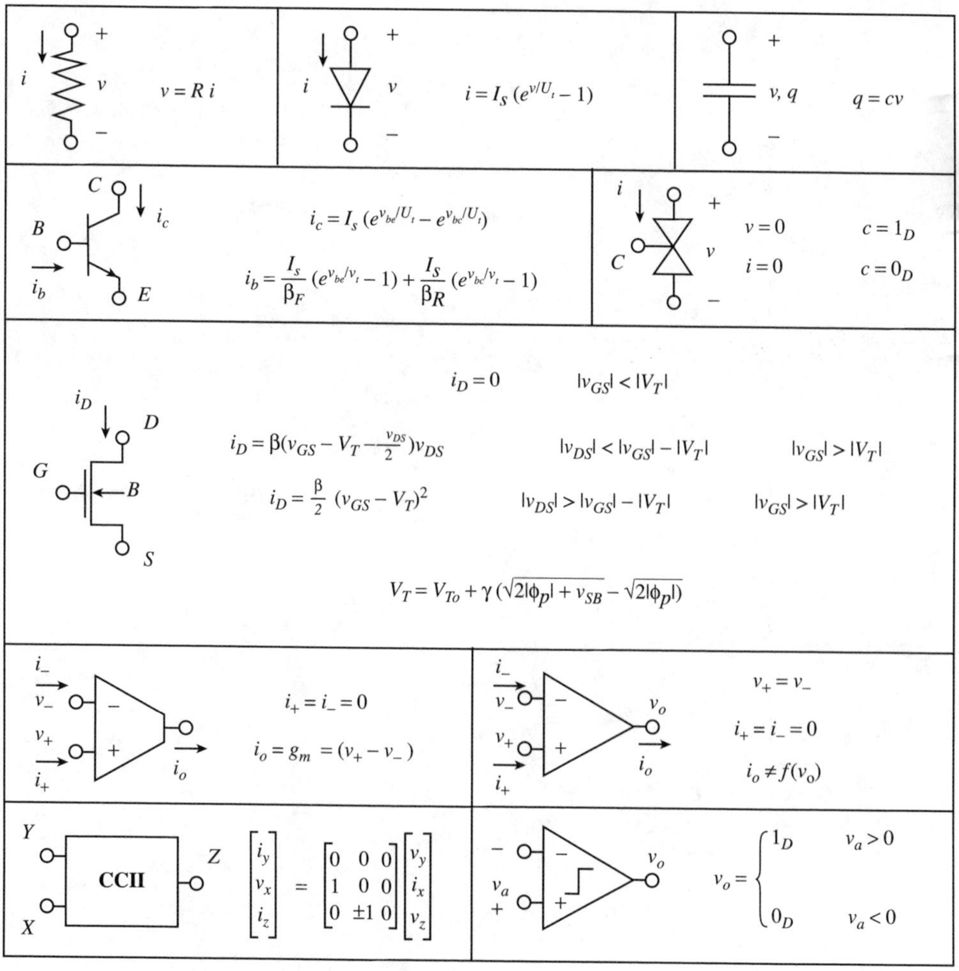

FIGURE 2.42 Section catalog of primitive circuit components.

2.9 Appendix B: Value and Slope Hermite Basis Functions

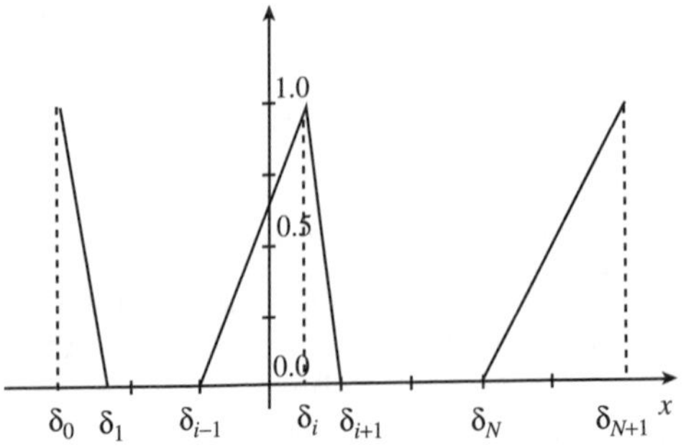

$$I_0(x) = \begin{cases} (\delta_1 - x)/\Delta_0 & \delta_0 \le x \le \delta_1 \\ 0 & \delta_1 \le x \le \delta_{N+1} \end{cases}$$

$$I_i(x) = \begin{cases} 0 & \delta_0 \le x \le \delta_{N+1} \\ (x - \delta_{i-1})/\Delta_{i-1} & \delta_{i-1} \le x \le \delta_i \\ (\delta_{i+1} - x)/\Delta_i & \delta_i \le x \le \delta_{i+1} \\ 0 & \delta_{i+1} \le x \le \delta_{N+1} \end{cases}$$

$$I_{N+1}(x) = \begin{cases} (x - \delta_N)/\Delta_N & \delta_N \le x \le \delta_{N+1} \\ 0 & \delta_0 \le x \le \delta_N \end{cases}$$

(a)

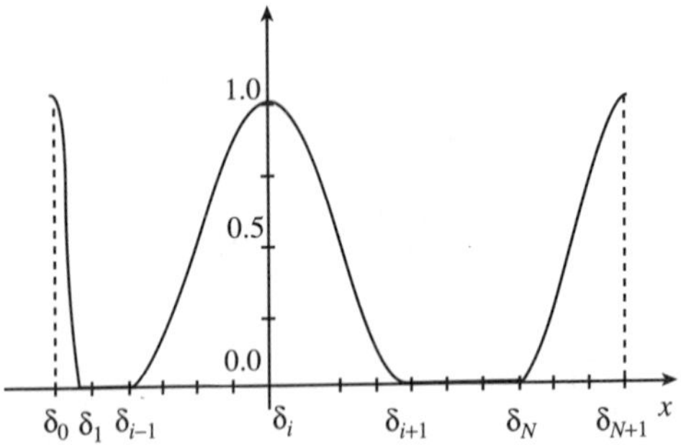

$$v_0(x) = \begin{cases} 1 - 2\beta_0^3 + 3\beta_0^2 & \delta_0 \le x \le \delta_1 \\ 0 & \delta_1 \le x \le \delta_{N+1} \end{cases}$$

$$v_i(x) = \begin{cases} 0 & \delta_0 \le x \le \delta_{N+1} \\ 1 - 3\alpha_i^2 - 2\alpha_i^3 & \delta_{i-1} \le x \le \delta_i \\ 1 - 3\beta_i^2 - 2\beta_i^3 & \delta_i \le x \le \delta_{i+1} \\ 0 & \delta_{i+1} \le x \le \delta_{N+1} \end{cases}$$

$$v_{N+1}(x) = \begin{cases} 1 - 3\alpha_{N+1}^2 - 2\alpha_{N+1}^3 & \delta_N \le x \le \delta_{N+1} \\ 0 & \delta_0 \le x \le \delta_N \end{cases}$$

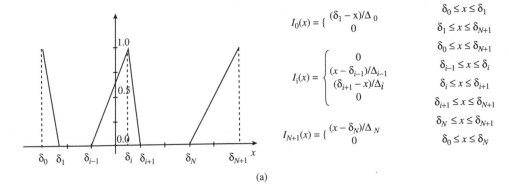

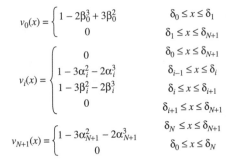

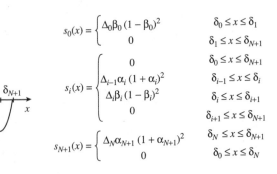

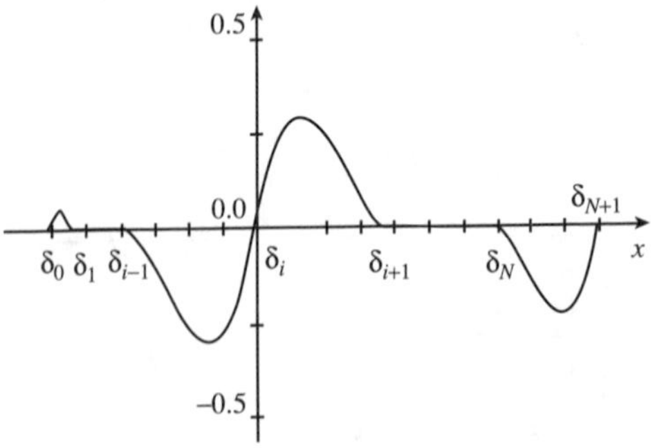

$$s_0(x) = \begin{cases} \Delta_0\beta_0 (1 - \beta_0)^2 & \delta_0 \le x \le \delta_1 \\ 0 & \delta_1 \le x \le \delta_{N+1} \end{cases}$$

$$s_i(x) = \begin{cases} 0 & \delta_0 \le x \le \delta_{N+1} \\ \Delta_{i-1}\alpha_i (1 + \alpha_i)^2 & \delta_{i-1} \le x \le \delta_i \\ \Delta_i\beta_i (1 - \beta_i)^2 & \delta_i \le x \le \delta_{i+1} \\ 0 & \delta_{i+1} \le x \le \delta_{N+1} \end{cases}$$

$$s_{N+1}(x) = \begin{cases} \Delta_N\alpha_{N+1} (1 + \alpha_{N+1})^2 & \delta_N \le x \le \delta_{N+1} \\ 0 & \delta_0 \le x \le \delta_N \end{cases}$$

$\Delta_i = \delta_{i+1} - \delta_i$ $\qquad$ $\alpha_i = (x - \delta_i)/\Delta_{i-1}$ $\qquad$ $\beta_i = (x - \delta_i)/\Delta_i$

(b)

FIGURE 2.43 Hermite basis functions: (a) PWL case; (b) PWC case.

References

[1] X. Arreguit, E. Vittoz, and M. Merz, "Precision compressor gain controller in CMOS technology," *IEEE J. Solid-State Circuits,* vol. 22, pp. 442–445, 1987.

[2] J. N. Babanezhad and G. C. Temes, "A 20-V four quadrant CMOS analog multiplier," *IEEE J. Solid-State Circuits,* vol. 20, pp. 1158–1168, 1985.

[3] K. Bult and H. Wallinga, "A class of analog CMOS circuits based on the square-law characteristics of an MOS transistor in saturation," *IEEE J. Solid-State Circuits*, vol. 22, pp. 357–365, 1987.

[4] J. Choi, B. J. Sheu, and J. C. F. Change, "A Gaussian synapse circuit for analog VLSI neural network," *IEEE Trans. Very Large-Scale Integration Syst.,* vol. 2, pp. 129–133, 1994.

[5] L. O. Chua, "The linear transformation converter and its applications to the synthesis of nonlinear networks," *IEEE Trans. Circuit Theory,* vol. 17, pp. 584–594, 1970.

[6] L. O. Chua and S. M. Kang, "Section-wise piecewise-linear functions: canonical representation, properties and applications," *Proc. IEEE,* vol. 67, pp. 915–929, 1977.

[7] L. O. Chua and S. Wong, "Synthesis of piecewise-linear networks," *Elect. Circuits Syst.,* vol. 2, pp. 102–108, 1978.

[8] L. O. Chua, "Nonlinear circuits," *IEEE Trans. Circuits Syst.,* vol. 31, pp. 69–87, 1984.

[9] L. O. Chua and G. Zhong, "Negative resistance curve tracer," *IEEE Trans. Circuits Syst.,* vol. 32, pp. 569–582, 1985.

[10] L. O. Chua, C. A. Desoer, and E. S. Kuh, *Linear and Nonlinear Circuits,* New York: McGraw-Hill, 1987.

[11] L. O. Chua and A. C. Deng, "Canonical piecewise linear representation," *IEEE Trans. Circuits Syst.,* vol. 35. pp. 101–111, 1988.

[12] G. Cybenko, "Approximation by superposition of a sigmoidal function," *Math. Control. Syst. Signals,* vol. 2, pp. 303–314, 1989.

[13] Z. Czarnul, "Modification of Banu-Tsividis continuous-time integrator structure," *IEEE Trans. Circuits Syst.,* vol. 33, pp. 714–716, 1986.

[14] R. W. Daniels, *Approximation Methods for Electronic Filter Design,* New York: McGraw-Hill, 1974.

[15] M. Delgado-Restituto and A. Rodríguez-Vázquez, "Switched-current chaotic neutrons," *Electr. Lett.,* vol. 30, pp. 429–430, 1994.

[16] M. Delgado-Restituto, "A chaotic switched-capacitor circuit for $1/f$ noise generation," *IEEE Trans. Circuits Syst.,* vol. 39, pp. 325–328, 1992.

[17] S. T. Dupuie and M. Ismail, "High-frequency CMOS transconductors," in *Analogue IC Design: The Current-Mode Approach,* C. Toumazou, F. J. Lidgey, and D. G. Haigh, Eds., London: Peter Peregrinus Ltd., 1990.

[18] T. Enomoto and M. Yasumoto, "Integrated MOS four-quadrant analog multiplier using switched capacitor technology for analog signal processor IC's," *IEEE J. Solid-State Circuits,* vol. 20, pp. 852–859, 1985.

[19] J. W. Fattaruso and R. G. Meyer, "MOS analog function synthesis," *IEEE J. Solid-State Circuits,* vol. 22, pp.1059–1063, 1987.

[20] I. M. Filanovsky and H. Baltes, "CMOS Schmitt trigger design," *IEEE Trans. Circuits Syst.,* vol. 41, pp. 46–49, 1994.

[21] K. Funahashi, "On the approximate realization of continuous mappings by neural networks," *Neural Networks,* vol. 2, pp. 183–192, 1989.

[22] R. L. Geiger and E. Sánchez-Sinencio, "Active filter design using operational transconductance amplifiers: a tutorial," *IEEE Circuits Devices Mag.,* vol. 1, pp. 20–32, 1985.

[23] B. Gilbert, "Current-mode circuits from a translinear view point: a tutorial," in *Analogue IC Design: The Current-Mode Approach,* C. Toumazou, F. J. Lidgey, and D. G. Haigh, Eds., London: Peter Peregrinus Ltd., 1990.

[24] J. Glover, "Basic T matrix patterns for 2-port linear transformation networks in the real domain," *IEEE Trans. Circuit Theory*, vol. 10, pp. 495–497, 1974.

[25] P. R. Gray and R. G. Meyer, *Analysis and Design of Analog Integrated Circuits*, 3rd ed., New York: John Wiley & Sons, 1993.

[26] A. Hausner, *Analog and Analog/Hybrid Computer Programming*, Englewood Cliffs, NJ: Prentice Hall, 1971.

[27] B. J. Hosticka, W. Brockherde, U. Kleine, and R. Schweer, "Design of nonlinear analog switched-capacitor circuits using building blocks," *IEEE Trans. Circuits Syst.*, vol. 31, pp. 354–368, 1984.

[28] J. L. Huertas, J. A. Acha, and A. Gago, "Design of general voltage or current controlled resistive elements and their applications to the synthesis of nonlinear networks," *IEEE Trans. Circuits Syst.*, vol. 27, pp. 92–103, 1980.

[29] J. L. Huertas, "DT-adaptor: Applications to the design of nonlinear *n*-ports," *Int. J. Circuit Theory Appl.*, vol. 8, pp. 273–290, 1980.

[30] J. L. Huertas and A. Rueda, "Sectionwise piecewise polynomial function: applications to the analysis and synthesis on nonlinear *n*-ports Networks," *IEEE Trans. Circuits Syst.*, vol. 31, pp. 897–905, 1984.

[31] J. L. Huertas, L. O. Chua, A. Rodríguez-Vázquez, and A. Rueda, "Nonlinear switched-capacitor networks: Basic principles and piecewise-linear design," *IEEE Trans. Circuits Syst.*, vol. 32, pp. 305–319, 1985.

[32] M. Ismail, "Four-transistor continuous-time MOS transconductance," *Electr. Lett.*, vol. 23, pp. 1099–1100, 1987.

[33] J. S. Jang, Neuro-Fuzzy Modeling: Architectures, Analyses and Applications, Ph.D. dissertation, University of California, Berkeley, 1992.

[34] C. Jansson, K. Chen, and C. Svensson, "Linear, polynomial and exponential ramp generators with automatic slope adjustments," *IEEE Trans. Circuits Syst.*, vol. 41, pp. 181–185, 1994.

[35] N. I. Khachab and M. Ismail, "Linearization techniques for *n*th-order sensor models in MOS VLSI technology," *IEEE Trans. Circuits Syst.*, vol. 38, pp. 1439–1449, 1991.

[36] Y. H. Kim and S. B. Park, "Four-quadrant CMOS analog multiplier," *Electr. Lett.*, vol. 28, pp. 649–650, 1992.

[37] K. Kimura, "Some circuit design techniques using two cross-coupled, emitter-coupled pairs," *IEEE Trans. Circuits Syst.*, vol. 41, pp. 411–423, 1994.

[38] O. Landolt, E. Vittoz, and P. Heim, "CMOS self-biased Euclidean distance computing circuit with high dynamic range," *Electr. Lett.*, vol. 28, pp. 352–354, 1992.

[39] N. R. Malik, G. L. Jackson, and Y. S. Kim, "Theory and applications of resistor, linear controlled resistor, linear controlled conductor networks," *IEEE Trans. Circuits Syst.*, vol. 31, pp. 222–228, 1976.

[40] A. I. Mees, M. F. Jackson, and L. O. Chua, "Device modeling by radial basis functions," *IEEE Trans. Circuits Syst.*, vol. 39, pp.19–27, 1992.

[41] C. Michael and M. Ismail, *Statistical Modeling for Computer-Aided Design of MOS VLSI Circuits*, Boston: Kluwer Academic, 1993.

[42] C. A. Micchelli, "Interpolation of scattered data: distance matrices and conditionally positive definite functions," *Constr. Approx*, vol. 2, pp. 11–22, 1986.

[43] J. Moody and C. Darken, "Fast learning in networks of locally-tuned processing units," *Neutral Comput.*, vol. 1, pp. 281–294, 1989.

[44] R. S. Muller and T. I. Kamins, *Device Electronics for Integrated Circuits*, New York: John Wiley & Sons, 1986.

[45] K. Nagaraj, K. Singhal, T. R. Viswanathan, and J. Vlach, "Reduction of finite-gain effect in switched-capacitor filters," *Electr. Lett.*, vol. 21, pp. 644– 645, 1985.

[46] K. Nagaraj, "A parasitic-insensitive area-efficient approach to realizing very large time constants in switched-capacitor circuits," *IEEE Trans. Circuit Syst.*, vol. 36, pp. 1210–1216, 1989.

[47] R. W. Newcomb, "Nonlinear differential systems: a canonic multivariable theory," *Proc. IEEE*, vol. 65, pp. 930–935, 1977.

[48] M. J. M. Pelgrom, A. C. J. Duinmaijer, and A. P. G. Welbers, "Matching properties of MOS transistors," *IEEE J. Solid-State Circuits*, vol. 24, pp. 1433–1440, 1989.

[49] J. S. Peña-Finol and J. A. Connelly, "A MOS four-quadrant analog multiplier using the quarter-square technique," *IEEE J. Solid-State Circuits*, vol. 22, pp. 1064–1073, 1987.

[50] J. Platt, "A resource allocating neural network for function interpolation," *Neural Comput.*, vol. 3, pp. 213–215, 1991.

[51] T. Poggio and F. Girosi, "Networks for approximation and learning," *Proc. IEEE*, vol. 78, pp. 1481–1497, 1990.

[52] M. J. D. Powell, "Radial basis functions for multivariate interpolation: a review," Rep 1985/NA12, Dept. Applied Mathematics and Theoretical Physics, Cambridge university, Cambridge, U.K., 1985.

[53] S. Qin and R. L. Geiger, "A 5-V CMOS analog multiplier," *IEEE J. Solid-State Circuits*, vol. 22, pp. 1143–1146, 1987.

[54] J. K. Roberge, *Operational Amplifiers: Theory and Practice*, New York: John Wiley, 1975.

[55] A. Rodríguez-Vázquez, J. L. Huertas, and A. Rueda, "Low-order polynomial curve fitting using switched-capacitor circuits," *Proc. IEEE Int. Symp. Circuits Syst.*, May, pp. 40, 1123–1125, 1984.

[56] A. Rodríguez-Vázquez and M. Delgado-Restituto, "CMOS design of chaotic oscillators using state variables: a monolithic Chua's circuit," *IEEE Trans. Circuits Syst.*, vol. 40, pp. 596–613, 1993.

[57] A. Rodríguez-Vázquez and M. Delgado-Restituto, "Generation of chaotic signals using current-mode techniques," *J. Intelligent Fuzzy Syst.*, vol. 2, pp. 15–37, 1994.

[58] A. Rodríguez-Vázquez, R. Domínguez-Castro, F. Mederio, and M. Delgado-Restituto, "High-resolution CMOS current comparators: design and applications to current-mode functions generation," *Analog Integrated Circuits Signal Process.*, in press.

[59] A. Rueda, J. L. Huertas, and A. Rodríguez-Vázquez, "Basic circuit structures for the synthesis of piecewise polynomial one-port resistors," *IEEE Proc.*, vol. 132 (part G), pp. 123–129, 1985.

[60] S. Sakurai and M. Ismail, "High frequency wide range CMOS analogue multiplier," *Electr. Lett.*, vol. 28, pp. 2228–2229, 1992.

[61] E. Sánchez-Sinencio, J. Ramírez-Angulo, B. Linares-Barranco, and A. Rodríguez-Vázquez, "Operational transconductance amplifier-based nonlinear function synthesis," *IEEE J. Solid-State Circuits*, vol. 24, pp. 1576–1586, 1989.

[62] E. Seevinck, *Analysis and Synthesis of Translinear Integrated Circuits*, Amsterdam: Elsevier, 1988.

[63] D. H. Sheingold, *Nonlinear Circuits Handbook*, Norwood, CA: Analog Devices Inc., 1976.

[64] J. Silva-Martínez and E. Sánchez-Sinencio, "Analog OTA multiplier without input voltage swing restrictions, and temperature compensated," *Electr. Lett.*, vol. 22, pp. 559–560, 1986.

[65] S. Soclof, *Analog Integrated Circuits*, Englewood Cliffs, NJ: Prentice Hall, 1985.

[66] B. Song, "CMOS RF circuits for data communication applications," *IEEE J. Solid-State Circuits*, vol. 21, pp. 310–317, 1986.

[67] L. Strauss, *Wave Generation and Shaping*, New York: McGraw-Hill, 1970.

[68] G. C. Temes and K. Haug, "Improved offset compensation schemes for switched-capacitor circuits," *Electr. Lett.*, vol. 20, pp. 508–509, 1984.

[69] Y. P. Tsividis, M. Banu, and J. Khoury, "Continuous-time MOSFET-C filters in VLSI," *IEEE J. Solid-State Circuits*, vol. 21, pp. 15–30, 1986.

[70] Y. P. Tsividis and J. O. Voorman, Eds., *Integrated Continuous-Time Filters*, New York: IEEE Press, 1993.

[71] C. Turchetti and M. Conti, "A general approach to nonlinear synthesis with MOS analog circuits," *IEEE Trans. Circuits Syst.*, vol. 40, pp. 608–612, 1993.

[72] R. Unbehauen and A. Cichocki, *MOS Switched-Capacitor and Continuous-Time Integrated Circuits and Systems*, Berlin: Springer-Verlag, 1989.

[73] G. N. Vanderplaats, *Numerical Optimization Techniques for Engineering Design: With Applications*, New York: McGraw-Hill, 1984.

[74] F. Vidal and A. Rodríguez-Vázquez, "A basic building block approach to CMOS design of analog neuro/fuzzy systems." *Proc. IEEE Int. Conf. Fuzzy Systems*, vol. 1, pp. 118–123, 1994.

[75] E. Vittoz, "Analog VLSI signal processing why, where and how?," *Analog Integrated Circuits Signal Process.*, vol. 6, pp. 27–44, 1994.

[76] J. Vlach and K. Singhal, *Computer Methods for Circuit Analysis and Design*, 2nd ed., New York: Van Nostrand Reinhold, 1994.

[77] Z. Wang, "Analytical determination of output resistance and DC matching errors in MOS current mirrors," *IEEE Proc.*, vol. 137 (Part G), pp. 397–404, 1990.

[78] G. A. Watson, *Approximation Theory and Numerical Methods*, Chichester, U.K.: John Wiley & Sons, 1980.

[79] R. Wiegerink, *Analysis and synthesis of MOS Translinear Circuits*, Boston: Kluwer Academic, 1993.

[80] Y. J. Wong and W. E. Ott, *Function Circuits; Design and Applications*, New York: McGraw-Hill, 1976.

[81] T. Yamakawa, "A fuzzy inference engineer in nonlinear analog mode and its application to fuzzy control," *IEEE Trans. Neural Networks*, vol. 4, pp. 496–522, 1993.

[82] J. M. Zurada, *Introduction to Artificial Neural Systems*, St. Paul, MN: West Publishing, 1992.

3

Representation, Approximation, and Identification

Guanrong Chen
City University of Hong Kong

3.1 Introduction

Representation, approximation, and identification of physical systems, linear or nonlinear, deterministic or random, or even chaotic, are three fundamental issues in systems theory and engineering. To describe a physical system, such as a circuit or a microprocessor, we need a mathematical formula or equation that can represent the system both qualitatively and quantitatively. Such a formulation is what we call a mathematical representation of the physical system. If the physical system is so simple that the mathematical formula or equation, or the like, can describe it perfectly without error, then the representation is ideal and ready to use for analysis, computation, and synthesis of the system. An ideal representation of a real system is generally impossible, so that system approximation becomes necessary in practice. Intuitively, approximation is always possible. However, the key issues are what kind of approximation is good, where the sense of "goodness" must first be defined, of course, and how to find such a good approximation. On the other hand, when looking for either an ideal or a approximate mathematical representation for a physical system, one must know the system structure (the form of the linearity or nonlinearity) and parameters (their values). If some of these are unknown, then one must identify them, leading to the problem of system identification.

This chapter is devoted to a brief description of mathematical representation, approximation, and identification of, in most cases, nonlinear systems. As usual, a linear system is considered to be a special case of a nonlinear system, but we do not focus on linear systems in this chapter on nonlinear circuits. It is known that a signal, continuous or discrete, is represented by a function of time. Hence, a signal can be approximated by other functions and also may be identified using its sampled data. These are within the context of "representation, approximation and identification," but at a lower level — one is dealing with functions. A system, in contrast, transforms input signals to output signals, namely, maps functions to functions, and is therefore at a higher level — it can only be represented by an operator (i.e., a mapping). Hence, while talking about representation, approximation, and identification in this chapter, we essentially refer to operators. However, we notice that two systems are considered to be equivalent over a set of input signals if and only if (iff) they map the same input signal from the set to the same output signal, regardless of the distinct structures of the two systems. From this point of view, one system is a good approximation of the other if the same input produces outputs that are approximately the same under certain measure. For this reason, we also briefly discuss the classical function approximation theory in this chapter.

The issue of system representation is addressed in Section 3.2, while approximation (for both operators and functions) is discussed in Section 3.3, leaving the system identification problem to Section 3.4. Limited by space, we can discuss only deterministic systems. Topics on stochastic systems are hence referred to some standard textbooks (e.g., [13, 17]).

It is impossible to cover all the important subjects and to mention many significant results in the field in this short and sketchy chapter. The selections made only touch upon the very elementary theories, commonly used methods, and basic results related to the central topics of the chapter, reflecting the author's personal preference. In order to simplify the presentation, we elected to cite only those closely related references known to us, which may or may not be the original sources. From our citations, the reader should be able to find more references for further reading.

3.2 Representation

The scientific term "representation" as used here refers to a mathematical description of a physical system. The fundamental issue in representing a physical system by a mathematical formulation, called a **mathematical model**, is its correct symbolization, accurate quantization, and strong ability to illustrate and reproduce important properties of the original system.

A circuit consisting of some capacitor(s), inductor(s), and/or resistors(s), and possibly driven by a voltage source or a current source, is a physical system. In order to describe this system mathematically for the purpose of analysis, design, and/or synthesis, a mathematical model is needed. Any mathematical model, which can correctly describe the physical behavior of the circuit, is considered a mathematical representation of the circuit. A lower level **mathematical representation** of a circuit can, for instance, be a signal flow chart or a circuit diagram like the nonlinear Chua's circuit shown in Figure 3.1, which is discussed next.

A circuit, such as that shown in Figure 3.1, can be used to describe a physical system, including its components and its internal as well as external connections. However, it is not convenient for carrying out theoretical analysis or numerical computations. This is because no qualitative or quantitative description exists about the relations among the circuit elements and their dynamic behavior. Hence, a higher level mathematical model is needed to provide a qualitative and quantitative representation of the real physical circuit.

Among several commonly used mathematical modeling approaches for various physical systems, differential equations, state-space formulations, I-O mappings, and functional series (particularly, the Volterra series) are the most important and useful, which have been very popular in the field of circuits and systems engineering. In the following, we introduce these mathematical representation methods,

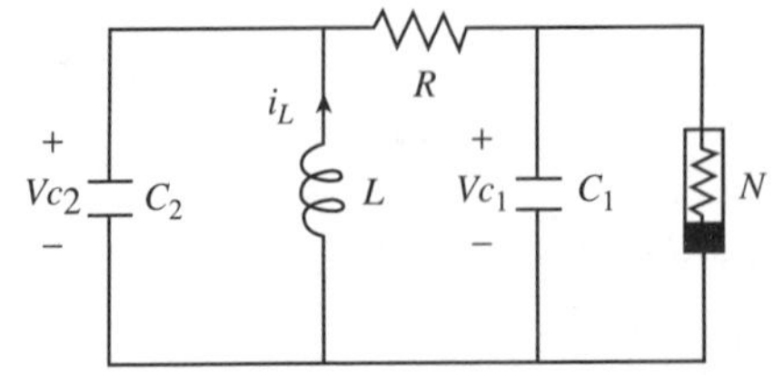

FIGURE 3.1 Chua's circuit.

along with some brief discussions of other related issues. Limited by space, detailed derivations are omitted.

Differential Equation and State-Space Representations

Mathematical modeling via differential equations and via state-space descriptions are the most basic mathematical representation methods. We illustrate the concept of mathematical modeling and the two representation methods by a simple, yet representative example: the nonlinear circuit in Figure 3.1. This circuit consists of one inductor L, two capacitors C_1 and C_2, one linear resistor R, and one nonlinear resistor, N, which is a nonlinear function of the voltage across its two terminals: $N = N(V_{C_1}(t))$. Let $i_L(t)$ be the current through the inductor L, and $V_{c_1}(t)$ and $V_{c_2}(t)$ be the voltages across C_1 and C_2, respectively. For the time being, let us remove the nonlinear resistor N from Figure 3.1 and consider the remaining linear circuit. This nonlinear resistor N is readded to the circuit with detailed discussions in (3.6).

For this linear circuit without the resistor N, it follows from Kirchhoff's laws that

$$C_1 \frac{d}{dt} V_{C_1}(t) = \frac{1}{R}\left[V_{C_2}(t) - V_{C_1}(t)\right] \tag{3.1}$$

$$C_2 \frac{d}{dt} V_{C_2}(t) = \frac{1}{R}\left[V_{C_1}(t) - V_{C_2}(t)\right] + i_L(t) \tag{3.2}$$

$$L \frac{d}{dt} i_L(t) = -V_{C_2}(t) \tag{3.3}$$

By simple calculation we can eliminate both $V_{c_2}(t)$ and i_L, leaving a single ordinary differential equation on the unknown voltage $V_{c_1}(t)$ as follows:

$$\frac{d^3}{dt^3} V_{C_1}(t) + \frac{1}{R}\left(\frac{1}{C_1} + \frac{1}{C_2}\right)\frac{d^2}{dt^2} V_{C_1}(t) + \frac{1}{C_2 L}\frac{d}{dt} V_{C_1}(t) + \frac{1}{C_1 C_2 R L} V_{C_1}(t) = 0 \tag{3.4}$$

Once $V_{c_1}(t)$ is obtained from (3.4), based on certain initial conditions, the other two unknowns, $V_{c_2}(t)$ and i_L, can be obtained by using (3.1) and (3.3), successively. Hence, this third-order ordinary differential equation describes both qualitatively and quantitatively the circuit shown in Figure 3.1 (without the nonlinear resistor N). For this reason, (3.4) is considered to be a mathematical representation, called a **differential equation representation**, of the physical linear circuit.

Very often, a higher-order, single-variable ordinary differential equation similar to (3.4) is not as convenient as a first-order multivariable system of ordinary differential equations as is the original system of (3.1) to (3.3), even when an analytic formulation of the solution is desired. Hence, a more suitable way for modeling a physical system is to introduce the concept of system state variables, which leads to a first-order higher dimensional system of ordinary differential equations.

If we introduce three **state variables** in (3.1) to (3.3):

$$x_1(t) = V_{C_1}(t) \quad x_2(t) = V_{C_2}(t) \quad x_3(t) = i_L(t)$$

then we can rewrite those equations in the following vector form:

$$\begin{cases} \dot{\mathbf{x}}(t) = A\mathbf{x}(t) + B\mathbf{u}(t) & t \ge 0 \\ \mathbf{x}(0) = \mathbf{x}_0 \end{cases} \tag{3.5}$$

with an initial value $\mathbf{x}_0$ (usually given), where

$$\mathbf{x}(t) = \begin{bmatrix} x_1(t) \\ x_2(t) \\ x_3(t) \end{bmatrix} \text{ and } A = \begin{bmatrix} -\frac{1}{RC_1} & \frac{1}{RC_1} & 0 \\ \frac{1}{RC_2} & -\frac{1}{RC_2} & \frac{1}{C_2} \\ 0 & -\frac{1}{L} & 0 \end{bmatrix}$$

in which $\mathbf{x}(t)$ is called the **state vector** of the system. Here, to be more general and for convenience in the discussions following, we formally added the term $B\mathbf{u}(t)$ to the system, in which B is a constant matrix and $\mathbf{u}(t)$ is called the **control input** of the system. In the present case, of course, $\mathbf{u} = 0$ and it is not important to specify B. However, note that $\mathbf{u}$ can be a nonzero external input to the circuit [19], which is discussed in more detail below.

This first-order, vector-valued linear ordinary differential equation is equivalent to the third-order differential equation representation, (3.4), of the same physical circuit. A special feature of this state vector formulation is that with different initial state vectors and with zero control inputs, all the possible system state vectors together constitute a linear space of the same dimension [31]. Hence, (3.5) is also called a **linear state-space representation** (or, a **linear state-space description**) for the circuit.

A few important remarks are in order.

First, if the circuit is nonlinear, its state vectors do not constitute a linear space in general. Hence, its mathematical model in the state vector form should not be called a "state-space" representation. Note, however, that some of the linear system terminology such as state variables and state vectors usually make physical sense for nonlinear systems. Therefore, we use the term **nonlinear state-variable representation** to describe a first-order, vector-valued nonlinear ordinary differential equation of the form $\dot{\mathbf{x}}(t) = \mathbf{f}(\mathbf{x}(t), \mathbf{u}(t), t)$, where $\mathbf{f}(\cdot, \cdot, t)$ is generally a vector-valued nonlinear function. This is illustrated in more detail shortly.

Second, a linear state-space representation for a given physical system is not unique because one can choose different state variables. For example, in (3.1) to (3.3) if we instead define $x_1 = V_{c_2}(t)$ and $x_2 = V_{c_1}(t)$, we arrive at a different linear state-space representation of the same circuit. However, we should note that if a linear nonsingular transformation of state vectors can map one state-space representation to another, then these two seemingly different representations are actually equivalent in the sense that the same initial values and control inputs will generate the same outputs (perhaps in different forms) through these two representations. Also worth noting is that not every circuit element can be used as a state variable, particularly for nonlinear systems. A basic requirement is that all the chosen state variables must be "linearly independent" in that the first-order, vector-valued ordinary differential equation has a unique solution (in terms of the control input) for any given initial values of the chosen state variables.

Finally, because A and B in the state-space representation (3.5) are both constant (independent of time), the representation is called a **linear time-invariant system**. If A or B is a matrix-valued function of time, then it will be called a **linear time-varying system**. Clearly, a time-invariant system is a special case of a time-varying system.

Now, let us return to the nonlinear circuit, with the nonlinear resistor N being connected to the circuit, as illustrated in Figure 3.1. Similar to (3.1) to (3.3), we have the following circuit equations:

$$C_1 \frac{d}{dt} V_{C_1}(t) = \frac{1}{R}\left[V_{C_2}(t) - V_{C_1}(t)\right] - N\left(V_{C_1}(t)\right) \tag{3.6}$$

$$C_2 \frac{d}{dt} V_{C_2}(t) = \frac{1}{R}\left[V_{C_1}(t) - V_{C_2}(t)\right] + i_L(t) \tag{3.7}$$

$$L \frac{d}{dt} i_L(t) = -V_{C_2}(t) \tag{3.8}$$

Note that if the nonlinear resistor N is given by

$$\begin{aligned} N\left(V_{C_1}(t)\right) &= N\left(V_{C_1}(t); m_0, m_1\right) \\ &= m_0 V_{C_1}(t) + \tfrac{1}{2}(m_1 - m_0)\left(\left|V_{C_1}(t)+1\right| - \left|V_{C_1}(t)-1\right|\right) \end{aligned} \tag{3.9}$$

with $m_0 < 0$ and $m_1 < 0$ being two appropriately chosen constant parameters, then this nonlinear circuit is the well-known Chua's circuit [24].

It is clear that compared with the linear case, it would be rather difficult to eliminate two unknowns, particularly $V_{C_1}(t)$, in order to obtain a simple third-order, nonlinear differential equation that describes the nonlinear circuit. That is, it would often be inconvenient to use a higher-order, single-variable differential equation representation for a nonlinear physical system in general. By introducing suitable state variables, however, one can easily obtain a nonlinear state-variable representation in a first-order, vector-valued, non-linear differential equation form. For instance, we may choose the following state variables:

$$x(\tau) = V_{C_1}(t) \quad y(\tau) = V_{C_2}(t) \quad \text{and} \quad z(\tau) = Ri_L(t) \quad \text{with} \quad \tau = t/RC_2$$

where the new variable $z(\tau) = Ri_L(t)$ and the rescaled time variable $\tau = t/RC_2$ are introduced to simplify the resulting representation of this particular circuit. Under this nonsingular linear transform, the previous circuit equations are converted to the following state-variable representation:

$$\begin{cases} \dot{x}(\tau) = p\left[-x(\tau) + y(\tau) - \tilde{N}(x(\tau))\right] \\ \dot{y}(\tau) = x(\tau) - y(\tau) + z(\tau) \\ \dot{z}(\tau) = -qy(\tau) \end{cases} \tag{3.10}$$

where $p = C_2/C_1$, $q = R^2C_2/L$, and

$$\begin{aligned} \tilde{N}(x(\tau)) &= N\left(x(\tau); \tilde{m}_0, \tilde{m}_1\right) \\ &= \tilde{m}_0 x(\tau) + \tfrac{1}{2}(\tilde{m}_1 - \tilde{m}_0)\left(\left|x(\tau)+1\right| - \left|x(\tau)-1\right|\right) \end{aligned} \tag{3.11}$$

with $\tilde{m}_0 = Rm_0$ and $\tilde{m}_1 = Rm_1$.

It is easy to see that this state-variable representation can be written as a special case in the following form, known as a **canonical representation** of Chua's circuit family:

$$\dot{\mathbf{x}}(\tau) = \mathbf{a} + A\mathbf{x}(\tau) + \sum_{i=1}^{k} \left|\mathbf{h}_i^{\mathrm{T}}\mathbf{x}(\tau) - \beta_i\right| \mathbf{c}_i + B\mathbf{u}(\tau) \tag{3.12}$$

namely, with $\mathbf{a} = 0$, $k = 2$, $\mathbf{h}_1 = \mathbf{h}_2 = [1\ 0\ 0]^{\mathrm{T}}$, $\beta_1 = -\beta_2 = -1$, $\mathbf{c}_1 = -\mathbf{c}_2 = ½(\tilde{m}_1 - \tilde{m}_0)$, $B\mathbf{u}(\tau)$ being a possible control input to the circuit [19], and

$$A = \begin{bmatrix} -\tilde{m}_0 - p & p & 0 \\ 1 & -1 & 1 \\ 0 & -q & 0 \end{bmatrix}$$

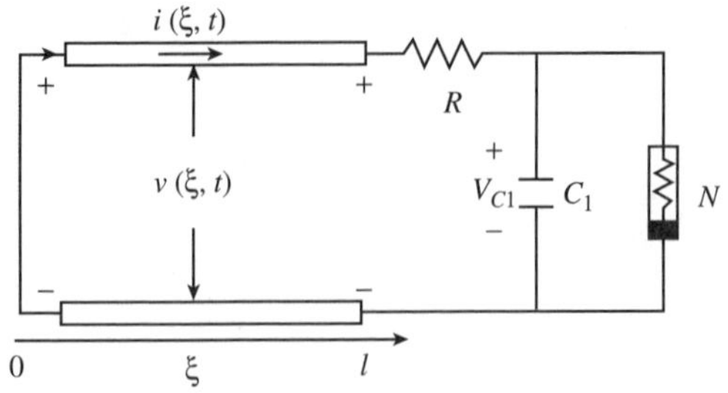

FIGURE 3.2 Time-delayed Chua's circuit.

The canonical (piecewise-linear) representation given by (3.12) describes a large class of circuits that have very rich nonlinear dynamics [25, 55].

Now, we return to (3.6) to (3.8) and Figure 3.1. If we replace the L-C_2 part of Chua's circuit by a lossless transmission line (with the spatial variable ξ) of length l terminated on its left-hand side (at $\xi = 0$) by a short circuit, as depicted in Figure 3.2, then we obtain a time-delayed Chua's circuit [89]. This circuit has a partial differential equation representation of the form

$$\begin{cases} \partial v/\partial\xi = -L\partial i(\xi,t)/\partial t \\ \partial i(\xi,t)/\partial\xi = -C_1\partial v(\xi,t)/\partial t \\ v(0,t) = 0 \\ i(l,t) = N(v(l,t) - e - Ri(l,t)) + C_1\partial\left[v(l,t) - Ri(l,t)\right]/\partial t \end{cases} \tag{3.13}$$

where $v(\xi,t)$ and $i(\xi,t)$ are the voltage and current, respectively, at the point $\xi \in [0, l]$ at time t, and $V_{C_1} = e > 0$ is a constant, with the nonlinear resistor N satisfying

$$N\left(V_{C_1} - e\right) = \begin{cases} m_0\left(V_{C_1} - e\right) & \left|V_{C_1} - e\right| < 1 \\ m_1\left(V_{C_1} - e\right) - \left(m_1 - m_0\right)\operatorname{sgn}\left(V_{C_1} - e\right) & \left|V_{C_1} - e\right| \geq 1 \end{cases}$$

In general, systems that are described by (linear or nonlinear) partial differential equations, with initial-boundary value conditions, are studied under a unified frame work of (linear or nonlinear) operator semigroup theory, and are considered to have an **infinite-dimensional system representation** [7].

Input-Output Representation

A state-variable representation of a nonlinear physical system generally can be written as

$$\begin{cases} \dot{\mathbf{x}}(t) = \mathbf{f}(\mathbf{x}(t), \mathbf{u}(t), t) & t \geq 0 \\ \mathbf{x}(0) = \mathbf{x}_0 \end{cases} \tag{3.14}$$

where $\mathbf{f}(\cdot, \cdot, t)$ is a nonlinear, vector-valued function, $\mathbf{x}_0$ is a (given) initial value for the state vector $\mathbf{x}$ at $t = 0$, and $\mathbf{u}$ is a control input to the system.

Because not all state variables in the state vector $\mathbf{x}$ can be measured (observed) in a physical system, let us suppose that what can be measured is only part of $\mathbf{x}$, or a mixture of its components, expressed by a vector-valued function of $\mathbf{x}$ in the form

$$\mathbf{y}(t) = \mathbf{g}(\mathbf{x}(t), t) \qquad t \geq 0 \tag{3.15}$$

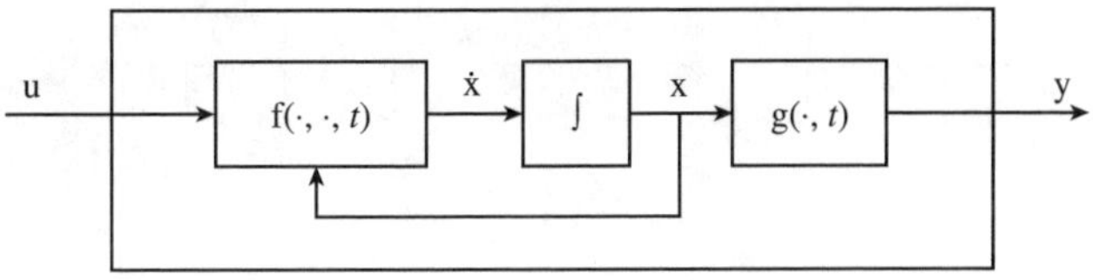

FIGURE 3.3 System I-O relationship.

where $\mathbf{y}$ is called a (**measurement** or **observation**) **output** of the physical system, and $\mathbf{g}$ is in general a lower dimensional vector-valued nonlinear function. As a particular case, $\mathbf{g}$ can be linear or, even more so, be $\mathbf{g}(\mathbf{x}(t), t) = \mathbf{x}(t)$ when all the components of the state vector are directly measurable.

If both $\mathbf{f} = \mathbf{f}(\mathbf{x}(t), \mathbf{u}(t))$ and $\mathbf{g} = \mathbf{g}(\mathbf{x}(t))$ are not explicit functions of the independent time variable t, the corresponding state-variable representation (3.14) and (3.15) is said to be **autonomous**.

It is clear that with both the system input $\mathbf{u}$ and output $\mathbf{y}$, one can simply represent the overall physical system by its input-output (I-O) relationship, as illustrated in Figure 3.3.

Now, under certain mild conditions on the nonlinear function $\mathbf{f}$, for a given control input $\mathbf{u}$, and an initial value $\mathbf{x}_0$, the state-variable representation (3.14) has a unique solution, $\mathbf{x}$, which depends on both $\mathbf{u}$ and $\mathbf{x}_0$. If we denote the solution as

$$\mathbf{x}(t) = \mathcal{F}\left(t; \mathbf{u}(t), \mathbf{x}_0\right) \tag{3.16}$$

where $\mathcal{F}$ is called an input-state mapping, then the overall I-O relationship shown in Figure 3.3 can be formulated as

$$\mathbf{y}(t) = \mathbf{g}\left(\mathcal{F}\left(t; \mathbf{u}(t), \mathbf{x}_0\right), t\right) \tag{3.17}$$

This is an **I-O representation** of the physical system having the state-variable representation (3.14) and (3.15).

As a simple example, let us consider the linear state-space representation (3.5), with a special linear measurement equation of the form $\mathbf{y}(t) = C\mathbf{x}(t)$, where C is a constant matrix. It is well known [31] that

$$\mathbf{y}(t) = C\mathcal{F}\left(t; \mathbf{u}(t), \mathbf{x}_0\right) = C\left\{e^{tA}\mathbf{x}_0 + \int_0^t e^{(t-\tau)A}\, B\mathbf{u}(\tau)d\tau\right\} \qquad t \geq 0 \tag{3.18}$$

yielding an explicit representation formula for the I-O relationship of the linear circuit (together with the assumed measurement equation).

Note that because the state-variable representation (3.14) is not unique, as mentioned previously, this I-O representation is not unique in general. However, we note that if two state-variable representations are equivalent, then their corresponding I-O relationships also will be equivalent.

It is also important to note that although the above I-O relationship is formulated for a finite-dimensional open-loop system, it can also be applied to infinite-dimensional [7] and closed-loop systems [39]. In particular, similar to linear systems, many finite-dimensional, closed-loop nonlinear systems possess an elegant **coprime factorization representation**. The (left or right) coprime factorization representation of a nonlinear feedback system is a general I-O relationship that can be used as a fundamental framework, particularly suitable for studies of stabilization, tracking, and disturbance rejection. The problem is briefly described as follows. Let a nonlinear system (mapping) P be given, not necessarily stable, and assume that it has a right-coprime factorization $P = ND^{-1}$, where both N and D are stable (D^{-1} usually has the same stability as P). One is looking for two stable, nonlinear subsystems (mappings), A and B^{-1}, representing feedback and feed-forward controllers, respectively, satisfying the Bezout identity

$$AN + BD = I$$

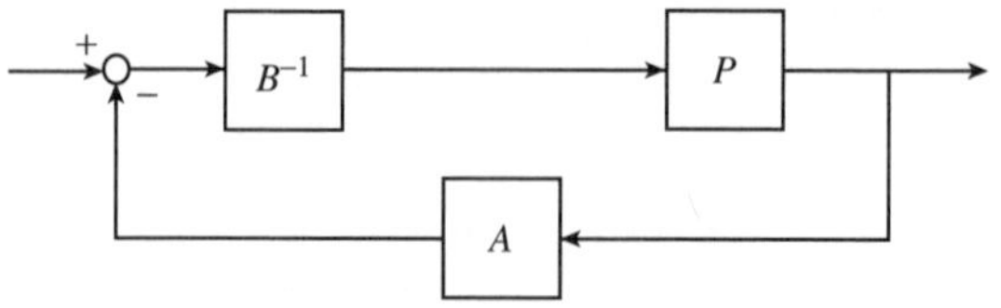

FIGURE 3.4 Right-coprime factorization of a nonlinear feedback system.

which are connected as shown in Figure 3.4, where B is also stable. If two controllers, A and B, can be found to satisfy such conditions, then even with an unstable P, the resulting closed-loop control system will be I-O, as well as internally, stable. In this sense, A and B together stabilize P.

For the left-coprime factorization, one simply uses formulas $P = D^{-1}N$ and $NA + DB = I$ instead and interchanges the two blocks of A and B^{-1} in Figure 3.4.

Taking into account causality and well-posedness of the overall closed-loop system, it is a technical issue as to how to construct the four subsystems A, B, D, and N, such that the preceding requirements can be satisfied. Some characterization results and construction methods are available in the literature [38, 45, 51, 95].

Volterra Series Representation

Recall from the fundamental theory of ordinary differential equations that an explicit I-O representation of the overall system still can be found, even if the linear state-space representation (3.5) is time varying, via the state transition matrix $\mathbf{\Phi}(t, \tau)$ determined by

$$\begin{cases} \dfrac{d}{dt}\mathbf{\Phi}(t,\tau) = A(t)\mathbf{\Phi}(t,\tau) & t \geq \tau \\ \mathbf{\Phi}(\tau,\tau) = I \end{cases} \tag{3.19}$$

where I is the identity matrix. The formula, for the simple case $\mathbf{y}(t) = C(t)\,\mathbf{x}(t)$, is

$$\mathbf{y}(t) = C(t)\left\{\mathbf{\Phi}(t,0)\mathbf{x}_0 + \int_0^t \mathbf{\Phi}(t,\tau)B(\tau)\mathbf{u}(\tau)d\tau\right\} \qquad t \geq 0 \tag{3.20}$$

For linear time-invariant systems, we actually have $\mathbf{\Phi}(t, \tau) = e^{(t-\tau)A}$, so that (3.20) reduces to the explicit formula (3.18).

For a nonlinear system, a simple explicit I-O representation with a single integral of the form (3.18) or (3.20) is generally impossible. A natural generalization of such an integral formulation is the Volterra series representation. For simplicity, let us consider the one-dimensional case in which $y(t) = g(x(t), t) = x(t)$ below. A Volterra series representation for a nonlinear I-O relationship $\mathcal{F}(\cdot)$, convergent in some measure, is an infinite sum of integrals in the following form:

$$\begin{aligned} \mathcal{F}(t,u(t)) &= \phi_0(t;x_0) + \int_0^t \phi_1(t,\tau_1)u(\tau_1)d\tau_1 + \cdots \\ &\quad + \int_0^t \cdots \int_0^{\tau_2} \phi_n(t,\tau_1,\ldots,\tau_n)u(\tau_1)\cdots u(\tau_n)d\tau_1\cdots d\tau_n + \cdots \end{aligned} \tag{3.21}$$

where $\{\phi_n\}_{n=0}^{\infty}$ are called the Volterra kernels of the series. Here, we note that this Volterra series representation can be extended easily to higher-dimensional systems.

For some representations $\mathcal{F}$, the corresponding Volterra series may have only finitely many nonzero terms in the above infinite sum. In this case, it is called a Volterra polynomial, which does not have convergence problem for bounded inputs, provided that all the integrals exist. In particular, when $\mathcal{F}$ is affine (or linear, if initial conditions are zero, so that $\phi_0 = 0$), its Volterra series has at most two nonzero terms, as given by (3.18) and (3.20), and is called a first-order Volterra polynomial. In general, however, the Volterra series (3.21) is an infinite sum. Hence, the convergence of a Volterra series is a crucial issue in formulating such a representation for a given nonlinear I-O relationship [5, 12, 59, 85].

In order to state a fundamental result about the convergence of a Volterra series, we must first recall that a mapping that takes a function to a (real or complex) value is called a **functional** and a mapping that takes a function to another function is called an **operator**. A functional may be considered to be a special operator if one views a value as a constant function in the image of the mapping. Clearly, the I-O relationship (3.17) and the Volterra series (3.21), including Volterra polynomials, are nonlinear operators. Recall also that an operator $\mathcal{T}: X \to Y$, where X and Y are normed linear spaces, is said to be continuous at $x \in X$ if $\|x_n - x\|_x \to 0$ implies $\|\mathcal{T}(x_n) - \mathcal{T}(x)\|_y \to 0$ as $n \to \infty$. Note that for a linear operator, if it is continuous at a point, then it is also continuous on its entire domain [34], but this is not necessarily true for nonlinear operators.

As usual, we denote by $C[0, T]$ and $L_p[0, T]$, respectively, the space of continuous functions defined on $[0, T]$ and the space of measurable functions f satisfying $\int_0^T |f(t)|^p\, dt < \infty$ for $1 \le p < \infty$ or $\sup_{t\in[0,T]} |f(t)| < \infty$ for $p = \infty$. The following result [5] is an extension of the classical Stone–Weierstrass theorem [22, 36, 40].

Theorem 3.1: *Let X be either $C[0,T]$ or $L_p[0,T]$, with $1 \le p < \infty$, and Ω be a compact subset in X. Then, for any continuous operator $\mathcal{F}: \Omega \to L_q[0, T]$, where $(1/p) + (1/q) = 1$, and for any $\varepsilon > 0$, a Volterra polynomial $P_n(\cdot)$ exists, with n determined by ε, such that*

$$\sup_{x\in\Omega} \left\| \mathcal{F}(x) - P_n(x) \right\|_{L_q} < \varepsilon$$

In other words, $P_n \to \mathcal{F}$ uniformly on the compact subset $\Omega \subset X$ as $n \to \infty$.

In the literature, many variants of this fundamental convergence theorem exist under various conditions in different forms, including the $L_\infty[0, T]$ case [45, 59, 84, 85]. We may also find different methods for constructing the Volterra kernels $\{\phi_n\}_{n=0}^{\infty}$ for $\mathcal{F}$ [83]. In addition, specially structured Volterra series representations abound for nonlinear systems, such as the Volterra series with finite memory [5], approximately finite memory [86], and fading memory [10].

Finally, it should be mentioned that in a more general manner, a few abstract functional series representations exist, including the **generating power series representation** for certain nonlinear systems [48], from which the Volterra series can be derived. Briefly, an important result is the following theorem [6, 54, 71, 91].

Theorem 3.2: *Consider a nonlinear control system of the form*

$$\begin{cases} \dot{x}(t) = g_0(x(t)) + \displaystyle\sum_{k=1}^{m} g_k(x(t))u_k(t) & t \in [0, T] \\ y(t) = h(x(t)) \end{cases}$$

where $h(\cdot)$ and $\{g_i(\cdot)\}_{i=0}^{m}$ are sufficiently smooth functionals, with an initial state x_0. If the control inputs satisfy $\max_{0 \le \tau \le T} |u_k(t)| < 1$, then the corresponding output of this nonlinear system has a convergent functional series of the form

$$y(t) = h(x_0) + \sum_{i=0}^{\infty} \sum_{k_0,\ldots,k_1=0}^{m} L_{g_{k_0}} \cdots L_{g_{k_i}} h(x_0) \int_0^t d\xi_{k_i} \cdots d\xi_{k_0} \tag{3.22}$$

where $L_g h(x_0) := [\partial h/\partial x] g(x)|_{x=x0}$ and ξ_k are defined by

$$\xi_0(t) = t \quad \xi_k(t) = \int_0^t u_k(\tau)d\tau \qquad k = 1, \ldots, m$$

with the notation

$$\int_0^t d\xi_{k_i} \cdots d\xi_{k_0} := \int_0^t d\xi_{k_i}(\tau) \int_0^t d\xi_{k_{i-1}} \cdots d\xi_{k_0}$$

Note that in order to guarantee the convergence of the functional series (3.22), in many cases it may be necessary for T to be sufficiently small.

Analogous to the classical Taylor series of smooth functions, a fairly general series representation for some nonlinear systems is still possible using **polynomial operators**, or the like [90]. As usual, however, the more general the presentation is, the less concrete the results. Moreover, a very general series expansion is likely to be very local, and its convergence is difficult to analyze.

3.3 Approximation

The mathematical term "approximation" used here refers to the theory and methodology of function (functional or operator) approximation. Mathematical approximation theory and techniques are important in engineering when one seeks to represent a set of discrete data by a continuous function, to replace a complicated signal by a simpler one, or to approximate an infinite-dimensional system by a finite-dimensional model, etc., under certain optimality criteria.

Approximation is widely used in system modeling, reduction, and identification, as well as in many other areas of control systems and signal processing [32]. A Volterra polynomial as a truncation of the infinite Volterra series (discussed earlier) serves as a good example of system (or operator) approximation, where the question "In what sense is this approximation good?" must be addressed further.

Best Approximation of Systems (Operators)

Intuitively, approximation is always possible. However, two key issues are the quality of the approximation and the efficiency of its computation (or implementation). Whenever possible, one would like to have the best (or optimal) approximation, based on the available conditions and subject to all the requirements.

A commonly used criterion for best (or optimal) approximations is to achieve a minimum norm of the approximation error using a norm that is meaningful to the problem. Best approximations of systems (operators) include the familiar least-squares technique, and various other uniform approximations.

Least-Squares Approximation and Projections

Let us start with the most popular "best approximation" technique (the least-squares method), which can also be thought of as a projection, and a special min-max approximation discussed in the next section. Discrete data fitting by a continuous function is perhaps the best-known example of least-squares. The special structure of Hilbert space, a complete inner-product space of functions, provides a general and convenient framework for exploring the common feature of various least-squares approximation techniques. Because we are concerned with approximation of nonlinear systems rather than functions, a higher-level framework, the Hilbert space of operators, is needed. We illustrate such least-squares system (or operator) approximations with the following two examples.

First, we consider the linear space, H, of certain nonlinear systems that have a convergent Volterra series representation (3.21) mapping an input space X to an output space Y. Note that although a nontrivial Volterra series is a nonlinear operator, together they constitute a linear space just like nonlinear functions.

To form a Hilbert space, we first need an inner product between any two Volterra series. One way to introduce an inner product structure into this space is as follows. Suppose that all the Volterra series,

$\mathcal{F}: X \to Y$, where both X and Y are Hilbert spaces of real-valued functions, have bounded admissible inputs from the set

$$\Omega = \left\{ x \in X \big| \|x\|_X \leq \gamma < \infty \right\}$$

For any two convergent Volterra series of the form (3.21), say $\mathcal{F}$ and $\mathcal{G}$, with the corresponding Volterra kernel sequences $\{\phi_n\}$ and $\{\psi_n\}$, respectively, we can define an inner product between them via the convergent series formulation

$$\langle \mathcal{F}, \mathcal{G} \rangle_H : = \sum_{n=0}^{\infty} \tfrac{\rho_n}{n!} |\phi_n \psi_n|$$

with the induced norm $\|\mathcal{F}\|_H = \langle \mathcal{F},\mathcal{F}\rangle^{1/2}{}_H$, where the weights $\{\rho_n\}$ satisfy

$$\sum_{n=o}^{\infty} \frac{1}{\rho_n} \frac{\gamma^{2n}}{n!} < \infty$$

Recall also that a **reproducing kernel Hilbert** space $\tilde{H}$ is a Hilbert space (of real-valued functions or operators) defined on a set S, with a **reproducing kernel** $K(x, y)$, which belongs to $\tilde{H}$ for each fixed x or y in S and has the property

$$\langle K(x, y), \mathcal{F}(y) \rangle_{\tilde{H}} = \mathcal{F}(x) \qquad \forall \mathcal{F} \in \tilde{H} \text{ and } \forall x, y \in S$$

Using the notation defined above, the following useful result was established [43, 45] and is useful for nonlinear systems identification (see Theorem 3.23).

Theorem 3.3: *The family of all the convergent Volterra series of the form* (3.21) *that maps the bounded input set* Ω *to* Y *constitutes a reproducing kernel Hilbert space with the reproducing kernel*

$$K(x, y) = \sum_{n=0}^{\infty} \frac{1}{n!} \frac{1}{\rho_n} \langle x, y \rangle_X^n \qquad x, y \in \Omega \subset X \tag{3.23}$$

The reproducing kernel Hilbert space H defined above is called a **generalized Fock space** [46]. For the special case in which $\rho_n \equiv 1$, its reproducing kernel has a nice closed-form formula as an exponential operator $K(x,y) = e^{\langle xy \rangle}$.

Now, suppose that a nonlinear system $\mathcal{F}$ is given, which has a convergent Volterra series representation (3.21) with infinitely many nonzero terms in the series. For a fixed integer $n \geq 0$, if we want to find an nth-order Volterra polynomial, denoted V_n^*, from the Hilbert space H such that

$$\|\mathcal{F} - V_n^*\|_H = \inf_{V_n \in H} \|\mathcal{F} - V_n\|_H \tag{3.24}$$

then we have a best approximation problem in the least-squares sense. To solve this optimization problem is to find the best Volterra kernels $\{\phi_k(t)\}_{k=0}^n$ over all the possible kernels that define the Volterra polynomial V_n, such that the minimization (3.24) is achieved.

Note that, if we view the optimal solution V_n^* as the projection of $\mathcal{F}$ onto the $(n + 1)$-dimensional subspace of H, then this least-squares minimization is indeed a projection approximation. It is then clear, even from the Hilbert space geometry (see Figure 3.5), that such an optimal solution, called a **best approximant**, always exists due to the norm-completeness of Hilbert space and is unique by the convexity of inner product space.

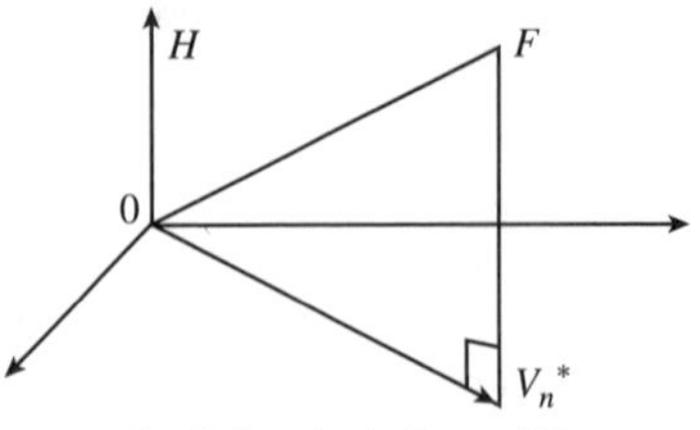

FIGURE 3.5 Projection in a Hilbert space.

As a second example, let H be a Hilbert space consisting of all the linear and nonlinear systems that have an nth-order Taylor series representation of the form

$$P_n(\cdot)=\sum_{k=0}^{n}\alpha_k(t)M_k(\cdot)=\alpha_0(t)+\alpha_1(t)(\cdot)(t)+\cdots+\alpha_n(t)(\cdot)^n(t) \quad (3.25)$$

where $M_k(\cdot) := (\cdot)^k$ is the monomial operator of degree k and $\{\alpha_k\}_{k=0}^{n}$ are continuous real-valued functions satisfying certain conditions arising from some basic properties of both the domain and the range of the operator. Suppose that the monomial operators $\{M_k\}_{k=0}^{\infty}$ are orthogonal under the inner product of H. Given an lth-order polynomial operator P_l with $\alpha_l(t) \neq 0$ almost everywhere, if for a fixed integer $n < l$ we want to find an nth-order polynomial operator P_n^* of the form (3.25) from H, such that

$$\left\|P_l-P_n^*\right\|_H=\inf_{P_n\in H}\left\|P_l-P_n\right\|_H \quad (3.26)$$

then we have a best approximation problem in the least-squares sense. To solve this optimization problem is to find the best coefficient functions $\{\alpha_k(t)\}_{k=0}^{n}$ over all possible functions that define the polynomial operator P_n. Again, because the optimal solution is the projection of P_l onto the $(n + 1)$-dimensional subspace H of a Hilbert space to which P_l belongs, it always exists and is unique.

We now state a general result of least-squares approximation for systems, which is a straightforward generalization of the classical result of least-squares approximation for functions [22, 36].

Theorem 3.4: *Let H be a Hilbert space of nonlinear operators, and let H_n be its n-dimensional subspace. Then, given an $\mathcal{F}\in H$, the least-squares approximation problem*

$$\left\|\mathcal{F}-\mathcal{N}_n^*\right\|_H=\inf_{\mathcal{N}_n\in H_n}\left\|\mathcal{F}-\mathcal{N}_n\right\|_H$$

is always uniquely solvable, with the optimal solution given by

$$\mathcal{N}^*(\cdot)=\sum_{k=1}^{n}\left\langle\mathcal{F},h_k\right\rangle_H h_k(\cdot)$$

where $\{h_k\}_{k=1}^{n}$ is an orthonormal basis of H_n.

A more general setting is to replace the Hilber space H by a Banach space (a complete normed linear space, such as L_1 and L_∞, which may not have an inner product structure). This extension includes the Hilbert space setting as a special case, but generally does not have so many special features. Even the existence and uniqueness of best approximants cannot be taken for granted in general — not even for the simpler case of best approximation of real-valued functions — if a Banach (non-Hilbert) space is considered [73]. Nevertheless, the following result is still convenient to use [22].

Theorem 3.5: *Let B be a uniformly convex Banach space and Ω be a closed convex set in B. Then, for any given* $\mathcal{F} \in B$, *the optimal approximation problem*

$$\left\|\mathcal{F}-\omega^*\right\|_B = \inf_{\omega\in\Omega}\left\|\mathcal{F}-\omega\right\|_B$$

has a unique solution.

Here, a space (or subset) B is said to be **uniformly convex** if, for any $\varepsilon > 0$, there exists a $\|f\|_B = \|g\|_B = 1$ and $\|\tfrac{1}{2}(f + g)\|_B > 1 - \delta$ together imply $\|f - g\|_B < \varepsilon$. Geometrically, a disk is uniformly convex while a triangle is only convex, but not uniformly so. It is then intuitively clear that for a given point outside (or inside) a disk, only a single point exists in the disk that has the shortest distance to the given point. However, this is not always true for a nonuniform case. In fact, a best approximation problem in the general Banach space setting has either a unique solution or has infinitely many solutions (if it is solvable), as can be seen from the next result [32].

Theorem 3.6: *Let Ω be a closed convex set in a Banach space B, and* ω_1^* *and* ω_2^* *be two optimal solutions of the best approximation problem*

$$\left\|\mathcal{F}-\omega^*\right\|_B = \inf_{\omega\in\Omega}\left\|\mathcal{F}-\omega\right\|_B$$

Then, any convex combination of ω_1^* *and* ω_2^* *in the form*

$$\omega^* = a\omega_1^* + (1-a)\omega_2^* \quad 0 \le a \le 1$$

is also an optimal solution of the problem.

Usually, a best approximant (if it exists) for an optimal approximation problem in a Banach space is also called a (**minimal**) **projection** of the given operator from a higher-dimensional subspace onto a lower-dimensional subspace. In this extension, the projection has no simple geometric meaning of "orthonormality" due to the lack of an inner product structure. However, a projection operator with a unity norm in the Banach space setting is a natural generalization of the orthonormal projection in the Hilbert space framework.

Min-Max (Uniform) Approximation

It is clear from the least-squares approximation formulation that if the given nonlinear representation (operator) $\mathcal{F}$ and the lower-order approximant (used to approximate $\mathcal{F}$) do not have the same structure (the same type of series), then the least-squares approximation cannot be applied directly or efficiently.

To introduce another approach, we first recall that for two given normed linear spaces X and Y and for a given bounded subset Ω of X, with $0 \in \Omega$, the operator norm of a nonlinear operator $\mathcal{N}: \Omega \to Y$ satisfying $\mathcal{N}(0) = 0$, can be defined as

$$\|\mathcal{N}\| = \sup_{\substack{x,y\in\Omega \\ x\neq y}} \frac{\left\|\mathcal{N}(x)-\mathcal{N}(y)\right\|_y}{\|x-y\|_X} \tag{3.27}$$

Thus, given a norm-bounded nonlinear operator $\mathcal{F}$, representing a given physical system, we may consider the problem of finding another norm-bounded nonlinear operator $\mathcal{N}^*$ from a certain class $\mathcal{N}$ of desired nonlinear operators (systems), not necessarily having the same structure as $\mathcal{F}$, to best approximate $\mathcal{F}$ in the sense that

$$\left\|\mathcal{F}-\mathcal{N}^*\right\| = \inf_{\mathcal{N}\in N}\left\|\mathcal{F}-\mathcal{N}\right\| \tag{3.28}$$

For example, N can be the family of nth-order Volterra polynomilas or nth-order polynomial operators discussed previously. Commonly used function spaces X and Y include the space of all continuous functions, the standard L_p space (or l_p for the discrete case), and the Hardy space H_p (for complex-variable functions [32]), with $1 \leq p \leq \infty$.

Because the nonlinear operator norm defined by (3.27) is a sup (max) norm and this optimization is an inf (min) operation, the best approximation problem (3.28) is called a **min-max approximation.** Note also that because the nonlinear operator norm (3.27) is defined over all the bounded inputs in the set Ω, this approximation is uniform, and thus independent of each individual input function of the set Ω. For this reason, this approximation is also called a **uniform approximation**, indicating that the best approximant is the optimal solution over *all* input functions.

It should be noted that both existence and uniqueness of best approximation solutions to the min-max approximation problem (3.28) must be investigated according to the choice of the operator family N and the I-O spaces X and Y, which generally cannot be taken for granted, as previously discussed.

An important and useful class of nonlinear operators which can be put into a Banach space setting with great potential in systems and control engineering is the family of **generalized Lipschitz operators** [45]. To introduce this concept, we first need some notation. Let X be a Banach space of real-valued functions defined on $[0, \infty)$ and, for any $f \in X$ and any $T \in [0, \infty)$, define

$$[f]_T(t) = \begin{cases} f(t) & t \leq T \\ 0 & t > T \end{cases}$$

Then, form a normed linear space X^e, called the **extended linear space associated with** $\boldsymbol{X}$, by

$$X^e = \left\{ f \in X \,\middle|\, \left\| [f]_T \right\|_X < \infty, \forall T < \infty \right\}$$

For a subset $D \subseteq X^e$, any (linear or nonlinear) operator $\mathcal{G}$: $D \rightarrow Y^e$ satisfying

$$\left\| [\mathcal{G}(x_1)]_T - [\mathcal{G}(x_2)]_T \right\|_Y \leq L \left\| [x_1]_T - [x_2]_T \right\|_X, \forall x_1, x_2 \in D, \forall T \in [0, \infty)$$

for some constant $L < \infty$, is called a generalized Lipschitz operator defined on D. The least of such constants L is given by the semi-norm of the operator $\mathcal{G}$:

$$\|\mathcal{G}\| \colon \sup_{T \in [0,\infty)} \sup_{\substack{x_1, x_2 \in D \\ [x_1]_T \neq [x_2]_T}} \frac{\left\| [\mathcal{G}(x_1)]_T - [\mathcal{G}(x_2)]_T \right\|_Y}{\left\| [x_1]_T - [x_2]_T \right\|_X}$$

and the operator norm of $\mathcal{G}$ is defined via this semi-norm by

$$\|\mathcal{G}\|_{\text{Lip}} = \|\mathcal{G}(x_0)\|_Y + \|\mathcal{G}\|$$

for an arbitrarily chosen and fixed $x_0 \in D$. The following result has been established [45].

Theorem 3.7: *The family of generalized Lipschitz operators*

$$\text{Lip}(D, Y^e) = \left\{ \mathcal{G} : D \subseteq X^e \rightarrow Y^e \,\middle|\, \|\mathcal{G}\|_{\text{Lip}} < \infty \text{ on } D \right\}$$

is a Banach space.

Based on this theorem, a best approximation problem for generalized Lipschitz operators can be similarly formulated, and many fundamental approximation results can be obtained. In addition, generalized Lipschitz operators provide a self-unified framework for both left- and right-coprime factorization representations of nonlinear feedback systems. Under this framework, the overall closed-loop system shown in Figure 3.4 can have a causal, stable, and well-posed coprime factorization representation, which can be applied to optimal designs such as tracking and disturbance rejection [45].

We now discuss briefly a different kind of min-max (uniform) approximation: the **best Hankel norm approximation,** where the norm (3.27) is replaced by the operator norm of a Hankel operator defined as follows [32, 77]. Consider, for instance, the transfer function

$$H(z) = \alpha_0 + \alpha_1 z^{-1} + \alpha_2 z^{-2} + \cdots$$

of a discrete time linear time-invariant system. The **Hankel operator** associated with this series is defined as the infinite matrix

$$\Gamma_\alpha := \left[\alpha_{|i-j|}\right] = \begin{bmatrix} \alpha_0 & \alpha_1 & \alpha_2 & \cdots \\ \alpha_1 & \alpha_2 & \cdots & \\ \alpha_2 & \cdots & & \\ \vdots & & & \end{bmatrix}$$

which is a linear operator on a normed linear space of sequences. The operator norm of Γ_α over the l_2-space is called the **Hankel norm** of Γ_α.

One important feature of the Hankel operators is reflected in the following theorem [32, 77].

Theorem 3.8: *An infinite Hankel matrix has a finite rank iff its corresponding functional series is rational (it sums up to a rational function); and this is true iff the rational series corresponds to a finite-dimensional bilinear system.*

Another useful property of Hankel operators in system approximation is represented in the following theorem [28].

Theorem 3.9: *The family of compact Hankel operators is an M-ideal in the space of Hankel operators that are defined on a Hilbert space of real-valued functions.*

Here, a compact operator is one that maps bounded sets to compact closures and an M-ideal is a closed subspace X of a Banach space Z such that $X^\perp$, the orthogonal complemental subspace of X in Z, is the range of the projection P from the dual space Z^* to $X^\perp$ that has the property

$$\|f\| = \|P(f)\| + \|f - P(f)\| \quad \forall f \in Z^*$$

The importance of the M-ideal is that it is a proximinal subspace with certain useful approximation characteristics, where the proximinal property is defined as follows. Let $L(X)$ and $C(X)$ be the classes of bounded linear operators and compact operators, respectively, both defined on a Banach space X. If every $\mathcal{L} \in L(X)$ has at least one best approximant from $C(X)$, then $C(X)$ is said to be proximinal in $L(X)$. A typical result would be the following: for any $1 < p < \infty$, $C(l_p)$ is proximinal in $L(l_p)$. However, $C(X)$ is not proximinal in $L(X)$ if $X = C[a, b]$, the space of continuous functions defined on $[a, b]$, or $X = L_p[a, b]$ for all $1 < p < \infty$ except $p = 2$.

Best (Uniform) Approximation of Signals (Functions)

Best approximations of signals for circuits and systems are also important. For example, two (different) systems (e.g., circuits) are considered to be equivalent over a set Ω of admissible input signals iff the same input from Ω yields the same outputs through the two systems. Thus, the problem of using a system to best approximate another may be converted, in many cases, to the best approximation problem for their output signals.

A signal is a function of time, usually real valued and one-dimensional. The most general formulation for best approximation of functions can be stated as follows. Let X be a normed linear space of real-valued functions and Ω be a subset of X. For a given f in X but not in Ω, find a $g^* \in \Omega$ such that

$$\|f - g^*\|_X = \inf_{g \in \Omega} \|f - g\|_X \tag{3.29}$$

In particular, if $X = L_\infty$, l_∞, or H_∞, the optimal solution is the best result for the worst case.

If such a g^* exists, then it is called a **best approximant** of f from the subset Ω. In particular, if $\Omega_1 \subset \Omega_2 \subset \cdots$ is a sequence of subspaces in X, such that $\overline{\cup \Omega_n} = X$, an important practical problem is to find a sequence of best approximants $g_n^* \in \Omega_n$ satisfying the requirement (3.29) for each $n = 1, 2, \ldots$, such that $\|g_n^* - g^*\|_X \to 0$ as $n \to \infty$. In this way, for each n, one may be able to construct a simple approximant g_n^* for a complicated (even unknown) function f, which is optimal in the sense of the min-max approximation (3.29).

Existence of a solution is the first question about this best approximation. The fundamental result is the following [22, 36].

Theorem 3.10: *For any $f \in X$, a best approximant g^* of f in Ω always exists, if Ω is a compact subset of X; or Ω is a finite-dimensional subspace of X.*

Uniqueness of a solution is the second question in approximation theory, but it is not as important as the existence issue in engineering applications. Instead, characterization of a best approximant for a specific problem is significant in that it is often useful for constructing a best approximant.

As a special case, the preceding best approximation reduces to the least-squares approximation if X is a Hilbert space. The basic result is the following (compare it with Theorem 3.4, and see Figure 3.5).

Theorem 3:11: *Let H be a Hilbert space of real-valued functions, and let H_n be its n-dimenstional subspace. Then, given an $f \in H$, the least-squares approximation problem*

$$\|f - h_n^*\|_H = \inf_{h_n \in H_n} \|f - h_n\|_H$$

is always uniquely solvable, with the optimal solution given by

$$h_n^*(t) = \sum_{k=1}^{n} \langle f, h_k \rangle_H h_k(t)$$

where $\{h_k\}_{k=1}^n$ is an orthonormal basis of H_n.

Here, the orthonormal basis of H_n is a **Chebyshev** system, a system of functions which satisfy the **Haar condition** that the determinant of the matrix $[h_i(t_j)]$ is nonzero at n distinct points $t_1 < \cdots < n_n$ in the domain. Chebyshev systems include many commonly used functions, such as algebraic and trigonometric polynomials, splines, and radial functions. Best approximation by these functions is discussed in more detail below.

We remark that the least-squares solution shown in Theorem 3.11 is very general, which includes the familiar truncations of the Fourier series [36] and the wavelet series [29] as best approximation.

Polynomial and Rational Approximations

Let π_n be the space of all algebraic polynomials $p_n(t)$ of degree not greater than n. For any continuous function $f(t)$ defined on $[a, b]$, one is typically looking for a best approximant $p_n^* \in \pi_n$ for a fixed n, such that

$$\|f - p_n^*\|_{L_\infty[a,b]} = \min_{p_n \in \pi_n} \|f - p_n\|_{L_\infty[a,b]} \tag{3.30}$$

This is a best (min-max and uniform) **algebraic polynomial approximation** problem. Replacing the algebraic polynomials by the nth-order trigonometric polynomials of the form $\sum_{k=0}^{n}(a_k \cos(kt) + b_k \sin(kt))$ changes the problem to the best **trigonometric polynomial approximation**, in the same sense as the best algebraic polynomial approximation, for a given function $f \in C[-\pi,\pi]$. This can be much further extended to any Chebyshev system, such as the radial basis functions and polynomial spline functions, which are discussed later. According to the second part of Theorem 3.10, the best uniform polynomial approximation problem (3.30) always has a solution that, in this case, is unique. Moreover, this best approximant is characterized by the following important sign-alternation theorem. This theorem is also valid for the best uniform approximation from any other Chebyshev system [22,36].

Theorem 3.12: *The algebraic polynomial p_n^* is a best uniform approximant of $f \in C[a, b]$ from π_n iff there exist $n + 2$ points $a \le t_0 < \cdots < t_{n+1} \le b$* such that

$$f(t_k) = p_n^*(t_k\_ = c(-1)^k \|f - p_n^*\|_{L_\infty[a,b]} \qquad k = 0, 1, \ldots, n+1$$

where $c = 1$ or -1.

An efficient **Remes (exchange) algorithm** is available for constructing such a best approximant [79].

Another type of function is related to algebraic polynomials: the algebraic rational functions of the form $r_{n,m}(t) = p_n(t)/q_m(t)$, which has finite values on $[a, b]$ with coprime $p_n \in \pi_n$ and $q_m \in \pi_m$. We denote by $R_{n,m}$ the family of all such rational functions, or a subset of them, with fixed integers $n \ge 0$ and $m \ge 1$. Although $R_{n,m}$ is not a compact set or a linear space, the following result can be established [22].

Theorem 3.13: *For any given function $f \in C\,[a, b]$, there exists a unique $r_{n,m}^*(t) \in R_{n,m}$ such that*

$$\|f - r_{n,m}^*\|_{L_\infty[a,b]} = \inf_{r_{n,m} \varepsilon R_{n,m}} \|f - r_{n,m}\|_{L\infty[a,b]} \tag{3.31}$$

The optimal solution $r_{n,m}^*(t)$ of (3.31) is called the **best uniform rational approximant** of $f(t)$ on $[a, b]$ from $R_{n,m}$.

Note that the unique best rational approximant may have different expressions unless it is coprime, as assumed previously. The following theorem [22] characterizes such a best approximant, in which we use $d(p_n)$ to denote the actual degree of p_n, $0 \le d\,(p_n) \le n$.

Theorem 3.14: *A rational function $r_{n,m}^* = p_n^*/q_m^*$ is a best uniform approximant of $f \in C\,[a, b]$ from $R_{n,m}$ iff there exist s points $a \le t_1 < \cdots < t_s \le b$, with $s = 2 + \min\{n + d(q_m), m + d(p_n)\}$,* such that

$$f(t_k) - r_{n,m}^*(t_k) = c(-1)^k \|f - r_{n,m}^*\|L_{\infty[a,b]} \qquad k = 1, \ldots, s$$

where $c = 1$ or -1.

The Remes (exchange) algorithm [79] also can be used for constructing a best rational approximant.

An important type of function approximation, which utilizes rational functions, is the **Padé approximation**. Given a formal power series of the form

$$f(t) = c_0 + c_1 t + c_2 t^2 + \cdots \qquad t \in [-1, 1]$$

not necessarily convergent, the question is to find a rational function $p_n(t)/q_m(t)$, where n and m are both fixed, to best approximate $f(t)$ on $[-1, 1]$, in the sense that

$$\left| f(t) - \frac{p_n(t)}{q_m(t)} \right| \le c|t|^l \qquad t \in [-1, 1] \tag{3.32}$$

for a "largest possible" integer l. It turns out that normally the largest possible integer is $l = n + m + 1$. If such a rational function exists, it is called the **[*n*, *m*]th-order Padé approximant** of $f(t)$ on [-1, 1]. The following result is important [22].

Theorem 3.15: *If $f(t)$ is $(n + m + 1)$ times continuously differntiable in a neighborhood of $t = 0$, then the $[n, m]$th-order Padé approximant of $f(t)$ exists, with $l > n$. If $l \leq n + m + 1$, then the coefficients $\{a_k\}_{k=0}^{n}$ and $\{b_k\}_{k=0}^{m}$ of $p_n(t)$ and $q_m(t)$ are determined by the following linear system of algebraic equations:*

$$\sum_{j=0}^{i} \frac{f^j(0)}{j!} b_{i-j} = a_i \qquad i = 0, 1, \ldots, l-1$$

with $a_{n+j} = b_{m+j} = 0$ for all $j = 1, 2, \ldots$. Moreover, if p_n/q_m is the $[n, m]$th-order Padé approximant of $f(t) = \sum^{\infty}\{a_k\}_{k=0}^{n} f_k t^k$, then the approximation error is given by

$$\left| f(t) - \frac{p_n(t)}{q_m(t)} \right| = \sum_{k=n+1}^{\infty} \left(\sum_{j=0}^{m} f_{k-j} b_j \right) \frac{t^k}{q_m(t)} \qquad t \in [-1, 1]$$

Padé approximation can be extended from algebraic polynomials to any other Chebyshev systems [22].

Approximation via Splines and Radial Functions

Roughly speaking, spline functions, or simply splines, are piecewise smooth functions that are structurally connected and satisfy some special properties. The most elementary and useful splines are polynomial splines, which are piecewise algebraic polynomials, usually continuous, with a certain degree of smoothness at the connections. More precisely, let

$$a = t_0 < t_1 < \cdots < t_n < t_{n+1} = b$$

be a partition of interval $[a, b]$. The **polynomial spline** of degree m with knots $\{t_k\}_{k=1}^{n}$ on $[a, b]$ is defined to be the piecewise polynomial $g_m(t)$ that is a regular algebraic polynomial of degree m on each subinterval $[t_k, t_{k+1}]$, $k = 0, \ldots, n$, and is $(m - 1)$ times continuously differentiable at all knots [41, 88]. We denote the family of these algebraic polynomial splines by $S_m(t_1, \ldots, t_n)$, which is an $(n + m + 1)$-dimensional linear space.

Given a continuous function $f(t)$ on $[a, b]$, the best uniform spline approximation problem is to find a $g_m^* \in S_m(t_1, \ldots, t_n)$ such that

$$\|f - g_m^*\|_{L_\infty[a,b]} = \inf_{g_m \in S_m} \|f - g_m\|_{L_\infty[a,b]} \tag{3.33}$$

According to the second part of Theorem 3.10, this best uniform approximation problem always has a solution. A best spline approximant can be characterized by the following sign-alteration theorem [72], which is a generalization of Theorem 3.12, from polynomials to polynomial splines.

Theorem 3.16: *The polynomial spline $g_m^*(t)$ is a best uniform approximant of $f \in C\,[a, b]$ from $S_m(t_1, \ldots, t_n)$ iff there exists a subinterval $[t_r, t_{r+s}] \subset [a, b]$, with integers r and $s \leq 1$, such that the maximal number γ of sign-alteration points on this subinterval $[t_r, t_{r+s}]$, namely,*

$$f(t_k) - g_m(t_k) = c(-1)^k \|f - g_m\|_{L_\infty[a,b]} \qquad t_k \in [t_r, t_{r+s}] \quad k = 1, \ldots, \gamma$$

satisfies $\gamma \geq m + s + 1$, where $c = 1$ or -1.

Polynomial splines can be used for least-squares approximation, just like regular polynomials, if the L_∞-norm is replaced by the L_2-norm in (3.33). For example, *B*-splines, i.e., basic splines with a compact support, are very efficient in least-squares approximation. The spline quasi-interpolant provides another type of efficient approximation, which has the following structure

$$g_m(t) = \sum_k f(t_k)\phi_k^m(t) \tag{3.34}$$

and can achieve the optimal approximation order, where $\{\phi_k^m\}$ is a certain linear combination of *B*-splines of order *m* [18].

Spline functions have many variants and generalizations, including natural splines, perfect splines, various multivariate splines, and some generalized splines defined by linear ordinary or partial differntial operators with initial-boundary conditions [27, 41, 42, 44, 88].

Splines are essentially local, in the sense of having compact supports, perhaps with the exception perhaps of the **thin-plate splines** [94], where the domains do not have a boundary.

Radial functions are global, with the property $\phi(r) \to \infty$ as $r \to \infty$ and, normally, $\phi(0) = 0$. Well-conditioned radial functions include $|r|^{2m+1}$, $r^{2m}\log(r)$, $(r^2 + a^2)^{\pm 1/2}$, $0 < a \ll 1$, etc. [80]. Many radial functions are good candidates for modeling nonlinear circuits and systems [63, 64]. For example, for l distinct points $\mathbf{t}_1, \ldots, \mathbf{t}_l$ in R^n, the radial functions $\{\phi(|\mathbf{t} - \mathbf{t}_k|)\}_{k=1}^{l}$ are linearly independent, and thus the minimization

$$\min_{\{c_k\}} \left| f(\mathbf{t}) - \sum_{k=1}^{l} c_k \phi(|\mathbf{t} - \mathbf{t}_k|) \right|^2 \tag{3.35}$$

at some scattered points can yield a best least-squares approximant for a given function $f(t)$, with some especially desirable features [81]. In particular, an affine plus radial function in the form

$$\mathbf{a} \cdot \mathbf{t} + b + \sum_{k=1}^{l} c_k \phi(|\mathbf{t} - \mathbf{t}_k|) \quad \mathbf{t} \in R^n \tag{3.36}$$

where $\mathbf{a}$, b, $\{c_k\}_{k=1}^{l}$ are constants, provides a good modeling framework for the canonical piecewise linear representation (3.12) of a nonlinear circuit [63].

Approximation by Means of Interpolation

Interpolation plays a central role in function approximation theory. The main theme of interpolation is this: suppose that an unknown function exists for which we are given some measurement data such as its function values, and perhaps some values of its derivatives, at some discrete points in the domain. How can we use this information to construct a new function that interpolates these values at the given points as an approximant of the unknown function, preferably in an optimal sense? Constructing such a function, called an **interpolant**, is usually not a difficult problem, but the technical issue that remains is what kind of functions should be used as the interpolant so that a certain meaningful and optimal objective is attained?

Algebraic polynomial interpolation is the simplest approach for the following **Lagrange interpolation** problem [22, 36].

Theorem 3.17: *For arbitrarily given* $n + 1$ *distinct points* $0 \le t_0 < t_1 < \cdots < t_n \le 1$ *and* $n + 1$ *real values* $v_0, v_1, \ldots, v_n$*, there exists a unique polynomial* $p_n(t)$ *of degree n, which satisfies*

$$p_n(t_k) = v_k \qquad k = 0, 1, \ldots, n$$

This polynomial is given by

$$p_n(t) = \sum_{k=0}^{n} v_k L_k(t)$$

with the Lagrange basis polynomials

$$L_k(t) := \frac{(t-t_0)\cdots(t-t_{k-1})(t-t_{k+1})(t-t_n)}{(t_k-t_0)\cdots(t_k-t_{k-1})(t_k-t_{k+1})(t_k-t_n)} \qquad k=0,\ldots,n$$

Moreover, if $f(t)$ is l ($\leq n+1$) times continuously differentiable on $[a, b]$, then the interpolation error is bounded by

$$\|f-p_n\|_{L_\infty[0,1]} \leq \frac{1}{n!}\left\|f^{(l)}\right\|_{L_\infty[0,1]}\|h\|_{L_\infty[0,1]}$$

where $h(t) = \prod_{k=0}^{n-1}(t - t_k)$, and $\|h\|_{L_\infty[0,1]}$ *attains its minimum at the Chebyshev points* $t_k = \cos(2k+1)\pi/2(n+1))$, $k = 0, 1,\ldots, n$.

Note that the set $\{L_k(t)\}_{k=0}^n$ is a Chebyshev system on the interval $[t_0, t_n]$, which guarantees the existence and uniqueness of the solution. This set of basis functions can be replaced by any other Chebyshev system to obtain a unique interpolant.

If not only functional values, but also derivative values, are available and required to be interpolated by the polynomial,

$$p_n^{(i_k)}(t_k) = v_{k,i_k} \qquad i_k = 0,\ldots,m_k \qquad k = 0, 1,\ldots, n$$

then we have a **Hermite interpolation** problem. An algebraic polynomial of degree $d = n + \sum_{k=0}^n m_k$ always exists as a Hermite interpolant. An explicit closed-form formula for the Hermite interpolant also can be constructed. For example, if only the functional values $\{v_k\}_{k=0}^n$ and the first derivative values $\{w_k\}_{k=0}^n$ are given and required to be interpolated, then the Hermite interpolant is given by

$$p_{2n}(t) = \sum_{k=0}^{n}\{v_k A_k(t) + w_k B_k(t)\}$$

where, with notation $L_k'(t_k) := (d/dt)L_k(t)\big|_{t=t_k}$,

$$A_k(t) = \left[1 - 2(t-t_k)L_k'(t_k)\right]L_k^2(t) \quad \text{and} \quad B_k(t) = (t-t_k)L_k^2(t)$$

in which $L_k(t)$ are Lagrange basis polynomials, $k = 0, 1, \ldots, n$.

However, if those derivative values are not consecutively given, we have a **Hermite–Birkhoff interpolation** problem, which is not always uniquely solvable [61].

The preceding discussions did not take into consideration any optimality. The unique algebraic polynomial interpolant obtained previously may not be a good result in many cases. A well-known example is provided by Runge, in interpolating the continuous and smooth function $f(t) = 1/(1 + 25t^2)$ at $n + 1$ equally spaced points on the interval $[-1, 1]$. The polynomial interpolant $p_n(t)$ shows extremely high oscillations near the two end-points ($|t| > 0.726,\ldots$). Hence, it is important to impose an additional optimality requirement (e.g., a uniform approximation requirement) on the interpolant. In this concern, the following result is useful [36].

Theorem 3.18: *Given a continuous function* $f \in C\,[-1, 1]$, let $\{t_k\}_{k=1}^{n}$ *be the Chebyshev points on* $[-1, 1]$; *namely,* $t_k = \cos\,((2k-1)\,\pi/(2n))$, $k = 1,\ldots, n$. *Let also* $P_{2n-1}(t)$ *be the polynomial of degree* $2n - 1$ *that satisfies the following special Hermite interpolation conditions:* $P_{2n-1}(t_k) = f(t_k)$ *and* $P_{2n-1}(t_k) = f(t_k)$ and $P'_{2n-1}(t_k) = 0$, $k = 1, \ldots, n$. *Then, the interpolant* $P_{2n}(t)$ *has the uniform approximation property*

$$\left\| f - P_{2n-1} \right\|_{L_\infty[-1,1]} \to 0 \quad \text{as} \quad n \to \infty$$

Because polynomial splines are piecewise algebraic polynomials, similar uniform approximation results for polynomial spline interpolants may be established [41, 72, 88].

Finally, a simultaneous interpolation and uniform approximation for a polynomial of a finite (and fixed) degree may be very desirable in engineering applications. The problem is that given and $f \in C\,[a, b]$ with $n+1$ points $a \le t_0 < t_1 < \cdots < t_n \le b$ and a given $\varepsilon > 0$, find a polynomial $p(t)$ of finite degree (usually, larger than n) that satisfies both

$$\left\| f - p \right\|_{L_\infty[a,b]} < \varepsilon \quad \text{and} \quad p(t_k) = f(t_k) \qquad k = 0, 1, \ldots, n$$

The answer to this question is the Walsh theorem, which states that this is always possible, even for complex polynomials [36]. Note that natural splines can also solve this simultaneous interpolation and uniform-approximation problem.

Best Approximation of Linear Functionals

As already mentioned, a functional is a mapping that maps functions to values. Definite integrals, derivatives evaluated at some points, and interpolation formulas are good examples of linear functionals.

The best approximation problem for a given bounded linear functional, L, by a linear combination of n independent and bounded linear functionals $L_1,\ldots, L_n$, all defined on the same normed linear space X of functions, can be similarly stated as follows: determine n constant coefficients $\{a^*_k\}_{k=1}^{n}$ such that

$$\left\| L - \sum_{k=1}^{n} a_k^* L_k \right\|_{X^*} = \min_{\{a_k\}} \left\| L - \sum_{k=1}^{n} a_k L_k \right\|_{X^*} \tag{3.37}$$

where X^* is the dual space of X, which is also a normed linear space. A basic result is described by the following theorem [36].

Theorem 3.19: *If* X *is a Hilbert space, then the best approximation problem* (3.37) *is uniquely solvable. Moreover, if* r *and* $\{r_k\}_{k=1}^{n}$, *are the functional representors of* L *and* $\{L_k\}_{k=1}^{n}$, *respectively, then*

$$\left\| r - \sum_{k=1}^{n} a_k r_k \right\|_{X^*} = \min \Rightarrow \left\| L - \sum_{k=1}^{n} a_k L_k \right\|_{X^*} = \min$$

It is important to note that for linear functionals, we have an interpolation problem: given bounded linear functionals L and $\{L_k\}_{k=1}^{n}$, all defined on a normed linear space X, where the last n functionals are linearly independent on X, and given also n points $x_k \in X$, $k = 1,\ldots, n$, determine n constant coefficients $\{a_k\}_{k=1}^{n}$, such that

$$\sum_{k=1}^{n} a_k L_k(x_i) = L(x_i) \qquad i = 1, \ldots, n$$

Obviously, this problem is uniquely solvable. Depending on the specific formulation of the linear functionals, a bulk of the approximation formulas in the field of numerical analysis can be derived from this general interpolation formulation.

Finally, convergence problems also can be formulated and discussed for bounded linear functionals in a manner similar to interpolation and approximation of functions. The following result is significant [36].

Theorem 3.20: *Let* L *and* $\{L_k\}_{k=1}^{\infty}$ *be bounded linear functionals defined on a Banach space* X. *A necessary and sufficient condition for*

$$\lim_{k\to\infty}\|L_k - L\|_{X^*} = 0$$

is that $\{L_k\}_{k=1}^{\infty}$ *are uniformly bounded:*

$$\|L_k\|_{X^*} \le M < \infty \qquad \forall k = 1, 2, \ldots$$

and there is a convergent sequence $\{x_i\}_{i=1}^{\infty} \in X$, *such that*

$$\lim_{k\to\infty} L_k(x_i) = L(x_i) \quad \text{for each } i = 1, 2, \ldots$$

Artificial Neural Network for Approximation

Artificial neural networks offer a useful framework for signal and system approximations, including approximation of continuous and smooth functions of multivariables. Due to its usually mutilayered structure with many weights, an artificial neural network can be "trained," and hence has a certain "learning" capability in data processing. For this reason, artificial neural networks can be very efficient in performing various approximations. The main concern with a large-scale artificial neural network is its demand on computational speed and computer memory.

Both parametrized and nonparametrized approaches to approximations use artificial neural networks. In the parametrized approach the activation function, basic function, and network topology are all predetermined; hence, the entire network structure is fixed, leaving only a set of parameters (weights) to be adjusted to best fit the available data. In this way, the network with optimal weights becomes a best approximant, usually in the least-squares sense, to a nonlinear system. Determining the weights from the data is called a **training process**. Back-propagation multilayered artificial neural networks are a typical example of the parametrized framework. The nonparametrized approach requires that the activation and/or basic functions also be determined, which turns out to be difficult in general.

To illustrate how an artificial neural network can be used as a system or signal approximant, we first describe the structure of a network. The term neuron used here refers to an operator or processing unit, which maps R^n to R, with the mathematical expression

$$o_i = f_a\left(f_b(\mathbf{i}_i, \mathbf{w}_i)\right) \tag{3.38}$$

where $\mathrm{i}_i = [i_1 \ldots i_n]^{\mathsf{T}}$ is the input vector, $\mathbf{w}_i = [w_{i1} \ \ldots \ w_{in}]^{\mathsf{T}}$ is the weight vector associated with the ith neuron, o_i the output of the ith neuron, f_a the activation function (usually sigmoidal or Gaussian), and f_b the basic function (which can be linear, affine, or radial). For example, if an affine basic function is used, (3.38) takes on the form

$$o_i = f_a(\mathbf{i}_i \cdot \mathbf{w}_i + b_i) \tag{3.39}$$

where b_i is a constant.

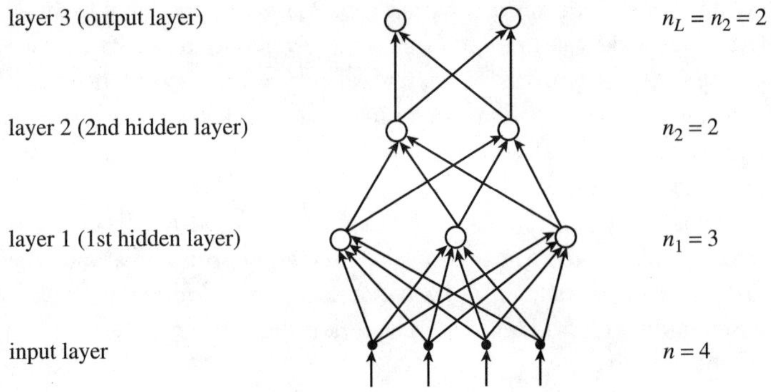

FIGURE 3.6 A two-hidden-layer, feed-forward artificial neural network.

A fully connected feed-forward artificial neural network is generally a multi-input/multi-output network, where the output from each neuron of each layer is an input to each neuron of the next layer. Such a network, arranged in one input layer, multiple hidden layers, and one output layer, can be constructed as follows (see Figure 3.6). Suppose we have n-inputs, n_L-outputs and $L - 1$ hidden layers, and a linear basic function is used with a sigmoidal activation function $f_a(t) = \sigma(t)$:

$$\sigma(t) \to \begin{cases} 1 & \text{as } t \to +\infty \\ 0 & \text{as } t \to -\infty \end{cases}$$

Also, let $o_{l,i}$ be the output of the ith neuron at the lth layer and $w_{l,i} = [w_{l,i,1} \cdots w_{l,i,s}]^{\mathrm{T}}$ be the weight vector associated with the same neuron connected to the neurons at the $(l - 1)$st layer. Then, we have

$$o_{l,i} = \sigma\left(\sum_{j=1}^{n_l} o_{l-1,j} w_{l,i,j} + w_{l,i,0}\right) \tag{3.40}$$

Inductively, the output of the ith neuron in the last (the Lth) layer is given by

$$o_{L,i} = \sigma\left(\sum_{j=1}^{n_{L-1}} w_{L,i,j}\left(\cdots\sigma\left(\sum_{q=1}^{n_i} w_{1,p,q} i_{n_0} + w_{1,p,o}\right) + w_{2,p,0}\right) + \cdots + w_{L,i,o}\right) \tag{3.41}$$

where $i = 1, \ldots, n_L$.

The following best uniform approximation property of an artificial neural network is a fundamental result in neural-network approximation [35].

Theorem 3.21: *Let $f(t)$ be a continuous function defined on a compact subset $\Omega \subset R^n$. Then, for any $\in >$ 0, there exists an integer $m \geq 1$ and real parameters $\{c_k, w_{ki}, b_k\}_{k=1}^{m}$ such that using any nonconstant, bounded, and monotonically increasing continuous function f_a as the activation function, the artificial neural network can uniformly approximate f on Ω, in the sense that*

$$\|f - N\|_{L_\infty(\Omega)} < \varepsilon$$

where the network has the form

$$N(\mathbf{t}) = \sum_{k=1}^{m} c_k f_a\left(\sum_{i=1}^{n} w_{k,i} t_i + b_k\right) \qquad \mathbf{t} = [t_1 \cdots t_n]^{\mathrm{T}} \in \Omega$$

Neural networks can also provide approximation for a mapping together with its derivatives [52]. On the other hand, neural networks can provide localized approximation, which is advantageous in that if a certain portion of the data is perturbed, only a few weights in the network need to be retrained. It was demonstrated that a single hidden layered network cannot provide localized approximation of continuous functions on any compact set of a Euclidean space with dimension higher than one; however, two hidden layers are sufficient for the purpose [33].

As mentioned previously, the basic function f_b in a network need not be linear. An artificial neural network, using a radial function for f_b, can also give very good approximation results [76]. Also, as a system approximation framework, stability of a network is very important [68]. Finally, a major issue that must be addressed in designing a large-scale network is the computer memory, which requires some special realization techniques [67].

3.4 Identification

System identification is a problem of finding a good mathematical model, preferably optimal in some sense, for an unknown physical system, using some available measurement data. These data usually include system outputs and sometimes also inputs. Very often, the available data are discrete, but the system to be identified is continuous [97].

A general formulation of the system identification problem can be described as follows. Let S be the family of systems under consideration (linear or nonlinear, deterministic or stochastic, or even chaotic), with input u and output γ, and let $R_{(u,y)}$ be the set of I-O data. Define a mapping $M: S \to R_{(u,y)}$. Then, a system $\mathcal{F} \in S$ is said to be (**exactly**) **identifiable** if the mapping M is invertible, and the problem is to find the $\mathcal{F} = M^{-1}(\tilde{u}, \tilde{y})$ using the available data $(\tilde{u}, \tilde{y}) \in R_{(u,y)}$. Here, how to define the mapping M, linear or not, is the key to the identification problem. Usually, we also want M^{-1} to be causal for the implementation purpose.

The first question about system identification is of course the identifiability [82]. Not all systems, not even linear deterministic systems, are exactly identifiable [21]. Because many physical systems are not exactly identifiable, system identification in a weaker sense is more realistic.

Suppose that some inputs and their corresponding outputs of an unknown system, $\mathcal{S}_1$, are given. We want to identify this unknown system by an approximate model, $\mathcal{S}_2$, using the available I-O data, such that the corresponding outputs produced by any input through $\mathcal{S}_1$ and $\mathcal{S}_2$, respectively, are "very close" under certain meaningful measure. If the structure of $\mathcal{S}_1$ (hence, $\mathcal{S}_2$) is known *a priori*, then what we need is to identify some system parameters. If the structure of $\mathcal{S}_1$ is not clear, the task becomes much more difficult because we must determine what kind of model to choose in approximating the unknown system [50]. This includes many crucial issues such as the linearity and dimension (or order) of the model used. In particular, if the system is nonlinear and contains uncertainties, special techniques from set-valued mapping and differential inclusion theories may be needed [58].

Usually, the basic requirement is that $\mathcal{S}_2$ should be a best approximant of $\mathcal{S}_1$ from a desired class of simple and realizable models under a suitably chosen criterion. For example, the least-squares operator approximation discussed previously can be thought of as an identification scheme. For this reason, identification in the weak sense is traditionally considered to be one of the typical best approximation problems in mathematics. If a minimal, worst-case model-matching error bound is required, the approximation is known as the optimal recovery problem, for either functions or functionals [65, 66], or for operators [15, 45]. In system engineering it usually refers to **system identification** or **reconstruction**, with an emphasis on obtaining an identified model or a reconstruction scheme.

Generally speaking, system identification is a difficult problem, often leading to nonunique solutions when it is solvable. This is typically true for nonlinear circuits and systems. In systems and control engineering, an unknown system is identified by a desired model such that they can produce "close enough" outputs from the same input, measured by a norm in the signal space, such as L_p, l_p or H_p ($1 \leq p \leq \infty$). For dynamic systems, however, this norm-measure is generally not a good choice because one is concerned with nonlinear dynamics of the unknown system, such as limit cycles, attractors, bifurcations, and chaos. Hence, it is preferable to have an identified model that preserves the same dynamic behavior. This is a very challenging research topic; its fundamental theories and methodologies are still open for further exploration.

Linear Systems Identification

Compared to nonlinear systems, linear systems, either ARMA (autoregressive with moving-average) or state-space models, can be relatively easily identified, especially when the system dimension (order) is fixed. The mainstream theory of linear system identification has the following characteristics [37]:

1. The model class consists of linear, causal, stable, finite-dimensional systems with constant parameters.
2. Both system inputs and their corresponding outputs are available as discrete or continuous data.
3. Noise, if any, is stationary and ergodic (usually with rational spectral densities), white and uncorrelated with state vectors in the past.
4. Criteria for measuring the closeness in model-matching are of least-squares type (in the deterministic case) or of maximum likelihood type (in the stochastic case).
5. Large-scale linear systems are decomposed into lower-dimensional subsystems, and nonlinear systems are decomposed into linear and simple (e.g., memoryless) nonlinear subsystems.

Because for linear systems, ARMA models and state-space models are equivalent under a nonsingular linear transformation [17, 32], we discuss only ARMA models here.

An (n, m, l)th-order ARMAX model (an ARMA model with exogenous noisy inputs) has the general form

$$a(z^{-1})\mathbf{y}(t) = b(z^{-1})\mathbf{u}(t)c(z^{-1})\boldsymbol{\varepsilon}(t) \qquad t = \ldots, -1, 0, 1, \ldots \tag{3.42}$$

in which z^{-1} is the time-delay operator defined by $z^{-1}\mathbf{f}(t) = \mathbf{f}(t - l)$, and

$$a(z^{-1}) = \sum_{i=0}^{n} A_i z^{-i} \qquad b(z^{-1}) = \sum_{j=0}^{m} B_j z^{-j} \qquad c(z^{-1}) = \sum_{k=1}^{l} C_k z^{-k}$$

with constant coefficient matrices $\{A_i\}$, $\{B_j\}$, $\{C_k\}$ of appropriate dimensions, where $A_0 = I$ (or, is nonsingular). In the ARMAX model (3.42) $\mathbf{u}(t)$, $\mathbf{y}(t)$, and $\varepsilon(t)$ are considered to be system input, output, and noise vectors, respectively, where the input can be either deterministic or random. In particular, if $l = 0$ and $n = 0$ (or $m = 0$), then (3.42) reduces to a simple moving-average (MA) (or autoregressive, AR) model. Kolmogorov [56] proved that every linear system can be represented by an infinite-order AR model. It is also true that every nonlinear system with a Volterra series representation can be represented by a nonlinear AR model of infinite order [53].

The system identification problem for the ARMAX model (3.42) can now be described as follows. Given the system I-O data $(\mathbf{u}(t), \mathbf{y}(t))$ and the statistics of $\varepsilon(t)$, determine integers (n, m, l) (system-order determination) and constant coefficient matrices $\{A_i\}$, $\{B_j\}$, $\{C_k\}$ (system-parameter identification). While many successful methods exist for system parameter identification [3, 23, 49, 60], system order determination is a difficult problem [47].

As already mentioned, the identifiability of an unknown ARMAX model using the given I-O data is a fundamental issue. We discuss the exact model identification problem here. The ARMAX model (3.42) is said to be **exactly identifiable** if $(\tilde{a}(z^{-1}), \tilde{b}(z^{-1}), \tilde{c}(z^{-1}))$ is an ARMAX model with $\tilde{n} \le n$, $\tilde{m} \le m$, and $\tilde{l} \le l$, such that

$$\begin{cases} \left[\tilde{a}(z^{-1})\right]^{-1}\tilde{b}(z^{-1}) = \left[a(z^{-1})\right]^{-1}b(z^{-1}) \\ \left[\tilde{a}(z^{-1})\right]^{-1}\tilde{c}(z^{-1}) = \left[a(z^{-1})\right]^{-1}c(z^{-1}) \end{cases}$$

Note that not all ARMAX models are exactly identifiable in this sense. A basic result about this identifiability is the following [21].

Theorem 3.22: *The ARMAX model* (3.42) (*with* $t \geq 0$) *is exactly identifiable iff* $a(z^{-1})$, $b(z^{-1})$, *and* $c(z^{-1})$ *have no common left factor and the rank of the constant matrix* $[A_n, B_m, C_l]$, *consisting of the highest-order coefficient terms in* $a(z^{-1})$, $b(z^{-1})$, $c(z^{-1})$, *respectively, is equal to the dimension of the system output* $\mathbf{y}$.

Even if an unknown system is exactly identifiable and its identification is unique, how to find the identified system is still a very technical issue. For simple AR models, the well-known Levinson–Durbin algorithm is a good scheme for constructing the identified model; for MA models, one can use Trench–Zohar and Berlekamp–Massey algorithms. There exist some generalizations of these algorithms in the literature [23]. For stochastic models with significant exogenous noise inputs, various statistical criteria and estimation techniques, under different conditions, are available [82]. Various recursive least-squares schemes, such as the least-mean-square (LMS) algorithm [96], and various stochastic searching methods, such as the stochastic gradient algorithm [49], are popular. Because of their simplicity and efficiency, the successful (standard and extended) Kalman filtering algorithms [16, 30] have also been widely applied in parameters identification for stochastic systems [13, 17, 62], with many real-world applications [92].

Finally, for linear systems, a new framework, called the **behavioral approach**, is proposed for mathematical system modeling and some other related topics [2, 98].

Nonlinear Systems Identification

Identifying a nonlinear system is much more difficult than identifying a linear system in general, whether it is in the exact or in the weak sense, as is commonly known and can be seen from its information-based complexity analysis [15, 45].

For some nonlinear systems with simple Volterra series representations, the least-squares approximation technique can be employed for the purpose of identification in the weak sense [8]. As a simple illustrative example, consider the cascaded nonlinear system with noise input shown in Figure 3.7. In this figure $h_1(t)$ and $h_2(t)$ are unit impulse responses of two linear subsystems, respectively, and $V_n(\cdot)$ is a memoryless nonlinear subsystem which is assumed to have an nth-order Volterra polynomial in the special form

$$y(t) = \sum_{k=1}^{n} c_k \int_0^t \cdots \int_0^t \phi_k(\tau_{1,\ldots,}\tau_k) x(\tau_1) \cdots x(\tau_k) d\tau_1 \cdots d\tau_k \tag{3.43}$$

where all the Volterra kernels $\{\phi_k\}_{k=0}^n$ are assumed to be known, but the constant coefficients $\{c_k\}_{k=0}^n$ must be identified.

It is clear from Figure 3.7 that the output of the cascaded system can be expressed via convolution-type integrals as

$$z(t) = c_1\left(\int h_2\phi_1 h_1 u\right)(t) + \cdots + c_n\left(\int \cdots \int h_2\phi_1 h_1 \cdots h_2 u \cdots u\right)(t) + \varepsilon(t) \tag{3.44}$$

Now, because all the integrals can be computed if the input function u(t) is given, the standard least-squares technique can be used to determine the unknown constant coefficients $\{c_k\}_{k=0}^n$, using the measured system output $z(t)$.

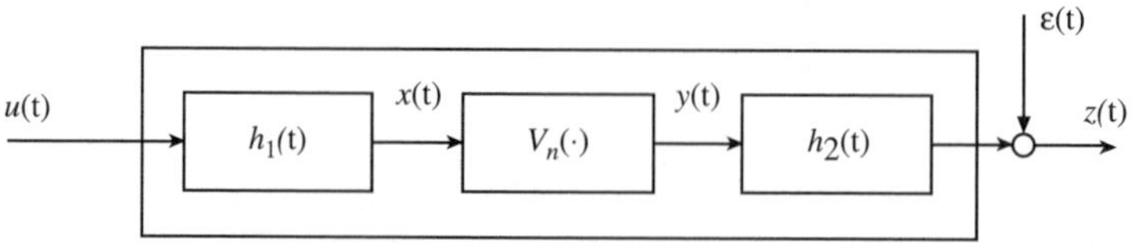

FIGURE 3.7 A cascaded linear–nonlinear system with noise input.

A neural network implementation of Volterra series model identification is described in [1]. Neural network for system identification has been used in many different cases, as can also be seen from [70].

Finally, we consider one approach to nonlinear systems identification which combines the special structure of the generalized Fock space of Volterra series (Theorem 3.3) and the "training" idea from neural networks discussed previously (Theorem 3.21). For simplicity, consider the scalar nonlinear system

$$y(t)+f\big(y(t-1),y(t-2),\dots,y(t-n)\big)=u(t) \quad t=0,1,\dots \tag{3.45}$$

where n is a fixed integer, with the given initial conditions $y(-1)=y_1,\dots,y(-n)=y_n$. Introducing a simple notation

$$y^{t-1}=\big(y(t-1),\dots,y(t-n)\big) \tag{3.46}$$

we first rewrite this system as

$$\gamma(t)+f\big(\gamma^{t-1}\big)=u(t) \tag{3.47}$$

Then, we denote by E^n as the n-dimensional Euclidean space of continuous functions and let $u^1,\dots,u^m\in E^n$, called the domain training samples, be given data vectors that are componentwise nonzero and distinct, namely, $u_i^k\neq u_j^k$ if $i\neq j$ for all $k=1,\dots,m$, $1\le i,j\le n$. Here, (3.46) also has been used for these domain training samples. Also, let $r_1,\dots,r_m$, be given real numbers, called the corresponding **range training samples**. The identification problem is to find an approximate system, $f^*(\cdot)$, among all Volterra series representations from the generalized Fock space formulated in Theorem 3.3, such that f^* maps all the domain training samples to their corresponding range training samples:

$$f^*\big(u^k\big)=r_k \qquad k=1,\dots,m \tag{3.48}$$

and f^* has the minimum operator-norm among all such candidates. The following theorem provides an answer to this problem [45].

Theorem 3.23: *There is a unique element f^* of minimum norm in the generalized Fock space defined in Theorem 3.3, with the domain $\Omega=E^n$ therein, that satisfies the constraint* (3.48). *Moreover, f^* has the following expression:*

$$f^*(v)=\sum_{k=1}^{m}a_kK\big(u^k,v\big) \qquad \forall v\in E^n$$

where $K(\cdot,\cdot)$ is the reproducing kernel defined in Theorem 3.3, and the system parameters are determined by

$$\begin{bmatrix}a_1\\ \vdots\\ a_m\end{bmatrix}=\begin{bmatrix}K(u^1,u^1) & \cdots & K(u^1,u^m)\\ \vdots & & \vdots\\ K(u^m,u^1) & \cdots & K(u^m,u^m)\end{bmatrix}^{-1}\begin{bmatrix}r_1\\ \vdots\\ r_m\end{bmatrix}$$

Here, it should be noted that because K is a reproducing kernel, the set of functions $\{K(u^k,\cdot)\}_{k=1}^{m}$ are linearly independent, so that the above inverse matrix exists.

Also, note that this system identification method can be applied to higher-dimensional systems and the continuous-time setting [45].

Nonlinear Dynamic Systems Identification from Time Series

Measurement (observation) data obtained from an unknown system are often available in the form of time series. There are some successful techniques for identification of linear and nonlinear systems from time series if the time series is generated from Gaussian white noise. For example, for linear systems we have the Box–Jenkins scheme and for nonlinear systems, a statistical method using a nonlinear filter [75].

Concerned with nonlinear dynamic systems, however, statistical methods and the commonly used norm-measure criterion may not be capable of identifying the system dynamics in general. This is because the main issue in this concern is the nonlinear dynamic behavior of the unknown system, such as limit cycles, attractors, bifurcations, and chaos. Hence, it is preferable that the identified model can preserve the nonlinear dynamics of the unknown system. This turns out to be a very challenging task; many fundamental theories and methodologies for this task remain to be developed.

When an unknown nonlinear dynamic system is measured to produce a set of continuous or discrete data (a time series), a natural approach for studying its dynamics from the available time series is to take an integral transform of the series, so as to convert the problem from the time domain to the frequency domain. Then, some well-developed engineering frequency domain methods can be applied to perform analysis and computation of the nonlinear dynamics [69].

One common approach formulated in the time domain is the **(delay-coordinate) embedding method** that can be applied to reconstruct (identify) an unknown nonlinear dynamic model from which only a set of discrete measurement data (a time series) is available.

Let us consider the problem of identifying a periodic trajectory of an unknown, nonlinear dynamic system using only an experimental time series measured from the system. Let $\{r_k\}$ be the available data. The embedding theory guarantees this can be done in the space R^m with the embedding dimension $m \geq 2n + 1$, where n is the dimension of the dynamic system [93], or $m \geq 2d_A$, where d_A is the dimension of the attractor [87]. A way to achieve this is to use the delay-coordinate technique, which approximates the unknown, nonlinear dynamics in R^m by introducing the embedding vector

$$\mathbf{r}_k = \left[r_k r_{k-\mu} \cdots r_{k-(m-1)\mu} \right]^{\mathrm{T}} \tag{3.49}$$

where μ is the time-delay step. This embedding vector provides enough information to characterize the essence of the system dynamics and can be used to obtain an experimental Poincaré map, which helps in understanding the dynamics. For example, one may let the map be the equation of the first component of the vector being equal to a constant: r_{k_i} = constant. This procedure yields the successive points

$$\xi_i := \left[r_{k_i-\mu} \cdots r_{k_i-(m-1)\mu} \right]^{\mathrm{T}} \tag{3.50}$$

at the ith piercing of the map by the trajectory (or the vector r_k), where k_i is the time index at the ith piercing. Then, one can locate the periodic trajectories of the unknown system using the experimental data [4, 14]. In this approach, however, determining a reasonably good time-delay step size, i.e., the real number μ in (3.49), remains an open technical problem.

Finally, we note that the embedding method discussed previously has been applied to the control of chaotic circuits and systems [19, 20, 74].

References

[1] A. Abdulaziz and M. Farsi, "Non-linear system identification and control based on neural and self-tuning control," *Int. J. Adapt. Control Signal Process.*, vol. 7, no. 4, pp. 297–307, 1993.

[2] A. C. Antoulas and J. C. Willems, "A behavioral approach to linear exact modeling," *IEEE Trans. Autom. Control*, vol. 38, no. 12, pp. 1776–1802, 1993.

[3] K. J. Aström and P. Eykhoff, "System identification: A survey," *Automatica*, vol. 7, pp. 123–162, 1971.

[4] H. Atmanspacher, H. Scheingraber, and W. Voges, "Attractor reconstruction and dimensional analysis of chaotic signals," *Data Analysis in Astronomy III*, V. di Gesú et al., Eds., New York: Plenum Press, 1988, pp. 3–19.
[5] I. Baesler and I. K. Daugavet, "Approximation of nonlinear operators by Volterra polynomials," *Am. Math. Soc. Transl.*, vol. 155, no. 2, pp. 47–57, 1993.
[6] S. P. Banks, *Mathematical Theories of Nonlinear Systems*, New York: Prentice Hall, 1988.
[7] A. Bensoussan, G. de Prado, M. C. Felfour, and S. K. Mitter, *Representation and Control of Infinite Dimensional Systems*, Boston: Birkhäuser, 1992.
[8] S. A. Billings, "Identification of nonlinear systems — a survey," *IEEE Proc., Pt. D*, vol. 127, pp. 272–285, 1980.
[9] S. Boyd and L. O. Chua, "Fading memory and the problem of approximating nonlinear operators with Volterra series," *IEEE Trans. Circuits Syst.*, vol. 32, no. 11, pp. 1150–1161, 1985.
[10] S. Boyd, L. O. Chua, and C. A. Desoer, "Analytic foundations of Volterra series," *IMA J. Info. Control*, vol. 1, pp. 243–282, 1984.
[11] D. Braess, *Nonlinear Approximation Theory*, New York: Springer-Verlag, 1986.
[12] R. W. Brockett, "Volterra series and geometric control theory," *Automatica*, vol. 12, pp. 167–176, 1976.
[13] P. E. Caines, *Linear Stochastic Systems*, New York: John Wiley & Sons, 1988.
[14] M. Casdagli, S. Eubank, J. D. Farmer, and J. Gibson, "State space reconstruction in the presence of noise," *Physica D*, vol. 51, pp. 52–98, 1991.
[15] G. Chen, "Optimal recovery of certain nonlinear analytic mappings," in *Optimal Recovery*, B. Bojanov and H. Wozniakowski, Eds., New York: Nova Science, 1992, pp. 141–144.
[16] G. Chen, Ed., *Approximate Kalman Filtering*, New York: World Scientific, 1993.
[17] G. Chen, G. Chen, and H. S. Hsu, *Linear Stochastic Control Systems*, Boca Raton, FL: CRC Press, 1995.
[18] G. Chen, C. K. Chui, and M. J. Lai, "Construction of real-time spline quasi-interpolation schemes," *Approx. Theory Appl.*, vol. 4, no. 4, pp. 61–75, 1988.
[19] G. Chen and X. Dong, *From Chaos to Order: Methodologies, Perspectives, and Applications*, Singapore: World Scientific, 1998.
[20] G. Chen and J. Moiola, "An overview of bifurcation, chaos and nonlinear dynamics in control systems," *The Franklin Institute Journal*, vol. 331B, pp. 819–858, 1994.
[21] H. F. Chen and J. F. Zhang, "On identifiability for multidimensional ARMAX model," *Acta Math. Appl. Sin.*, vol. 9, no. 1, pp. 1–8, 1993.
[22] E. W. Cheney, *Introduction to Approximation Theory*, New York: McGraw-Hill, 1966.
[23] B. Choi, *ARMA Model Identification*, New York: Springer-Verlag, 1992.
[24] L. O. Chua, "The genesis of Chua's circuit," *AEU Int. J. Electr. Commun.*, vol. 46, pp. 187–257, 1992.
[25] L. O. Chua and A. C. Deng, "Canonical piecewise-linear modeling," *IEEE Trans. Circuits Syst.*, vol 33, no. 5, pp. 511–525, 1986.
[26] L. O. Chua and A. C. Deng, "Canonical piecewise-linear representation," *IEEE Trans. Circuits Syst.*, vol. 35, no. 1, pp. 101–111, 1988.
[27] C. K. Chui, *Multi-Variate Splines*, Philadelphia: SIAM, 1988.
[28] C. K. Chui, "Approximations and expansions," in *Encyclopedia of Physical Science and Technology*, vol. 2, New York: Academic Press, 1992, pp. 1–29.
[29] C. K. Chui, *An Introduction to Wavelets*, New York: Academic Press, 1992.
[30] C. K. Chui and G. Chen, *Kalman Filtering with Real-Time Applications*, New York: Springer-Verlag, 1987.
[31] C. K. Chui and G. Chen, *Linear Systems and Optimal Control*, New York: Springer-Verlag, 1989.
[32] C. K. Chui and G. Chen, *Signal Processing and Systems Theory: Selected Topics*, New York: Springer-Verlag, 1992.
[33] C. K. Chui, X. Li, and H. N. Mhaskar, "Neural networks for localized approximation," *Math. Comput.*, vol. 63, pp. 607–623, 1994.

[34] J. B. Conway, *A Course in Functional Analysis*, New York: Springer-Verlag, 1985.
[35] G. Cybenko, "Approximation by superposition of sigmoidal functions," *Math. Control, Signals Syst.*, vol. 2, pp. 303–314, 1989.
[36] P. J. Davis, *Interpolation and Approximation*, New York: Dover, 1975.
[37] M. Deistler, "Identification of Linear Systems (a survey)," in *Stochastic Theory and Adaptive Control*, T. E. Duncan and B. P. Duncan, Eds., New York: Springer-Verlag, 1991, pp. 127–141.
[38] C. A. Desoer and M. G. Kabuli, "Right factorizations of a class of time-varying nonlinear systems," *IEEE Trans. Autom. Control*, vol. 33, pp. 755–757, 1988.
[39] C. A. Desoer and M. Vidyasagar, *Feedback Systems: Input–Output Properties*, New York: Academic Press, 1975.
[40] R. A. DeVore and G. G. Lorentz, *Constructive Approximation*, New York: Springer-Verlag, 1993.
[41] C. de Boor, *A Practical Guide to Splines*, New York: Springer-Verlag, 1978.
[42] C. de Boor, K. Höllig, and S. Riemenschneider, *Box Splines*, New York: Springer-Verlag, 1993.
[43] R. J. P. de Figueiredo and G. Chen, "Optimal interpolation on a generalized Fock space of analytic functions," in *Approximation Theory VI*, C. K. Chui, L. L. Schumaker, and J. D. Ward, Eds., New York: Academic Press, 1989, pp. 247–250.
[44] R. J. P. de Figueiredo and G. Chen, "PDLG splines defined by partial differential operators with initial and boundary value conditions," *SIAM J. Numerical Anal.*, vol. 27, no. 2, pp. 519–528, 1990.
[45] R. J. P. de Figueiredo and G. Chen, *Nonlinear Feedback Control Systems: An Operator Theory approach*, New York: Academic Press, 1993.
[46] R. J. P. de Figueiredo and T. A. Dwyer, "A best approximation framework and implementation for simulation of large scale nonlinear systems," *IEEE Trans. Circuits Syst.*, vol. 27, no. 11, pp. 1005–1014, 1980.
[47] J. G. de Gooijer, B. Abraham, A. Gould, and L. Robinson, "Methods for determining the order of an autoregressive-moving average process: a survey," *Int. Statist. Rev.*, vol. 53, pp. 301–329, 1985.
[48] M. Fliess, M. Lamnabhi, and F. L. Lagarrigue, "An algebraic approach to nonlinear functional expansions," *IEEE Trans. Circuits Syst.*, vol. 39, no. 8, pp. 554–570, 1983.
[49] G. C. Goodwin and K. S. Sin, *Adaptive Filtering, Prediction and Control*, Englewood Cliffs, NJ. Prentice Hall, 1984.
[50] R. Haber and H. Unbehauen, "Structure identification of nonlinear dynamic systems — a survey on input/output approaches," *Automatica*, vol. 26, no. 4, pp. 651–677, 1990.
[51] J. Hammer, "Fraction representations of nonlinear systems: a simplifed approach," *Int. J. Control*, vol. 46, pp. 455–472, 1987.
[52] K. Hornik, M. Stinchcombe, and H. White, "Universal approximation of an unknown mapping and its derivatives using multilayer feed-forward networks," *Neural Networks*, vol. 3, pp. 551–560, 1990.
[53] L. R. Hunt, R. D. DeGroat, and D. A. Linebarger, "Nonlinear AR modeling," *Circuits, Syst. Signal Process.*, in press.
[54] A. Isidori, *Nonlinear Control Systems*, New York: Springer-Verlag, 1985.
[55] C. Kahlert and L. O. Chua, "A generalized canonical piecewise-linear representation," *IEEE Trans. Circuits Syst.*, vol. 37, no. 3, pp. 373–383, 1990.
[56] A. N. Kolmogorov, "Interpolation and extrapolation von stationaren zufallingen folgen," *Bull. Acad. Sci. USSR Ser. Math.*, vol. 5, pp. 3–14, 1941.
[57] S. Y. Kung, *Digital Neural Networks*, Englewood Cliffs, NJ: Prentice Hall, 1993.
[58] A. B. Kurzhanski, "Identification — a theory of guaranteed estimates," in *From Data to Model*, J. C. Willems, Ed., New York: Springer-Verlag, 1989, pp. 135–214.
[59] C. Lesiak and A. J. Krener, "The existence and uniqueness of Volterra series for nonlinear systems," *IEEE Trans. Autom. Control*, vol. 23, no. 6, pp. 1090–1095, 1978.
[60] L. Ljung, *System Identification: Theory for the User*, Englewood Cliffs, NJ: Prentice Hall, 1987.
[61] G. G. Lorentz, K. Jetter, and S. D. Riemenschneider, *Birkhoff Interpolation*, Reading, MA: Addison-Wesley, 1983.
[62] H. Lütkepohl, *Introduction to Multiple Time Series Analysis*, New York: Springer-Verlag, 1991.

[63] A. I. Mees, M. F. Jackson, and L. O. Chua, "Device modeling by radial basis functions," *IEEE Trans. Circuits Syst.*, vol. 39, no. 11, pp. 19–27, 1992.
[64] A. I. Mees, "Parsimonious dynamical reconstruction," *Int. J. Bifurcation Chaos*, vol. 3, no. 3, pp. 669–675, 1993.
[65] C. A. Micchelli and T. J. Rivlin, "A survey of optimal recovery," in *Optimal Estimation in Approximation Theory*, C. A. Micchelli and T. J. Rivlin, Eds., New York: Plenum Press, 1977, pp. 1–54.
[66] C. A. Micchelli and T. J. Rivlin, "Lectures on optimal recovery," in *Numerical Analysis*, P. R. Turner, Ed., New York: Springer-Verlag, 1984, pp. 21–93.
[67] A. N. Michel, J. A. Farrell, and W. Porod, "Associative memories via artificial neural networks," *IEEE Control Syst. Mag.*, pp. 6–17, Apr. 1990.
[68] A. N. Michel, J. A. Farrell, and W. Porod, "Qualitative analysis of neural networks," *IEEE Trans. Circuits Syst.*, vol. 36, no. 2, pp. 229–243, 1989.
[69] J. L. Moiola and G. Chen, *Hopf Bifurcation Analysis: A Frequency Domain Approach*, Singapore: World Scientific, 1996.
[70] K. S. Narendra and K. Parthasarathy, "Identification and control of dynamic systems using neural networks," *IEEE Trans. Neural Networks*, vol. 1, no. 1, pp. 4–27, 1990.
[71] H. Nijmeijer and A. J. van der Schaft, *Nonlinear Dynamical Control Systems*, New York: Springer-Verlag, 1990.
[72] G. Nürnberger, *Approximation by Spline Functions*, New York: Springer-Verlag, 1989.
[73] W. Odyniec and G. Lewicki, *Minimal Projections in Banach Spaces*, New York: Springer-Verlag, 1990.
[74] E. Ott, C. Grebogi, and J. A. Yorke, "Controlling chaos," *Phys. Rev. Lett.*, vol. 64, pp. 1196–1199, 1990.
[75] T. Ozaki, "Identification of nonlinearities and non-Gaussianities in time series," in *New Directions in Time Series Analysis, Part I*, D. Brillinger et al., Eds., New York: Springer-Verlag, 1993, pp. 227–264.
[76] J. Park and I. W. Sandberg, "Approximation and radial-basic-function networks," *Neural comput.*, vol. 5, pp. 305–316, 1993.
[77] J. R. Partington, *An Introduction to Hankel Operators*, London: Cambridge University Press, 1988.
[78] P. P. Petrushev and V. A. Popov, *Rational Approximation of Real Functions*, London: Cambridge University Press, 1987.
[79] M. J. D. Powell, *Approximation Theory and Methods*, London: Cambridge University Press, 1981.
[80] M. J. D. Powell, "Radial basis functions for multi-variable interpolation: A review," in *IMA Conf. Algorithms for Approximation of Functions and Data*, RMCS, Shrivenham, U.K., 1985.
[81] E. Quak, N. Sivakumar, and J. D. Ward, "Least-squares approximation by radial functions," *SIAM J. Math. Anal.*, vol. 24, no. 4, pp. 1043–1066, 1993.
[82] B. L. S. P. Rao. *Identifiability in Stochastic Models*, New York: Academic Press, 1992.
[83] W. J. Rugh, *Nonlinear Systems Theory: The Volterra/Wiener Approach*, Baltimore: Johns Hopkins University Press, 1981.
[84] I. W. Sandberg, "A perspective on system theory," *IEEE Trans. Circuits Syst.*, vol. 31, no. 1, pp. 88–103, 1984.
[85] I. W. Sandberg, "Criteria for the global existence of functional expansions for input/output maps," *AT&T Tech, J.*, vol. 64, no. 7, pp. 1639–1658, 1985.
[86] I. W. Sandberg, "Approximately finite memory and input–output maps," *IEEE Trans. Circuits Syst.*, vol. 39, no. 7, pp. 549–556, 1992.
[87] T. Sauer, J. A. Yorke, and M. Casdagli, "Embedology," *J. Stat. Phys.*, vol. pp. 579–616, 1991.
[88] L. L. Schumaker, *Spline Functions: Basic Theory*, New York: John Wiley & Sons, 1981.
[89] A. N. Sharkovsky, "Chaos from a time-delayed Chua's circuit," *IEEE Trans. Circuits Syst.*, vol. 40, no. 10, pp. 781–783, 1993.
[90] E. D. Sontag, *Polynomial Response Maps*, New York: Springer-Verlag, 1979.
[91] E. D. Sontag, *Mathematical Control Theory*, New York: Springer-Verlag, 1990.
[92] H. W. Sorenson, Ed., *Kalman Filtering: Theory and Applications*, New York: IEEE Press, 1985.
[93] F. Takens, "Detecting strange attractors in turbulence," in *Lecture Notes in Mathematics*, vol. 898, D. A. Rand and L. S. Yong, Eds., New York: Springer-Verlag, 1981, pp. 366–381.

[94] F. I. Utreras, "Positive thin plate splines," *J. Approx. Theory Appl.*, vol. 1, pp. 77–108, 1985.
[95] M. S. Verma and L. R. Hunt, "Right coprime factorizations and stabilization for nonlinear systems," *IEEE Trans. Autom. Control*, vol. 38, pp. 222–231, 1993.
[96] B. Widrow and S. D. Stearns, *Adaptive Signal Processing*, Englewood Cliffs, NJ: Prentice Hall, 1985.
[97] J. C. Willems, Ed., *From Data to Model*, New York: Springer-Verlag, 1989.
[98] J. C. Willems, "Paradigms and puzzles in the theory of dynamical systems," *IEEE Trans. Autom. Control*, vol. 36, pp. 259–294, 1991.

4

Transformation and Equivalence

Wolfgang Mathis
University of Hannover, Germany

4.1 General Equivalence Theorems for Nonlinear Circuits

One of the basic problems in the study of linear and nonlinear dynamical electrical networks is the analysis of the underlying descriptive equations and their solution manifold. In the case of linear or affine networks, the constitutive relations of network elements are restricted to classes of linear or affine functions and, therefore, possess rather restricted types of solutions. In contrast, the solution manifold of nonlinear networks may consist of many different types. Naturally, it is useful to decompose nonlinear networks into classes that possess certain similarities. One approach, for example, is to consider the solution manifold and to decompose solutions into similar classes. Furthermore, if the descriptive differential equations of dynamic networks are considered to be mathematical sets, their decompositions will be of interest.

The technique of equivalence relations is the preferred method used to decompose a set of mathematical objects into certain classes. A well-known approach to define equivalence relations uses transformation groups. For example, real symmetric n x n matrices $\mathbb{R}_s^{n\times n}$ can be decomposed into equivalence classes by using the general linear transformation group $GL(n; \mathbb{R})$, and by applying the following similarity transformation:

$$\mathbf{M} \mapsto \mathbf{U}^{-1}\mathbf{M}\mathbf{U} \tag{4.1}$$

where $\mathbf{U} \in GL(n; \mathbb{R})$. By applying $GL(n; \mathbb{R})$, the set $\mathbb{R}_s^{n\times n}$ is decomposed into similarity classes that are characterized by certain eigenvalues. Furthermore, each class $\mathbb{R}_s^{n\times n}$ contains a diagonal matrix $\mathbf{D}$ with the eigenvalues on the main diagonal [33]. These eigenvalues are invariants of the group and characterize the classes. These and other related results can be applied to classify linear and affine dynamical networks [20]. Thus, properties of the A-matrix of the state space equations are used for the classification. Note that each linear or affine network can be described in state-space form. We will discuss the theory of equivalence of linear and affine dynamical networks only as special cases of nonlinear networks. A fine reformulation of the classical material of the decomposition of real matrices by using similarity transformations in the framework of one-parameter groups in $GL(n; \mathbb{R})$ is given by [22].

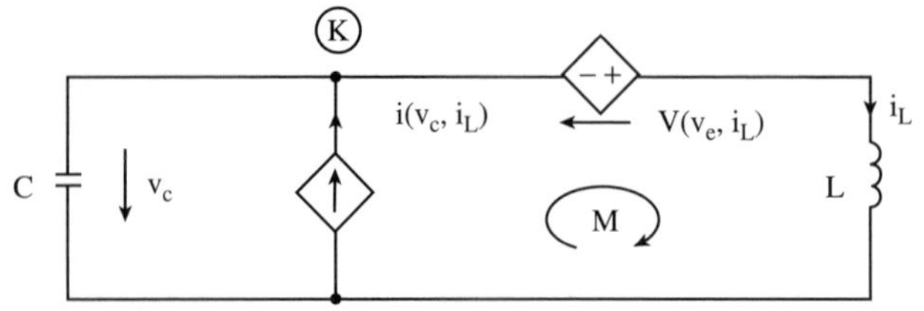

FIGURE 4.1 A nonlinear network.

A classification of the vector fields is needed in order to classify differential equations of the type $\dot{\mathbf{x}} = \mathbf{f}(\mathbf{x})$, where $\mathbf{x} \in \mathbb{R}^n$ and $\mathbf{f}: \mathbb{R}^n \to \mathbb{R}^n$ is a vector field on $\mathbb{R}^n$. A first concept is established by a k-times differentiable change of coordinates that transforms the differential equation $\dot{\mathbf{x}} = \mathbf{f}(\mathbf{x})$ into $\dot{\mathbf{y}} = \mathbf{g}(\mathbf{y})$ by a function h ε C^k. C^k is the set of k-fold continuously differentiable functions h: $\mathbb{R}^n \to \mathbb{R}^n$. In other words, two vector fields are called C^k conjugate if there exists a C^k diffeomorphism h ($k \geq 1$) such that $h \circ f = g \circ h$. An equivalent formulation uses the concept of flows associated with differential equations $\dot{x} = f(x)$. A flow is a continuously differentiable function φ: $\mathbb{R} \times \mathbb{R}^n \to \mathbb{R}^n$ such that, for each t ε $\mathbb{R}$, the restriction $\varphi(t, \cdot) =: \varphi_t(\cdot)$ satisfies $\varphi_0 = \mathrm{id}_{\mathbb{R}^n}$ and $\varphi_t \circ \varphi_s = \varphi_{t+s}$ for all t, s ε $\mathbb{R}$. The relationship to a differential equation is given by

$$\mathbf{f}(\mathbf{x}) := \frac{d\varphi_t}{dt}(\mathbf{x})\Big|_{t=0} = \lim_{\varepsilon \to 0} \left\{ \frac{\varphi(\varepsilon, \mathbf{x}) - \varepsilon(0, \mathbf{x})}{\varepsilon} \right\}.$$

For more details see, for example, [3]. Two flows φ_t and ψ_t (associated with **f** and **g**, respectively) are called C^k conjugate if there exists a C^k diffeomorphism **h** ($k \geq 1$) such that $\mathbf{h} \circ \varphi_t = \psi_t \circ \mathbf{h}$. In the case when $k = 0$, the term C^k conjugate needs to be replaced by C^0 or topologically conjugate and **h** is a homeomorphism. Clearly, differential equations, vector fields, and flows are only alternative ways of presenting the same dynamics.

By the previous definitions, equivalence relations can be generated and the set of differential equations $\dot{\mathbf{x}} = \mathbf{f}(\mathbf{x})$ (as well as vector fields and flows) can be decomposed in certain classes. Although C^k conjugacy sems to be a natural concept for classifying differential equations, vector fields, and flows, this approach leads to a very refined classification (up to a diffeomorphism). In other words, too many systems become inequivalent. The following two examples illustrate this statement.

Consider the nonlinear dynamical circuit (see Figure 4.1) with the descriptive equations (dimensionless form)

$$C\frac{dv_C}{dt} = i(v_C, i_L) - i_L \tag{4.2}$$

$$L\frac{di_L}{dt} = v_C + v(v_C, i_L) \tag{4.3}$$

If the nolinear controlled sources are defined by

$$\begin{aligned} i = i(v_C, i_L) &:= -\frac{1}{2}\left(\sqrt{v_C^2 + i_L^2} - 1\right)v_C, \\ v = v(v_C, i_L) &:= -\frac{1}{2}\left(\sqrt{v_C^2 + i_L^2} - 1\right)i_L \end{aligned} \tag{4.4}$$

the following equations are derived:

$$C\frac{dv_C}{dt} = -\frac{1}{2}\left(\sqrt{v_C^2 + i_L^2} - 1\right)v_C - i_L \tag{4.5}$$

$$L\frac{di_L}{dt} = v_C - \frac{1}{2}\left(\sqrt{v_C^2 + i_L^2} - 1\right)i_L \tag{4.6}$$

These equations can be transformed to polar coordinates (r, ϕ) by $v_C := r\cos(\phi t)$ and $i_L := r\sin(\phi t)$ $(C = 1$ and $L = 1)$

$$\frac{dr}{dt} = \frac{1}{2}(1 - r)r \quad and \quad \dot{\phi} = 1 \tag{4.7}$$

If we consider the same circuit with parameters $C/2$ and $L/2$, the descriptive equations formulated in polar coordinates become

$$\frac{dr}{dt} = (1 - r)r \quad and \quad \dot{\phi} = 2 \tag{4.8}$$

Note that both differential equations differ only by a time rescaling $(t \mapsto 2t)$.

It can be demonstrated (see [1], [28]) that if a diffeomorphism converts a singular point of a vector field into a singular point of another vector field, then the derivative of the diffeomorphism converts the Jacobian matrix of the first vector field at its singular point into the Jacobian matrix of the second field at its singular point. Consequently, these two Jacobian matrices are in the same similarity class and, therefore, have the same eigenvalues. In other words, the eigenvalues of the Jacobian matrices are invariants with respect to diffeomorphism, and the corresponding decomposition of the set of vector fields (differential equations and flows) is continuous rather than discrete. Obviously, the eigenvalues of (4.7) and (4.8) $(\lambda_1 = 1/2, \lambda_2 = 1$ and $\tilde{\lambda}_1 = 1, \lambda_2 = 2$, respectively) are different and, in conclusion, the two vector fields are not C^1 conjugate. Moreover, these two vector fields are not topologically or C^0 conjugate. A more "coarse" equivalence relation is needed in order to classify those vector fields, differential equations, and flows. As mentioned above, a time rescaling transforms the differential equations (4.7) and (4.8) into one another. This motivates the following definition.

Definition 1. Two flows φ_t and ψ_t are called C^k equivalent $(k \geq 1)$ if there exists a C^k diffeomorphism $\mathbf{h}$ that takes each orbit of φ_t into an orbit of ψ_t, preserving their orientation. In the case of $k = 0$, the flows are called C^0 or topologically equivalent.

Because C^k equivalence preserves the orientation of orbits, the relation $\mathbf{h}[\varphi_t(\mathbf{x})] = \varphi_{\tau_y(t)}(y)$ with $\mathbf{y} = \mathbf{h}(\mathbf{x})$ between φ_t and ψ_t is allowed, where τ_y is an increasing function of t for every $\mathbf{y}$.

It can be demonstrated (see [28]) that the eigenvalues of the Jacobian matrices of two vector fields must be in the same ratio if a monotonic time rescaling is allowed. Therefore, the two vector fields (4.7) and (4.8) are C^1 equivalent. But, even the two linear vector fields of the equations

$$\begin{pmatrix}\dot{x}\\ \dot{y}\end{pmatrix} = \begin{pmatrix}1 & 0\\ 0 & 1\end{pmatrix}\begin{pmatrix}x\\ y\end{pmatrix} \quad \begin{pmatrix}\dot{x}\\ \dot{y}\end{pmatrix} = \begin{pmatrix}1 & 0\\ 0 & 1+\varepsilon\end{pmatrix}\begin{pmatrix}x\\ y\end{pmatrix} \tag{4.9}$$

are not C^1 equivalent for any $\varepsilon \neq 0$, although the solutions are very close for small ε in a finite time interval. In conclusion, topological equivalence is the appropriate setting for classifying differential equations, vector fields, and flows. Note that the decomposition of the set of linear vector fields into equivalence classes using the topological equivalence does not distinguish between nodes, improper nodes, and foci, but does distinguish between sinks, saddles, and sources. This suggests the following theorem [3].

Theorem 1: *Let* $\dot{\mathbf{x}} = \mathbf{Ax}$ ($\mathbf{x} \in \mathbb{R}^n$ and $\mathbf{A} \in \mathbb{R}^{n\times n}$) *define a hyperbolic flow on* $\mathbb{R}^n$, *i.e., the eigenvalues of* A *have only nonzero real parts, with* n_s *eigenvalues with a negative real part. Then,* $\dot{\mathbf{x}} = \mathbf{Ax}$ *is topologically equivalent to the system* ($n_u := n - n_s$)

$$\dot{\mathbf{x}}_s = -\mathbf{x}_s, \quad \mathbf{x}_s \in \mathbb{R}^{n_s} \tag{4.10}$$

$$\dot{\mathbf{x}}_u = +\mathbf{x}_u, \quad \mathbf{x}_u \in \mathbb{R}^{n_u} \tag{4.11}$$

Therefore, it follows that hyperbolic linear flows can be classified in a finite number of types using topological equivalence.

A local generalization of this theorem to nonlinear differential equations is known as the theorem of Hartman and Grobman (see [3]).

Theorem 2: *Let* $\mathbf{x}^*$ *be a hyperbolic fixed point of* $\mathbf{x} = \mathbf{f}(\mathbf{x})$ *with the flow* φ_t: $U \subseteq \mathbb{R}^n \rightarrow \mathbb{R}^n$, *i.e., the eigenvalues of the Jacobian matrix* $J_f(\mathbf{x}^*)$ *have only nonzero real parts. Then, there is a neighborhood N of* $\mathbf{x}^*$ *on which* $\dot{\mathbf{x}} = \mathbf{f}(\mathbf{x})$ *is topologically equivalent to* $\dot{\mathbf{x}} = \mathbf{J}_f(\mathbf{x}^*)\mathbf{x}$.

The combination of the two theorems implies that a very large set of differential equations can be classified in an isolated hyperbolic fixed point by a finite number of types [namely, n_s (or n_u)]. The reason behind this interesting result is that the theorem of Hartman and Grobman, based on homeomorphisms, leads to a coarse decomposition of the set under consideration.

As a consequence of the preceding theorems, the behavior of nonlinear differential equations near a hyperbolic fixed point is equivalent up to a homeomorphism to the behavior of a simple system of linear differential equations. In the theory of nonlinear circuits, these mathematical results can be interpreted in the following way: The behavior of nonlinear circuits near an operational point where the Jacobian matrix of descriptive equations (in state-space form) has only eigenvalues with nonzero real parts is "similar" to that of a suitable linear circuit (so-called small-signal behavior).

In the following section, we discuss more general methods for the analysis of vector fields that have a least one nonhyperbolic fixed point.

4.2 Normal Forms

In Section 4.1, we presented theorems useful for classifying the "local" behavior of nonlinear differential equations near hyperbolic fixed points by using "global" results from the theory of linear differential equations. A main remaining problem is to calculate a homeomorphic transformation **h** in a concrete case. An alternative way to circumvent some of the difficulties is to apply the theory of normal forms that goes back to the beginning of this century and is based on classical ideas of Poincaré and Dulac. Detailed investigations of this subject are beyond the scope of this section and an interested reader should consult the monographs of [1], [3], [10], as well as [28], where further references to the theory of normal forms can be found. In this section, we present only the main ideas to illustrate its areas of application.

In contrast to theory described in Section 4.1, which is dedicated to hyperbolic cases, the theory of normal forms applies diffeomorphisms instead of homoemorphisms. This is necessary in order to distinguish the dynamical behavior in more detail. To classify the topological types of fixed points of the nonlinear differential equation $\dot{\mathbf{x}} = \mathbf{f}(\mathbf{x})$ one proceeds in two steps:

a) construction of a "normal form" in which the nonlinear terms of the vector field **f** take their "most simple" form, and b) determination of the topological type of the fixed point (under consideration) from the normal form. We present the main aspect of this "algorithm" without a proof.

First, we suppose that the vector field $\mathbf{f}(\mathbf{x})$ of the nonlinear differential equation $\dot{\mathbf{x}} = \mathbf{f}(\mathbf{x})$ satisfies $\mathbf{f}(\mathbf{0}) = \mathbf{0}$ (using a suitable transformation) and that it is represented by a formal Taylor expansion

$$\mathbf{f}(\mathbf{x}) = \mathbf{Ax} + \tilde{\mathbf{f}}(\mathbf{x}) \tag{4.12}$$

where $\mathbf{A} = \mathbf{J}_f(0)$ and $\tilde{\mathbf{f}}(\mathbf{x}) = 0(\|\mathbf{x}\|^2)$ is of class C^r. Power (Taylor) series with no assumptions about convergence are called *formal series*. In practice, we begin with the formal series and then we determine the corresponding region of converence. Then, we apply a diffeomorphic C^r change of coordinates $\mathbf{h}: \mathbb{R}^n \to \mathbb{R}^n$ with $\mathbf{y} \mapsto \mathbf{x} = \mathbf{h}(\mathbf{y})$ ($\mathbf{h}(0) = 0$) in the form of a near identity transformation

$$\mathbf{h}(\mathbf{y}) := \mathbf{y} + \mathbf{h}^k(\mathbf{y}) \tag{4.13}$$

where $\mathbf{h}^k(\mathbf{y})$ is a homogeneous polynomial of order k in $\mathbf{y}$ ($k \geq 2$). The result of the transformation is

$$\dot{\mathbf{y}} = \left\{\mathrm{id} + \mathbf{h}_y^k(\mathbf{y})\right\}^{-1} \mathrm{A}\left\{\mathbf{y} + \mathbf{h}^k(\mathbf{y})\right\} + \left\{\mathrm{id} + \mathbf{h}_y^k(\mathbf{y})\right\}^{-1} \tilde{\mathbf{f}}\left[\mathbf{y} + \mathbf{h}^k(\mathbf{y})\right] \tag{4.14}$$

$$= \mathbf{Ay} + \mathbf{g}(\mathbf{y}) \tag{4.15}$$

where $\mathbf{h}_y^k$ is the Jacobina matrix of $\mathbf{h}^k$ with respect to $\mathbf{y}$, and $\mathbf{g}(\mathbf{y}) = O(\|\mathbf{y}\|^2)$ is of class C^r. It is useful to define $\mathbf{f}^k$ of a function $\mathbf{f}: \mathbb{R}^n \to \mathbb{R}^n$ as the "truncated" Taylor series in $\mathbf{0}$ expansion that have a degree less or equal k; the ith-order terms are denoted by $\mathbf{f}_k$. The set of f^k's forms a real vector space of functions where the components are homogeneous polynomials in n variables of degree less or equal k. The vector space of f_k's, the homogeneous polynomials in n variables of degree k, is denoted by H_k^n. Using the k-jet notation and expanding $\mathbf{g}$ into a formal Taylor series, (4.15) can be reformulated as

$$\dot{\mathbf{y}} = \mathbf{g}^{k-1}(\mathbf{y}) + \overset{k}{\mathbf{g}}(\mathbf{y}) \tag{4.16}$$

where $\overset{k}{\mathbf{g}}(\mathbf{y})$ contains all terms of degree k or higher. Expanding $\tilde{\mathbf{f}}$ into a formal Taylor series

$$\tilde{\mathbf{f}} = \tilde{\mathbf{f}}_2 + \tilde{\mathbf{f}}_3 + \ldots, \tag{4.17}$$

(4.16) and (4.17) can be represented by

$$\dot{\mathbf{y}} = \mathbf{g}^{k-1}(\mathbf{y}) + \left\{\mathbf{f}_k - \left[\mathbf{Ay}, \mathbf{h}^k\right]\right\} + O\left(\|\mathbf{y}\|^{k-1}\right) \tag{4.18}$$

where $[\mathbf{Ay}, \mathbf{h}^k]$ is the so-called *Lie bracket* of the linear vector field $\mathbf{Ay}$ and $\mathbf{h}^k(\mathbf{y})$ is defined by

$$\left[\mathbf{Ay}, \mathbf{h}^k\right] := \mathbf{h}_y^k(\mathbf{y})\mathbf{Ay} - \mathbf{Ah}^k(\mathbf{y}) \tag{4.19}$$

Define a linear operator L_A^k: $H_k^n \to H_k^n$ on H_k^n by

$$L_A^k \mathrm{h}^k : \mathbf{y} \mapsto \left[\mathbf{Ay}, \mathbf{h}^k\right] \tag{4.20}$$

with the range $\mathcal{R}_k$, and let $\mathcal{C}_k$ be any comlementary subspace to $\mathcal{R}_k$ in H_k^n, that is $H_k^n = \mathcal{R}_k \oplus \mathcal{C}_k (k \geq 2)$. Then, the following theorem implies a simplification of a nonlinear differntial equation $\dot{\mathbf{x}} = \mathbf{f}(\mathbf{x})$.

Theorem 3: *Let* $\mathbf{f}: \mathbb{R}^n \to \mathbb{R}^n$ be *a* C^r *vector field with* $\mathbf{f}(\mathbf{0}) = \mathbf{0}$ *and* $\mathbf{A} \in \mathbb{R}^{n \times n}$, *and let the decomposition* $H_k^n = \mathcal{R}_k \oplus \mathcal{C}_k$ of H_k^n *be given. Then, there exists a series of near identify transformations* $\mathbf{x} = \mathbf{y} + \mathbf{h}_k(\mathbf{y}) \in \Omega$, $k = 2, 3, \ldots, r$ *in some region* Ω, *where* $\mathbf{h}_k \in H_k^n$ and Ω *is a neighborhood of the origin, such that the equation* $\dot{\mathbf{x}} = \mathbf{f}(\mathbf{x})$ *is transformed to*

$$\dot{\mathbf{y}} = \mathbf{Ay} + \mathrm{g}_2(\mathbf{y}) + \ldots + \mathbf{g}_k(\mathbf{y}) + O\left(\|\mathbf{y}\|^{r+1}\right) \quad \mathbf{y} \in \Omega \tag{4.21}$$

where $\mathrm{g}_k \in \mathcal{C}_k$ *for* $k = 2, 3, \ldots$.

The proof of this theorem and the following definition can be found in [4].

Definition 2. Let $\mathcal{R}_k \oplus \mathcal{C}_k$ be decompositions of H_k^n for $k = 2, 3, \ldots, r$. The truncated equation (4.21)

$$\dot{\mathbf{y}} = \mathbf{Ay} + \mathbf{g}_2(\mathbf{y}) + \ldots + \mathbf{g}_r(\mathbf{y}) \tag{4.22}$$

where $\mathbf{g}_k \in \mathcal{C}_k$ (k = 2, 3,..., r), is called normal form of $\dot{\mathbf{x}} = \mathbf{Ax} + \tilde{\mathbf{f}}(\mathbf{x})$ associated with matrix $\mathbf{A}$ up to order r $\geq$ 2 (with respect to the decomposition $\mathcal{R}_k \oplus \mathcal{C}_k$, or an A-normal form of $\dot{\mathbf{x}} = \mathbf{f}(\mathbf{x})$.

Theorem 3 suggests an equivalence relation in the set of vector fields **f** that decomposes the set into equivalence classes. Each class can be represented by using definition of normal forms. Because a concrete normal form depends on the choice of complementary subspaces $\mathcal{C}_k$, it is not unique. In practical problems a constructive method for finding these subspaces is needed. An elegant way to find the subspaces is to start with the introduction of a suitable inner product $\langle \cdot | \cdot \rangle_n$ in H_k^n that is needed to define the adjoint operator $(L_A^k)^*$ of L_A^k (in a usual way) by

$$\left\langle \eta \middle| L_A^k(\xi) \right\rangle_n := \left\langle (L_A^k)^*(\eta) \middle| \xi \right\rangle_n, \quad \text{for all } \eta, \xi \in H_k^n \tag{4.23}$$

It can be shown that $(L_A^k)^* = L_{A^*}^k$, where $\mathbf{A}^* = \mathbf{A}^T$, is the transposed matrix of $\mathbf{A}$. The desired construction is available as an application of the following theorem.

Theorem 4: *Vector space* ker $\{L_{A^*}^k\}$ *that is the solution space of the equation* $L_{A^*}^k \xi = 0$ *is a complementary subspace of* $\mathcal{R}_k$ in H_k^n, i.e.,

$$H_k^n = \mathcal{R}_k \oplus \ker\{L_{A^*}^k\} \tag{4.24}$$

The interested reader is referred to [4] for a detailed discussion of this subject. As a consequence, finding a normal form in the above defined sense up to the order r, requires solving the partial differential equation. From the algebraic point of view this means that a base of ker $\{L_{A^*}^k\}$ has to be chosen, but this can be done, again, with some degree of freedom. For example, the two sets of differential equations are distinct normal forms of $\dot{\mathbf{x}} = \mathbf{Ax} + \tilde{\mathbf{f}}(\mathbf{x})$ associated with the same matrix $\mathbf{A}$ (see [4]):

$$\frac{d}{dt}\begin{pmatrix} x_1 \\ x_2 \end{pmatrix} = \begin{pmatrix} 0 & 1 \\ 0 & 0 \end{pmatrix}\begin{pmatrix} x_1 \\ x_2 \end{pmatrix} + \begin{pmatrix} ax_1^2 \\ bx_2^2 + ax_1x_2 \end{pmatrix} \tag{4.25}$$

$$\frac{d}{dt}\begin{pmatrix} x_1 \\ x_2 \end{pmatrix} = \begin{pmatrix} 0 & 1 \\ 0 & 0 \end{pmatrix}\begin{pmatrix} x_1 \\ x_2 \end{pmatrix} + \begin{pmatrix} 0 \\ ax_1^2 + bx_1x_2 \end{pmatrix} \tag{4.26}$$

To reduce the number of nonlinear monomials in the normal form, a more useful base of $\mathcal{C}_k$ must be determined. If a nonlinear differential equation of the form $\dot{\mathbf{x}} = \mathbf{Ax} + \tilde{\mathbf{f}}(\mathbf{x})$ is given with an arbitrary matrix **A**, several partial differential equations need to be solved. This, in general, is not an easy task. If **A** is diagonal or has an upper triangular form, methods for constructing a base are available. For this purpose we introduce the following definition.

Definition 3. Let A ε $\mathbb{R}^{n\times n}$ possess the eigenvalues $\lambda_1, \ldots, \lambda_n$ and let $x_1^{\alpha_1} x_1^{\alpha_2} \cdots x_2^{\alpha} x_n^{\alpha_n} \mathbf{e}_j$ be a monomial in n variables. It is called a resonant monomial so-called resonance condition

$$\alpha^T \lambda - \lambda_j = 0 \tag{4.27}$$

is satisfied ($\alpha^T := (\alpha_1 \ldots \alpha_n)$, $\lambda^T := (\lambda_1 \ldots \lambda_n)$]. If the resonance condition holds for some $\alpha_1 + \alpha_2 + \alpha_n \geq$ 2 and some $j \in \{1, \ldots, n\}$, we say that A has a resonant set of eigenvalues.

The next theorem proves that if **A** is diagonal, a minimal normal form exists (in certain sense).

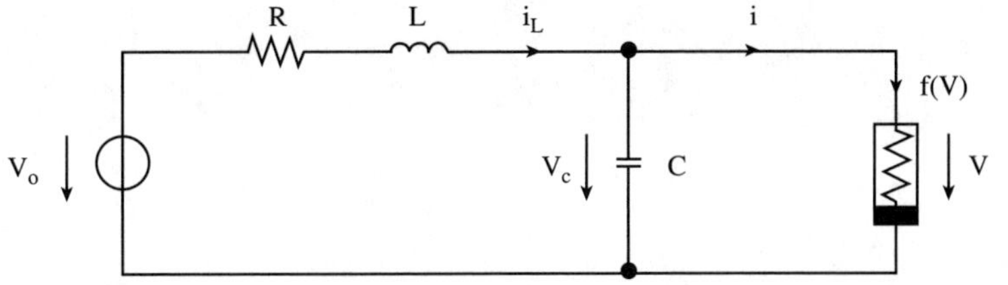

FIGURE 4.2 Nonlinear network with tunnel diode.

Theorem 5: *Let* **A** = diag $(\lambda_1,\ldots,\lambda_n)$. *Then, an A-normal form equation up to order r can be chosen to contain all resonant monomials up to order r.*

If some eigenvalues of **A** are complex, a linear change to complex coordinates is needed to apply this theorem. Furthermore, theorems and definitions need to be modified to complex cases. In the case of differential equations,

$$\frac{d}{dt}\begin{pmatrix} x_1 \\ x_2 \end{pmatrix} = \begin{pmatrix} 0 & -1 \\ +1 & 0 \end{pmatrix}\begin{pmatrix} x_2 \\ x_2 \end{pmatrix} + O\left(\|x\|^2\right) \tag{4.28}$$

that can be used to describe oscillatory circuits, the coordinates are transformed to

$$\dot{z}_1 := x_1 + jx_2 \tag{4.29}$$

$$\dot{z}_2 := x_1 - jx_2 \tag{4.30}$$

with the resonant set of eigenvalues $\{-j, +j\}$ that can be written in a complex normal form (z presents one of the z_i's):

$$\dot{z} = j_z + a_1|z|^2 z + \ldots + a_k|z|^{2k} z \tag{4.31}$$

This last normal form equation — Poincaré's normal form — is used intensively in the theory of the Poincaré–Andronov–Hopf bifurcation. The book of Hale and Kocak [30], is worth reading for the illustration of this phenomenon in nonlinear dynamical systems. A proof of the Poincaré–Andronov–Hopf theorem can be found in [32]. These authors emphasize that further preparation of the system of nonlinear differential equations is needed before the normal form theorem is applicable. In general, the linearized part of $\dot{\mathbf{x}} = \mathbf{Ax} + \tilde{\mathbf{f}}(\mathbf{x})$ (in a certain fixed point of the vector field **f**) has two classes of eigenvalues: *central* eigenvalues that lie on the imaginary axis and *noncentral* eigenvalues. The dynamical behavior that is associated with the noncentral eigenvalues is governed by the theorem of Grobman and Hartman (see Section 4.1). A systematic procedure to eliminate these noncentral eigenvalues from the system is based on the so-called center manifold theorem (see [11]) that is used also in the paper of Hassard and Wan [32]. A numerical algorithm is presented by Hassard, Kazarinoff, and Wan [31]. Discussion of this essential theorem goes beyond the scope of this section. A detailed presentation of the theory of normal forms of vector fields and its applications can be found in [21], [22].

Until now, we have discussed differential equations of the type $\dot{\mathbf{x}} = \mathbf{f}(\mathbf{x})$. Descriptive equations of circuits consist, in general, of linear and nonlinear algebraic equations, as well as differential equations. Therefore, so-called constrained differential equations need to be considered for applications in network theory. A typical example is the well-known circuit shown in Figure 4.2, containing a model of a tunnel diode (see [9], [41]). Circuit equations can be written as

$$C\frac{dv_C}{dt} = i_L - f(v_C) \tag{4.32}$$

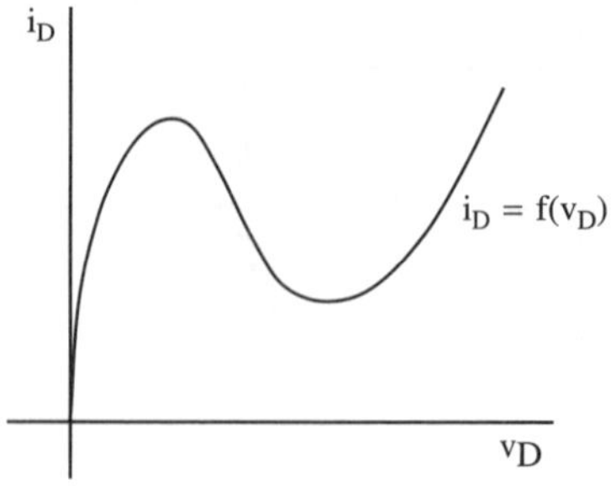

FIGURE 4.3 Characteristic of a tunnel diode.

$$L\frac{di_L}{dt} = -v_C - Ri_L + V_0 \tag{4.33}$$

where $i_D = f(v_D)$ is the characteristics of the tunnel diode (see Figure 4.3).

If the behavior of the circuit is considered with respect to the time scale $\tau := RC$, then the differential equations (4.32) and (4.33) can be reformulated as

$$\frac{dv}{d\theta} = i - F(v) \tag{4.34}$$

$$\varepsilon\frac{di}{d\theta} = -v - i + 1 \tag{4.35}$$

with $\theta := t/\tau$, $i := (r/V_0)i_L$, $v := v_C/V_0$, $\varepsilon := L/(CR^2)$, and $F(v) := (R/V_0)f(V_0 \cdot v)$. The behavior of the *idealized* circuit, where $\varepsilon = 0$ is often of interest. These types of descriptive equations are called **constrained equations**. The theory of normal forms cannot be applied on descriptive equations of this type. Fortunately, a generalized theory is available, as presented (with applications) in [24], [25].

4.3 Dimensionless Form

In Section 4.1, the problem of equivalence is considered in a rather complex manner by adopting the mathematical point of view of dynamical systems. We will use this approach in Section 4.6 to classify nonlinear dynamical networks. In contrast to this approach, another kind of equivalence of descriptive equations for linear time-invariant circuits and for certain nonlinear circuits is well known. It is illustrated by means of a simple example. The *RLC* parallel circuit in Figure 4.4, analyzed in the frequency domain, can be described by its impedance

$$Z = \frac{1}{\frac{1}{R} + j\left(\omega C - \frac{1}{\omega L}\right)} \tag{4.36}$$

In the case of small damping, this formula can be simplified using the so-called *Q*-factor that is defined as the ratio of the current flowing through *L* (or *C*) to the current through the circuit in the case of

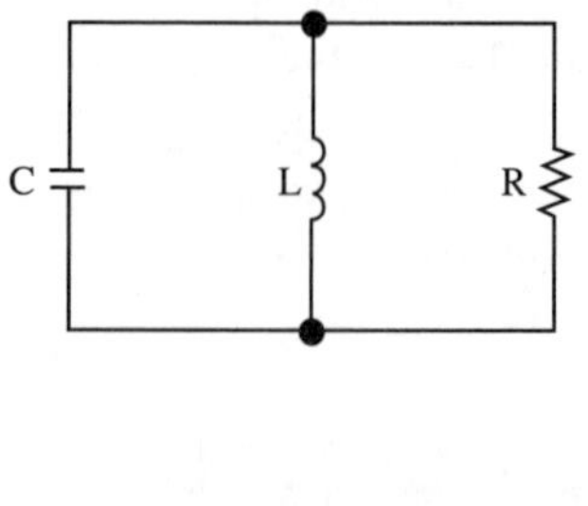

FIGURE 4.4 RLC network.

resonance. The case of small damping is characterized by the condition $R/(2L) \ll \omega_0$, where $\omega_0 := 1/\sqrt{LC}$ (Thompson's formula for LC circuits with $R = 0$). The Q-factor is given by $Q := 1/(\omega_0 CR) = \omega_0 L/R$. Simple calculations lead to a so-called normalized impedance

$$\tilde{Z} := \frac{Z}{R} = \left[1 - jQ^{-1}\left(\frac{\omega}{\omega_0} - \frac{\omega_0}{\omega}\right)\right]^{-1} \tag{4.37}$$

that contains only two instead of three parameters. By using this method, a whole class of RLC circuits may be described by the same normalized impedance. Therefore, an equivalence relation is defined in this manner. Handbooks written for practical electrical engineers contain diagrams of those and of similar curves of normalized impedances and admittances. Note that the formula is exact, although the interpretations of the parameters Q and ω_0 depend on the condition $R/(2L) \ll \omega_0$.

Methods for normalizing descriptive equations of circuits and for reducing the number of parameters are known in linear and nonlinear circuit theory (see [20]). Unfortunately, these methods are stated without a presentation of their mathematical foundations. The main ideas for justification of normalization procedures are based on so-called dimensional analysis. Their first applications in physics and the development of their mathematical foundations can be traced to the end of the last century. In this section, we discuss only a few aspects of this subject. Interested readers may find more details about the theory and the applications of dimensional analysis in the paper by Mathis and Chua [38]. They demonstrated that for a complete mathematical discussion of physical quantities, several algebraic introductions are necessary. In this section, a concise introduction is preferred and therefore, an intuitive introduction based on Lie groups is presented. The following presentation uses ideas from the book of Ovsiannikov [42].

For describing the physical arrangements, we require descriptive quantities that can be measured. To perform a physical measurement, we need at least one measuring instrument that provides a value on its scale. Intuitively, a mathematical model of a physical quantity ϕ consists of two parts: a real number $|\phi|$ that characterizes its value, and symbol E_ϕ that is due to the measuring arrangement of ϕ. In general, a measuring arrangement is composed of elementary measuring instruments to evaluate, for example, time or frequency, length, voltage, and charge. Each elementary instrument will have associated a symbol E_k. Therefore, a physical quantity ϕ is defined by

$$\phi := |\phi| E_1^{\lambda_1} E_2^{\lambda_2} \ldots E_r^{\lambda_r} \tag{4.38}$$

where integers λ_k determine how many times an instrument is applied and whether the value on the scale needs to be multiplied or divided. The *dimensionality* of a physical quantity ϕ is defined by

$$[\phi] := E_1^{\lambda_1} E_2^{\lambda_2} \ldots E_r^{\lambda_r} \tag{4.39}$$

where $r \leq n$. A quantity is called **dimensionless** if its dimensionality is null, that is if $\lambda_1 = 0, \lambda_2 = 0, \ldots, \lambda_r = 0$. Moreover, a set of n physical quantities $\phi_1, \phi_2, \ldots, \phi_n$ is called *dependent* (in the sense of their dimensionality) if there exist integers $\chi_1, \chi_2, \ldots, \chi_n$ (not all equal zero) such that the product of these quantities

$$\phi_1^{\chi_1} \phi_2^{\chi_2} \ldots \phi_n^{\chi_n} \tag{4.40}$$

is dimensionless. Otherwise, $\phi_1, \phi_2, \ldots, \phi_n$ are called **independent**.

The main problem of dimension theory is to determine how many independent physical quantities are in a given set $\phi_1, \phi_2, \ldots, \phi_n$, to find them, and then to express the other quantities in terms of these independent quantities. As an application, a systematic procedure to normalize physical descriptive equations can be derived.

In order to solve the main problem, the change of measuring instruments and measuring scales needs to be introduced. Obviously, in terms of modeling physical quantities, the same ϕ can be represented in different ways. Using two sets of measuring instruments, denoted by $E_1, E_2, \ldots, E_r$ and $\tilde{E}_1, \tilde{E}_2, \ldots, \tilde{E}_r$, respectively, ϕ is given by

$$\phi = |\phi| E_1^{\lambda_1} E_2^{\lambda_2} \ldots E_r^{\lambda_r} = |\tilde{\phi}| \tilde{E}_1^{\tilde{\lambda}_1} \tilde{E}_2^{\tilde{\lambda}_2} \ldots \tilde{E}_p^{\tilde{\lambda}_p} \tag{4.41}$$

where in general $r \neq \rho$. This suggests the so-called **analogy transformation** (see [38]).

$$E_k = a_1^{\alpha_1} \cdot a_{a_k}^{\alpha_k} \tilde{E}_1^{\alpha_1} \cdots \tilde{E}_k^{\alpha_k} \qquad k = 1, \ldots, r \tag{4.42}$$

Transformations of the scales of measuring instruments (in the following *scale transformation*) are special analogy transformations

$$E_k = a_k \tilde{E}_k, \qquad k = 1, \ldots, r \tag{4.43}$$

It can be demonstrated that analogy transformations of dimension theory are special cases of so-called *extension groups*. These groups belong to the *r-parameter Lie groups*, and, subsequently, all results of dimension theory can be interpreted by theorems from this mathematical theory. Further details are contained in [42]. To introduce the main ideas, we consider in this section only scale transformations.

Let $Z := \mathbb{R}^n \times \mathbb{R}^n$ be the Cartesian product of the set of n-column vectors $\mathbf{x}$ and m-column vectors $\mathbf{y}$, and let $\{\mathbf{e}_1, \ldots, \mathbf{e}_n\}$ and $\{\mathbf{f}_1, \ldots, \mathbf{f}_m\}$, be, respectively, arbitrary (but fixed) bases of this vector spaces. Endowing Z with the structure of a direct sum $\mathbb{R}^n \oplus \mathbb{R}^m$, each $\mathbf{z} \in Z$ can be represented by (with respect to the bases)

$$\mathbf{z} = \sum_{i=1}^{n} x^i \mathbf{e}_i + \sum_{k=1}^{n} y^k \mathbf{f}_k \tag{4.44}$$

An *extension* of Z (with respect to the bases) is defined by the transformation

$$h: \mathbf{z} \mapsto \sum_{i=1}^{n} c^i x^i \mathbf{e}_i + \sum_{k=1}^{n} d^k y^k \mathbf{f}_k \qquad \left(c^i > 0, \mathrm{d}^k > 0\right) \tag{4.45}$$

Obviously, the set of all extensions generates Abel's group of transformations on Z that is a $(n + m)$-parameter Lie group. This group is denoted by $\mathrm{Diag}\{\boldsymbol{e}_i, \boldsymbol{f}_k\}$. Any subgroup $H \subset \mathrm{Diag}\{\boldsymbol{e}_i, \boldsymbol{f}_k\}$ is called an *extension group* of Z. We now consider extension groups H^r, with $0 < \mathrm{r} \leq n + m$.

Ovsiannikov demonstrated that extensions of H^r can be represented, choosing a parametric group, in the form

$$\tilde{x}^i = x^i \prod_{\alpha=1}^{r} (a_\alpha)^{\lambda_\alpha^i} \qquad \tilde{y}^k = y^k \prod_{\alpha=1}^{r} (a_\alpha)^{\mu_\alpha^k} \tag{4.46}$$

where $i = 1, \ldots, n$ and $k = 1, \ldots, m$.

The main property of transformation groups is that they induce equivalence relations decomposing the subjects into equivalence classes on which the group acts. If h_p acts on elements $x \in X$, and $\mathbf{p} \in \mathbb{R}^p$ is the vector of parameters, an *orbit U of a point* $x \in X$ is defined by the set $U := \{\xi \in X | = h_p(x, \mathbf{p})$, for all $\mathbf{p} \in \mathbb{R}^p$. In this sense, the points of an orbit can be identified by a transformation group. A transformation group acts *transitive* on X if there exists an orbit U that is an open subset of X, with $\overline{U} = X$.

To study so-called local Lie groups with actions that are defined near a null neighborhood of the parameter space (includes its vector $\mathbf{0}$), we can discuss the Lie algebra that characterizes the local behavior of the associated local Lie group. In finite dimensional parameter spaces, a Lie algebra is generated by certain partial differential operators. Using the representations (4.46) of H^r, the operators are of the form

$$\sum_{i=1}^{n} \lambda_\alpha^i x^i \frac{\partial}{\partial x^i} + \sum_{k=1}^{m} \mu_\alpha^k y^k \frac{\partial}{\partial y^k} \tag{4.47}$$

where $i = 1, \ldots, n$ and $k = 1, \ldots, m$. These operators can be represented in a matrix form

$$\mathbf{M}(z) := \mathbf{M}_1 \circ \mathrm{diag}\{x^1, \ldots, x^n; y^1, \ldots, y^m\} \tag{4.48}$$

where

$$\mathbf{M}_1 := \begin{pmatrix} \lambda_1^1 & \lambda_1^n, \mu_1^1 & \mu_1^m \\ \\ \lambda_r^1 & \lambda_r^n, \mu_r^1 & \mu_r^m \end{pmatrix} \tag{4.49}$$

Obviously, H^r is intransitive if $r < n + m$.

In order to solve the main problem of dimension theory, we need to introduce invariants of a Lie group. Let $F: X \to Y$ be a function on X and let transformation h_p of a transformation group act on X, then F is an invariant of the group if $F[h_a(x)] = F(x)$ holds for any $x \in X$ and $\mathbb{R}^p$. The invariant $J: X \to Y$ is called a *universal invariant* if there exists, for any invariant $F: X \to Y$ of the group, a function Φ such that $F = \Phi \circ J$. The following main theorem can be proved for the extension group.

Theorem 6: *For the extension group H^r on Z, there exists a universal invariant $J: Z \to \mathbb{R}^{n+m-r}$ if the condition $r < n + m$ is satisfied. The independent components of J have the monomial form*

$$J^\tau(z) = \prod_{i=1}^{n} \left(x^i\right)^{\theta_i^\tau} \cdot \prod_{k=1}^{m} \left(y^k\right)^{\sigma_k^\tau} \tag{4.50}$$

where $\tau = 1, \ldots, n + m - r$.

If dimensional analysis considers only scale transformations (4.43), this theorem contains the essential result of the so-called **Pi-theorem**. For this purpose we present a connection between the dimensionalities and the extension group H^r (see [42]). The group H^r of the space $\mathbb{R}^n$, defined only by the dimensions of the physical quantities ϕ_k with respect to the set of symbols $\{E_\alpha\}$, has a one-to-one correspondence with every finite set $\{\phi_k\}$ of n physical quantities, which can be measured in the system of symbols $\{E_\alpha\}$ consisting of r independent measurement units [see (4.41)]. The transformations belonging to the group H^r give the rule of change, in the form

$$|\tilde{\phi}| = |\phi| \prod_{\alpha=1}^{r} \left(a^\alpha\right)^{\lambda_\alpha} \tag{4.51}$$

of the numerical values $|\phi_k|$ as a result of the transition from the units $\{E_\alpha\}$ to $\{\tilde{E}_\alpha\}$ by means of (4.43).

As a consequence of this relationship, a quantity ϕ is dimensionless if and only if its numerical value is an invariant of the group H^r. Thus, the problem to determine the independent physical quantities of a given set of quantities is solved by the construction of a universal invariant of H^r stated by the Pi-theorem (see also [5]). Normalization, as well as the popular method of *dimension comparison*, are consequences of the invariance of physical equations with respect to the group of analogy transformations. In applications of dimensional theory, a normal form that has certain advantageous properties is desired. For example, it is useful to reduce the number of parameters in physical equations. Normal forms of this type are used very often in practice, but with no clarification of their mathematical foundation.

Network equations, similar to other physical equations, contain numerous parameters. In applications, it is often desired to suppress some of these properties and they should be replaced by the numerical value 1. For this purpose Desloge [27], chooses a new system of units $\{E_\alpha\}$. A theory of Desloge's method, based on analogy transformations (4.42) instead of scale transformations (4.43), was presented by Mathis and Chua [38]. The main idea behind this method is that, beside the foundation units time $[T]$, voltage $[E]$, and charge $[Q]$ that are useful in network theory, the units of other parameters are considered foundational units. We denote the units by $[A_\alpha]$ instead of E_α. For example, in the case of the tunnel-diode circuit (see Figure 4.2), $[T]$, $[E]$, and $[Q]$, as well as $[R]$, $[C]$, and $[L]$ need to be discussed. As a consequence of Desloge's method, three of the four parameters can be suppressed and the other variables will be normalized. The method works in the case of linear as well as nonlinear networks.

The method is illustrated using the tunnel-diode circuit (see (4.32) and (4.43). At first, the dimensional matrix is determined by

$$\begin{array}{c} \\ [R] \\ [L] \\ [C] \end{array} \begin{array}{c} \begin{matrix} [T] & [E] & [Q] \end{matrix} \\ \begin{pmatrix} 1 & 1 & -1 \\ 2 & 1 & -1 \\ 0 & -1 & 1 \end{pmatrix} \end{array} \tag{4.52}$$

that characterizes the relation between the dimensions of t, v, q, R, C, L.

Desloge now considers another set of power independent dimensional scalars A_1, A_2, A_3 with

$$[A_i] = [T]^{a_i^1}[E]^{a_i^2}[Q]^{a_i^3} \qquad (i = 1,2,3) \tag{4.53}$$

These relations are interpreted as an analogy transformation (4.42). Applying the map $L(\cdot)$ that has the same properties as the logarithmic function (see [38]) to (4.53), the symbols $L([A_1])$, $L([A_2])$, $L([A_3])$ are represented by linear combinations of $L([T])$, $L([E])$, $L([Q])$. The coefficient matrix in (4.53) is regular and contains the exponents. Solving these linear equations using "antilog," the $[T]$, $[E]$, $[Q]$ are products of powers of $[A_1]$, $[A_2]$, $[A_3]$. In this manner, dimensionless versions of differential equations of the tunnel diode can be derived.

By using the independent units $A_1 := L$, $A_2 := C$, $A_3 := V_0$ to replace $|V_0|, |L|, |C| \to 1$ (with respect to the new units), the following equation is derived by the approach sketched previously:

$$\begin{array}{c} \\ [V_0] \\ [L] \\ [C] \end{array} \begin{array}{c} \begin{matrix} [T] & [E] & [Q] \end{matrix} \\ \begin{pmatrix} 0 & 1 & 0 \\ 2 & 1 & -1 \\ 0 & -1 & 1 \end{pmatrix} \end{array} \begin{pmatrix} \mathrm{In}([T]) \\ \mathrm{In}([E]) \\ \mathrm{In}([Q]) \end{pmatrix} = \begin{pmatrix} \mathrm{In}([V_0]) \\ \mathrm{In}([L]) \\ \mathrm{In}([C]) \end{pmatrix} \tag{4.54}$$

Multiplying (4.54) with the inverse of the dimensional matrix

$$\begin{array}{c} \\ [T] \\ [E] \\ [Q] \end{array} \begin{array}{c} \begin{matrix} [V_0] & [L] & [C] \end{matrix} \\ \begin{pmatrix} 0 & 1/2 & 1/2 \\ 1 & 0 & 0 \\ 1 & 0 & 1 \end{pmatrix} \end{array} \tag{4.55}$$

and applying "antilog" to the result, we obtain

$$[T]=[L]^{1/2}[C]^{1/2} \quad [E]=[V_0] \quad [Q]=[V_0][C] \tag{4.56}$$

From these equations, the relations between the old and the new units can be derived (see [38]). *T, E,* and *Q* are expressed by the new units *L, C,* and V_0 and the parameters and variables in (4.34) and (4.35) can be reformulated if the numerical values of V_0, *L*, and *C* are added:

$$T=|L|^{-1/2}|C|^{-1/2}L^{1/2}C^{1/2}, \quad E=|V_0|^{-1}V_0, \quad Q=|V_0|^{-1}|C|^{-1}V_0C \tag{4.57}$$

These relations represent parameters and variables of the tunnel-diode network with respect to the new V_0, *L* and *C*.

$$R=\frac{|R||C|^{1/2}}{|L|^{1/2}}L^{1/2}C^{-1/2}, \quad V_0=1\cdot V_0, \quad L=1\cdot L, \quad C=1\cdot C \tag{4.58}$$

$$i_L=\frac{|i_L||L|^{1/2}}{|V_0||C|^{1/2}}V_0L^{-1/2}C^{1/2}, \quad v_C=\frac{|v_C|}{|V_0|}V_0, \quad t=\frac{|t|}{|L|^{1/2}|C|^{1/2}}L^{1/2}C^{1/2} \tag{4.59}$$

The dimensional exponents for these quantities can be found by finding the inverse dimensional matrix (4.55):

1. *T, E, Q*: their exponents correspond the associated rows of (4.55).
2. V_0, *L, C, R*: premultiply (4.55) with the corresponding row (4.52).

For example, taking $[C] \triangleq (0 - 1 \ \ 1)$ results in

$$\begin{array}{c} \\ [C] \end{array}\!\!\begin{array}{ccc} [T] & [E] & [Q] \\ (0 & -1 & 1) \end{array} \begin{array}{c} \\ [T] \\ [E] \\ [Q] \end{array}\!\!\begin{array}{c} \begin{array}{ccc}[V_0] & [L] & [C]\end{array} \\ \begin{pmatrix} 0 & 1/2 & 1/2 \\ 1 & 0 & 0 \\ 1 & 0 & 1 \end{pmatrix} \end{array} = \begin{array}{c} \\ [C] \end{array}\!\!\begin{array}{ccc} [V_0] & [L] & [C] \\ (0 & 0 & 1) \end{array} \tag{4.60}$$

or with (4.52) $[R] \triangleq (1 \ \ 1 \ -1)$

$$\begin{array}{c} \\ [R] \end{array}\!\!\begin{array}{ccc} [T] & [E] & [Q] \\ (1 & 1 & -1) \end{array} \begin{array}{c} \\ [T] \\ [E] \\ [Q] \end{array}\!\!\begin{array}{c} \begin{array}{ccc}[V_0] & [L] & [C]\end{array} \\ \begin{pmatrix} 0 & 1/2 & 1/2 \\ 1 & 0 & 0 \\ 1 & 0 & 1 \end{pmatrix} \end{array} = \begin{array}{c} \\ [R] \end{array}\!\!\begin{array}{ccc} [V_0] & [L] & [C] \\ (0 & 1/2 & -1/2) \end{array} \tag{4.61}$$

With these representations of the dimensional quantities, we can obtain a dimensionless representation of (4.34) and (4.35)

$$\frac{d\bar{v}_C}{d\bar{t}}=\bar{i}_L-\bar{f}(\bar{v}_C) \tag{4.62}$$

$$\frac{d\bar{i}_L}{d\bar{t}}=1-\sqrt{\varepsilon}\,\bar{i}_L-\bar{v}_C \tag{4.63}$$

where

$$\bar{v}_c := \frac{|v_C|}{|V_0|}, \quad \bar{t} := \frac{|t|}{\sqrt{|L||C|}} \quad \sqrt{\varepsilon} := \frac{|R||C|^{1/2}}{|L|^{1/2}} \tag{4.64}$$

$$\bar{i}_L := \frac{|i_L||L|^{1/2}}{|V_0||C|^{1/2}} \tag{4.65}$$

Furthermore, the dimensionless tunnel-diode current $\bar{f}$ is defined by

$$\bar{f}(\bar{v}_C) := V_0^{-1} L^{1/2} C^{-1/2} f(V_0 \bar{v}_C) \tag{4.66}$$

The associated dimensionless form of the (4.34) and (4.35) can be derived by another scaling of the current $\bar{\bar{i}}_L := \sqrt{\varepsilon}\,\bar{i}_L$. Obviously, the dimensionless normal form is not unique.

The classical dimensional analysis shows that R_2C/L is the only dimensionless constant of (4.32) and (4.33). Because the parallel *RLC* circuit includes the same constants and variables, the results of the previous dimensional analysis of the tunnel-diode circuit can be used to normalize (4.37).

Further interesting applications of Desloge's approach of suppressing superfluous parameters can be found in the theory of singular perturbations. The reader is referred to the monograph of Smith [43] for further details. Miranker [41] demonstrated that the differential equations of the tunnel-diode circuit can be studied on three time scales $\tau_1 = L/R$, $\tau_2 = RC$, and $\tau_3 = \sqrt{LC}$ with different phenomena arising. The corresponding dimensionless equations can be derived in a systematic manner by Desloge's method. In this way, normalized differential equations describing Chua's circuit (see [39]) can be obtained but other representations of these differential equations are possible using dimensional analysis.

4.4 Equivalence between Nonlinear Resistive Circuits

In this section, we consider equivalence of nonlinear resistive *n*-ports. (We do not discuss resistive networks without accessible ports.) Although the explanations that follow are restricted to resistive *n*-ports, this theory can be extended to capacitive and inductive *n*-ports (see [23]). In Section 4.5, we give a definition of those *n*-ports.

At first, we consider linear resistive 1-ports that contain Ohmic resistors described by $v_k = R_k i_k$ or/and $i_k = G_k i_k$, and independent current and voltages sources $v_k = V_0^k$ and $i_k = I_0^k$. We can use Thevenin's or Norton's theorem to compare any two of those 1-ports and reduce a complex 1-port to a simple "normal" form. Therefore, two of those 1-ports are called equivalent if they have the same Thévenin (or Norton) 1-port. Clearly, by this approach, an equivalence relation is defined in the set linear resistive 1-ports and it is decomposed into "rich" classes of 1-ports. To calculate these normal forms, Δ-*Y* and/or *Y*-Δ transformations are needed (see [20, 47]). It is known that this approach is not applicable to nonlinear resistive networks because Δ–*Y* and *Y*–Δ transformations generally do not exist for nonlinear networks. (This was observed by Millar [40] for the first time.) Certain networks where these transformations can be performed were presented by Chua [14]. More recently, Boyd and Chua [6, 7] clarified the reasons behind this difficulty from the point of view of a Volterra series. As a conclusion, the set of nonlinear resistive 1-ports can be decomposed into equivalence classes, but, no reasonably large class of equivalent 1-ports exists. More general studies of this subject are based on the well-known substitution theorem, which can be extended to a certain class of nonlinear networks (see [26], [29]). Some results applicable to 1-ports can be generalized to linear resistive *n*-ports ("extraction of independent sources"), but this point of view is not suitable for nonlinear resistive *n*-ports.

Better understanding of nonlinear resistive *n*-ports and the problem of equivalence cannot be based on the "operational" approach mentioned earlier. Instead, a geometric approach that was developed by

Brayton and Moser [9] is more useful. These authors (see also [8]) characterize a resistive *n*-port in a generic manner by *n* independent relations between the 2*n* port variables, *n*-port currents $i_1, \ldots, i_n$ and *n*-port voltages $v_1, \ldots, v_n$. Geometrically, this means that in the 2*n*-dimensional space of port variables the external behavior of a resistive *n*-port can be represented generically by an *n*-dimensional surface. The classical approach formulates a system of equations

$$\begin{gathered} y_1 - f_1(x_1, \ldots, x_n) = 0 \\ \vdots \\ y_n - f_n(x_1, \ldots, x_n) = 0 \end{gathered} \tag{4.67}$$

where *x*'s and *y*'s are the port variables. The zero set of equations (4.67) corresponds to the *n*-dimensional surface. Therefore, two *n*-ports are called equivalent if they are different parameterizations of the same surface. As an application of this point of view, Brayton and Moser [9] demonstrated that a 2-port consisting of a *Y*-circuit and a circuit consisting of a Δ-circuit cannot be equivalent, in general. For example, they proved by means of Legendre transformations that a *Y*-circuit with two ohmic resistors and a third resistor can be equivalent to a Δ-circuit if and only if the third resistor is also linear. Therefore, the operational approach is not a very useful concept for nonlinear *n*-ports.

The subject of synthesizing a prescribed input–output behavior of nonlinear resistive *n*-ports is closely related to the problem of the equivalence. Several results were published in this area using ideal diodes, concave and convex resistors, dc voltage and current sources, ideal op amps, and controlled sources. Therefore, we give a short review of some results. We do not consider here the synthesis of resistive *n*-ports.

Although the synthesis of nonlinear resistive *n*-ports was of interest to many circuit designers since the beginning of this century, the first systematic studies of this subject were published by Chua [13], [14]. Chua's synthesis approach is based on the introduction of new linear 2-ports (*R*-rotator, *R*-reflector, and scalors) as well as their electronic realizations. Now, curves in the *i*–*v* space of port current *i* and port voltage *v* that characterize a (nonlinear) resistive 1-port can be reflected and scaled in a certain manner. Chua suggested that a prescribed behavior of an active or passive nonlinear resistive 1-port can be reduced essentially to the realization of passive *i*–*v* curves. Piecewise-linear approximations of characteristics of different types of diodes, as well as the previously mentioned 2-ports, are used to realize a piecewise-linear approximation of any prescribed passive *i*-*v* curve. In a succeeding article, Chua [15] discussed a unified procedure to synthesize a nonlinear dc circuit mode that represents a prescribed family of input and output curves of any strongly passive 3-terminal device (e.g, transistor). It was assumed that the desired curves are piecewise-linear. Since then, this research area has grown very rapidly and piecewise-linear synthesis and modeling has become an essential tool in the simulation of nonlinear circuits. (see [19], [35], [37] for further references.)

4.5 Equivalence of Lumped *n*-Port Networks

In this section, we consider more general *n*-ports that can be used in for device modeling (see [16]). Although many different *lumped* multiterminal and multiport networks are used, a decomposition into two mutually exclusive classes is possible: *algebraic* and *dynamic* multiterminal and multiport networks. Adopting the definition of Chua [16], an (n + 1)-terminal or *n*-port network is called an *algebraic element* if and only if its constitutive relations can be expressed symbolically by algebraic relationships involving at most two dynamically independent variables for each port. In the case of a 1-port, a so-called memristor is described by flux and charge, a resistor by voltage and current, a inductor by flux and current, and a capacitor by voltage and charge. An element is called a *dynamic element* if and only if it is not an algebraic element.

Despite the fact that the class of all dynamic elements is much larger than that of algebraic ones, the following theorem of Chua [16] suggests that resistive multiports are essential for dynamic elements, too.

Theorem 7: *Every lumped* $(n + 1)$*-terminal or* n*-port element can be synthesized using only a finite number* m *of linear 2-terminal capacitors (or inductors) and one (generally nonlinear)* $(n + m)$*-port resistor with* n *accessible ports and* m *ports for the capacitors.*

Theorem 7 demonstrates that any n-port made of lumped multiterminal and/or multiport elements is **equivalent** to a multiterminal network where all of its nonlinear elements are **memoryless.** This fact offers a possibility to classify $(n+1)$-terminal and n-port elements in an operational manner.

The proof of this theorem provides the answer of a fundamental question: what constitutes a *minimal* **set** of network elements from which **all** lumped elements can be synthesized?

Theorem 8: *The following set* $\mathcal{M}$ *of network elements constitutes the minimal basic building blocks in the sense that any lumped multiterminal or multiport element described by a continuous constitutive relation on any closed and bounded set can be synthesized using only a finite number of elements of* $\mathcal{M}$*, and that this statement is false if even one element is deleted from* $\mathcal{M}$:

1. *Linear 2-terminal capacitors (or inductors)*
2. *Nonlinear 2-terminal resistors*
3. *Linear 2-port current-controlled voltage sources* (CCVS) *defined by* $v_1 = 0$ *and* $v_2 = ki_1$
4. *Linear 2-port current-controlled current sources* (CCCS) *defined by* $i_1 = 0$ *and* $i_2 = kv_1$

The proof of Theorem 8 (see [16]), is based on a remarkable theorem of Kolmogoroff, which asserts that a continuous function $\mathbf{f}: \mathbb{R}^n \to \mathbb{R}$ can always be decomposed over the unit cube of $\mathbb{R}^n$ into a certain sum of functions of a *single* variable. Although the proof of Theorem 8 is constructive, it is mainly of theoretical interest because the number of controlled sources needed in the realization is often excessive.

4.6 Equivalence between Nonlinear Dynamic Circuits

As already mentioned in Section 4.1, a set of networks can be decomposed into classes of equivalent networks by some type of equivalence relation. Such equivalence relations are introduced in a direct manner with respect to the descriptive equations, using a transformation group or classifying the behavior of the solution of the descriptive equations. In the last three sections, several useful ideas for defining equivalence relations were discussed that can be suitable for circuit theory. In this section, equivalent dynamic circuits are discussed in more detail. It should be emphasized again that equivalence has a different meaning depending of the applied equivalence relation.

As the so-called state-space equations in network and system theory arose in the early 1960s, a first type of equivalence was defined because various networks can be described by the same state-space equations that induced an equivalence relation. (For further references, see [46].) Although this approach is interesting, in some cases different choices of variables for describing nonlinear networks exist that need not lead to equivalent state-space equations (see [17]). In other words, the transformations of coordinates are not well conditioned. This approach was applied also to nonlinear input–output systems.

A study of equivalence of a subclass of nonlinear input–output networks was presented by Verma [45] and Varaiya and Verma [44]. These authors discussed nonlinear reciprocal networks that can be formulated by a so-called mixed potential function. This approach was developed by Brayton and Moser [9]. If $\mathbf{x} \in \mathbb{R}^n$ is the state-space vector, $\mathbf{u} \in \mathbb{R}^m$ the input vector, and $\mathbf{e} \in \mathbb{R}^m$ is the output vector, then the state-space equations can be generated by a matrix-valued function $\mathbf{A}(\mathbf{x}): \mathbb{R}^n \to \mathbb{R}^{n \times n}$ and a real-valued function $\mathbf{P}: \mathbb{R}^n \times \mathbb{R}^m \to \mathbb{R}$

$$\mathbf{A}(\mathbf{x})\frac{d\mathbf{x}}{dt} = -\frac{\partial P}{\partial \mathbf{x}}(\mathbf{x}, \mathbf{u}) \tag{4.68}$$

$$\mathbf{e} = \frac{\partial P}{\partial \mathbf{u}}(\mathbf{x}, \mathbf{u}) \tag{4.69}$$

For two such networks $N_1 = \{\mathbf{A}_1, P_1\}$ and $N_2 = \{\mathbf{A}_2, P_2\}$, Varaiya and Verma defined the following equivalence.

Definition 4. Networks N_1 and N_2

$$\mathbf{A}_1(\mathbf{x})\frac{d\mathbf{x}}{dt} = -\frac{\partial P_1}{\partial \mathbf{x}}(\mathbf{x}, \mathbf{u}) \tag{4.70}$$

$$\mathbf{e}_1 = \frac{\partial P_1}{\partial \mathbf{u}}(\mathbf{x}, \mathbf{u}) \tag{4.71}$$

and

$$\mathbf{A}_2(\mathbf{y})\frac{d\mathbf{y}}{dt} = -\frac{\partial P_2}{\partial \mathbf{y}}(\mathbf{y}, \mathbf{u}) \tag{4.72}$$

$$\mathbf{e}_2 = \frac{\partial P_2}{\partial \mathbf{u}}(\mathbf{y}, \mathbf{u}) \tag{4.73}$$

are equivalent if there exists a diffeomorphism $\mathbf{y} = \phi(\mathbf{x})$, such that for all $\mathbf{x}_0 \in \mathbb{R}^n$, all input functions $\mathbf{u}$, and all $t \geq 0$:

1. $\phi[\xi(t, \xi_0, \upsilon)] = \psi(t, \mathbf{j}(\xi_0), \upsilon)$
2. $\mathbf{e}_1(t, \mathbf{x}_0, \mathbf{u}) = \mathbf{e}_2(t, \phi(\mathbf{x}_0), \mathbf{u})$

The diffeomorphism ϕ is called the equivalance map.

Thus, two networks are equivalent if their external behavior is identical, i.e., if for the same input and corresponding states they yield the same output. It is clear that this definition yields an equivalence relation on the set of all dynamical networks under consideration. In their paper, Varaiya and Verma showed that, under an additional assumption of controllability, the diffeomorphism ϕ establishes an isometry between the manifold with the (local) pseudo-Riemannian metric $(d\mathbf{x}, d\mathbf{x}) := \langle d\mathbf{x}, \mathrm{A}_1 d\mathbf{x}\rangle$ and the manifold with the (local) pseudo-Riemannian metric $(d\mathbf{y}, d\mathbf{y}) := \langle d\mathbf{y}, \mathrm{A}_2 d\mathbf{y}\rangle$ in many interesting cases of reciprocal nonlinear networks. This statement has an interesting interpretation in the network context. It can be proven that ϕ must relate the reactive parts of the networks N_1 and N_2 in such a way that, if N_1 is in the state $\mathbf{x}$ and N_2 is in the state $\mathbf{y} = \phi(\mathbf{x})$, and if the input $\mathbf{u}$ is applied, then

$$\left\langle \frac{d\mathbf{i}}{dt}, \mathbf{L}(\mathbf{i})\frac{d\mathbf{i}}{dt}\right\rangle - \left\langle \frac{d\mathbf{v}}{dt}, \mathbf{C}(\mathbf{v})\frac{d\mathbf{v}}{dt}\right\rangle = \left\langle \frac{d\tilde{\mathbf{i}}}{dt}, \tilde{\mathbf{L}}(\tilde{\mathbf{i}})\frac{d\tilde{\mathbf{i}}}{dt}\right\rangle - \left\langle \frac{d\tilde{\mathbf{v}}}{dt}, \tilde{\mathbf{C}}(\tilde{\mathbf{v}})\frac{d\tilde{\mathbf{v}}}{dt}\right\rangle \tag{4.74}$$

The concept of equivalence defined in a certain subset of nonlinear dynamic networks with input and output terminals given by Varaiya and Verma is based on diffeomorphic coordinate transformations (the transformation group of diffeomorphisms). Unfortunately, the authors presented no ideas about the kind of "coarse graining" produced in the set of networks by their equivalence relation. However, a comparison to C^k conjugacy or C^k equivalence of vector fields in Section 4.1 implies that input–output equivalence leads to a "fine" decomposition in the set of these networks. To classify the main features of the dynamics of networks, the concept of topological equivalence (the transformation group of homeomorphisms) is useful. On the other hand, in the case of networks with nonhyperbolic fixed points, the group of diffeomorphisms is needed to distinguish the interesting features. An interesting application of C^1 equivalence of vector fields is given by Chua [18]. To compare nonlinear networks that generate chaotic signals, Chua applied the concept of equivalence relation and concluded that the class of networks and systems that are C^1 equivalent to Chua's circuit (Figure 4.5) is relatively small. The nonlinearity in this network is described by a piecewise linear *i*–*v* characteristic. (See [39] for further details.) The equations describing the circuit are

$$\frac{dv_{C_1}}{dt} = \frac{1}{C_1}\left[G\left(v_{C_2} - v_{C_1}\right) - f\left(v_{C_1}\right)\right] \tag{4.75}$$

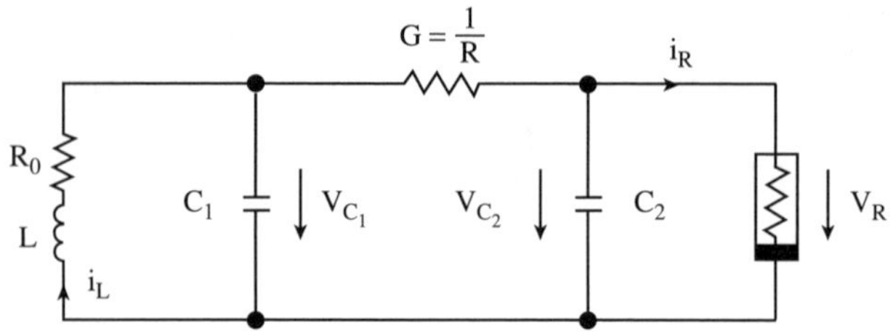

FIGURE 4.5 Modified Chua's circuit.

$$\frac{dv_{C_2}}{dt} = \frac{1}{C_2}\left(G\left(v_{C_1} - v_{C_2}\right) + i_L\right) \tag{4.76}$$

$$\frac{di_L}{dt} = \frac{I}{L}\left(v_{C_2} - R_0 i_L\right) \tag{4.77}$$

where $R_0 = 0$ and the piecewise linear function is defined by

$$f\left(v_{C_1}\right) := G_b v_{C_1} + \frac{1}{2}(G_a - G_b)\left(\left|v_{C_1} + E\right| - \left|v_{C_1} - E\right|\right) \tag{4.78}$$

Chua's extended approach to study the set of the piecewise linear networks that include Chua's circuit introduces the concept of global unfoldings. This concept can be considered as an analogy to the theory of "local unfoldings" of nonhyperbolic systems in a small neighborhood of singularities [3], [30]. Heuristically, a minimum number of parameters in a given nonhyperbolic system is introduced, and, as the parameters are varied "any other system" near the nonhyperbolic system is obtained. Chua demonstrated that Chua's circuit with arbitrary $(R_0 \neq 0)$ can be considered as an "unfolding" of the original circuit. Furthermore, he proved that a class of networks that can be described without loss of generality by

$$\dot{\mathbf{x}} = \mathbf{A}\mathbf{x} + \mathbf{b}, \quad x_1 \le -1 \tag{4.79}$$

$$= \mathbf{A}_0\mathbf{x}, \quad -1 \le x_1 \le 1 \tag{4.80}$$

$$= \mathbf{A}\mathbf{x} + \mathbf{b}, \quad x_1 \ge 1 \tag{4.81}$$

is equivalent to the unfolded Chua's circuit if certain conditions are satisfied. In the associated parameter space, these conditions define a set of measure zero. The proof of this theorem as well as some applications are included in [18].

The ideas of normal forms presented in Section 4.2 can be applied to nonlinear networks with hyperbolic and nonhyperbolic fixed points. A similar theory of normal forms of maps can be used to study limit cycles, but this subject is beyond our scope. (See [3] for further details.) In any case the vector field has to be reduced to lower dimensions and that can be achieved by the application of the so-called center manifold theorem. Altman [2] illustrated this approach by calculating the center manifold and a normal form of Chua's circuit in a tutorial style. To perform the analytical computations the piecewise nonlinearity (4.78) is replaced by a cubic function $f(x) = c_0x + c_1x^3$. Based on this normal form, Altman studied bifurcations of Chua's circuit.

Another application of normal forms in nonlinear dynamical networks is discussed by Keidies and Mathis [36]. In this approach, nonlinear dynamical networks with constant sources are considered and are described by nonlinear differential equations in state-space form:

$$\dot{\mathbf{x}} = \mathbf{f}(\mathbf{x}), \quad \mathbf{f}: \mathbb{R}^n \to \mathbb{R}^n \tag{4.82}$$

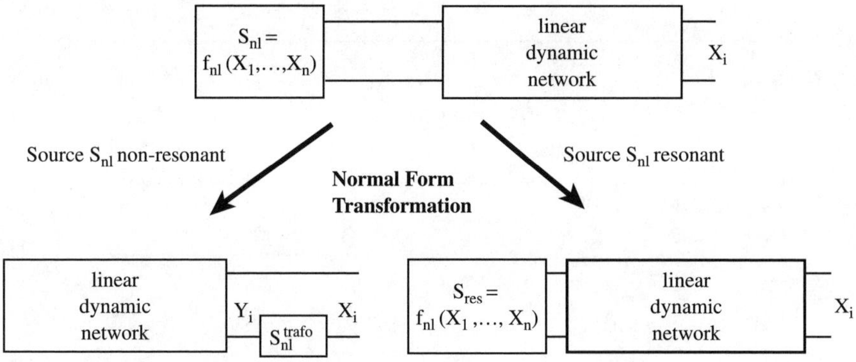

FIGURE 4.6 Decomposition of nonlinear dynamic networks.

where all nonlinear reactances are replaced by linear reactances, nonlinear resistors, and linear controlled sources. The nonlinearities are interpreted as nonlinear controlled sources. The network is decomposed into a linear part that consists of linear reactances and resistive elements, and the nonlinear sources that are used as input sources (Figure 4.6). The network is described by the vector of state-space variables **x**. Now, normal form theorems are used to transform the nonlinear sources to the input. In other words, if the RHS **f** of (4.82) is decomposed into a linear and a nonlinear part, $f(\mathbf{x}) = \mathbf{A}\mathbf{x} + \tilde{\mathbf{f}}(\mathbf{x})$, where $\tilde{\mathbf{f}}$ corresponds the nonlinear sources, the system can be decomposed into two equations:

$$\dot{\mathbf{y}} = \mathbf{A}\mathbf{y} \tag{4.83}$$

$$\mathbf{x} = \mathbf{y} + \mathbf{F}(\mathbf{y}) \tag{4.84}$$

We now have to define the nonresonant and resonant terms of vector fields that depend on the eigenvalues of the linear part **A** of **f** and the degrees of the polynomial nonlinearities. Under certain conditions, a finite recursive process exists, such that all nonlinear sources can be transformed to the input of the linear part of a network. In these cases, the networks are described by (4.82) and (4.83). In other cases, a number of new sources are generated during the recursive process that cannot transform sources to the input. This effect is shown in Figure 4.7(a) and (b). It should be mentioned that this idea is related

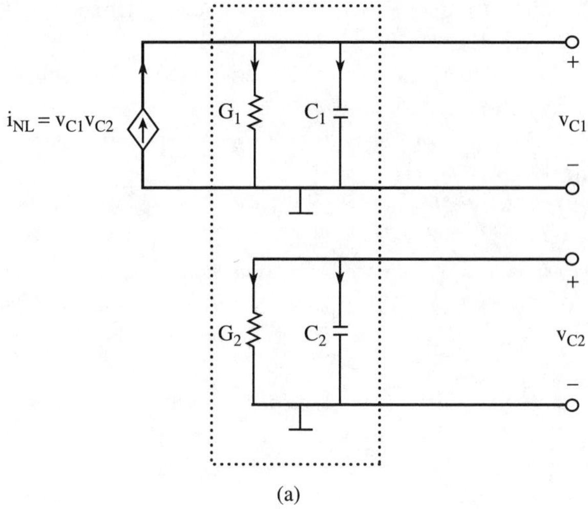

FIGURE 4.7 Decomposition of a simple nonlinear network.

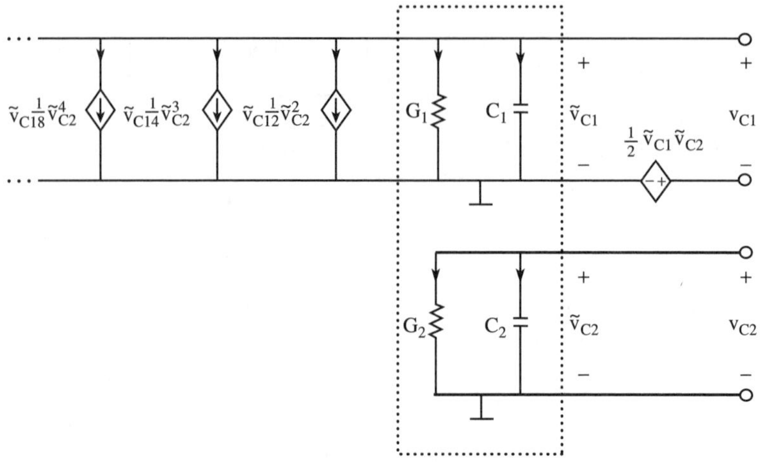

FIGURE 4.7 (continued).

in certain sense to the so-called exact linearization that is studied in the theory of nonlinear control systems (see [34]). Therefore, this application of normal form theorems can be interpreted as a kind of extraction of nonlinear controlled sources from a nonlinear dynamic network.

References

[1] V. I. Arnold, *Geometrical Methods in the Theory of Ordinary Differential Equations,* New York: Springer-Verlag, 1983.

[2] E. J. Altman, "Bifurcation analysis of Chua's circuit with application for low-level visual sensing," in *Chua's Circuit: A Paradigm for Chaos,* R. N. Madan, Ed., Singapore: World Scientific, 1993.

[3] D. K. Arrowsmith and C. M. Place, *An Introduction to Dynamical Systems,* Cambridge: Cambridge Univ., 1993.

[4] M. Ashkenazi and S.-N. Chow, "Normal forms near critical points for differential equations and maps," *IEEE Trans. Circuits Syst.,* vol. 35, pp. 850–862, 1988.

[5] G. W. Bluman and S. Kumei, *Symmetries and Differential Equations,* New York: Springer-Verlag, 1989.

[6] S. Boyd and L. O. Chua, "Uniqueness of a Basic Nonlinear Structure," *IEEE Trans. Circuits Syst.,* vol. CAS-30, pp. 648–651, 1983.

[7] S. Boyd and L. O. Chua, "Uniqueness of circuits and systems containing one nonlinearity," *IEEE Trans. Automat. Control,* vol. AC-30, pp. 674–681, 1985.

[8] R. K. Brayton, "Nonlinear reciprocal networks," in *Mathematical Aspects of Electrical Network Analysis,* Providence, RI: AMS, 1971.

[9] R. K. Brayton and J. K. Moser, "Nonlinear networks, I, II," *Quart. Appl. Math.,* vol. 23, pp. 1–33, 81–104, 1964.

[10] A. D. Bruno, *Local Methods in Nonlinear Differential Equations,* New York: Springer-Verlag, 1989.

[11] J. Carr, *Applications of Center Manifold Theorem,* New York: Springer-Verlag, 1981.

[12] L. O. Chua, "Δ–*Y* and *Y*–Δ transformation for nonlinear networks," *Proc. IEEE,* vol. 59, pp. 417–419, 1971.

[13] L. O. Chua, "The rotator — a new network element," *Proc IEEE.* vol. 55, pp. 1566–1577, 1967.

[14] L. O. Chua, "Synthesis of new nonlinear network elements" *Proc. IEEE,* vol. 56 pp. 1325–1340, 1968.

[15] L. O. Chua, "Modeling of three terminal devices: A black box approach," *IEEE Trans. Circuit Theory,* vol. CT-19, pp. 555–562, 1972.

[16] L. O. Chua, "Device modeling via basic nonlinear circuit elements," *IEEE Trans. Circuits Syst.* vol. CAS-27, pp. 1014–1044, 1980

[17] L. O. Chua "Dynamical nonlinear networks: State of the Art," *IEEE Trans. Circuits Syst.*, vol. CAS-27, pp. 1059–1087, 1980.
[18] L. O. Chua, "Global unfolding of Chua's circuit" *IEICE Trans. Fundam.*, vol. E76-A, pp. 704–734, 1993.
[19] L. O. Chua and A. C. Deng, "Canonical piecewise linear representation," *IEEE Trans. Circuits Syst.*, vol. 33, pp. 101–111, 1988.
[20] L. O. Chua, C. A. Desoer, and E. S. Kukh, *Linear and Nonlinear Circuits,* New York: McGraw-Hill, 1987.
[21] L. O. Chua and H. Kokubo, "Normal forms for nonlinear vector fields — Part I: Applications," *IEEE Trans. Circuits Syst.*,vol. 36, pp. 51–70, 1989.
[22] L. O. Chua and H. Kokubo, "Normal forms for nonlinear vector fields — Part II: Theory and algorithm," *IEEE Trans. Circuits Syst.*, vol. 35, pp. 863–880, 1988.
[23] L. O. Chua and Y.-F. Lam, "A theory of algebraic *n*-ports," *IEEE Trans. Circuit Theory,* vol. CT-20, pp. 370–382, 1973.
[24] L. O. Chua and H. Oka, "Normal forms of constrained nonlinear differential equations — Part I: Theory," *IEEE Trans. Circuits Syst.*, vol. 35, pp. 881–901, 1988.
[25] L. O. Chua and H. Oka, "Normal forms of constrained nonlinear differential equations — Part II: Bifurcation, *IEEE Circuits Syst.*, vol. 36, pp. 71–88, 1989.
[26] C. A. Desoer and E. S. Kuh, *Basic Circuit Theory,* New York: McGraw-Hill, 1969.
[27] E. A. Desloge, "Suppression and restoration of constants in physical equations," *Amer. J. Phys.*, vol. 52, pp. 312–316, 1984.
[28] J. Guckenheimer and P. Holmes, *Nonlinear Oscillations, Dynamical Systems, and Bifurcations of Vector Fields,* New York: Springer-Verlag, 1990.
[29] J. Haase, "On generalizations and applications of the substitution theorem," in *Proc. ECCTD '85*, Praha, Sep. 2–6, 1985, pp. 220–223.
[30] S. Hale and H. Kocak, *Dynamics and Bifurcations,* New York: Springer-Verlag, 1991.
[31] B. D. Hassard, N. D. Kazarinoff, and Y.-H. Wan, *Theory and Applications of the Hopf Bifurcation,* Cambridge: Cambridge Univ., 1980.
[32] B. D. Hassard and Y.-H. Wan, "Bifurcation formulae derived from center manifold theorem," *J. Math. Anal. Applicat.*, vol. 63, pp. 297–312, 1978.
[33] R. A. Horn and C. R. Johnson, *Matrix Analysis,* Cambridge: Cambridge Univ., 1992.
[34] A. Isidori, *Nonlinear Control Systems,* Berlin: Springer-Verlag, 1989.
[35] J. Jess, "Piecewise Linear Models for Nonlinear Dynamic Systems," *Frequenz,* vol. 42, pp. 71–78, 1988.
[36] C. Keidies and W. Mathis, "Applications of normal forms to the analysis of nonlinear circuits," in *Proc. 1993 Int. Symp. Nonlinear Theory, Applicat.*, Hawaii, Dec. 1993.
[37] T. A. M. Kevenaar and D. M. W. Leenaerts, "A comparison of piecewise-linear model descriptions," *IEEE Trans. Circuits Syst. I,* vol. 39, pp. 996–1004, 1992.
[38] W. Mathis and L. O. Chua "Applications of dimensional analysis to network theory," *Proc. ECCTD '91*, Copenhagen, Sep. 4–6, 1991.
[39] R. N. Madan, Ed., *Chua's Circuit: A Paradigm for Chaos,* Singapore: World Scientific, 1993.
[40] W. Millar, "The nonlinear resistive 3-pole: Some general concepts," in *Proc. Symp. Nonlinear Circuit Anal.*, Polytech. Instit., Brooklyn, NY: Interscience, 1957.
[41] W. L. Miranker, *Numerical Methods for Stiff Equations and Singular Perturbation Problems,* Dordrecht: D. Reidel, 1981.
[42] L. V. Ovsiannikov, *Group Analysis of Differential Equations,* New York: Academic Press, 1982.
[43] D. R. Smith, *Singular-Perturbation Theory,* Cambridge: Cambridge University, 1985.
[44] P. P. Varaiya and J. P. Verma, "Equivalent nonlinear reciprocal networks," *IEEE Trans. Circuit Theory,* vol. CT-18, pp. 214–217, 1971.
[45] J. P. Verma, "Equivalence of nonlinear networks," Ph.D. dissertation, University of California, Berkeley, 1969.
[46] A. N. Willson, Jr., *Nonlinear Networks: Theory and Analysis,* New York: IEEE, 1975.
[47] W. Mathis and R. Pauli, "Networks Theorems," in *Wiley Encyclopedia of Electrical and Electronics Engineering,* vol. 14, New York: John Wiley & Sons, 1999.

5

Piecewise-Linear Circuits and Piecewise-Linear Analysis

J. Vandewalle
Katholieke Universiteit

L. Vandenberghe
University of California, Los Angeles

5.1 Introduction and Motivation

In this chapter, we present a comprehensive description of the use of piecewise-linear methods in modeling, analysis, and structural properties of nonlinear circuits. The main advantages of piecewise linear circuits are fourfold. (1) Piecewise-linear circuits are the easiest in the class of nonlinear circuits to analyze exactly, because many methods for linear circuits can still be used. (2) The piecewise-linear approximation is an adequate approximation for most applications. Moreover, certain op amp, operational transconductance amplifier, diode and switch circuits are essentially piecewise linear. (3) Quite a number of methods exist to analyze piecewise-linear circuits. (4) Last, but not least, piecewise-linear circuits exhibit most of the phenomena of nonlinear circuits while still being manageable. Hence, PWL circuits provide unique insight in nonlinear circuits.

The section consists of six parts. First, the piecewise-linear models will be presented and interrelated. A complete hierarchy of models and representations of models is presented. Rather than proving many relations, simple examples are given. Second, the piecewise-linear models for several important electronic components are presented. Third, since many PWL properties are preserved by interconnection, a short discussion on the structural properties of piecewise-linear circuits is given in Section 5.4. Fourth, analysis methods of PWL circuits are presented, ranging from the Katzenelson algorithm to the linear complementarity methods and the homotopy methods. Fifth, we discuss PWL dynamic circuits, such as the famous Chua circuit, which produces chaos. Finally, in Section 5.7, efficient computer-aided analysis of PWL circuits and the hierarchical mixed-mode PWL analysis are described. A comprehensive reference list is included. For the synthesis of PWL circuits, we refer to Chapter 1.

In order to situate these subjects in the general framework of nonlinear circuits, it is instructive to interrelate the PWL circuit analysis methods (Figure 5.1). In the horizontal direction of the diagrams, one does the PWL approximation of the dc analysis from left to right. In the vertical direction, we show the conversion from a circuit to a set of equations by network equation formulation and the conversion

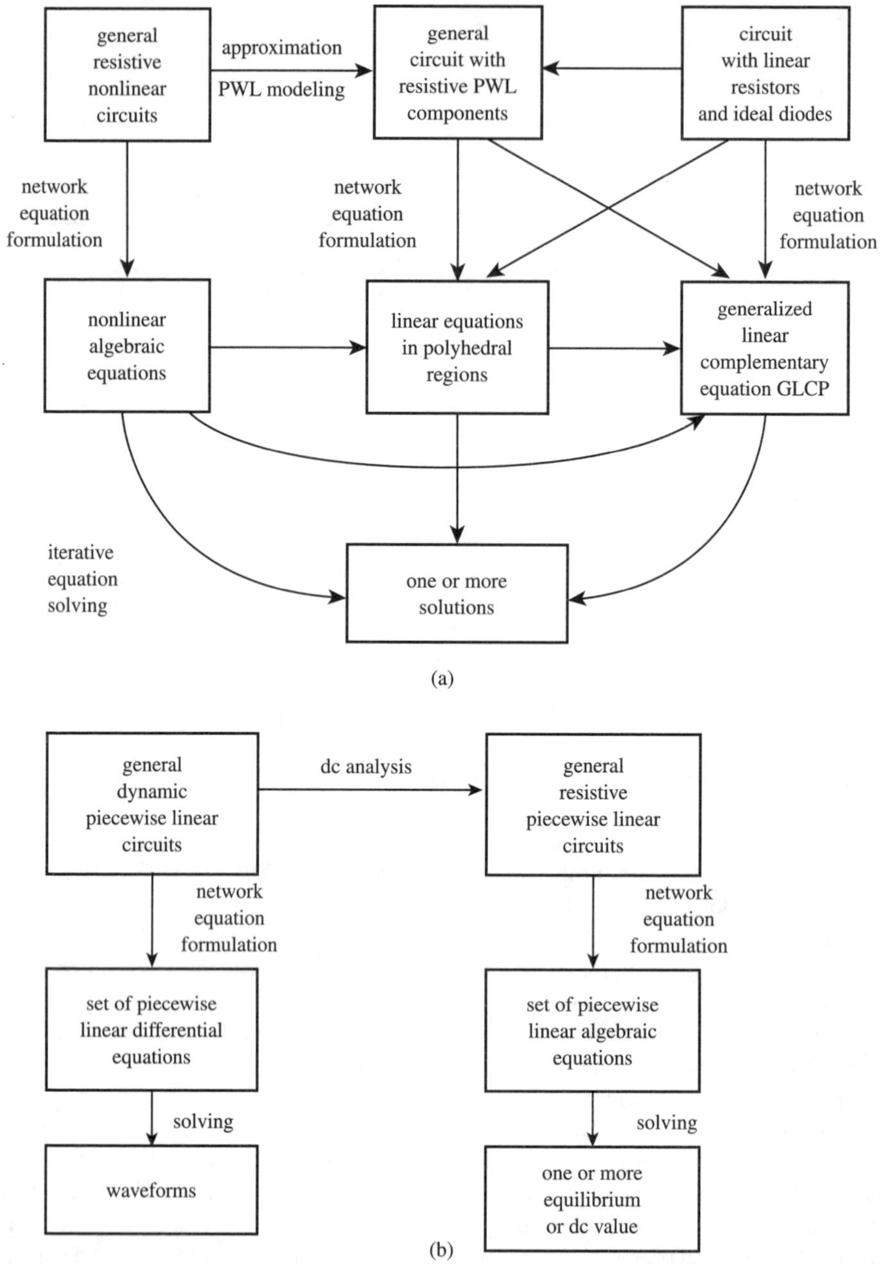

FIGURE 5.1 Interrelation of PWL circuit analysis methods: (a) resistive and (b) dynamic nonlinear circuits.

from equations to solutions (waveforms or dc values) by solution methods. The specific methods and names used in the figure are described in detail in the different parts.

5.2 Hierarchy of Piecewise-Linear Models and Their Representations

In the past 25 years, much progress has been achieved in the representations of piecewise-linear resistive multiports and their relationships (see references). From a practical point of view, a clear trade-off exists

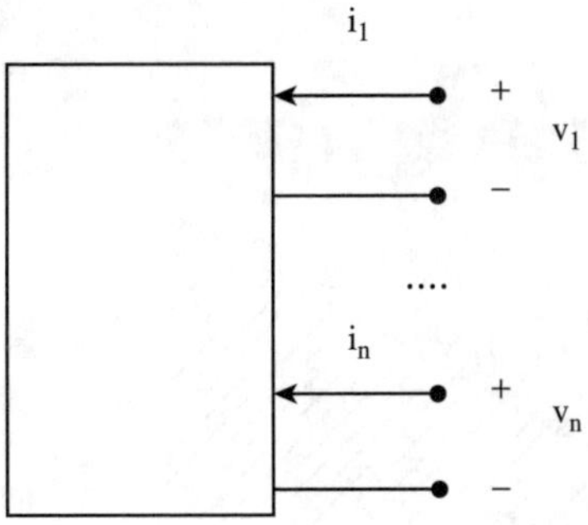

FIGURE 5.2 Resistive n-port.

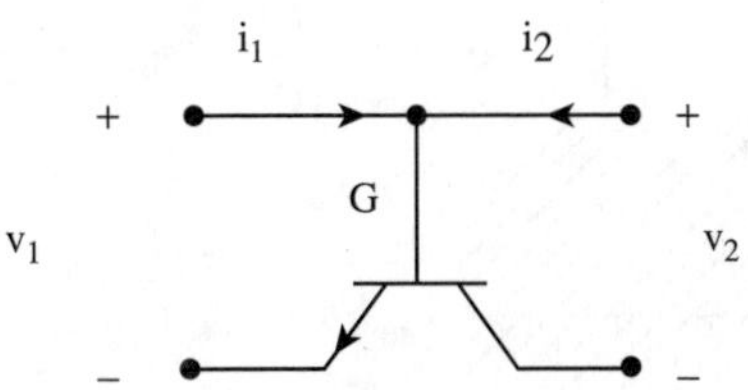

FIGURE 5.3 2-port configuration of a bipolar transistor.

between the efficiency of a representation in terms of the number of parameters and the ease of evaluation (explicit versus implicit models) on the one hand and the generality or accuracy on the other hand. Here, we go from the easiest and most efficient to the most general representations.

We define here a **resistive multiport** (Figure 5.2) as an n-port whose port variables (the vector of port currents $i = [i_1 \dots i_n]^T$ and the vector of port voltages $v = [v_1 \dots v_n]^T$) are related by m algebraic equations called **constitutive equations**

$$\varphi(i, v) = 0 \tag{5.1}$$

where $i, v \in \mathbb{R}^n$ and $\phi(.,.)$ maps $\mathbb{R}^{2n}$ into $\mathbb{R}^m$.

For example, for a bipolar transistor (Figure 5.3), one obtains the explicit form $i_1 = f_1(v_1, v_2)$ and $i_2 = f_2(v_1, v_2)$ and $i = [i_1, i_2]^T$ and $v = [v_1, v_2]^T$. These relations can be measured with a curve tracer as dc characteristic curves. Clearly, here $\phi(.,.)$ is a map from $\mathbb{R}^4 \rightarrow \mathbb{R}^2$ in the form

$$i_1 - f_1(v_1, v_2) = 0 \tag{5.2}$$

$$i_2 - f_2(v_1, v_2) = 0 \tag{5.3}$$

It is easy to see that a complete table of these relationships would require an excessive amount of computer storage already for a transistor. Hence, it is quite natural to describe a resistive n-port with a piecewise-linear map f over polyhedral regions P_k by

$$v = f(i) = a_k + B_k i, \quad i \in P_k, \quad k \in \{0, 1, \dots, 2^l - 1\} \tag{5.4}$$

where the Jacobian $B_k \in \mathbb{R}^{n \times n}$ and the offset vector $a_k \in \mathbb{R}^n$ are defined over the polyhedral region P_k, separated by hyperplanes $c_i^T x - d_i = 0$, $\mathrm{i} = 1, \dots, l$ and defined by

$$P_k = \left\{ x \in \mathbb{R}^n \middle| c_j^T x - d_j \geq 0, \ \mathrm{j} \in \mathrm{I}_k, c_j^T x - d_j \leq 0, j \notin I_k \right\} \tag{5.5}$$

where $k = \Sigma_{j \in I_k} 2^{j-1}$, $I_k \subseteq \{1, 2, \dots, l\}$ and $c_j \in \mathbb{R}^n$, $d_j \in \mathbb{R}^n$.

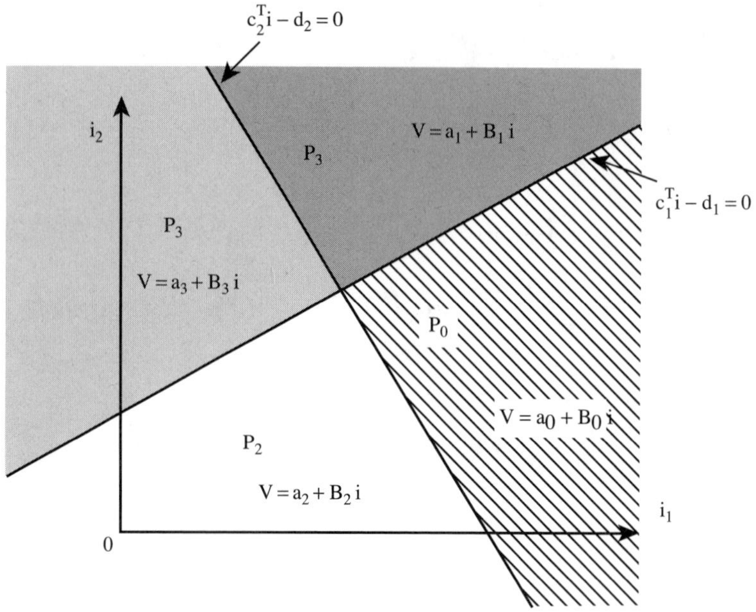

FIGURE 5.4 A PWL function defined in four polyhedral regions in $\mathbb{R}^n$ defined by $c_1^T i - d_1 \lessgtr 0$ and $c_2^T i \gtrless d_2 > 0$.

In other words, the hyperplanes $c_i^T x - d_i = 0,\ i = 1, \ldots, l$ separate the space $\mathbb{R}^n$ into 2^l polyhedral regions P_k (see Figure 5.4) where the constitutive equations are linear.

The computer storage requirements for this representation is still quite large, especially for large multiports. A more fundamental problem with this rather intuitive representation is that it is not necessarily continuous at the boundaries between two polyhedral regions. In fact, the continuity of the nonlinear map is usually desirable for physical reasons and also in order to avoid problems in the analysis.

The canonical PWL representation [6] is a very simple, attractive, and explicit description for a resistive multiport that solves both problems:

$$v = f(i) = a + Bi + \sum_{j=1}^{l} e_j \left| c_j^T i - d_j \right| \tag{5.6}$$

One can easily understand this equation by looking at the wedge form of the modulus map (see Figure 5.5). It has two linear regions: in the first $x \geq 0$ and $y = x$, while in the second $x \leq 0$ and $y = -x$. At the boundary the function is clearly continuous. Equation (5.6) is hence also continuous and is linear in each of the polyhedral regions P_k described by (5.5). If l modulus terms are in (5.6), there are 2^l polyhedral regions where the map (5.6) is linear. Because the map is represented canonically with $n + n^2 + l(n + 1)$ real parameters, this is a very compact and explicit representation.

Several examples of canonical PWL models for components are given in Section 5.3.

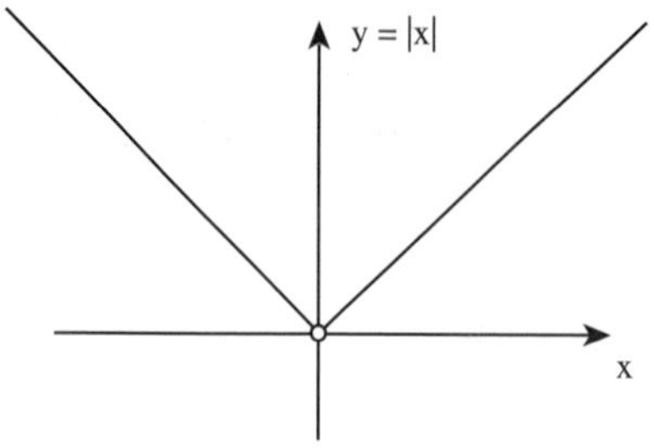

FIGURE 5.5 The absolute value function $y = |x|$.

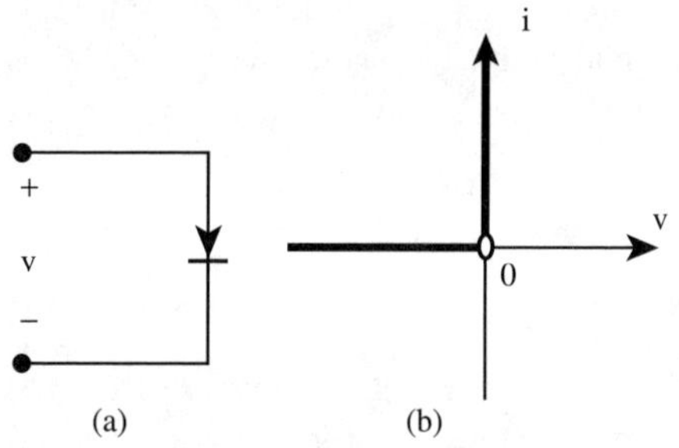

FIGURE 5.6 (a) The ideal diode and (b) the (i–v) relation of an ideal diode.

From Figure 5.5, it should be clear that the right and left derivative of $y = |x|$ at 0 are different, their difference being 2. Hence, the Jacobian J_+ and J_- of (5.6) will be different on the boundary between the two neighboring polyhedral regions where $(c_j i - d_j) \geq 0$ and $(c_j i - d_j) \leq 0$

$$J_+ - J_- = 2e_j c_j^T \tag{5.7}$$

Observe that this difference is a rank 1 matrix, which is also called a dyadic or outer vector product of e_j and c_j. Moreover, this difference is independent of the location of the independent variable i on the boundary. This important observation is made in [24], and is called the consistent variation property [10] and essentially says that the variation of the Jacobian of a canonical PWL representation is independent of the place where the hyperplane $c_j i - d_j = 0$ is crossed. Of course, this implies that the canonical PWL representation (5.6) is not the most general description for a continuous explicit PWL map. In [26] and [29] two more general representations, which include nested absolute values, are presented. These are too complicated for our discussion.

Clearly, the canonical PWL representation (5.6) is valid only for single-valued functions. It can clearly not be used for an important component: the ideal diode (Figure 5.6) characterized by the multivalued (i, v) relation. It can be presented analytically by introducing a real scalar parameter ρ [31].

$$i = \frac{1}{2}(\rho + |\rho|) \tag{5.8}$$

$$v = \frac{1}{2}(\rho - |\rho|) \tag{5.9}$$

This parametric description can easily be seen to correspond to Figure 5.6(b) because $i = \rho$ and $v = 0$ for $\rho \geq 0$, while $i = 0$ and $v = \rho$ when $\rho \leq 0$. Such a parametric description $i = f(\rho)$ and $v = g(\rho)$ with f and g PWL can be obtained for a whole class of unicursal curves (see[6]).

When we allow implicit representations between v and i for a multiport, we obtain an LCP (linear complementarity problem) model (5.10)–(5.12) with an interesting state space like form [55]:

$$v = Ai + Bu + f \tag{5.10}$$

$$s = Ci + Du + g \tag{5.11}$$

$$u \geq 0,\ s \geq 0,\quad u^T s = 0 \tag{5.12}$$

where $A \in \mathbb{R}^{n\times n}$, $B \in \mathbb{R}^{n\times l}$, $f \in \mathbb{R}^{n\times n}$, $c \in \mathbb{R}^{l\times n}$, $D \in \mathbb{R}^{l\times l}$ are the parameters that characterize the relationship between v and i. In the model, u and s are called the state vectors and we say that $u \geq 0$ when all its components are nonnegative. Clearly, (5.12) dictates that all components of u and s should be nonnegative and that, whenever a component u_j satisfies $u_j > 0$, then $s_j = 0$ and, vice versa, when $s_j > 0$, then $u_j = 0$.

This is called the linear complementarity property, which we have seen already in the ideal diode (5.8) and (5.9) where $i \geq 0, v \geq 0$ and $iv = 0$. Hence, an implicit or LCP model for the ideal diode (5.8) and (5.9) is

$$v = u \tag{5.13}$$

$$s = i \tag{5.14}$$

$$u \geq 0 \quad s \geq 0 \quad us = 0 \tag{5.15}$$

In order to understand that the general equations (5.10)–(5.12) describe a PWL relation such as (5.4)–(5.5) between i and v over polyhedral regions, one should observe first that $v = Ai + f$ is linear when $u = 0$ and $s = Ci + g \geq 0$. Hence, the relation is linear in the polyhedral region determined by $Ci + g \geq 0$. In general, one can consider 2^l possibilities for u and s according to

$$\left(u_j \geq 0 \text{ and } s_j = 0\right) \text{ or } \left(u_j = 0 \text{ and } s_j = 0\right), \quad \text{for } j = 1, 2, \ldots, l$$

Denote sets of indexes U and S for certain values of u and s satisfying (5.12)

$$U = \left\{ j \middle| u_j \geq 0 \text{ and } s_j = 0 \right\} \tag{5.16}$$

$$S = \left\{ j \middle| u_j = 0 \text{ and } s_j \geq 0 \right\} \tag{5.17}$$

then, clearly, U and S are complementary subsets of $\{1, 2, \ldots, l\}$ when for any j, u_j, and s_j cannot be both zero. Clearly, each of these 2^l possibilities corresponds to a polyhedral region P_U in $\mathbb{R}^n$, which can be determined from

$$u_j \geq 0, \quad (Ci + Du + g)_j = 0 \quad \text{for } j \in U \tag{5.18}$$

$$u_j = 0, \quad (Ci + Du + g)_j \geq 0 \quad \text{for } j \in S \tag{5.19}$$

The PWL map in region P_U is determined by solving the u_j for $j \in U$ from (5.18) and substituting these along with $u_j = 0$ for $j \in S$ into (5.10). This generates, of course, a map that is linear in the region P_U.

When (5.11) is replaced by the implicit equation

$$Es + Ci + Du + g\alpha = 0 \quad \alpha \geq 0$$

in (5.10)–(5.13), we call the problem a generalized linear complementarity problem (GLCP).

A nontrivial example of an implicit PWL relation (LCP model) is the hysteresis one port resistor (see Figure 5.7). Its equations are:

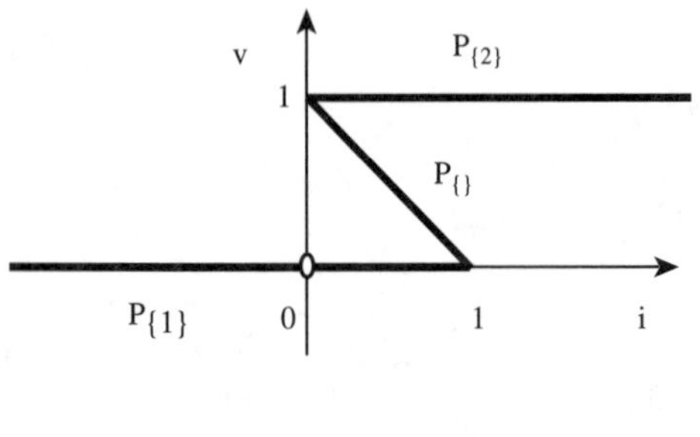

FIGURE 5.7 The hysteresis nonlinear resistor.

$$v = -i + \begin{bmatrix} -1 & 1 \end{bmatrix} \begin{bmatrix} u_1 \\ u_2 \end{bmatrix} + 1 \tag{5.20}$$

$$\begin{bmatrix} s_1 \\ s_2 \end{bmatrix} = \begin{bmatrix} -1 \\ 1 \end{bmatrix} i + \begin{bmatrix} -1 & 1 \\ 1 & -1 \end{bmatrix} \begin{bmatrix} u_1 \\ u_2 \end{bmatrix} + \begin{bmatrix} 1 \\ 0 \end{bmatrix} \tag{5.21}$$

$$s_1 \geq 0,\quad s_2 \geq 0,\quad u_1 \geq 0,\quad u_2 \geq 0,\quad u_1 s_1 + u_2 s_2 = 0 \tag{5.22}$$

In the first region P, we have

$$s_1 = -i + 1 \geq 0,\ s_2 = i \geq 0, \text{ and } v = -i + 1 \tag{5.23}$$

The region $P_{\{1,2\}}$, on the other hand, is empty because the following set of equations is contradictory:

$$s_1 = s_2 = 0, \qquad -i - u_1 + u_2 + 1 = 0, \qquad i + u_1 - u_2 = 0 \tag{5.24}$$

The region $P_{[1]}$ is

$$u_1 \geq 0, \qquad s_1 = -i - u_1 + 1 = 0, \qquad u_2 = 0, \qquad s_2 = i + u_1 \geq 0 \tag{5.25}$$

Hence, $u_1 = -i + 1$ and $s_2 = 1$ and $v = -i + i - 1 + 1 = 0$, while $i \leq 1$.

Finally, the region $P_{[2]}$ is

$$u_1 \geq 0, \qquad s_1 = -i + u_2 + 1 \geq 0, \quad u_2 \geq 0, \qquad s_2 = i - u_2 = 0$$

Hence,

$$u_2 = i \text{ and } s_1 = 1 \text{ and } v = -i + i + 1 = 1, \text{ while } i \geq 0 \tag{5.26}$$

It is now easy to show in general that the canonical PWL representation is a special case of the LCP model. Just choose $u_j \geq 0$ and $s_j \geq 0$ for all j as follows:

$$\left| c_j^T i - d_j \right| = \frac{1}{2}(u_j + s_j) \tag{5.27}$$

$$c_j^T i - d_j = \frac{1}{2}(u_j - s_j) \tag{5.28}$$

then, u and s are complementary vectors, i.e.,

$$u \geq 0 \qquad s \geq 0 \qquad u^T s = 0$$

Observe that the moduli in (5.6) can be eliminated with (5.27) to produce an equation of the form (5.10) and that (5.28) produces an equation of the form (5.11).

More generally, it has been proven [36] that the implicit model includes all explicit models. Because it also includes the parametric models, one obtains the general hierarchy of models as depicted in Figure 5.8.

A general remark should be made about all models that have been presented until now. Although the models have been given for resistive multiports where the voltages v at the ports are expressed in terms of the currents i, analogous equations can be given for the currents i in terms of the voltages, or hybrid

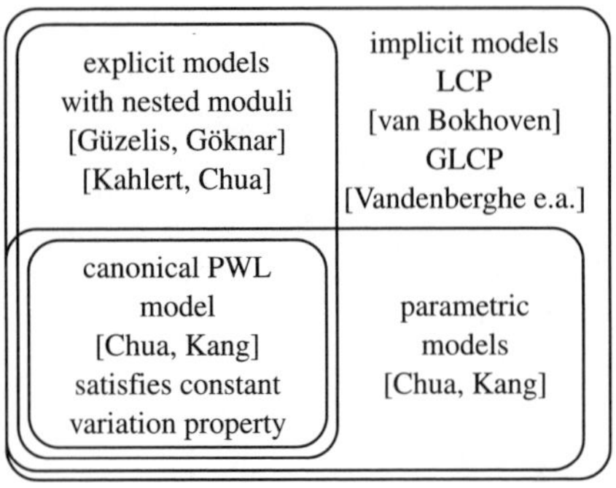

FIGURE 5.8 The interrelation of the PWL models.

variables. It can even be adapted for piecewise linear capacitors, inductors, or memristors, where the variables are, respectively, q, v for capacitors, φ, i for inductors, and q, φ for memristors.

5.3 Piecewise-Linear Models for Electronic Components

In order to simulate nonlinear networks with a circuit or network simulator, the nonlinear behavior of the components must be modeled first. During this modeling phase, properties of the component that are not considered important for the behavior of the system may be neglected. The nonlinear behavior is often important, therefore, nonlinear models have to be used. In typical simulators such as SPICE, nonlinear models often involve polynomials and transcendental functions for bipolar and MOS transistors. These consume a large part of the simulation time, so table lookup methods have been worked out. However, the table lookup methods need much storage for an accurate description of multiports and complex components.

The piecewise-linear models constitute an attractive alternative that is both efficient in memory use and in computation time. We discuss here the most important components. The derivation of a model usually requires two steps: first, the PWL approximation of constitutive equations, and second, the algebraic representation.

Two PWL models for an ideal diode (Figure 5.6) have been derived, that is, a parametric model (5.8) and (5.9) and an implicit model (5.13)–(5.15), while a canonical PWL model does not exist.

The piecewise-linear models for operational amplifiers (op amps) and operational transconductance amplifiers (OTA's) are also simple and frequently used. The piecewise-line approximation of op amps and OTA's of Figure 5.9 is quite accurate. It leads to the following representation for the op amp, which is in the linear region for $-E_{sat} \leq v_0 \leq E_{sat}$ with voltage amplification A_v and positive and negative saturation E_{sat} and $-E_{sat}$

$$v_0 = \frac{A_v}{2}\left(\left|v_i + \frac{E_{sat}}{A_v}\right| - \left|v_i - \frac{E_{sat}}{A_v}\right|\right) \tag{5.29}$$

$$i_- = i_+ = 0 \tag{5.30}$$

This is called the op amp finite-gain model. In each of the three regions, the op amp can be replaced by a linear circuit.

For the OTA, we have similarly in the linear region for $-I_{sat} \leq i_0 \leq I_{sat}$ with transconductance gain g_m and positive and negative saturation I_{sat} and $-I_{sat}$

$$i_0 = \frac{g_m}{2}\left(\left|v_i + \frac{I_{sat}}{g_m}\right| - \left|v_{i-} \frac{I_{sat}}{g_m}\right|\right) \tag{5.31}$$

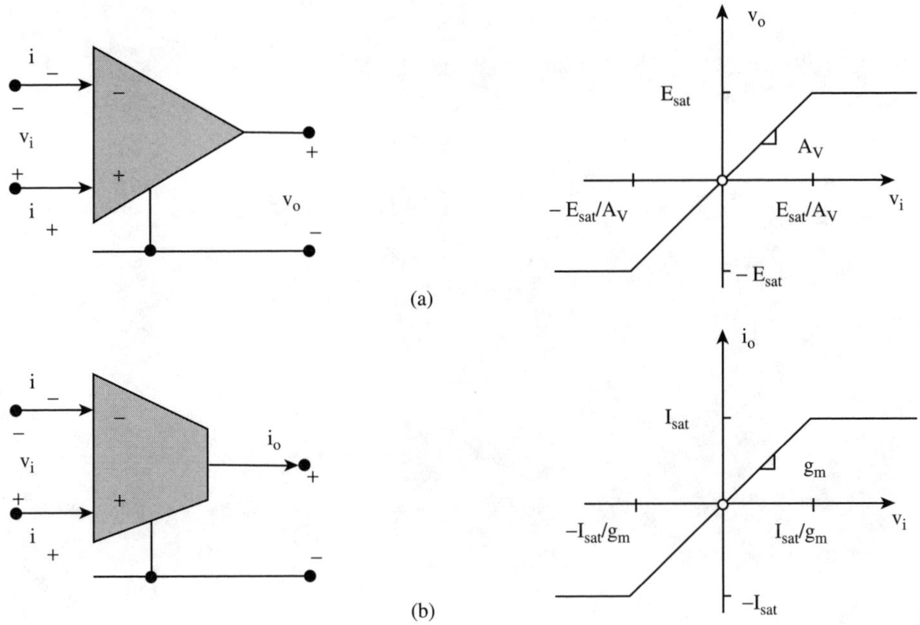

FIGURE 5.9 (a) Op amp and PWL model and (b) OTA and PWL model.

$$i_- = i_+ = 0 \tag{5.32}$$

Next, for a tunnel diode, one can perform a piecewise-linear approximation for the tunnel-diode characteristic as shown in Figure 5.10. It clearly has three regions with conductances g_1, g_2, and g_3. This PWL characteristic can be realized by three components [(Figure 5.10(b)] with conductances, voltage sources, and diodes. The three parameters G_0, G_1, and G_2 of Figure 5.10(b) must satisfy

$$\text{in Region 1:} \qquad G_0 = g_1 \tag{5.33}$$

$$\text{in Region 2:} \qquad G_0 + G_1 = g_2 \tag{5.34}$$

$$\text{in Region 3:} \qquad G_0 + G_1 + G_2 = g_3 \tag{5.35}$$

Thus, $G_0 = g_1$, $G_1 = -g_1 + g_2$, and $G_2 = -g_2 + g_3$. We can derive the canonical PWL representation as follows:

$$i = -\frac{1}{2}(G_1E_1 + G_2E_2) + \left(G_0 + \frac{1}{2}G_1 + \frac{1}{2}G_2\right)v + \frac{1}{2}G_1|v - E_1| + \frac{1}{2}G_2|v - E_2| \tag{5.36}$$

Next, we present a canonical piecewise-linear bipolar transistor model [12]. Assume a *npn* bipolar transistor is connected in the common base configuration with $v_1 = v_{BE}$, $v_2 = v_{BC}$, $i_1 = i_E$, and $i_2 = i_C$, as shown in Figure 5.3. We consider data points in a square region defined by $0.4 \le v_1 \le 0.7$ and $0.4 \le v_2 \le 0.7$, and assume the terminal behavior of the transistor follows the Ebers–Moll equation; namely,

$$i_1 = \frac{I_s}{\alpha_f}\left(e^{v_1/V_T} - 1\right) - I_s\left(e^{v_2/V_T} - 1\right) \tag{5.37}$$

$$i_2 = \frac{I_s}{\alpha_r}\left(e^{v_2/V_T} - 1\right) - I_s\left(e^{v_1/V_T} - 1\right) \tag{5.38}$$

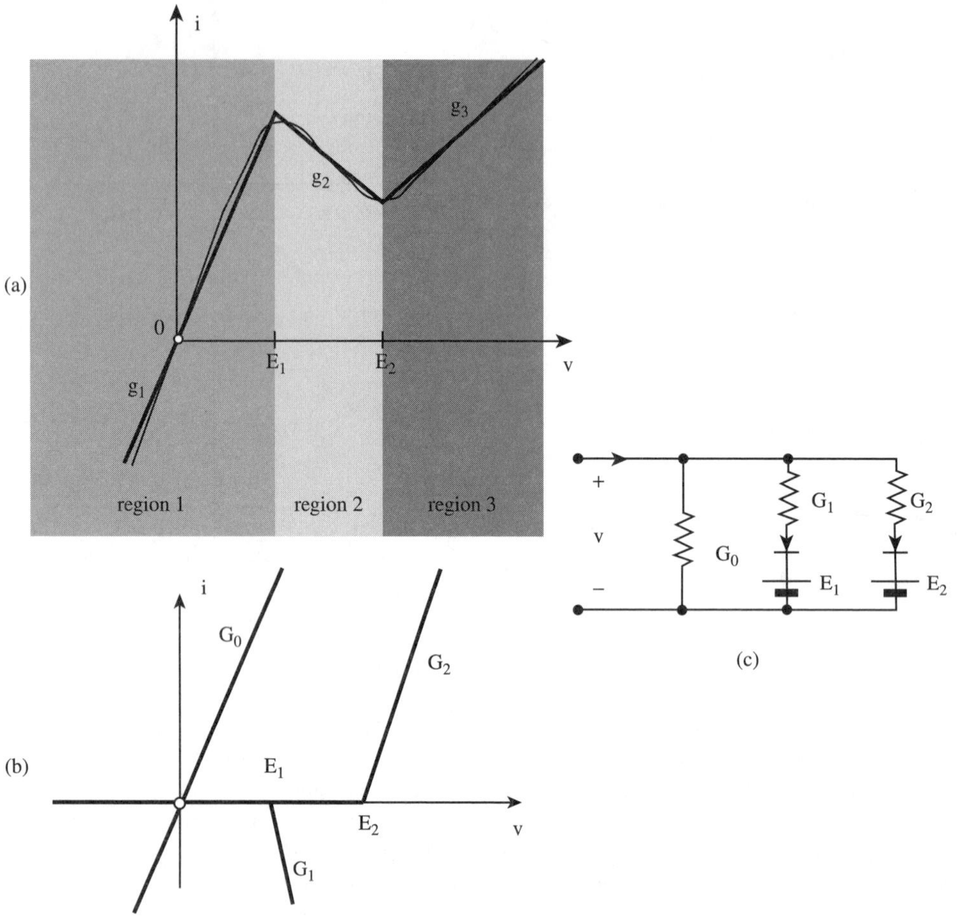

FIGURE 5.10 (a) Piecewise-linear approximation of the tunnel-diode characteristic. The three-segment approximation defines the three regions indicated. (b) Decomposition of the piecewise-linear characteristic (a) into three components, and (c) the corresponding circuit.

with $I_s = 10^{-14}$ A, $V_T = 26$ mV, $\alpha_f = 0.99$, and $\alpha_r = 0.5$. In [12], the following canonical piecewise-linear model is obtained, which optimally fits the data points (Figure 5.11)

$$\begin{bmatrix} i_1 \\ i_2 \end{bmatrix} = \begin{bmatrix} a_1 \\ a_2 \end{bmatrix} + \begin{bmatrix} b_{11} & b_{21} \\ b_{12} & b_{22} \end{bmatrix} \begin{bmatrix} v_1 \\ v_2 \end{bmatrix} + \begin{bmatrix} c_{11} \\ c_{21} \end{bmatrix} |m_1 v_1 - v_2 + t_1| + \begin{bmatrix} c_{12} \\ c_{22} \end{bmatrix} |m_2 v_1 - v_2 + t_2| + \begin{bmatrix} c_{13} \\ c_{23} \end{bmatrix} |m_3 v_1 - v_2 + t_3| \tag{5.39}$$

where

$$\begin{bmatrix} a_1 \\ a_2 \end{bmatrix} = \begin{bmatrix} 5.8722 \times 10^{-3} \\ -3.2652 \times 10^{-2} \end{bmatrix} \quad \begin{bmatrix} b_{11} \\ b_{21} \end{bmatrix} = \begin{bmatrix} 3.2392 \times 10^{-2} \\ -3.2067 \times 10^{-2} \end{bmatrix}$$

$$\begin{bmatrix} b_{12} \\ b_{22} \end{bmatrix} = \begin{bmatrix} -4.0897 \times 10^{-2} \\ 8.1793 \times 10^{-2} \end{bmatrix} \quad \begin{bmatrix} c_{11} \\ c_{21} \end{bmatrix} = \begin{bmatrix} 3.1095 \times 10^{-6} \\ -3.0784 \times 10^{-6} \end{bmatrix}$$

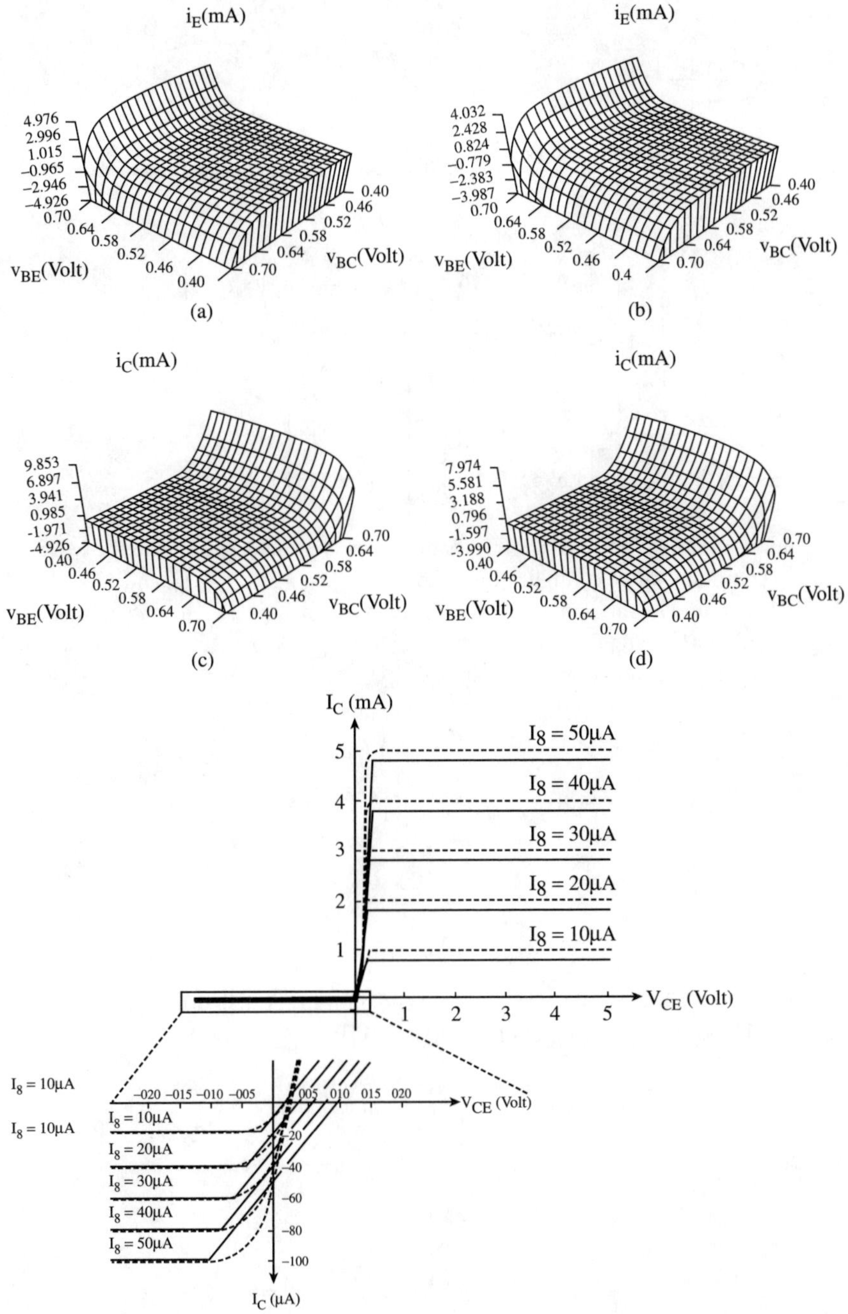

FIGURE 5.11 Three-dimensional plots for the emitter current in the Ebers-Moll model given by (5.37) and (5.38). (b) Three-dimensional plot for the emitter current in the canonical piecewise-linear model given by [10, (B.1)] (low-voltage version). (c) Three-dimensional plot for the collector current in the Ebers-Moll model given by (5.37) and (5.38). (d) Three-dimensional plot for the collector current in the canonical piecewise-linear model given by [10, (B.1)] (low-voltage version). (e) Comparison between the family of collector currents in the Ebers-Moll model (dashed line) and the canonical piecewise-linear model (solid line). *Source*: L.O. Chua and A. Deng, "Canonical piecewise linear modeling," *IEEE Trans. Circuits Syst.*, vol. CAS-33, p. 519, ® 1986, IEEE.

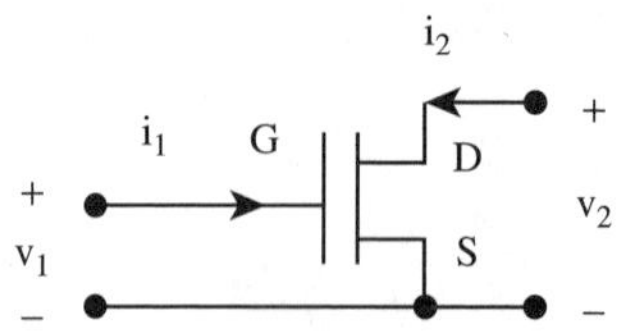

FIGURE 5.12 2-port configuration of the MOSFET.

$$\begin{bmatrix} c_{12} \\ c_{22} \end{bmatrix} = \begin{bmatrix} -9.9342\times10^{-3} \\ 1.9868\times10^{-2} \end{bmatrix} \begin{bmatrix} c_{13} \\ c_{23} \end{bmatrix} = \begin{bmatrix} -3.0471\times10^{-2} \\ 6.0943\times10^{-2} \end{bmatrix}$$

$$\begin{bmatrix} m_1 \\ m_2 \\ m_3 \end{bmatrix} = \begin{bmatrix} 1.002\times10^{4} \\ -1.4\times10^{-4} \\ 1.574\times10^{-6} \end{bmatrix} \begin{bmatrix} t_1 \\ t_2 \\ t_3 \end{bmatrix} = \begin{bmatrix} -6472 \\ 0.61714 \\ 0.66355 \end{bmatrix}$$

Next, a canonical piecewise-linear MOS transistor model is presented. Assume the MOS transistor is connected in the common source configuration with $v_1 = v_{GS}$, $v_2 = v_{DS}$, $i_1 = i_G$, and $i_2 = i_D$, as illustrated, in Figure 5.12, where both v_1, v_2 are in volts, and i_1, i_2 are in microamperes. The data points are uniformly spaced in a grid within a rectangular region defined by $0 \le v_1 \le 5$, and $0 \le v_2 \le 5$. We assume the data points follow the Shichman–Hodges model, namely,

$$i_1 = 0$$

$$i_2 = k\left[(v_1 - V_t)v_2 - 0.5v_2^2\right], \text{ if } v_1 - V_t \ge v_2$$

or

$$i_2 = 0.5k(v_1 - V_t)^2\left[1 + \lambda(v_2 - v_1 + V_t)\right], \text{ if } v_1 - V_t < v_2 \tag{5.40}$$

with $k = 50\ \mu A/V^2$, $V_t = 1$ V, $\lambda = 0.02\ V^{-1}$. Applying the optimization algorithm of [11], we obtain the following canonical piecewise-linear model (see Figure 5.13):

$$\begin{aligned} i_2 &= a_2 + b_{21}v_1 + b_{22}v_2 + c_{21}|m_1v_1 - v_2 + t_1| \\ &\quad + c_{22}|m_2v_1 - v_2 + t_2| + c_{23}|m_3v_1 - v_2 + t_3| \end{aligned} \tag{5.41}$$

where

$$a_2 = -61.167, \quad b_{21} = 30.242, \quad b_{22} = 72.7925$$

$$c_{21} = -49.718, \quad c_{22} = -21.027, \quad c_{23} = 2.0348$$

$$m_1 = 0.8175, \quad m_2 = 1.0171, \quad m_3 = -23.406$$

$$t_1 = -2.1052, \quad t_2 = -1.4652, \quad t_3 = 69$$

Finally, a canonical piecewise-linear model of GaAs FET is presented. The GaAs FET has become increasingly important in the development of microwave circuits and high-speed digital IC's due to its fast switching speed.

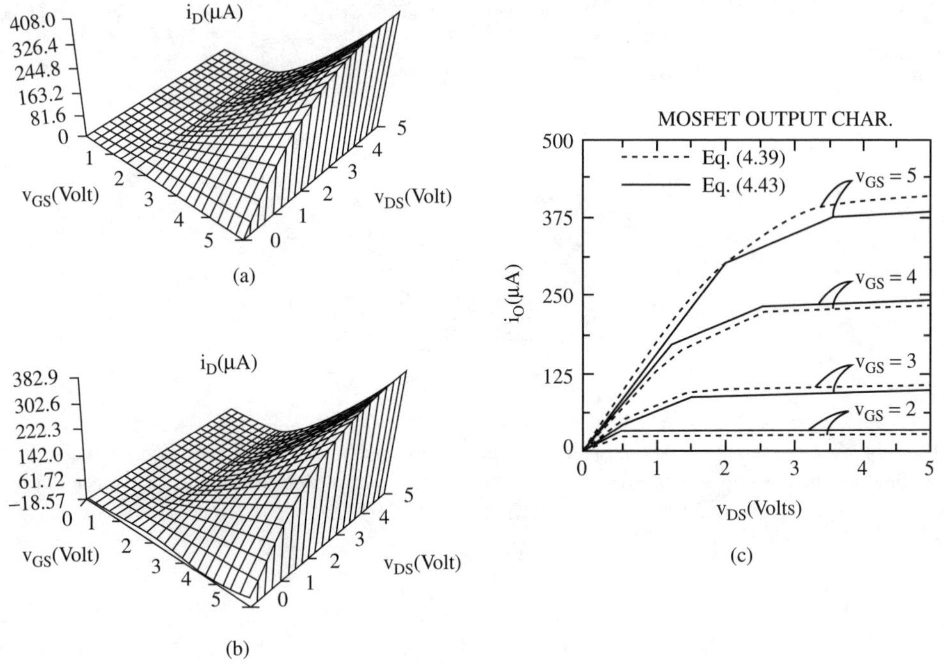

FIGURE 5.13 (a) Three-dimensional plot of drain current from the Shichman–Hodges model. (b) Three-dimensional plot of the drain current from the canonical piecewise-linear model. (c) Family of drain currents modeled by (5.40) (dashed line) and (5.41) (solid line). *Source*: L. O. Chua and A. Deng, "Canonical piecewise-linear modeling," *IEEE Trans. Circuits Syst.*, vol. CAS-33, p. 520, 1986. ® 1986 IEEE.

$$i_2 = a_2 + b_{21}v_1 + b_{22}v_2 + c_{21}|m_1v_1 - v_2 + t_2| + c_{22}|m_2v_1 - v_2 + t_2| + c_{23}|m_3v_1 - v_2 + t_3| \quad (5.42)$$

where $v_1 = v_{GS}$ (volt), $v_2 = v_{DS}$ (volt), $i_2 - i_D$ (mA), and

$$a_2 = 6.3645, \quad b_{21} = 2.4961, \quad b_{22} = 32.339$$
$$c_{21} = 0.6008, \quad c_{22} = 0.9819, \quad c_{23} = -29.507$$
$$m_1 = -19.594, \quad m_2 = -6.0736, \quad m_3 = 0.6473$$
$$t_1 = -44.551, \quad t_2 = -8.9962, \quad t_3 = 1.3738$$

Observe that this model requires only three absolute-value functions and 12 numerical coefficients and compares rather well to the analytical model (Figure 5.14).

More piecewise-linear models for timing analysis of logic circuits can be found in [21]. In the context of analog computer design, even PWL models of other nonlinear relationships have been derived in [51].

5.4 Structural Properties of Piecewise-Linear Resistive Circuits

When considering interconnections of PWL resistors (components), it follows from the linearity of KVL and KCL that the resulting multiport is also a piecewise-linear resistor. However, if the components have a canonical PWL representation, the resulting multiport may not have a canonical PWL representation. This can be illustrated by graphically deriving the equivalent one port of the series connection of two

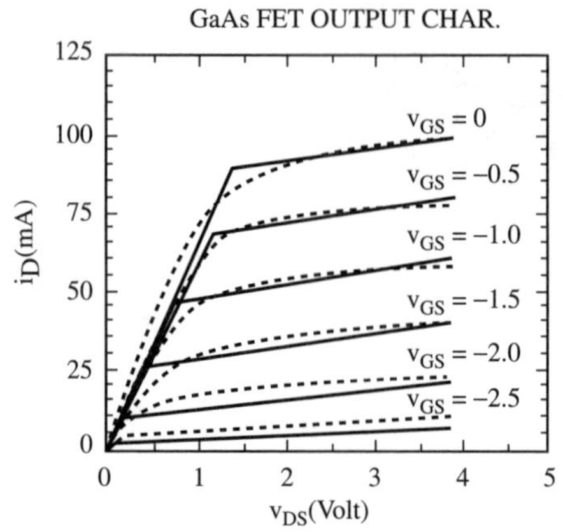

FIGURE 5.14 Comparison of the canonical piecewise-linear described by (5.42) (solid line) and the analytical model (dashed line) for the ion-implanted GaAs FET. *Source*: L. O. Chua and A. Deng, "Canonical piecewise-linear modeling," *IEEE Trans. Circuits Syst.*, vol. CAS-33, p. 522, 1986, ® 1986 IEEE.

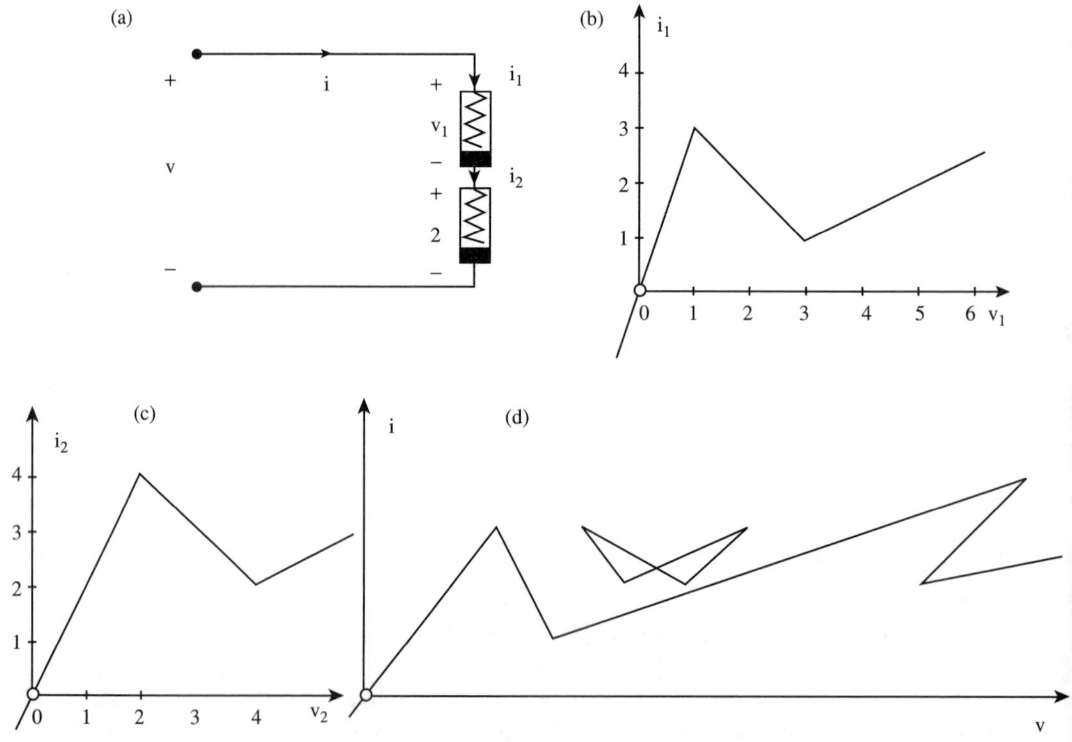

FIGURE 5.15 (a) The series connection of two tunnel diodes, (b) and (c), their *i-v* characteristics, and (d) the composite *i–v* plot, which consists of two unconnected parts.

tunnel diodes [3] (Figure 5.15). Both resistors have the same current, so we have to add the corresponding voltages $v = v_1 + v_2$ and obtain an $i - v$ plot with two unconnected parts. Values of i correspond to 3 values of v_1 for R_1 and 3 values of v_2 for R_2, and hence to 9 values of the equivalent resistor [Figure 5.15(d)]. This illustrates once more that nonlinear circuits may have more solutions than expected at first sight.

Although the two tunnel diodes R_1 and R_2 have a canonical PWL representation, the equivalent one port of their series connection has neither a canonical PWL voltage description, nor a current one. It, however, has a GLCP description because KVL, KCL, and the LCP of R_1 and R_2 constitute a GLCP. If the *v-i* PWL relation is monotonic, the inverse *i-v* function exists and then some uniqueness properties hold.

These observations are, of course, also valid for the parallel connection of two PWL resistors and for more complicated interconnections.

In Section 5.3 we illustrated with an example how a PWL one-port resistor can be realized with linear resistors and ideal diodes. This can be proven in general. One essentially needs a diode for each breakpoint in the PWL characteristic. Conversely, each one port with diodes and resistors is a PWL one port resistor.

This brings us to an interesting class of circuits composed of linear resistors, independent sources, linear controlled sources, and ideal diodes. These circuits belong to the general class of circuits with PWL components [see Figure 5.1(a)] and can be described by GLCP equations. Such networks have not only shown their importance in analysis but also in the topologic study of the number of solutions and more general qualitative properties. When only short-circuit and open-circuit branches are present, one independent voltage source with internal resistance and ideal diodes, an interesting loop cut set exclusion property holds that is also called the colored branch theorem or the arc coloring theorem. It says that the voltage source either forms a conducting loop with forward-oriented diodes and some short circuits or there is a cut set of the voltage source, some open circuits, and blocking diodes. Such arguments have been used to obtain [23] topologic criteria for upper bounds of the number of solutions of PWL resistive circuits. In fact, diode resistor circuits have been used extensively in PWL function generators for analog computers [51]. These electrical analogs can also be used for mathematical programming problems (similar to linear programming) and have reappeared in the neural network literature.

5.5 Analysis of Piecewise-Linear Resistive Circuits

It is first demonstrated that all conventional network formulation methods (nodal, cut, set, hybrid, modified nodal, and tableau) can be used for PWL resistive circuits where the components are described with canonical or with LCP equations. These network equations may have one or more solutions. In order to find solutions, one can either search through all the polyhedral regions P_k by solving the linear equations for that region or by checking whether its solution is located inside that region P_k.

Because many regions often exist, this is a time-consuming method, but several methods can be used to reduce the search [28], [61]. If one is interested in only one solution, one can use solution tracing methods, also called continuation methods or homotopy methods, of which the Katzenelson method is best known. If one is interested in all solutions, the problem is more complicated, but some algorithms exist.

Theorem Canonical PWL (Tableau Analysis) [8]

Consider a connected resistive circuit N containing only linear two-terminal resistors, dc independent sources, current-controlled and voltage-controlled piecewise-linear two-terminal resistors, linear- and piecewise-linear-controlled sources (all four types) and any linear multiterminal resistive elements. A composite branch of this circuit is given in Figure 5.16. If each piecewise-linear function is represented in the canonical form (5.6), then the *tableau formulation* also has the canonical PWL form

$$f(x) = a + Bx + \sum_{i+1}^{p} c_i \left|\alpha_i^T x - \beta_i\right| = 0 \tag{5.43}$$

where $x = [i^T, v^T, v_n^T]^T$ and i, respectively v, is the branch current voltage vector (Figure 5.16) and v_n is the node-to-datum voltage vector.

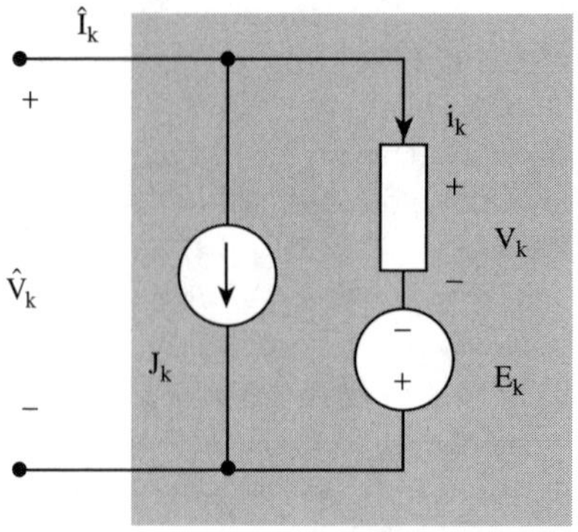

FIGURE 5.16 A composite branch.

PROOF. Let A be the reduced incidence matrix of N relative to some datum node, then KCL, KVL, and element constitutive relations give

$$Ai = AJ \tag{5.44}$$

$$v = A^T v_n + E \tag{5.45}$$

$$f_I(i) + f_v(v) = S \tag{5.46}$$

where we can express $f_1(\cdot)$ and $f_v(\cdot)$ in the canonical form (5.6)

$$f_I(i) = a_1 + B_I i + C_I \text{abs}\left(D_I^T e - e_1\right) \tag{5.47}$$

$$f_v(v) = a_v + B_v v + C_v \text{abs}\left(D_v^T v - e_v\right) \tag{5.48}$$

Substituting (5.47) and (5.48) into (5.46), we obtain

$$\begin{bmatrix} -AJ \\ -E \\ a_I + a_v - S \end{bmatrix} + \begin{bmatrix} A & 0 & 0 \\ 0 & 1 & A^T \\ B_I & B_v & 0 \end{bmatrix} \begin{bmatrix} i \\ v \\ v_n \end{bmatrix} = \begin{bmatrix} 0 & 0 & 0 \\ 0 & 0 & 0 \\ C_I & C_v & 0 \end{bmatrix}$$
$$\text{abs}\left[\begin{bmatrix} D_I & 0 & 0 \\ 0 & D_V & 0 \\ 0 & 0 & 0 \end{bmatrix} \begin{bmatrix} i \\ v \\ v_n \end{bmatrix} - \begin{bmatrix} e_I \\ e_V \\ 0 \end{bmatrix}\right] = 0 \tag{5.49}$$

Clearly, (5.49) is in the canonical form of (5.43).

Of course, an analogous theorem can be given when the PWL resistors are given in LCP form. Then the tableau constitute a GLCP. Moreover, one can derive nodal, cut set, loop, hybrid, and modified nodal analysis from the tableau analysis by eliminating certain variables. Alternatively, one can also directly derive these equations.

Whatever the description for the PWL components may be, one can always formulate the network equations as linear equations

$$0 = f(x) = a_k + B_k x, \qquad x \in P_k \tag{5.50}$$

in the polyhedral region P_k defined by (5.50). The map f is a continuous PWL map. A solution x of (5.50) can then be computed in a finite number of steps with the Katzenelson algorithm [4], [33], by tracing the map f from an initial point $(x^{(1)},y^{(1)})$ to a value $(x^*,0)$ (see Figure 5.18).

Algorithm

STEP 1. Choose an initial point $x^{(1)}$ and determine its polyhedral region $P^{(1)}$, and compute

$$y^{(1)} = f\left(x^{(1)}\right) = a^{(1)} + B^{(1)}x \quad \text{and set} \quad j=1$$

STEP 2. Compute

$$\hat{x} = x^{(j)} + \left(B^{(j)}\right)^{-1}\left(0 - y^{(j)}\right) \tag{5.51}$$

STEP 3. If $\hat{x} \in P^{(j)}$, we have obtained a solution $\hat{x}$ of $f(\hat{x}) = 0$. Stop.

STEP 4. Otherwise, compute

$$x^{(j+1)} = x^{(j)} + \lambda^{(j)}\left(\hat{x} - x^{(j)}\right) \tag{5.52}$$

where $\lambda^{(j)}$ is the largest number such that $x^{(j+1)} \in P^{(j)}$, i.e., $x^{(j+1)}$ is on the boundary between $P^{(j)}$ and $P^{(j+1)}$ (see Figure 5.17).

STEP 5. Identify $P^{(j+1)}$ and the linear map $y = a^{(j+1)} + B^{(j+1)}x$ in the polyhedral region $P^{(j+1)}$ and compute

$$y^{(j+1)} = y^{(j)} + \lambda^{(j)}\left(y^{*} - y^{(j)}\right) \tag{5.53}$$

Set $j = j + 1$. Go step 2.

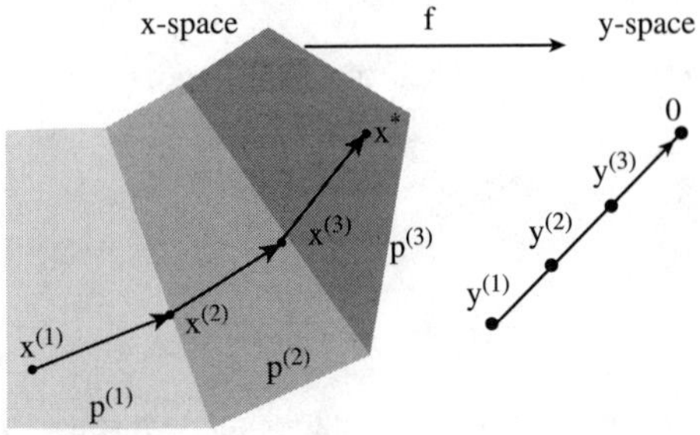

FIGURE 5.17 The iteration in the Katzenelson algorithm for solving $y = f(x) = 0$.

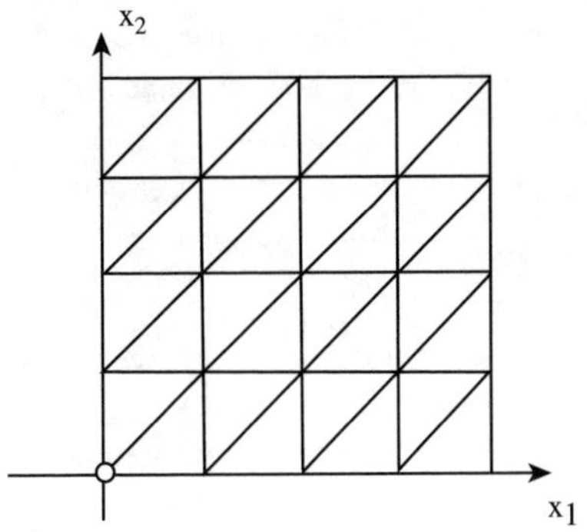

FIGURE 5.18 Simplicial subdivision.

This algorithm converges to a solution in a finite number of steps if the determinants of all matrices $B^{(j)}$ have the same sign. This condition is satisfied when the *i-v* curves for the PWL one port resistors are monotonic. The Katzenelson algorithm was extended in [45] by taking the sign of the determinants into account in (5.52) and (5.53). This requires the PWL resistors to be globally coercive. If by accident in the iteration the point $x^{(j+1)}$ is not on a single boundary and instead is located on a corner, the region $P^{(j+1)}$ is not uniquely defined. However, with a small perturbation [1], one can avoid this corner and still be guaranteed to converge.

This algorithm was adapted to the canonical PWL equation (5.49) in [8]. It can also be adapted to the GLCP. However, there exist circuits where this algorithm fails to converge. For the LCP problem, one can then use other algorithms [20], [40], [56]. One can also use other homotopy methods [43], [57], [60], which can be shown to converge based on eventual passivity arguments. In fact, this algorithm extends the rather natural method of source stepping, where the PWL circuit is solved by first making all sources zero and then tracing the solution for increasing (stepping up) the sources. It is instructive to observe here that these methods can be used successfully in another sequence of the steps in Figure 5.1(a). Until now, we always first performed the horizontal step of PWL approximation or modeling and then the vertical step of network equation formulation. With these methods, one can first perform the network equation formulation and then the PWL approximation. The advantage is that one can use a coarser grid in the simplicial subdivision far away from the solution, and hence dynamically adapt the accuracy of the PWL approximation.

In any case, if all solutions are requested, all these homotopy-based methods are not adequate, because not all solutions can be found even if the homotopy method is started with many different $x^{(1)}$. Hence, special methods have been designed. It is beyond the scope of this text to give a complete algorithm [39], [59], but the solution of the GLCP basically involves two parts. First, calculate the solution set of all nonnegative solutions to (5.10) and (5.11). This is a polyhedral cone where extremal rays can be easily determined [44], [54]. Second, this solution set is intersected with a hyperplane and the complementarity condition $u^T s = 0$ implies the elimination of vertices (respectively, convex combinations) where this complementarity (respectively, cross complementarity) is not satisfied. This allows for the complete solution set for the circuit of Figure 5.15 and for circuits with infinitely many solutions.

A more recent method [46] covers the PWL *i-v* characteristic with a union of polyhedra and hierarchically solves the circuit with finer and finer polyhedra.

An important improvement in efficiency for the methods is possible when the PWL function $f(\cdot)$ is separable, i.e., there exist $f^i\colon \mathbb{R} \to \mathbb{R}^n$ i = 1, 2, ..., n such that

$$f(x) = \sum_{i=1}^{n} f^i(x_i) \tag{5.54}$$

This happens when there are only two terminal PWL resistors, linear resistors, and independent sources, and if the bipolar transistors are modeled by the Ebers–Moll model [see (5.39)]. Then, the subdivision for x is rectangular and each rectangle is subdivided into simplices (see Figure 5.18). This property can be used to eliminate certain polyhedral regions without solutions [62] and also to speed up the Katzenelson-type algorithm [60], [62]. If there are MOS transistors, the map f is not separable but one can apply the extended concept of pairwise separable map [62].

5.6 Piecewise-Linear Dynamic Circuits

As mentioned at the end of Section 5.2, the piecewise linear descriptions of Section 5.2 can be used also for PWL capacitors, respectively, inductors and memristors, by replacing the port voltages v and currents i by q, v, respectively, φ, i and φ, q. Whenever we have a network obtained by interconnecting linear and/or PWL resistors, inductors, capacitors, and memristors, we have a dynamic piecewise-linear circuits.

Of course, such networks are often encountered because they include the networks with linear R, L, C, and linear dependent sources, diodes, switches, op amps, and components such as bipolar and MOS transistors, and GaAs FET's with PWL resistive models. This includes several important and famous nonlinear circuits such as Chua's circuit [18], [19], and the cellular neural networks (CNN's) [48], which are discussed in Chapter 7 and Chapter 8, Section 8.2.

Of course, PWL dynamic circuits are much more interesting and much more complicated and can exhibit a much more complex behavior than resistive circuits and hence this subject is much less explored. It is clear from the definition of a PWL dynamic circuit that it can be described by linear differential equations over polyhedral regions. Hence, it can exhibit many different types of behavior. It may have many equilibria, which can essentially be determined by solving the resistive network (see Section 5.5 and Figure 5.1) obtained by opening the capacitive ports and short circuiting the inductive ports (dc analysis). When there is no input waveform, the circuit is said to be autonomous and has transients. Some transients may be periodic and are called limit cycles but they may also show chaotic behavior. Next, one may be interested in the behavior of the circuit for certain input waveforms (transient analysis). This can be performed by using integration rules in simulations.

For the analysis of limit cycles, chaos, and transients, one can, of course, use the general methods for nonlinear circuits, but some improvements can be made based on the PWL nature of the nonlinearities. Here, we only describe the methods briefly. If one is interested in the periodic behavior of a PWL dynamic circuit (autonomous or with a periodic input), then one can, for each PWL nonlinearity, make some approximations.

First, consider the case that one is only interested in the dc and fundamental sinusoidal contributions in all signals of the form $i(t) = A_0 + A_1 \cos \omega t$. The widely used describing function method [6] for PWL resistors $v = f(i)$ consists of approximating this resistor by an approximate resistor where $\hat{v}(t) = D_0 + D_1 \cos \omega t$ has only the dc and fundamental contribution of $v(t)$. This is often a good approximation since the remainder of the circuit often filters out all higher harmonics anyway. Using a Fourier series, one can then find D_0 and D_1 as

$$D_0(A_0, A_1) = \frac{1}{2\pi} \int_0^{2\pi} f(A_0 + A_1 \cos\phi) d\phi$$

$$D_1(A_0, A_1) = \frac{1}{\pi A_1} \int_0^{2\pi} f(A_0 + A_1 \cos\phi) d\phi$$

By replacing all PWL components by their describing functions, one can use linear methods to set up the network equations in the Laplace–Fourier domain. When this approximation is not sufficient, one can include more harmonics. Then, one obtains the famous harmonic balance method, because one is balancing more harmonic components.

Alternatively, one can calculate the periodic solution by simulating the circuit with a certain initial condition and considering the map F: $x_0 \to x_1$ from the initial condition x_0 to the state x_1 one period later. Of course, a fixed point $x^* = F(x^*)$ of the map corresponds to a periodic solution. It has been demonstrated [27] that the map F is differentiable for PWL circuits. This is very useful in setting up an efficient iterative search for a fixed point of F. This map is also useful in studying the eventual chaotic behavior and is then called the Poincaré return map.

In transient analysis of PWL circuits, one is often interested in the sensitivity of the solution to certain parameters in order to optimize the behavior. As a natural extension of the adjoint network for linear circuits in [22], the adjoint PWL circuit is defined and used to determine simple sensitivity calculations for transient analysis.

Another important issue is whether the PWL approximation of a nonlinear characteristic in a dynamic circuit has a serious impact on the transient behavior. In [63], error bounds were obtained on the differences of the waveforms.

5.7 Efficient Computer-Aided Analysis of PWL Circuits

Transient analysis and timing verification are an essential part of the VLSI system design process. The most reliable way of analyzing the timing performance of a design is to use analog circuit analysis methods. Here as well, a set of algebraic-differential equations has to be solved. This can be done by using implicit integration formulas that convert these equations into a set of algebraic equations, which can be solved by iterative techniques like Newton–Raphson (see Chapter 6). The computation time then becomes excessive for large circuits. It mainly consists of linearizations of the nonlinear component models and the solution of the linear equations. In addition, the design process can be facilitated substantially if this simulation tool can be used at many different levels from the top level of specifications over the logic and switch level to the circuit level. Such a hierarchical simulator can support the design from top to bottom and allow for mixtures of these levels. In limited space, we describe here the main simulation methods for improving the efficiency and supporting the hierarchy of models with piecewise-linear methods.

It is clear from our previous discussion that PWL models and circuit descriptions can be used at many different levels. An op amp, for example, can be described by the finite gain model [see Figure 5.9 and (5.29) and (5.30)], but when it is designed with a transistor circuit it can be described by PWL circuit equations as in Section 5.5. Hence, it is attractive to use a simulator that can support this top-down design process [35]. One can then even incorporate logic gates into the PWL models. One can organize the topological equations of the network hierarchically, so that it is easy to change the network topology. The separation between topological equations and model descriptions allows for an efficient updating of the model when moving from one polyhedral region into another. Several other efficiency issues can be built into a hierarchical PWL simulator.

An important reduction in computation time needed for solving the network equations can be obtained by using the consistent variation property. In fact, only a rank one difference exists between the matrices of two neighboring polyhedral regions, and hence, one inverse can be easily derived from the other [8], [35]. In the same spirit, one can at the circuit level take advantage of the PWL transistor models (see [62] and separability discussion in Section 2.5). In [53], the circuit is partitioned dynamically into subcircuits during the solution process, depending on the transistor region of operation. Then, the subcircuits are dynamically ordered and solved with block Gauss–Seidel for minimal or no coupling among them.

Interesting savings can be obtained [34] by solving the linear differential equations in a polyhedral region with Laplace transformations and by partitioning the equations. However, the computation of the intersection between trajectories in neighboring polyhedral regions can be a disadvantage of this method.

Acknowledgment

This work was supported by the Research Council Kuleuven Project MEFISTO666GOA.

References

[1] M. J. Chien, "Piecewise-linear homeomorphic resistive networks," *IEEE Trans. Circuits Syst.*, vol. CAS-24, pp. 118–127, Mar. 1977.

[2] M. J. Chien and E. S. Kuh,"Solving nonlinear resistive network using piecewise-linear analysis and simplicial subdivision," *IEEE Trans. Circuits Syst.*, vol. CAS-24, pp. 305–317, 1977.

[3] L. O. Chua, "Analysis and synthesis of multivalued memoryless nonlinear networks," *IEEE Trans. Circuit Theory*, vol. CT-14, pp. 192–209, June 1967.

[4] L. O. Chua and P. M. Lin, *Computer-Aided Analysis of Electronic Circuits: Algorithms and Computational Techniques*, Englewood Cliffs, NJ: Prentice Hall, 1975.

[5] L. O. Chua and P. M. Lin, "A switching-parameter algorithm for finding multiple solutions of nonlinear resistive circuits," *Int. J. Circuit Theory, Appl.*, vol. 4, pp. 215–239, 1976.

[6] L. O. Chua and S. M. Kang, "Section-wise piecewise-linear functions: Canonical representation properties and applications," *Proc. IEEE,* vol. 65, pp. 915–929, June 1977.

[7] L. O. Chua and D. J. Curtin, "Reciprocal *n*-port resistor represented by continuous *n*-dimensional piecewise-linear function realized by circuit with 2-terminal piecewise-linear resistor and $p+g$ port transformer," *IEEE Trans. Circuits Syst.*, vol. CAS-27, pp. 367–380, May 1980.

[8] L. O. Chua and R. L. P. Ying, "Finding all solutions of piecewise-linear circuits," *Int. J. Circuit Theory, Appl.*, vol. 10, pp. 201–229, 1982.

[9] L. O. Chua and R. L. P. Ying, "Canonical piecewise-linear analysis," *IEEE Trans. Circuits Syst.*, vol. CAS-30, pp. 125–140, 1983.

[10] L. O. Chua and A C. Deng, "Canonical piecewise-linear analysis — Part II: Tracing driving-point and transfer characteristics," *IEEE Trans. Circuits Syst.*, vol. CAS-32, pp. 417–444, May 1985.

[11] L. O. Chua and A. C. Deng, "Canonical piecewise-linear modeling, unified parameter optimization algorithm: Application to pn junctions, bipolar transistors, MOSFETs, and GaAs FETs, *IEEE Trans. Circuits Syst.*,vol. CAS-33, pp. 511–525, May 1986.

[12] L. O. Chua and A. C. Deng, "Canonical piecewise linear modeling," *IEEE Trans. Circuits Syst.*, vol. CAS-33, pp. 511–525, May 1986.

[13] L. O. Chua and A. C. Deng, "Canonical piecewise linear analysis: Generalized breakpoint hopping algorithm," *Int. J. Circuit Theory, Appl.*, vol. 14, pp. 35–52, 1986.

[14] L. O. Chua and A. C. Deng, "Canonical piecewise linear representation," *IEEE Trans. Circuits Syst.*, vol. 35, pp. 101–111, Jan. 1988.

[15] L. O. Chua and A. C. Deng, "Canonical piecewise-linear modeling," ERL Memo. UCB/ERL M85/35, Univ. California, Berkeley, April 26, 1985.

[16] L. O. Chua and G. Lin, "Canonical realization of Chua's circuit family," *IEEE Trans. Circuits Syst.*, vol. 37, pp. 885–902, July 1990.

[17] L. O. Chua and G. Lin, "Intermittency in piecewise-linear circuit," *IEEE Trans. Circuits Syst.*, vol. 38, pp. 510–520, May 1991.

[18] L. O. Chua, "The genesis of Chua's circuit," *Archiv Elektronik Übertragungstechnik*, vol. 46, no. 4, pp. 250–257, 1992.

[19] L. O. Chua, C.-W. Wu, A.-S. Huang, and G.-Q. Zhong, "A Universal circuit for studying and generating chaos," *IEEE Trans. Circuits Syst. I*, vol. 40, no. 10, pp. 732–744, 745–761, Oct. 1993.

[20] R. Cottle, J.-S. Pang, and R. Stone, *The Linear Complementarity Problem,* New York: Academic Press, 1992.

[21] A. C. Deng, "Piecewise-linear timing model for digital CMOS circuits," *IEEE Trans. Circuits Syst.*, vol. 35, pp. 1330–1334, Oct. 1988.

[22] Y. Elcherif and P. Lin, "Transient analysis and sensitivity computation in piecewise-linear circuits," *IEEE Trans. Circuits Syst.*, vol. 36, pp. 1525–1533, Dec. 1988.

[23] M. Fossepréz, M. J. Hasler, and C. Schnetzler, "On the number of solutions of piecewise-linear resistive circuits," *IEEE Trans. Circuits Syst.*, vol. 36, pp. 393–402, March 1989.

[24] T. Fujisawa, E. S. Kuh, "Piecewise-linear theory of nonlinear networks," *SIAM J. Appl. Math.*, vol. 22, no. 2, pp. 307–328, March 1972.

[25] T. Fujisawa, E. S. Kuh, and T. Ohtsuki, "A sparse matrix method for analysis of piecewise-linear resistive circuits," *IEEE Trans. Circuit Theory*, vol.19, pp. 571–584, Nov. 1972.

[26] G. Güzelis and I. Göknar, "A canonical representation for piecewise affine maps and its applications to circuit analysis," *IEEE Trans. Circuits Syst.*, vol. 38, pp. 1342–1354, Nov. 1991.

[27] I. N. Hajj and S. Skelboe, "Dynamic systems: Steady-state analysis," *IEEE Trans. Circuits Syst.*, vol. CAS-28, pp. 234–242, March 1981.

[28] Q. Huang and R. W. Liu, "A simple algorithm for finding all solutions of piecewise-linear networks," *IEEE Trans. Circuits Syst.*, vol. 36, pp. 600–609, April 1989.

[29] C. Kahlert and L. O. Chua, "A generalized canonical piecewise-linear representation," *IEEE Trans. Circuits Syst.*, vol. 37, pp. 373–383, March 1990.
[30] C. Kahlert and L. O. Chua, "Completed canonical piecewise-linear representation: Geometry of domain space," *IEEE Trans. Circuits Syst.*, vol. 39, pp. 222–236, March 1992.
[31] S. M. Kang and L. O. Chua, "A global representation of multidimensional piecewise-linear functions with linear partitions," *IEEE Trans. Circuits Syst.*, vol. CAS-25, pp. 938–940, Nov. 1978.
[32] S. Karamardian, "The complementarity problem," *Mathemat. Program.*, vol. 2, pp. 107–129, 1972.
[33] S. Karamardian and J. Katzenelson, "An algorithm for solving nonlinear resistive networks," *Bell Syst. Tech. J.*, vol. 44, pp. 1605–1620, 1965.
[34] R. J. Kaye and A. Sangiovanni-Vincentelli, "Solution of piecewise-linear ordinary differential equations using waveform relaxation and Laplace transforms," *IEEE Trans. Circuits Syst.*, vol. CAS-30, pp. 353–357, June 1983.
[35] T. A. M. Kevenaar and D. M. W. Leenaerts, "A flexible hierarchical piecewise-linear simulator," *Integrat., VLSI J.*, vol. 12, pp. 211–235, 1991.
[36] T. A. M. Kevenaar and D. M. W. Leenaerts, "A comparison of piecewise-linear model descriptions," *IEEE Trans. Circuits Syst.*, vol. 39, pp. 996–1004, Dec. 1992.
[37] M. Kojima and Y. Yamamoto, "Variable dimension algorithms: Basic theory, interpretations and extensions of some existing methods," *Mathemat. Program.*, vol. 24, pp. 177–215, 1982.
[38] S. Lee and K. Chao, "Multiple solution of piecewise-linear resistive networks," *IEEE Trans. Circuits Syst.*, vol. CAS-30, pp. 84–89, Feb. 1983.
[39] D. M. W. Leenaerts and J. A. Hegt, "Finding all solutions of piecewise-linear functions and the application to circuits design," *Int. J. Circuit Theory, Appl.*, vol. 19, pp. 107–123, 1991.
[40] C. E. Lemke, "On complementary pivot theory," in *Nonlinear Programming*, J. B. Rosen, O. L. Mangasarian, and K. Ritten, Eds., New York: Academic Press, 1968, pp. 349–384.
[41] J. Lin and R. Unbehauen, "Canonical piecewise-linear approximations," *IEEE Trans. Circuits Syst.*, vol. 39, pp. 697–699, Aug. 1992.
[42] R. Lum and L. O. Chua, "Generic properties of continuous piecewise-linear vector fields in 2-D space," *IEEE Trans. Circuits Syst.*, vol. 38, pp. 1043–1066, Sep. 1991.
[43] R. Melville, L. Trajkovic, S.-C. Fang, and L. Watson, "Artificial homotopy methods for the DC operating point problem," *IEEE Trans Comput.-Aided Design Integrat. Circuits Syst.*, vol. 12, pp. 861–877, June 1993.
[44] T. S. Motzkin, H. Raiffa, G. L. Thompson, and R. M. Thrall, "The double description method," in *Contributions to the Theory of Games, Ann. Mathemat. Studies*, H.W. Kuhn and A.W. Tucker, Eds., Princeton: Princeton Univ. Press, 1953, pp. 51–73.
[45] T. Ohtsuki, T. Fujisawa, and S. Kumagai, "Existence theorem and a solution algorithm for piecewise-linear resistor circuits," *SIAM J. Math. Anal.*, vol. 8, no. 1, pp. 69–99, 1977.
[46] S. Pastore and A. Premoli, "Polyhedral elements: A new algorithm for capturing all the equilibrium points of piecewise-linear circuits," *IEEE Trans. Circuits Syst. I.*, vol. 40, pp. 124–132, Feb. 1993.
[47] V. C. Prasad and V. P. Prakash, "Homeomorphic piecewise-linear resistive networks," *IEEE Trans. Circuits Syst.*, vol. 35, pp. 251–253, Feb. 1988.
[48] T. Roska and J. Vandewalle, *Cellular Neural Networks*, New York: John Wiley & Sons, 1993.
[49] I. W. Sandberg, "A note on the operating-point equations of semiconductor-device networks," *IEEE Trans. Circuits Syst.*, vol. 37, p. 966, July 1990.
[50] A. S. Solodovnikov, *System of Linear Inequalities*, translated by L. M. Glasser and T. P. Branson, Chicago: Univ. Chicago, 1980.
[51] T. E. Stern, *Theory of Nonlinear Networks and Systems: An Introduction*, Reading, MA: Addison-Wesley, 1965.
[52] S. Stevens and P.-M. Lin, "Analysis of piecewise-linear resistive networks using complementary pivot theory," *IEEE Trans Circuits Syst.*, Vol. CAS-28, pp. 429–441, May 1981.
[53] O. Tejayadi and I. N. Hajj, "Dynamic partitioning method for piecewise-linear VLSI circuit simulation," *Int. J. Circuit Theory, Appl.*, vol. 16, pp. 457–472,1988.

[54] S. N. Tschernikow, *Lineare Ungleichungen*. Berlin: VEB Deutscher Verlag der Wissenschaften, 1971; translation from H. Weinert and H. Hollatz, *Lineinye Neravenstva*, 1968, into German.

[55] W. M. G. van Bokhoven, *Piecewise-Linear Modelling and Analysis.* Deventer: Kluwer Academic, 1980.

[56] C. Van de Panne, "A complementary variant of Lemke's method for the linear complementary problem," *Mathemat. Program.*, vol. 7, pp. 283–310, 1974.

[57] L. Vandenberghe and J. Vandewalle, "Variable dimension algorithms for solving resistive circuits," *Int. J. Circuit Theory Appl.*, vol. 18, pp. 443–474, 1990.

[58] L. Vandenberghe and J. Vandewalle, "A continuous deformation algorithm for DC-analysis of active nonlinear circuits," *J. Circuits, Syst., Comput.*, vol. 1, pp. 327–351, 1991.

[59] L. Vandenberghe, B. L. De Moor, and J. Vandewalle, "The generalized linear complementarity problem applied to the complete analysis of resistive piecewise-linear circuits," *IEEE Trans. Circuits Syst.*, vol. 36, pp. 1382–1391, 1989.

[60] K. Yamamura and K. Horiuchi, "A globally and quadratically convergent algorithm for solving resistive nonlinear resistive networks," *IEEE Trans. Computer-Aided Design Integrat. Circuits Syst.*, vol. 9, pp. 487–499, May 1990.

[61] K. Yamamura and M. Ochiai, "Efficient algorithm for finding all solutions of piecewise-linear resistive circuits," *IEEE Trans. Circuits Syst.*, vol. 39, pp. 213–221, March 1992.

[62] K. Yamamura, "Piecewise-linear approximation of nonlinear mappings containing Gummel-Poon models or Shichman-Hodges models," *IEEE Trans. Circuits Syst.*, vol. 39, pp. 694–697, Aug. 1992.

[63] M. E. Zaghloul and P. R. Bryant, "Nonlinear network elements: Error bounds," *IEEE Trans. Circuits Syst.*, vol. CAS-27, pp. 20–29, Jan. 1980.

6

Simulation

Erik Lindberg
Technical University of Denmark

This chapter deals with the **simulation** or analysis of a nonlinear electrical circuit by means of a computer program. The program creates and solves the differential-algebraic equations of a **model** of the circuit. The basic tools in the solution process are *linearization, difference approximation,* and *the solution of a set of linear equations.* The output of the analysis may consist of (1) all node and branch voltages and all branch currents of a bias point (dc analysis); (2) a linear small-signal model of a bias point that may be used for analysis in the frequency domain (ac analysis); or (3) all voltages and currents as functions of time in a certain time range for a certain excitation (transient analysis). A model is satisfactory if there is good agreement between measurements and simulation results. In this case, simulation may be used instead of measurement for obtaining a better understanding of the nature and abilities of the circuit. The crucial point is to set up a model that is as simple as possible, in order to obtain a fast and inexpensive simulation, but sufficiently detailed to give the proper answer to the questions concerning the behavior of the circuit under study. **Modeling is the bottleneck of simulation**.

The model is an equivalent scheme — "schematics-capture" — or a branch table — "net-list" — describing the basic components (n-terminal elements) of the circuit and their connection. It is always possible to model an n-terminal element by means of a number of 2-terminals (branches). These internal 2-terminals may be coupled. By pairing the terminals of an n-terminal element, a port description may be obtained. The branches are either admittance branches or impedance branches. All branches may be interpreted as controlled sources. An **admittance branch** is a current source primarily controlled by its own voltage or primarily controlled by the voltage or current of another branch (transadmittance). An **impedance branch** is a voltage source primarily controlled by its own current or primarily controlled by the current or voltage of another branch (transimpedance). Control by signal (voltage or current) and control by time-derivative of signal is allowed. Control by several variables is allowed. Examples of admittance branches are (1) the conductor is a current source controlled by its own voltage; (2) the capacitor is a current source sontrolled by the time-derivative of its own voltage; and (3) the open circuit is a zero-valued current source (a conductor with value zero). Examples of impedance branches are (1) the resistor is a voltage source controlled by its own current; (2) the inductor is a voltage source controlled by the time-derivative of its own current; and (3) the short circuit is a zero-valued voltage source (a resistor with value zero).

A component may often be modeled in different ways. A diode, for example, is normally modeled as a current source controlled by its own voltage such that the model can be linearized into a **dynamic conductor** in parallel with a current source during the iterative process of finding the bias point of the diode. The diode may also be modeled as (1) a voltage source controlled by its own current (a dynamic resistor in series with a voltage source); (2) a **static conductor** being a function of the voltage across the diode; or (3) a static resistor being a function of the current through the diode. Note that in the case where a small signal model is wanted, for frequency analysis, only the dynamic model is appropriate.

The **primary variables** of the model are the currents of the impedance branches and the node potentials. The current law of Kirchhoff (the sum of all the currents leaving a node is zero) and the current-voltage relations of the impedance branches are used for the creation of the equations describing the relations between the primary variables of the model. The contributions to the equations from the branches are taken one branch at a time based on the question: will this branch add new primary variables? If yes, then a new column (variables) and a new row (equations) must be created and updated, or else the columns and rows corresponding to the existing primary variables of the branch must be updated. This approach to equation formulation is called the **extended nodal approach** or the **modified nodal approach (MNA)**.

In the following, some algorithms for solving a set of nonlinear algebraic equations and nonlinear differential equations are briefly described. Because we are dealing with physical systems and because we are responsible for the models, we assume that at least one solution is possible. The zero solution is, of course, always a solution. It might happen that our models become invalid if we, for example, increase the amplitudes of the exciting signals, diminish the risetime of the exciting signals, or by mistake create unstable models. It is important to define the range of validity for our models. What are the consequences of our assumptions? Can we believe in our models?

6.1 Numerical Solution of Nonlinear Algebraic Equations

Let the equation system to be solved be $f(x,u) = 0$, where x is the vector of primary variables and u is the the excitation vector. Denote the solution by x_s. Then, if we define a new function $g(x) = \alpha(f(x,u)) + x$, where α may be some function of $f(x,u)$, which is zero for $f(x,u) = 0$, then we can define an **iterative scheme** where $g(x)$ converges to the solution x_s by means of the iteration: $x_{k+1} = g(x_k) = \alpha(f(x_k,u)) + x_k$ where k is the iteration counter.

If for all x in the interval $[x_a, x_b]$ the condition $\|g(x_a) - g(x_b)\| \leq L * \|x_a - x_b\|$ for some $L < 1$ is satisfied, the iteration is called a **contraction mapping**. The condition is called a **Lipschitz condition**. Note that a function is a contraction if it has a derivative less than 1.

For $\alpha = -1$, the iterative formula becomes: $x_{k+1} = g(x_k) = -f(x_k, u) + x_k$. This scheme is called the **Picard method**, the **functional method**, or the **contraction mapping algorithm**. At each step, each nonlinear component is replaced by a linear **static component** corresponding to the solution x_k. A nonlinear conductor, for example, is replaced by a linear conductor defined by the straight line through the origin and the solution point. Each iterative solution is calculated by solving a set of linear equations. All components are updated and the next iteration is made. When two consecutive solutions are within a prescribed tolerance, the solution point is accepted.

For $\alpha = -1/(df/dx)$, the iterative formula becomes: $x_{k+1} = g(x_k) = -f(x_k, u)/(df(x_k, u)/dx) + x_k$. This scheme is called the **Newton–Raphson method** or the **derivative method**. At each step, each nonlinear component is replaced by a linear **dynamic component** plus an independent source corresponding to the solution x_k. A nonlinear conductor, for example, is replaced by a linear conductor defined by the derivative of the branch current with respect to the branch voltage (the slope of the nonlinearity) in parallel with a current source corresponding to the branch voltage of the previous solution. A new solution is then caculated by solving a set of linear equations. The components are updated and the next iteration is made. When the solutions converge within a prescribed tolerance, the solution point is accepted.

It may, of course, happen that the previously mentioned iterative schemes do not **converge** before the iteration limit k_{max} is reached. One reason may be that the nonlinearity $f(x)$ changes very rapidly for a small change in x. Another reason could be that $f(x)$ possess some kind of symmetry that causes cycles in the Newton–Raphson iteration scheme. If **convergence problems** are detected, the iteration scheme can be *modified* by introducing a limiting of the actual step size. Another approach may be to change the modeling of the nonlinear branches from voltage control to current control or vice versa. Often, the user of a circuit analysis program may be able to solve convergence problems by means of proper modeling and adjustment of the program options [1, 2, 3, 8, 11].

6.2 Numerical Integration of Nonlinear Differential Equations

The dynamics of a nonlinear electronic circuit may be described by a set of coupled first order differential equations–algebraic equations of the form: $dx/dt = f(x, y, t)$ and $g(x, y, t) = 0$, where x is the vector of **primary variables** (node potentials and impedance branch currents), y is the vector of variables that cannot be explicitly eliminated, and f and g are nonlinear vector functions. It is always possible to express y as a function of x and t by inverting the function g and inserting it into the differential equations such that the general differential equation form $dx/dt = f(x, t)$ is obtained. The task is then to obtain a solution $x(t)$ when an initial value of x is given. The usual methods for solving differential equations reduce to the solution of **difference equations**, with either the derivatives or the integrals expressed *approximately* in terms of finite differences.

Assume, at a given time t_0, we have a known solution point $x_0 = x(t_0)$. At this point, the function f can be expanded in Taylor series: $dx/dt = f(x_0, t) + A(x_0)(x - x_0) + \ldots$ where $\mathrm{A}(\mathrm{x}_0)$ is the **Jacobian** of f evaluated at x_0. Truncating the series, we obtain a linearization of the equations such that the small-signal behavior of the circuit in the neighborhood of x_0 is described by $dx/dt = A * x + k$, where A is a constant matrix equal to the Jacobian and k is a constant vector.

The most simple scheme for the approximate solution of the differential equation $dx/dt = f(x, t) = Ax + k$ is the **forward Euler formula** $x(t) = x(t_0) + hA(t_0)$ where $h = t - t_0$ is the integration time step. From the actual solution point at time t_0, the next solution point at time t is found along the tangent of the solution curve. It is obvious that we will rapidly leave the vicinity of the exact solution curve if the integration step is too large. To guarantee stability of the computation, the time step h must be smaller than $2/|\lambda|$ where λ is the largest **eigenvalue** of the Jacobian A. Typically, h must not exceed $0.2/|\lambda|$.

The forward Euler formula is a linear *explicit* formula based on forward Taylor expansion from t_0. If we make backward Taylor expansion from t we arrive at the **backward Euler** formula: $\mathrm{x(t)} = \mathrm{x(t_0)} + \mathrm{hA(t)}$. Because the unknown appears on both sides of the equation, it must in general be found by iteration so the formula is a linear *implicit* formula. From a stability point of view, the backward Euler formula has a much larger stability region than the forward Euler formula. The truncation error for the Euler formulas is of order h^2.

The two Euler formulas can be thought of as polynomials of degree one that approximate $x(t)$ in the interval $[t_0, t]$. If we compute $x(t)$ from a second-order polynomial $p(t)$ that matches the conditions that $p(t_0) = x(t_0)$, $dp/dt(t_0) = dx/dt(t_0)$ and $dp/dt(t) = dx/dt(t)$, we arrive at the **trapezoidal** rule: $x(t) = x(t_0) + 0.5hA(t_0) + 0.5hA(t)$. In this case, the truncation error is of order h^3.

At each integration step, the size of the local truncation error can be estimated. If it is too large, the step size must be reduced. An explicit formula such as the forward Euler may be used as a **predictor** giving a starting point for an implicit formula like the trapezoidal, which in turn is used as a **corrector**. The use of a predictor-corrector pair provides the base for the estimate of the local truncation error. The trapezoidal formula with varying integration step size is the main formula used in the **SPICE program**.

The two Euler formulas and the trapezoidal formula are special cases of a general linear multistep formula $\sum(a_i x_{n-i} + b_i h(dx/dt)_{n-i})$, where i goes from -1 to $m - 1$ and m is the degree of the polynomial used for the approximation of the solution curve. The trapezoidal rule, for example, is obtained by setting $a_{-1} = -1$, $a_0 = +1$ and $b_{-1} = b_0 = 0.5$, all other coefficients being zero. The formula can be regarded as being derived from a polynomial of degree r which matches $r + 1$ of the solution points x_{n-i} and their derivatives $(dx/dt)_{n-i}$.

Very fast transients often occur together with very slow transients in electronic circuits. We observe widely different time constants. The large spread in component values, for example, from large decoupling capacitors to small parasitic capacitors, implies a large spread in the modules of the eigenvalues. We say that the circuits are **stiff.** A family of implicit **multistep methods** suitable for stiff differential equations has been proposed by C.W. Gear. The methods are stable up to the polynomial of order 6. For example, the second-order **Gear formula** for fixed integration step size h may be stated as: $x_{n+1} = -(1/3)x_{n-1} + (4/3)x_n + (2/3)h(dx/dt)_{n+1}$.

By changing both the *order* of the **approximating polynomial** and the integration *step size*, the methods adapt themselves dynamically to the performance of the solution curve. The family of Gear formulas is modified into a "stiff-stable variable-order variable-step predictor-corrector" method based on implicit approximation by means of **backward difference formulas (BDFs)**. The resulting set of nonlinear equations is solved by **modified Newton–Raphson iteration**. Note that numerical integration, in a sense, is a kind of low-pass filtering defined by means of the minimum integration step [1, 2, 3, 8, 11].

6.3 Use of Simulation Programs

Since 1960, a large number of circuit-simulation programs have been developed by universities, industrial companies, and commercial software companies. In particular, the SPICE program has become a standard simulator both in the industry and in academia. Here, only a few programs, which together cover a very large number of simulation possibilities, are presented. Due to competition, there is a tendency to develop programs that are supposed to cover any kind of analysis so that only one program should be sufficient ("the Swiss Army Knife Approach"). Unfortunately this implies that the programs become very large and complex to use. Also, it may be difficult to judge the correctness and accuracy of the results of the simulation having only one program at your disposal. If you try to make the same analysis of the same model with different programs, you will frequently see that the results from the programs may not agree completely. By comparing the results, you may obtain a better feel for the correctness and accuracy of the simulation. The programs SPICE and APLAC supplemented with the programs NAP2, ESACAP, and DYNAST have proven to be a good choice in the case where a large number of different kinds of circuits and systems are to be modeled and simulated ("the Tool Box Approach"). The programs are available in inexpensive evaluation versions running on IBM compatible personal computers. The "net-list" input languages are very close, making it possible to transfer input data easily between the programs. In order to make the programs more "user-friendly" graphics interphase language "schematics-capture," where you draw the circuit on the screen, has been introduced. Unfortunately, this approach makes it a little more difficult for the user to transfer data between the programs. In the following, short descriptions of the programs are given and a small circuit is simulated in order to give the reader an idea of the capabilities of the programs.

SPICE

The first versions of **SPICE** (**Simulation Program with Integrated Circuit Emphasis version 2**), based on the modified nodal approach, were developed in 1975 at the Electronics Research Laboratory, College of Engineeering, University of California, Berkeley, CA.

SPICE is a general-purpose circuit analysis program. Circuit models may contain resistors, capacitors, inductors, mutual inductors, independent sources, controlled sources, transmission lines, and the most common semiconductor devices: diodes, bipolar junction transistors, and field effect transistors. SPICE has very detailed built-in models for the semiconductor devices, which may be described by about 50 parameters. Besides the normal dc, ac, and transient analyses, the program can make sensitivity, noise, and distorsion analysis and analysis at different temperatures. In the various commercial versions of the program many other possibilities have been added; for example, **analog behavior modeling** (poles and zeros) and statistical analysis.

In order to give an impression of the "net-list" input language, the syntax of the statements describing controlled sources is the following:

```
Voltage Controlled Current Source: Gxxx N+ N- NC+ NC- VALUE
Voltage Controlled Voltage Source: Exxx N+ N- NC+ NC- VALUE
Current Controlled Current Source: Fxxx N+ N- VNAM VALUE
Current Controlled Voltage Source: Hxxx N+ N- VNAM VALUE
```

where the initial characters of the branch name G, E, F, and H indicate the type of the branch, N+ and N– are integers ("node numbers") indicating the placement and orientation of the branch, respectively, NC+ NC– and VNAM indicate from where the control comes (VNAM is a dummy dc voltage source with value 0 inserted as an ammeter!), and VALUE specifies the numerical value of the control, which may be a constant or a polynomial expression in case of nonlinear dependent sources. Independent sources are specified with Ixxx for current and Vxxx for voltage sources.

The following input file describes an analysis of the **Chua oscillator circuit**. It is a simple harmonic oscillator with losses (C2, L2, and RL2) loaded with a linear resistor (R61) in series with a capacitor (C1) in parallel with a nonlinear resistor. The circuit is influenced by a sinusoidal voltage source VRS through a coil L1. Comments may be specified either as lines starting with an asterisk "*" or by means of a semicolon ";" after the statement on a line. A statement may continue by means of a plus "+" as the first character on the following line.

```
PSpice input file CRC-CHUA.CIR, first line, title line                :
* *:  The Chua Oscillator, sinusoidal excitation, F=150mV >            :
* :               RL2=1 ohm, RL1=0 ohm f=1286.336389 Hz                :
* : ref. K. Murali and M. Lakshmanan,                                  :
* :     Effect of Sinusoidal Excitation on the Chua's Circuit,         :
* :     IEEE Transactions on Circuits and Systems — 1:                 :
* :     Fundamental Theory and Applications,                           :
* :     vol.39, No.4, April 1992, pp. 264-270                          :
* : input source; : - - - - - - - - - - - - - - - - - - - - - - - - - -:
    VRS    7    0   sin(0    150m     1.2863363889332e+3  0  0)        :
* : choke                                                              :
    L1     6    17   80e-3    ; mH                                     :
    VRL1  17     7   DC  0    ; ammeter for measure of IL1             :
* : harmonic oscillator; : - - - - - - - - - - - - - - - - - - - - - - :
    L2     6    16   13m                                               :
    RL2   16     0   1                                                 :
    C2     6     0   1.250u                                            :
* : load; : - - - - - - - - - - - - - - - - - - - - - - - - - - - - - -:
    r61     6    10    1310                                            :
    vrrC1  10    11    DC     0                                        :
    C1     11     0    0.017u                                          :
*   i(vrr10)=current of nonlinear resistor                             :
    vrr10  10     1    DC  0                                           :
* : non-linear circuit;  : - - - - - - - - - - - - - - - - - - - - - - :
    .model n4148 d (is=0.1p rs=16 n=1); vt=n*k*T/q                     :
    d13     1     3    n4148                                           :
    d21     2     1    n4148                                           :
    rm9     2    22    47k                                             :
    vrm9   22     0    DC     -9        ; negative power supply        :
    rp9     3    33    47k                                             :
    vrp9   33     0    DC     +9                                       :
    r20     2     0    3.3k                                            :
    r30     3     0    3.3k                                            :
* : ideal op. amp.;  : - - - - - - - - - - - - - - - - - - - - - - - - :
    evop  4 0     1   5     1e+20                                      :
    r14   1 4   290                                                    :
    r54   5 4   290                                                    :
    r50   5 0  1.2k                                                    :
* : - - - - - - - - - - - - - - - - - - - - - - - - - - - - - - - - - -:
    .TRAN 0.05m 200m 0 0.018m UIC                                      :
    .plot tran v(11)                                                   :
```

```
      .probe                                                     :
      .options acct nopage opts gmin=1e-15 reltol=1e-3           :
+     abstol=1e-12 vntol=1e-12 tnom=25 itl5=0                    :
+     limpts=15000                                               :
      .end                                                       :
```

The analysis is controlled by means of the statements: .TRAN, where, for example, the maximum integration step is fixed to 18 μ sec, and .OPTIONS, where, for example, the relative truncation error is set to 1e-3. The result of the analysis is presented in Figure 6.1. It is seen that transition from **chaotic behavior** to a period 5 **limit cycle** takes place at about 100 m sec. A very important observation is that **the result of the analysis may depend on: (1) the choice of the control parameters; and (2) the order of the branches in the "net-list,"** for example, if the truncation error is set to 1e-6 instead of 1e-3 previously, the result becomes quite different. This observation is valid for all programs [4–7, 9–11].

APLAC

The program **APLAC** (originally **Analysis Program for Linear Active Circuits**) [11] has been under constant development at the Helsinki University of Technology, Finland, since 1972. Over time it has developed into an object-oriented analog circuits and systems simulation and design tool. Inclusion of a new model into APLAC requires only the labor of introducing the parameters and equations defining the model under the control of "C-macros." The code of APLAC itself remains untouched. The APLAC *Interpreter* immediately understands the syntax of the new model. APLAC accepts SPICE "net-lists" by means of the program Spi2a (SPICE to APLAC netlist converter).

APLAC is capable of carrying out dc, ac, transient, noise, oscillator, and multitone harmonic steady-state analyses and measurements using **IEEE-488 bus**. Transient analysis correctly handles, through convolution, components defined by frequency-dependent characteristics. **Monte Carlo analysis** is available in all basic analysis modes and **sensitivity analysis** in dc and ac modes. *N*-port *z*, *y*, and *s* parameters, as well as two-port *h* parameters, are available in ac analysis. In addition, APLAC includes a versatile collection of system level blocks for the simulation and design of analog and digital communication systems. APLAC includes seven different **optimization methods**. Any parameter in the design problem can be used as a variable, and any user-defined function may act as an objective. Combined time and frequency domain optimization is possible.

The file below is the APLAC "net-list" of the **Chua oscillator circuit** created by the Spi2a converter program with the PSpice file CRC-CHUA.CIR above as input. Comments are indicated by means of the dollar sign $ or the asterisk *. Unfortunately, it is necessary to manually change the file. Comments semicolon ";" and colon ":" must be replaced with "$;" and "$:". Also, Spi2a indicates a few statements as "$ not implemented."

```
$$$$$$$$$$$$$$$$$$$$$$$$$$$$$$$$$$$$$$$$$$$$$$$$$$$$$$$$$$$$$$
$$                                                          $$
$$ Spi2a - SPICE to APLAC netlist converter, version 1.26   $$
$$                                                          $$
$$ This file is created at Tue Jul 17 14:48:02 2001 $$
$$ with command: spi2a C:\WINDOWS\DESKTOP\crc-chua.cir $$
$$                                                          $$
$$$$$$$$$$$$$$$$$$$$$$$$$$$$$$$$$$$$$$$$$$$$$$$$$$$$$$$$$$$$$

$PSpice input file CRC-CHUA.CIR, first line, title line        $:

  Prepare gmin=1e-15 ERR=1e-3 ABS_ERR=1e-12 TNOM=(273.15+(25))
$ .options acct nopage opts gmin=1e-15 reltol=1e-3             $:
$+    abstol=1e-12 vntol=1e-12 tnom=25 itl5=0                  $:
```

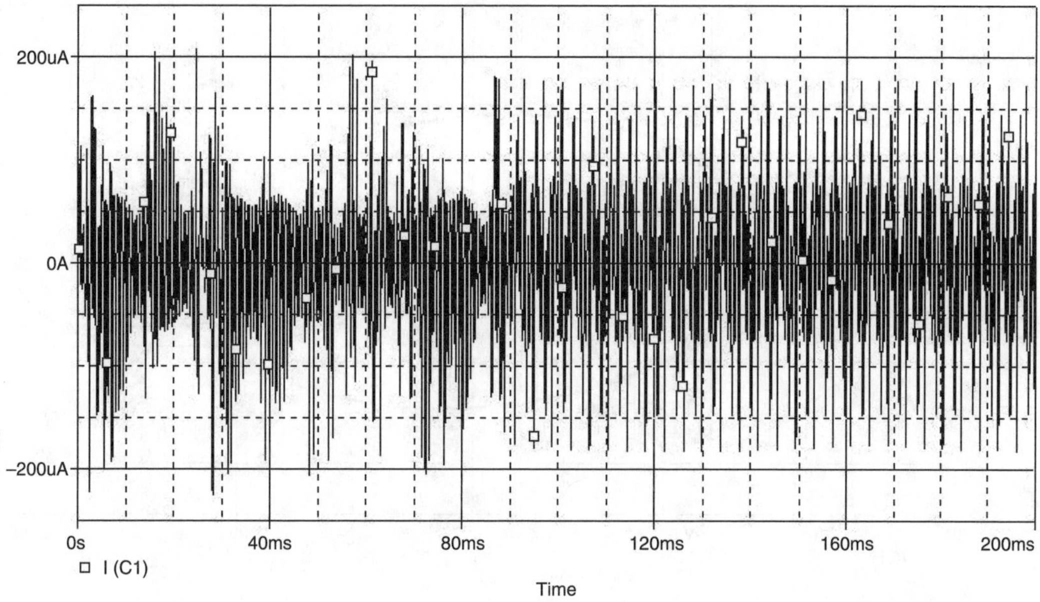

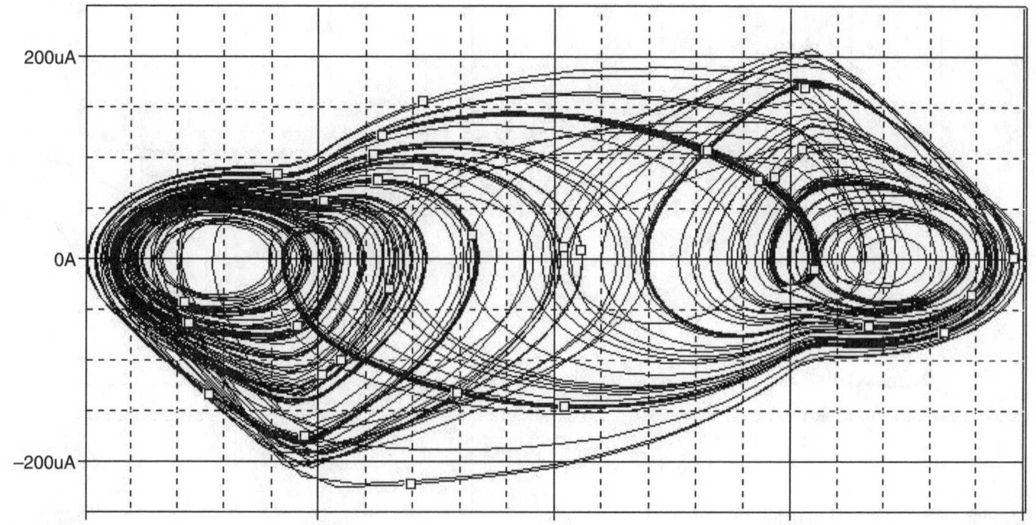

FIGURE 6.1 (top) PSPICE analysis. The current of C1: I(C1) as function of time in the interval 0 200 ms. (bottom) The current of C1: I(C1) as function of the voltage of C1: V(11). (next page, top) The current of C1: I(C1) as function of the voltage of C1: V(11) in the time interval 100 to 200 ms. (next page, bottom) The voltage of C2: V(6) as function of the voltage of C1: V(11) in the time interval 100 to 200 ms.

```
$+    limpts=15000                                            $:
$ .MODEL and .PARAM definitions                               $:
  Model "n4148" is=0.1p rs=16 n=1
+ $;=vt n*k*T/q                                               $:
$ Circuit definition                                          $:
$ Not implemented                                             $:
$ VRS 7 0 sin(0 150m 1.2863363889332e+3 0 0)                  $:
```

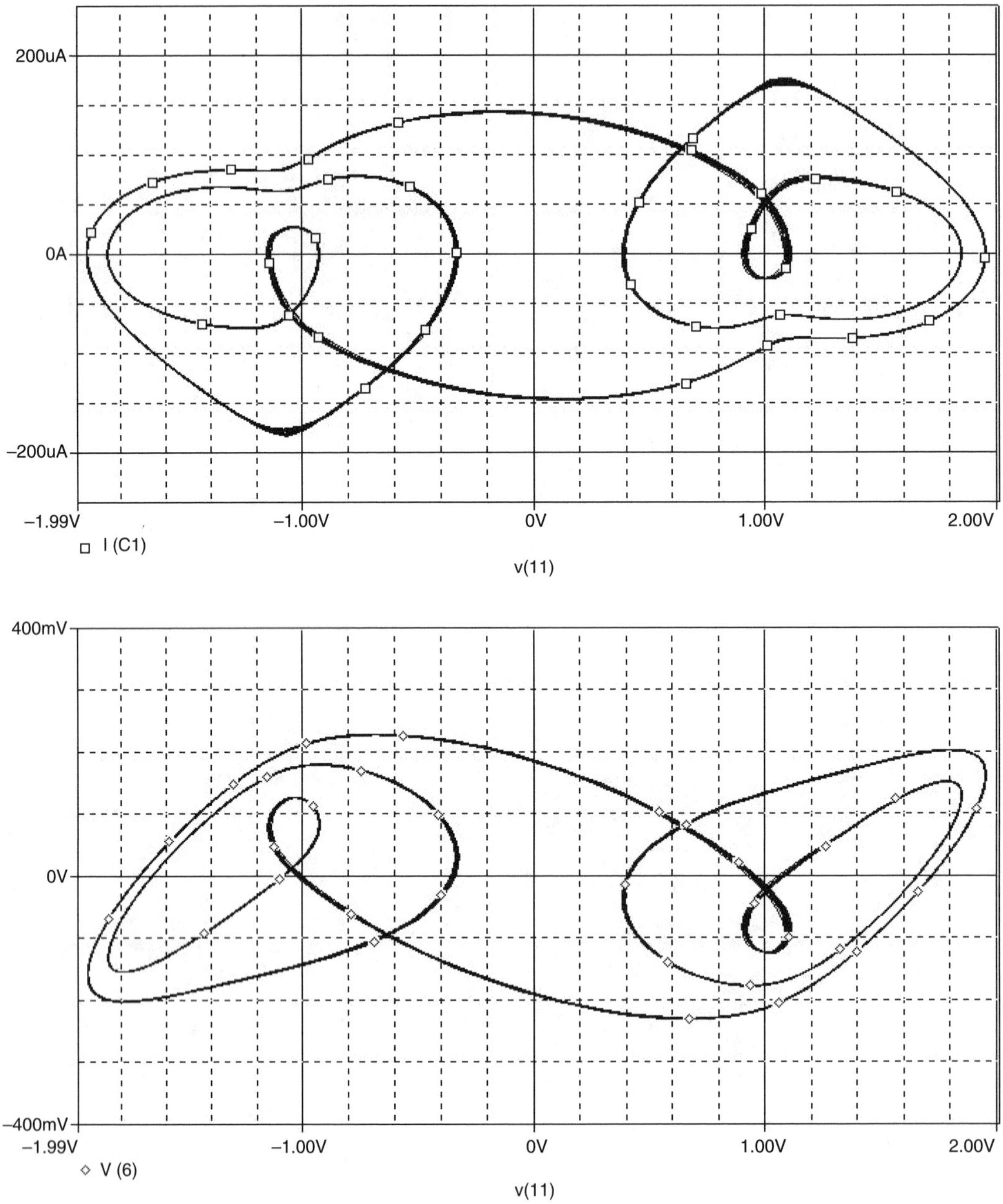

FIGURE 6.1 (continued).

```
  Volt VRS 7 0 sin=[0, 150m, 1.2863363889332e+3, 0, 0]
* $: choke                                                        $:
  Ind L1 6 17 80e-3 $; mH                                         $:
  Volt VRL1 17 7 DC={VRL1=0} $ $; ammeter for measure of IL1      $:
* $: harmonic oscillator$; $: - - - - - - - - - - - - - - - - - - - - - - - $:
+ I=I_VRL1
  Ind L2 6 16 13m                                                 $:
  Res RL2 16 0 1                                                  $:
  Cap C2 6 0 1.250u                                               $:
  Res r61 6 10 1310                                               $:
```

```
$ Not implemented                                                    $:
$ vrrC1  10  11  DC  0                                               $:
  Volt  vrrC1  10  11  DC={vrrC1=0}
+ I=IC1
  Cap  C1  11  0  0.017u                                             $:
$ Not implemented                                                    $:
$ vrr10  10  1  DC  0                                                $:
  Volt  vrr10  10  1  DC={vrr10=0}
+ I=IRNL
* $: non-linear circuit$; $: - - - - - - - - - - - - - - - - - - - - -$:
  Diode  d13  1  3  MODEL="n4148"                                    $:
  Diode  d21  2  1  MODEL="n4148"                                    $:
  Res  rm9  2  22  47k                                               $:
  Volt  vrm9  22  0  DC={vrm9=-9}  $ $; negative power supply        $:
+ I=I_vrm9
  Res  rp9  3  33  47k                                               $:
$ Not implemented                                                    $:
$ vrp9  33  0  DC  +9                                                $:
  Volt  vrp9  33  0  DC={vrp9=9}  $ +9 must be 9
  Res  r20  2  0  3.3k                                               $:
  Res  r30  3  0  3.3k                                               $:
  VCVS  evop  4  0  1  1  5  [1e+20]  LINEAR                         $:
  Res  r14  1  4  290                                                $:
  Res  r54  5  4  290                                                $:
  Res  r50  5  0  1.2k                                               $:

$$ Analysis commands                                                 $:
$$ .TRAN  0.05m  200m  0  0.018m  UIC                                $:
$ Sweep  "TRAN Analysis  1"
$+ LOOP  (1+(200m-(0))/(0.05m))  TIME  LIN  0  200m  TMAX=0.018m
$+ NW=1  $ UIC                                                       $:
$$ .plot  tran  v(11)                                                $:
$ Show  Y  Vtran(11)  $                                              $:
$ EndSweep

$ the following lines are added and the sweep above is commented
Sweep  "TRAN Analysis  2"
+ LOOP  (4001)  TIME  LIN  0  200m  TMAX=0.018m
$+ NW=1  $ UIC                                                       $:
$ .plot  tran  v(11)                                                 $:
Show  Y  Itran(IC1)  X  Vtran(11)  $                                 $:
EndSweep

$.probe                                                              $:
```

The result of the analysis is presented in Figure 6.2. It is observed that limit cycle behavior is not obtained in the APLAC analysis in the time interval from 0 to 200 ms.

NAP2

The first versions of **NAP2** (**Nonlinear Analysis Program version 2**) [9], based on the extended nodal equation formulation were developed in 1973 at the Institute of Circuit Theory and Telecommunication, Technical University of Denmark, Lyngby, Denmark.

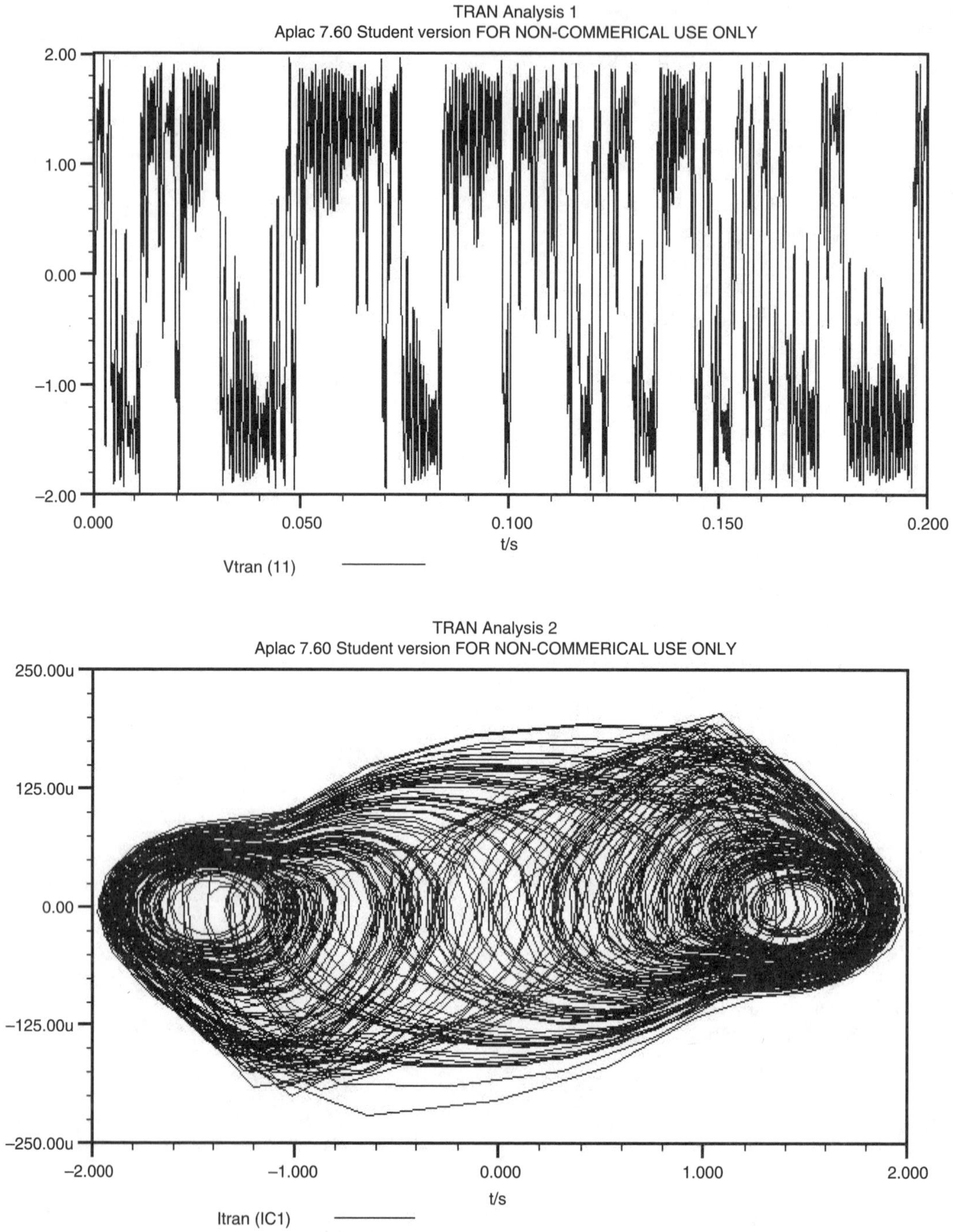

FIGURE 6.2 (top) The voltage of C1: V(11) as function of time in the interval 0 to 200 ms. (bottom) The current of C1: IC1 as function of the voltage of C1: V(11).

NAP2 is a general-purpose circuit analysis program. Circuit models may contain resistors, conductors, capacitors, inductors, mutual inductors, ideal operational amplifiers, independent sources, controlled sources, and the most common semiconductor devices: diodes, bipolar junction transistors, and field effect transistors. NAP2 has only simple built-in models for the semiconductor devices, which require about 15 parameters. Besides the normal dc, ac, and transient analyses, the program can make parameter variation analysis. Any parameter (e.g., component value or temperature) may be varied over a range in an arbitrary way and dc, ac, or transient analysis may be performed for each value of the parameter. **Optimization** of dc bias point ("given: voltages, find: resistors") is possible. **Event detection** is included so that it is possible to interrupt the analysis when a certain signal, for example, goes from a positive to a negative value. The results may be combined into one output plot. It is also possible to calculate the **poles** and **zeros** of driving point and transfer functions for the linearized model in a certain bias point. **Eigenvalue technique** (based on the **QR algorithm** by J. G. F. Francis) is the method behind the calculation of poles and zeros. **Group delay** (i.e., the derivative of the phase with respect to the angular frequency) is calculated from the poles and zeros. This part of the program is available as an independent program named **ANP3** (**Analytical Network Program version 3**).

In order to give an impression of the "net-list" input language, the syntax of the statements describing controlled sources is as follows:

```
Voltage Controlled Current Source:   Ixxx   N+   N-   VALUE VByyy
Voltage Controlled Voltage Source:   Vxxx   N+   N-   VALUE VByyy
Current Controlled Current Source:   Ixxx   N+   N-   VALUE IByyy
Current Controlled Voltage Source:   Vxxx   N+   N-   VALUE Ibyyy
```

where the initial characters of the branch name I and V indicate the type of the branch, N+ and N- are integers ("node numbers") indicating the placement and orientation of the branch, respectively, and VALUE specifies the numerical value of the control, which may be a constant or an arbitrary functional expression in case of nonlinear control. IB and VB refer to the current or voltage of the branch, respectively, from where the control comes. If the control is the time derivative of the branch signal, SI or SV may be specified. Independent sources must be connected to a resistor R or a conductor G as follows: Rxxx N+ N- VALUE E = VALUE and Gxxx N+ N- VALUE J = VALUE, where VALUE may be any function of time, temperature, and components.

The input file "net-list" below describes the same analysis of the **Chua oscillator circuit** as performed by means of SPICE and APLAC. The circuit is a simple harmonic oscillator with losses (C2, L2 and RL2) loaded with a linear resistor (R61) in series with a capacitor (C1) in parallel with a nonlinear resistor. The circuit is excited by a sinusoidal voltage source through a coil L1. The frequency is specified as angular frequency in rps. It is possible to specify more than one statement on one line. Colon ":" indicate start of a comment statement and semicolon ";" indicates end of a statement. The greater than character ">" indicates continuation of a statement on the following line. It is observed that most of the lines are comment lines with the PSPICE input statements.

```
*circuit; *list 2, 9;  : file CRC-CHUA.NAP
*:   PSpice input file CRC-CHUA.CIR, first line, title line >         :
:    translated into NAP2 input file
: The Chua Oscillator, sinusoidal excitation, F=150mV >               :
:             RL2=1 ohm, RL1=0 ohm f=1286.336389 Hz                   :
: ref. K. Murali and M. Lakshmanan,                                   :
:      Effect of Sinusoidal Excitation on the Chua's Circuit,         :
:      IEEE Transactions on Circuits and Systems — 1:                 :
:      Fundamental Theory and Applications,                           :
:      vol.39, No.4, April 1992, pp. 264-270                          :
: input source;  : - - - - - - - - - - - - - - - - - - - - - - - - - :
:   VRS  7  0   sin(0   150m   1.2863363889332e+3  0  0)              :
  sin/sin/; Rs  7  0  0  e=150m*sin(8.0822898994674e+3*time)          :
: choke  ;    L1   6   17     80mH;  RL1  17  7  0                    :
```

```
: L1     6    17  80e-3         ;:mH                                     :
: VRL1   17    7  DC    0       ;:ammeter for measure of IL1             :
:                               :- - - - - - - - - - - - - - - - - - - - :
: harmonic oscillator;  L2   6    16   13mH;    RL2  16   0    1         :
                        C2   6    0    1.250uF                           :
: L2     6    16  13m                                                    :
: RL2    16    0  1                                                      :
: C2     6     0  1.250u                                                 :
:              :- - - - - - - - - - - - - - - - - - - - - - - - - - - -  :
: load;  r61   6  10   1310;  rrc1  10   11    0;  c1   11   0    0.017uF :
       rr10  10   1    0:     irr10=current of nonlinear resistor        :
: r61      6  10  1310                                                   :
: vrrC1 10    11  DC   0                                                 :
: C1      11   0  0.017u                                                 :
: i(vrr10)=current of nonlinear resistor                                 :
: vrr10   10   1  DC   0                                                 :
: non-linear circuit;           :- - - - - - - - - - - - - - - - - - - - :
: .model   n4148   d   (is=0.1p  rs=16  n=1);    vt=n*k*T/q              :
  n4148    /diode/      is=0.1p gs=62.5m         vt=25mV;                :
  td13   1  3 n4148;   td21  2  1 n4148;                                 :
: d13    1    3 n4148                                                    :
: d21    2    1  n4148                                                   :
  rm9    2    0   47k  e=-9; rp9  3  0  47k  E=+9;                       :
: rm9    2    22  47k                                                    :
: vrm9   22   0   DC   -9; negative power supply                         :
: rp9    3    33  47k                                                    :
: vrp9   33   0   DC   +9                                                :
  r20    2    0   3.3k;  r30  3  0  3.3k;                                :
: r20    2    0   3.3k                                                   :
: r30    3    0   3.3k                                                   :
: ideal op. amp.;  :  - - - - - - - - - - - - - - - - - - - - - - - - -  :
  gop    1    5   0;    vop  4  0  vgop: no value means infinite value;  :
: evop   4    0   1    5      1e+20                                      :
  r14    1    4   290; r54  5  4  290; r50  5  0  1.2k;                  :
: r14    1    4   290                                                    :
: r54    5    4   290                                                    :
: r50    5    0   1.2k                                                   :
: - - - - - - - - - - - - - - - - - - - - - - - - - - - - - - - - - - -  :
  *time time 0 200m  : variable order, variable step                     :
: .TRAN 0.05m   200m  0  0.018m UIC                                      :
  *tr vnall *plot(50 v6) v1 *probe                                       :
: .plot tran v(11)                                                       :
: .probe                                                                 :
  *run cycle=15000 minstep=1e-20 >                                       :
       trunc=1e-3 step=50n                                               :
: .options acct nopage opts gmin=1e-15 reltol=1e-3                       :
:+abstol=1e-12 vntol=1e-12 tnom=25 itl5=0                                :
:+limpts=15000                                                           :
: .end                                                                   :
  *end                                                                   :
```

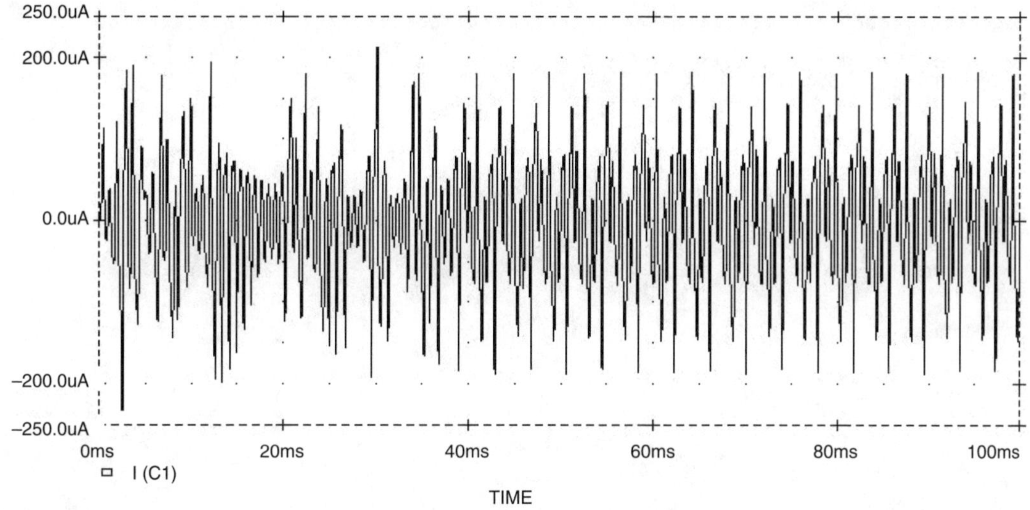

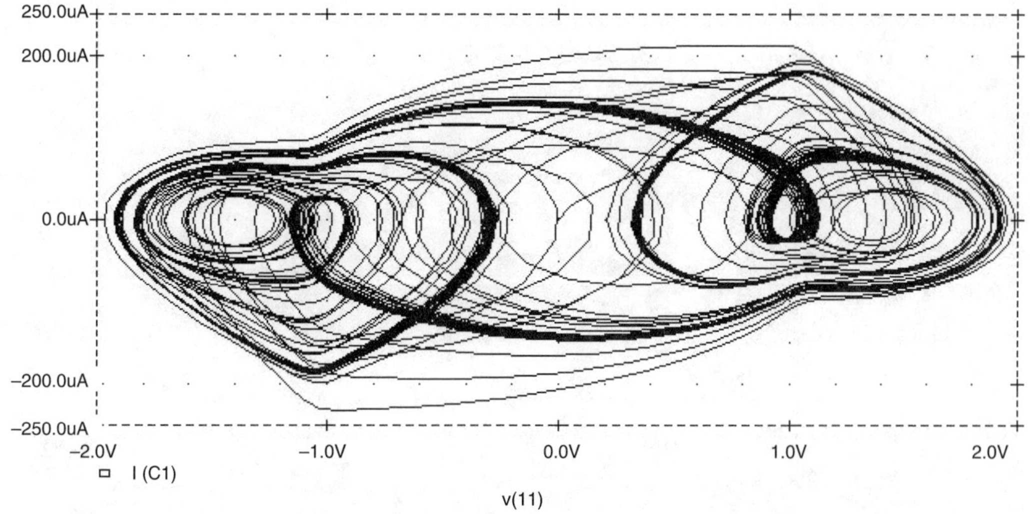

FIGURE 6.3 (top) NAP2 analysis. The current of C1: I(C1) as function of time in the interval 0 to 100 ms. (bottom) The current of C1: I(C1) as function of the voltage of C1: V(11) in the time interval 0 to 100 ms. (next page, top) The current of C1: I(C1) as function of time in the interval 180 to 200 ms. (next page, bottom) The current of C1: I(C1) as function of the voltage of C1: V(11) in the time interval 100 to 200 ms.

The program options are set by means of the statement *RUN, where, for example, the minimum integration step is set to 1e-20 s and the relative truncation error is set to 1e-6. The result of the analysis is presented in Figure 6.3. It can be observed that transition from **chaotic behavior** to a period 5 **limit cycle** takes place at about 50 ms. If we compare to the results obtained above by means of SPICE and APLAC, we see that although the three programs are "modeled and set" the same way, for example, with the same relative tolerance 1e-3, the results are different due to the chaotic nature of the circuit and possibly also due to the different strategies of equation formulation and solution used in the three programs. For example, SPICE uses the **trapezoidal integration method** with variable step; APLAC and NAP2 use the **Gear integration methods** with variable order and variable step.

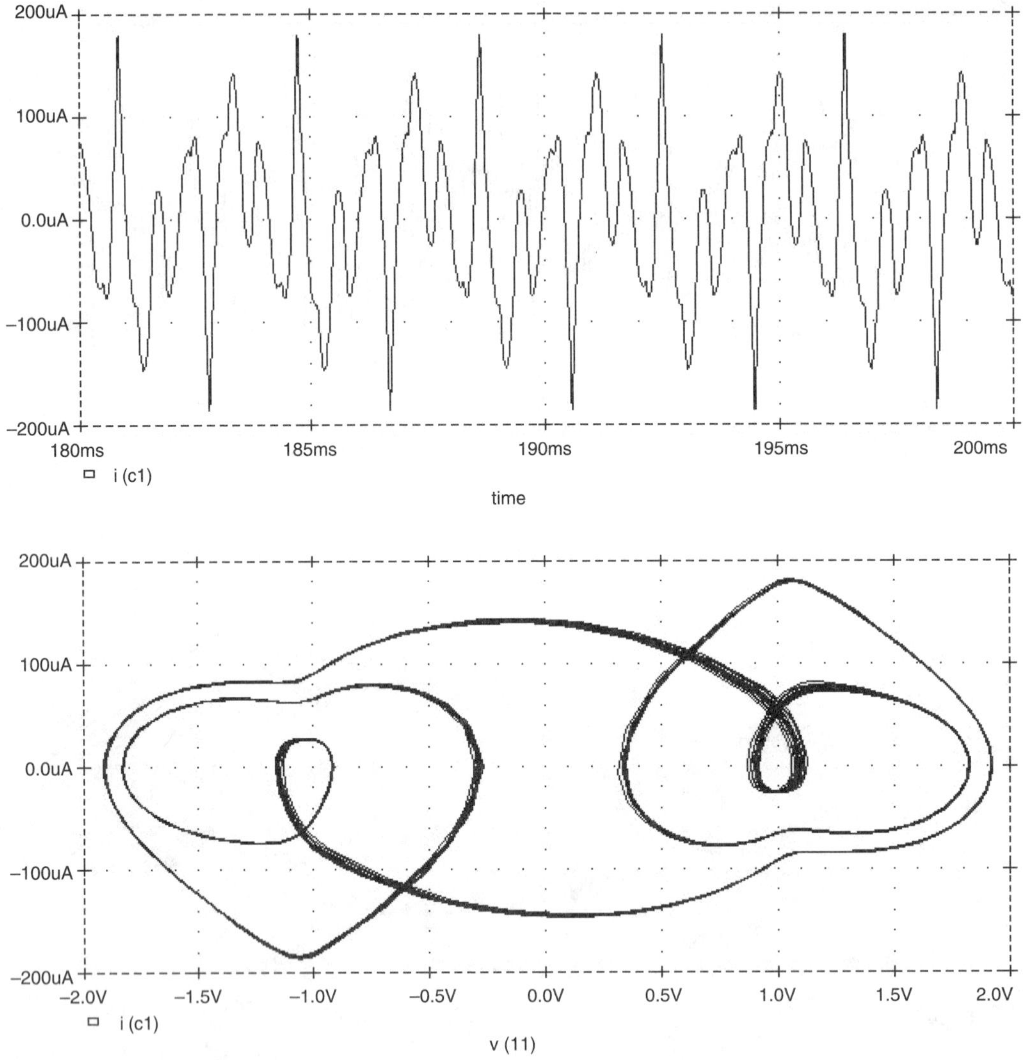

FIGURE 6.3 (continued).

ESACAP

The first versions of **ESACAP** (**Engineering System and Circuit Analysis Program**) based on the extended nodal equation formulation were developed in 1979 at Elektronik Centralen, Hoersholm, Denmark, for ESA (the European Space Agency) as a result of a strong need for a **simulation language** capable of handling interdisciplinary problems (e.g., coupled electrical and thermal phenomena). ESACAP was therefore born with facilities that have only recently been implemented in other simulation programs (e.g., facilities referred to as behavioral or functional modeling).

ESACAP carries out analyses on nonlinear systems in dc and in the time domain. The nonlinear equations are solved by a hybrid method combining the robustness of the gradient method with the good convergence properties of the **Newton–Raphson method.** The **derivatives** required by the Jacobian matrix are **symbolically evaluated** from arbitrarily complex arithmetic expressions and are therefore exact. The symbolic evaluation of derivatives was available in the very first version of ESACAP. It has now become a general numerical discipline known as **automatic differentiation**. The time-domain solution is found by numerial integration implemented as backward difference formulas, **BDFs**, of variable step and orders 1 through 6 (**modified Gear method**). An efficient **extrapolation method** (the **epsilon algorithm**) accelerates the asymptotic solution in the periodic steady-state case.

Frequency-domain analyses may be carried out on linear or linearized systems (e.g., after a dc analysis). Besides complex transfer functions, special outputs such as **group delay** and **poles/zeros** are available. The group delay is computed as the sum of the frequency sensitivities of all the reactive components in the system. Poles and zeros are found by a numerical interpolation of transfer functions evaluated on a circle in the complex frequency plane. ESACAP also includes a complex number postprocessor by means of which any function of the basic outputs can be generated (e.g., stability factor, s-parameters, complex ratios).

Sensitivities of all outputs with respect to all parameters are available in all analysis modes. The automatic differentiation combined with the adjoint network provides exact partial derivatives in the frequency domain. In the time domain, integration of a sensitivity network (using the already LU-factorized Jacobian) provides the partial derivatives as functions of time.

The ESACAP language combines procedural facilities, such as if-then-else, assignment statements, and do-loops, with the usual description by structure (nodes/branches). Arbitrary expressions containing system variables and their derivatives are allowed for specifying branch values thereby establishing any type of nonlinearity. System variables of non-potential and non-current type may be defined and used everywhere in the description (e.g., for defining power, charge). The language also accepts the specification of nonlinear differential equations. Besides all the standard functions known from high-level computer languages, ESACAP provides a number of useful functions. One of the most important of these functions is the **delay function**. The delay function returns one of its arguments delayed by a specified value, which in turn may depend on system variables. Another important function is the **threshold switch** — the ZEROREF function — used in if-then-else constructs for triggering discontinuities. The ZEROREF function interacts with the integration algorithm that may be reinitialized at the exact threshold crossing. The ZEROREF function is an efficient means for separating cause and action in physical models thereby eliminating many types of causality problems. **Causality problems** are typical examples of bad modeling techniques and the most frequent reason for divergence in the simulation of dynamic systems.

Typical ESACAP applications include electronics as well as **thermal and hydraulic systems**. The frequency domain facilities have been a powerful tool for designing stable control systems including nonelectronics engineering disciplines.

In order to give an idea of the input language, the syntax of the statements describing sources is as follows:

```
Current Source:   Jxxx(N+, N-)=VALUE;
Voltage Source:   Exxx(N+, N-)=VALUE;
```

where the initial characters of the branch name: J and E indicate the type of the branch, N+ and N- are node identifiers (character strings), which, as a special case, may be integer numbers ("node numbers"). The node identifiers indicate the placement and orientation of the branch. The VALUE specifies the numerical value of the source, which may be an arbitrary function of time, temperature, and parameters as well as system variables (including their time derivatives). Adding an apostrophe references the time derivative of a system variable. V(N1,N2)', for example, is the time derivative of the voltage drop from node N1 to node N2.

The next input file — actually, a small program written in the ESACAP language — describes an analysis of a tapered transmission line. The example shows some of the powerful tools available in the ESACAP language such as: (1) the delay function; (2) the do-loop; and (3) the sensitivity calculation. The description language of ESACAP is a genuine simulation and modeling language. However, for describing simple systems, the input language is just slightly more complicated than the languages of SPICE, APLAC, and NAP2. Data is specified in a number of blocks ("chapters" and "sections") starting with $$ and $. Note how the line model is specified in a do-loop where ESACAP creates nodes and branches of a ladder network [10].

```
Example.155 Tapered line in the time domain.
# Calculation of sensitivities.
# This example shows the use of a do-loop for the discretization of a
# tapered transmission line into a chain of short line segments. The
# example also demonstrates how to specify partial derivatives of any
# parameter for sensitivity calculations.
$$DESCRIPTION # chapter - - - - - - - - - - - - - - - - - - - - -
$CON: n_sections=60; END; # section - - - - - - - - - - - - - -
# n_sections is defined as a globally available constant.
# Only this constant needs to be modified in order to change
# the resolution of discretization
# Transmission line specified by characteristic impedance and length.
# Modelled by the ESACAP delay function (DEL).
$MODEL: LineCell(in,out): Z0,length;
 delay=length/3e8;
 J_reverse(0,in)=DEL(2*V(out)/Z0-I(J_forward), delay);
 J_forward(0,out)=DEL(2*V(in)/Z0-I(J_reverse), delay);
 G1(in,0)=1/Z0; G2(out,0)=1/Z0;
END; # end of section - - - - - - - - - - -
# Tapered line specified by input and output impedance and length
# This model calls LineCell n_sections times in a do-loop.
$MODEL: TaperedLine(in,out): Z1,Z2,length;
 ALIAS_NODE(in,1);                # Let node in and 1 be the same.
 ALIAS_NODE(out,[n_sections+1]);  # Let node out and n_sections+1 be
                                  # the same.
# Notice that values in square brackets become part of an identifier
FOR (i=1, n_sections) DO
X[i]([i],[i+1])=LineCell(Z1+i*(Z2-Z1)/n_sections, length/n_sections);
ENDDO;
END; # end of section - - - - - - - - - - -
# Main network calls the model of the tapered line and terminates
# it by a 50 ohm source and 100 ohm load.
$NETWORK:    # section - - - - - - - - - - -
IF(TIME.LT.1n) THEN
 Esource(source,0)=0;
ELSE
 Esource(source,0)=1;
ENDIF;
# Esource(source,0)=TABLE(TIME,(0,0),(1n,0),(1.001n,1),(10n,1));
 Rsource(source,in)=50;
 Rload(out,0)=100;
 Z1=50; Z2=100; length=1;
 X1(in,out)=TaperedLine(Z1,Z2,length);
END;   # end of section - - - - - - - - - - -
# Time-domain analysis
$$TRANSIENT   # chapter - - - - - - - - - - - - - - - - -
# Analysis parameters
$PARAMETERS:  # section - - - - - - - - - - - - - - - - -
 TIME=0,20n;                      # Total time sweep
 HMAX=2p;                         # Max integration step
END; # end of section - - - - - - - - - - - - - -
```

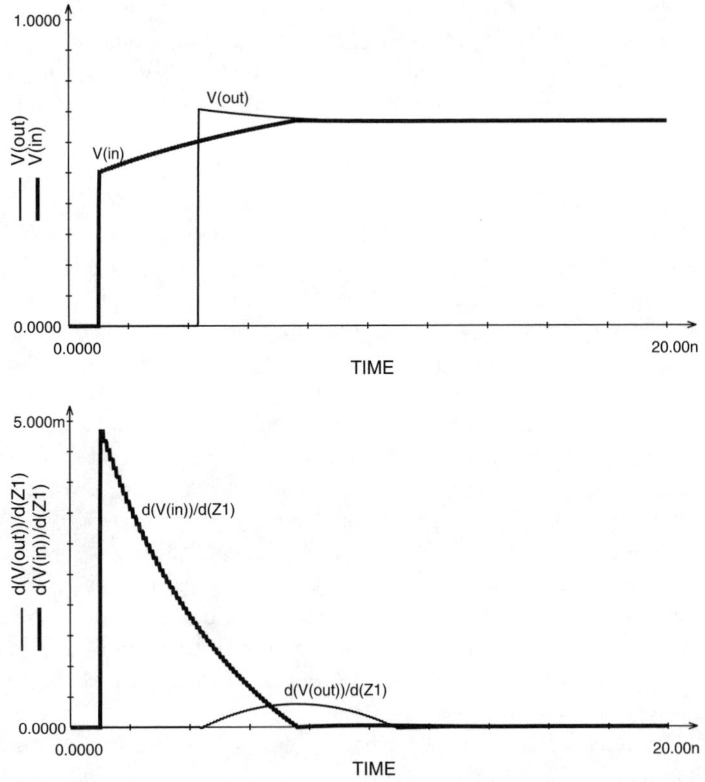

FIGURE 6.4 (top) ESACAP analysis. The input voltage of the tapered line: V(in) and the output voltage of the tapered line: V(out) as functions of time in the interval from 0 to 20 ns. (bottom) The sensitivities of V(in) and V(out) with respect to Z1.

```
# Specification of desired results. Adding an exclamation mark (!) to an
# output will show the value on the ESACAP real-time graphics display.
$DUMP: # section – - - - - - - - - - - - - - - - - - -
 FILE=<dump.155>; TIME=0,20n,20p;
 TIME; V(in)!; V(out)!;
 (V(in),DER(Z1))!;           # Partial derivatives with respect
 (V(out),DER(Z1))!;          # to Z1
END;    # end of section - - - - - - - - - -
$$STOP      # chapter - - - - - - - - - - - - - - - -
```

The result of the analysis is presented in Figure 6.4.

DYNAST

DYNAST (**DYNAmic Simulation Tool**) [5] was developed in 1992 in a joint venture between The Czech Technical University, Prague, the Czech Republic and Katholieke Universiteit Leuven, Heverlee, Belgium. The program was developed as an interdisciplinary simulation and design tool in the field of "mechatronics" (mixed mechanical/electrical systems).

The main purpose of DYNAST is to simulate dynamic systems decomposed into subsystems defined independently of the system structure. The structure can be hierarchical. DYNAST is a versatile software tool for modeling, simulation, and analysis of general linear as well as nonlinear dynamic systems, both in time and frequency domain. **Semisymbolic analysis** is possible (poles and zeros of network functions, inverse Laplace transformation using closed-form formulas).

Three types of subsystem models are available in DYNAST. The program admits systems descriptions in the form of (1) a multipole diagram respecting physical laws; (2) a causal or an acausal block diagram; (3) a set of equations; or (4) in a form combining the above approaches.

1) In DYNAST the physical-level modeling of dynamic systems is based on subsystem **multipole models** or **multiterminal models**. These models respect the continuity and compatibility postulates that apply to all physical energy-domains. (The former postulate corresponds to the laws of conservation of energy, mass, electrical charge, etc.; the latter is a consequence of the system connectedness). The multipole poles correspond directly to those subsystem locations in which the actual energetic interactions between the subsystems take place (such as shafts, electrical terminals, pipe inlets, etc.). The interactions are expressed in terms of products of complementary physical quantity pairs: the **through variables** flowing into the multipoles via the individual terminals, and the **across variables** identified between the terminals.

2) The **causal blocks**, specified by explicit functional expressions or transfer functions, are typical for any simulation program. But the variety of basic blocks is very poor in DYNAST, as its language permits definition of the block behavior in a very flexible way. Besides the built-in basic blocks, user specified multi-input multi-output macroblocks are available as well. The causal block interconnections are restricted by the rule that only one block output may be connected to one or several block inputs. In the DYNAST block variety, however, causal blocks are also available with no restrictions imposed on their inteconnections, as they are defined by implicit-form expressions.

3) DYNAST can also be used as an **equation solver** for systems of nonlinear first-order algebro-differential and algebraic equations in the implicit form. The equations can be submitted in a natural way (without converting them into block diagrams) using a rich variety of functions including the Boolean, event-dependent, and tabular ones. The equations, as well as any other input data, are directly interpreted by the program without any compilation.

The equation formulation approach used for both multipoles and block diagrams evolved from the **extended method of nodal voltages** (**MNA**) developed for electrical systems. Because all the equations of the diagrams are formulated simultaneously, no problems occur with the *algebraic loops*. As the formulated equations are in the implicit form, it does not create any problems with the *causality* of the physical models.

The integration method used to solve the nonlinear algebro-differential and algebraic equations is based on a stiff-stable implicit backward-differentiation formula (a **modified Gear method**). During the integration, the step length as well as the order of the method is varied continuously to minimize the computational time while respecting the admissible computational error. Jacobians necessary for the integration are computed by **symbolic differentiation**. Their evaluation as well as their LU decomposition, however, is not performed at each iteration step if the convergence is fast enough. Considerable savings of computational time and memory are achieved by a consistent **matrix sparsity** exploitation.

To accelerate the computation of periodic responses of weakly damped dynamic systems, the iterative **epsilon-algorithm** is utilized. Also, fast-Fourier transformation is available for **spectral analysis** of the periodic steady-state responses.

DYNAST runs under DOS- or WINDOWS-control on IBM-compatible PCs. Because it is coded in FORTRAN 77 and C-languages, it is easily implemented on other platforms. It is accompanied by a menu-driven *graphical environment*. The block and multiport diagrams can be submitted in a graphical form by a **schematic capture editor**. DYNAST can be easily augmented by various pre- and postprocessors because all its input and output data are available in the ASCII code. Free "net-list" access to DYNAST is possible by means of e-mail or online over the Internet [5].

References

[1] Calahan, D. A., *Computer-Aided Network Design*, New York: McGraw-Hill, 1972.

[2] Chua, L. O. and P.-M. Lin, *Computer-Aided Analysis of Electronic Circuits*, Englewood Cliffs, NJ: Prentice Hall, 1975.

[3] Dertouzos, M. L. et al., *Systems, Networks, and Computation: Basic Concepts*, New York: McGraw-Hill, 1972.

[4] Intusoft, *IsSpice3 — ICAPS System Packages*, San Pedro, CA: Intusoft, 1994, http://www.intusoft.com/.
[5] Mann, H., *DYNAST — A Multipurpose Engineering Simulation Tool*, Prague, Czech Republic: The Czech Technical University, 1994, http://www.it.dtu.dk/ecs/teacher.htm, http://icosym.cvut.cz/cacsd/msa/onlinetools.html, http://icosym.cvut.cz/dyn/download/public/.
[6] Meta-Software, *HSPICE User's Manual H*9001, Campbell, CA: Meta-Software, 1990.
[7] MicroSim, *PSpice — The Design Center*, Irvine, CA: MicroSim, 1994, http://www.cadencepcb.com and http://www.pspice.com/.
[8] Ostrowski, A.M., *Solution of Equations and Systems of Equations*, New York: Academic Press, 1966.
[9] Rübner-Petersen, T., *NAP2 — A Nonlinear Analysis Program for Electronic Circuits, version 2, Users Manual* 16/5-73, Report IT-63, ISSN-0105-8541, Lyngby, Denmark: Institute of Circuit Theory and Telecommunication, Technical University of Denmark, 1973, http://www.it.dtu.dk/ecs/programs.htm#nnn, http://www.it.dtu.dk/ecs/napanp.htm.
[10] Stangerup, P., *ESACAP User's Manual*, Nivaa, Denmark: StanSim Research Aps., 1990, http://www.it.dtu.dk/ecs/esacap.htm.
[11] Valtonen, M. et.al., *APLAC — An Object-Oriented Analog Circuit Simulator and Design Tool*, Espoo, Finland: Circuit Theory Lab., Helsinki University of Technology and Nokia Corporation Research Center, 1992, http://www.aplac.hut.fi/aplac/general.html, http://www.aplac.com/.
[12] Vlach, J. and K. Singhal, *Computer Methods for Circuit Analysis and Design*, New York: Van Nostrand Reinhold, 1983.
[13] Funk, D. G. and Christiansen, D. Eds., *Electronic Engineers' Handbook*, 3rd ed., New York: McGraw-Hill, 1989.

7

Cellular Neural Networks

Tamás Roska
Computer and Automation Research Institute of the Hungarian Academy of Sciences and the Pázmány P. Catholic University, Budapest

Ákos Zarándy
Hungarian Academy of Science

Csaba Rekeczky
Hungarian Academy of Sciences

7.1 Introduction: Definition and Classification

Current VLSI technologies provide for the fabrication of chips with several million transistors. With these technologies a single chip may contain one powerful digital processor, a huge memory containing millions of very simple units placed in a regular structure, and other complex functions. A powerful combination of a simple logic processor placed in a regular structure is the cellular automaton invented by John von Neumann. The cellular automaton is a highly parallel computer architecture. Although many living neural circuits resemble this architecture, the neurons do not function in a simple logical mode: they are analog "devices." The cellular neural network architecture, invented by Leon O. Chua and his graduate student Lin Yang [1] has both properties: the cell units are nonlinear continuous time dynamic elements placed in a cellular array. Of course, the resulting nonlinear dynamics in space could be extremely complex. The inventors, however, showed that these networks can be designed and used for a variety of engineering purposes, while maintaining stability and keeping the dynamic range within well-designed limits. Subsequent developments have uncovered the many inherent capabilities of this architecture (IEEE conferences: CNNA-90, CNNA-92, CNNA-94, 96, 98, 00, 02; Special issues: *Int. J. Circuit Theory and Applications*, 1993, 1996, 1998, 2002 and *IEEE Transactions on Circuits and Systems, I and II*, 1993, 1999, etc.). In the circuit implementation, unlike analog computers or general neural networks, the CNN cells are not the ubiquitous high-gain operational amplifiers. In most practical cases, they are either simple unity gain amplifiers or simple second- or third-order simple dynamic circuits with one to two simple nonlinear components. Tractability in the design and the possibility for exploiting the complex nonlinear dynamic phenomena in space, as well as the trillion operations per second computing speed in a single chip are but some of the many attractive properties of cellular neural networks. The trade-off is in the accuracy; however, in many cases, the accuracy achieved with current technologies is enough to solve a lot of real-life problems.

The cellular neural/nonlinear network (henceforth called CNN) is a new paradigm for multidimensional, nonlinear, dynamic processor arrays [1], [23]. The mainly uniform *processing elements*, called *cells* or artificial neurons, are placed on a regular *geometric grid* (with a square, hexagonal, or other pattern).

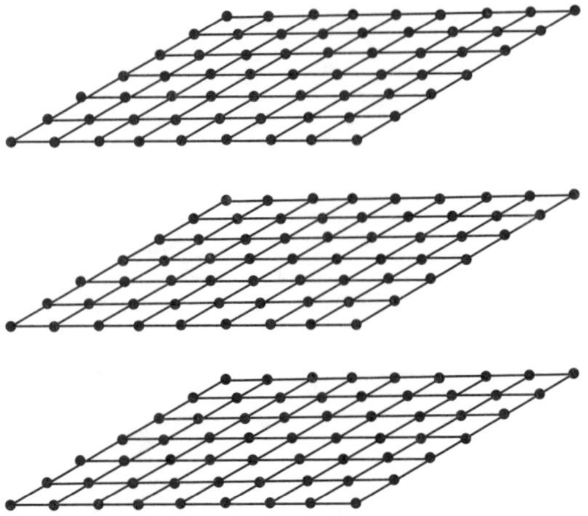

FIGURE 7.1 A CNN grid structure with the processing elements (cells) located at the vertices.

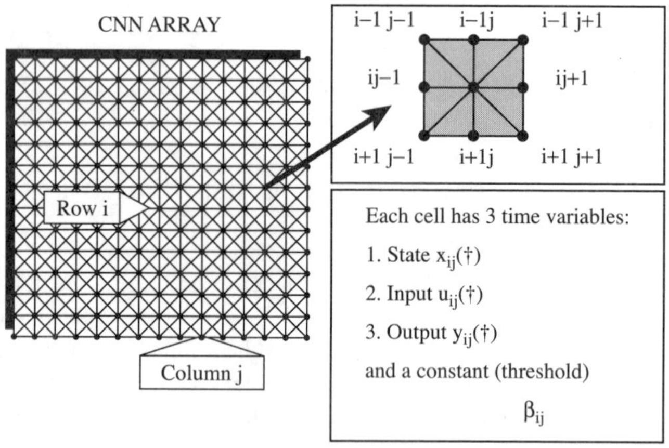

FIGURE 7.2 A single, two-dimensional CNN layer and a magnified cell with its neighbor cells with the normal neighborhood radius $r = 1$.

This grid may consist of several two-dimensional layers packed upon each other (Figure 7.1). Each processing element or cell is an analog dynamical system, the state (x), the input (u), and the output (y) signals are analog (real-valued) functions of time (both continuous-time and discrete-time signals are allowed). The *interconnection and interaction pattern* assumed at each cell is mainly local within a neighborhood N_r, where N_r denotes the first "r" circular layers of surrounding cells. Figure 7.2 shows a two-dimensional layer with a square grid of interconnection radius of 1 (nearest neighborhood). Each vertex contains a cell and the edges represent the interconnections between the cells. The pattern of interaction strengths between each cell and its neighbors is the "program" of the CNN array. It is called a *cloning template* (or just template).

Depending on the types of grids, processors (cells), interactions, and modes of operation, several classes of CNN architectures and models have been introduced. Although the summary below is not complete, it gives an impression of vast diversities.

Typical CNN Models

- Grid type:
  - square
  - hexagonal

planar
circular
equidistant
logarithmic

- Processor type:

linear
sigmoid
first order
second order
third order

- Interaction type:

linear memoryless
nonlinear
dynamic
delay-type

- Mode of operation:

continuous-time
discrete-time
equilibrium
oscillating
chaotic

7.2 The Simple CNN Circuit Structure

The simplest first-order dynamic CNN cell used in the seminal paper [1] is illustrated in Figure 7.3. It is placed on the grid in the position ij (row i and column j). It consists of a single state capacitor with a parallel resistor and an amplifier $[f(x_{ij})]$. This amplifier is a voltage-controlled current source (VCCS), where the controlling voltage is the state capacitor voltage. To make the amplifier model self-contained, a parallel resistor of unit value is assumed to be connected across the output port. Hence, the voltage transfer characteristic of this amplifier is also equal to $f(\cdot)$. In its simplest form this amplifier has a unity gain saturation characteristic (see Figure 7.7 for more details).

The aggregate feed-forward and feedback *interactions* are represented by the current sources i_{input} and i_{output}, respectively. Figure 7.4 shows these interactions in more detail. In fact, the feedforward interaction term i_{input} is a weighted sum of the input voltages (u_{kl}) of all cells in the neighborhood (N_r). Hence, the feedforward template, the so-called ***B** template*, is a small matrix of size $(2r + 1) \times (2r + 1)$ containing the template elements b_{kl}, which can be implemented by an array of linear voltage-controlled current sources. The controlling voltages of these controlled sources are the input voltages of the cells within the neighborhood of radius r. This means, for example, that b_{12} is the VCCS controlled by the input voltage of the cell lying north from the cell ij. In most practical cases the **B** template is translation invariant, i.e., the interaction pattern (the **B** template) is the same for all cells. Hence, the chip layout will be very regular (as in memories or PLAs). The feedback interaction term i_{output} is a weighted sum of the output voltages

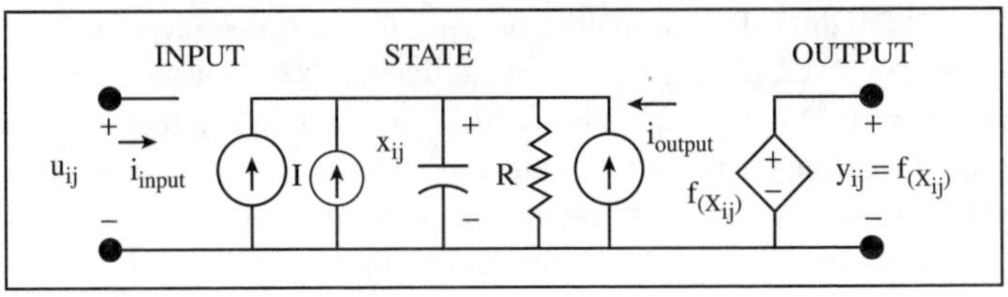

FIGURE 7.3 The simple first-order CNN cell.

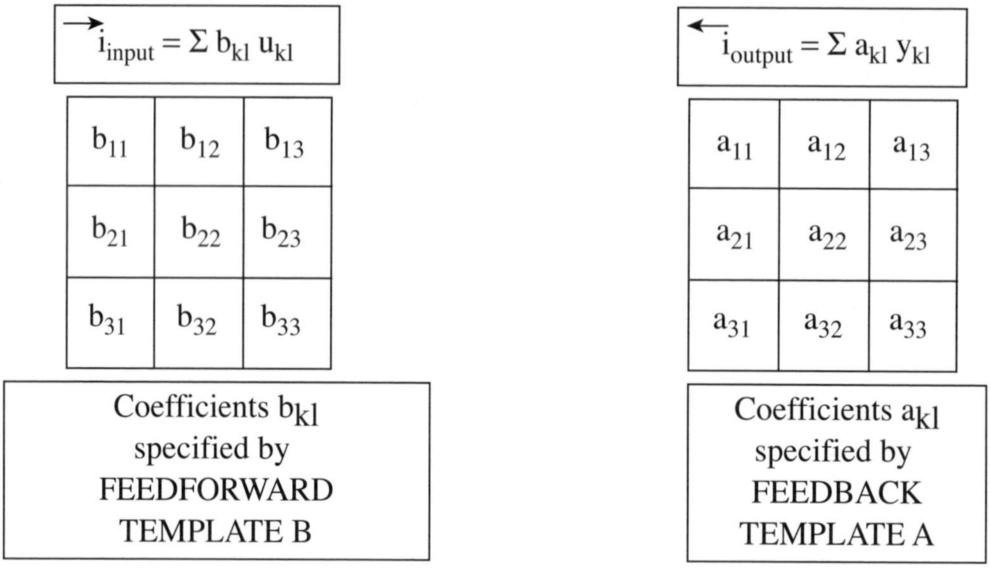

FIGURE 7.4 The 19 numbers (a program) that govern the CNN array (the 19th number is the constant bias term *I*, but it is not shown in the figure) define the cloning template (***A***, ***B***, and ***I***).

(y_{kl}) of all cells in the neighborhood (N_r). The weights are the elements of a small matrix ***A*** called the ***A*** template (or feedback template). Similar arguments apply for the ***A*** template as for the ***B*** template discussed previously. If the constant threshold term is translation invariant as denoted by the constant current source ***I***, then in the case of $r = 1$, the complete cloning template contains only 19 numbers (***A*** and ***B*** and ***I***, i.e., 9 + 9 + 1 terms), irrespective of the size of the CNN array. These 19 numbers define the task which the CNN array can solve.

What kind of tasks are we talking about? The simplest, and perhaps the most important, are image-processing tasks. In the CNN array computer, the input and output images are coded as follows. For each picture element (called pixel) in the image, a single cell is assigned in the CNN. This means that a one-to-one correspondence exists between the pixels and the CNN cells. Voltages in the CNN cells code the gray-scale values of the pixels. Black is coded by +1 V, white is –1 V, and the gray-scale values are in between. Two independent input images can be defined pixel by pixel: the input voltages u_{ij} and the initial voltage values of the capacitors x_{ij} (0) (cell by cell). Placing these input images onto the cell array and starting the transient, the steady state outputs y_{ij} will encode the output image. The computing time is equal to the settling time of the CNN array. This time is below one microsecond using a CNN chip made with a 1.0–1.5 μm technology containing thousands of CNN processing elements, i.e., pixels, in an area of about 2 cm^2. This translates to a computing power of several hundred billion operations per second (GXPS). The first tested CNN chip [4] was followed by several others implementing a discrete-time CNN model [6] and chips with on-chip photosensors in each cell [5].

For example, if we place the array of voltage values defined by the image shown in Figure 7.5(b) as the input voltage and the initial state capacitor voltage values in the CNN array with the cloning template shown in Figure 7.5(a), then after the transients have settled down, the output voltages will encode the output image of Figure 7.5(c). Observe that the vertical line has been deleted. Since the image contains 40 × 40 pixels, the CNN array contains 40 × 40 cells. It is quite interesting that if we had more than one vertical line, the computing time would be the same. Moreover, if we had an array of 100 × 100 cells on the chip, the computing time would remain the same as well. This remarkable result is due to the fully parallel nonlinear dynamics of the CNN computer. Some propagating-type templates induce wave-like phenomena. Their settling times increase with the size of the array.

$$A = \begin{bmatrix} 0 & 0 & 0 \\ 0 & 2 & 0 \\ 0 & 0 & 0 \end{bmatrix} \qquad B = \begin{bmatrix} 0 & -0.25 & 0 \\ 0 & 0 & 0 \\ 0 & -0.25 & 0 \end{bmatrix} \qquad I = -1.5$$

(a)

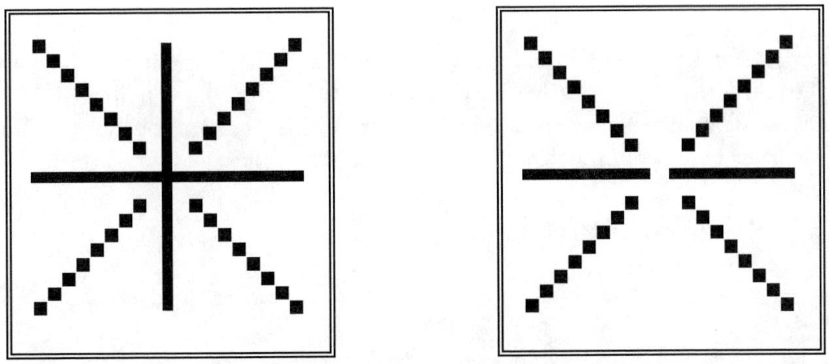

(b) (c)

FIGURE 7.5 An input and output image where the vertical line was deleted.

For other image-processing tasks, processing form, motion, color, and depth, more than 100 cloning templates have been developed to date and the library of new templates is growing rapidly. Using the Cellular Neural Network Workstation Tool Kit [10], they can be called in from a CNN Template Library (CTL). New templates are being developed and published continually.

The dynamics of the CNN array is described by the following set of differential equations:

$$dx_{ij}/dt = -x_{ij} + I + i_{\text{output}} + i_{\text{input}}$$

$$y_{ij} = f(x_{ij})$$

$$i = 1, 2, \ldots, N \quad \text{and} \quad j = 1, 2, \ldots, M \text{ (the array has } N \times M \text{ cells)}$$

where the last two terms in the state equation are given by the sums shown in Figure 7.4.

We can generalize the domain covered by the original CNN defined via linear and time-invariant templates by introducing the "nonlinear" templates (denoted by ^) and the "delay" templates (indicated by τ in the superscript) as well, to obtain the generalized state equation shown below. The unity-gain nonlinear sigmoid characteristics f are depicted in Figure 7.6.

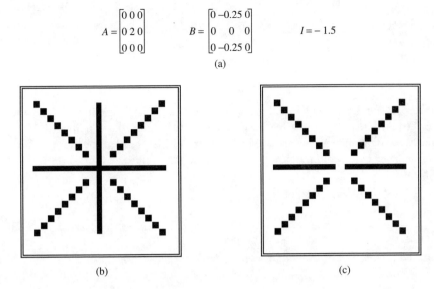

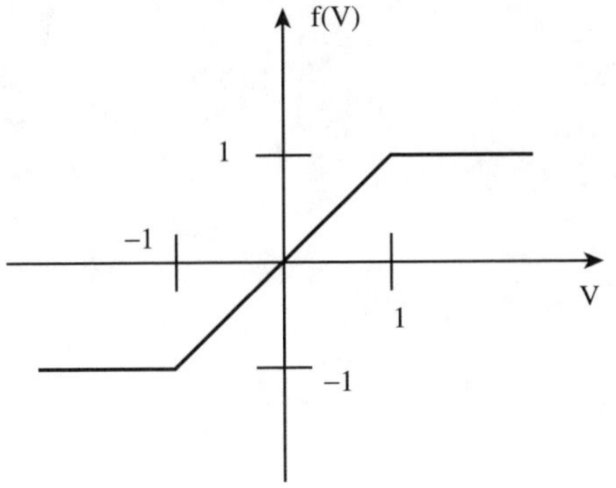

FIGURE 7.6 The simple unity-gain sigmoid characteristics.

$$\frac{dvx_{ij}}{dt} = -vx_{ij} + I_{ij} + \sum_{kl \in N_r(ij)} \hat{A}_{ij;kl}\left(vy_{kl}(t), vy_{ij}(t)\right) + \sum_{kl \in N_r(ij)} \hat{B}_{ij;kl}\left(vu_{kl}(t), vu_{ij}(t)\right)$$
$$+ \sum_{kl \in N_r(ij)} A^{\tau}_{ij;kl} vy_{kl}(t-\tau) + \sum_{kl \in N_r(ij)} B^{\tau}_{ij;kl} vu_{kl}(t-\tau)$$

Several strong results have been proved that assure stable and reliable operations. If the ***A*** template is symmetric, then the CNN is stable. Several other results have extended this condition [6,7]. The sum of the absolute values of all the 19 template elements plus one defines the dynamic range within which the state voltage remains bounded during the entire transient, if the input and initial state signals are less than 1 V in absolute value [1].

In a broader sense, the CNN is defined [2] as shown in Figure 7.7.

- 1-, 2-, 3-, or n-dimensional *array* of mainly identical *dynamical systems*, called cells or processor units, which satisfies two properties:
- most *interactions are local* within a finite radius *r*, and
- all *state variables are continuous valued* signals

FIGURE 7.7 The CNN definition.

7.3 The Stored Program CNN Universal Machine and the Analogic Supercomputer Chip

For different tasks, say image-processing, we need different cloning templates. If we want to implement them in hardware, we would need different chips. This is inefficient except for dedicated, mass-production applications.

The invention of the CNN universal machine [8] has overcome the problem above. It is the first stored program array computer with analog nonlinear array dynamics. One CNN operation, for example, solving thousands of nonlinear differential equations in a microsecond, is just one single instruction. In addition, a single instruction is represented by just a few analog (real) values (numbers). In the case when the nearest neighborhood is used, only 19 numbers are generated. When combining several CNN templates, for example extracting first contours in a gray-scale image, then detecting those areas where the contour has holes, etc., we have to design a flowchart-logic that satisfies the correct sequence of the different templates. The simple flowchart for the previous example is shown in Figure 7.8. One key point

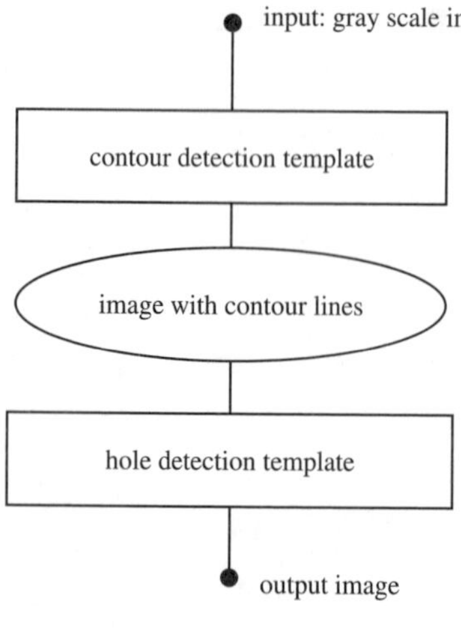

FIGURE 7.8 A flowchart representing the logic sequence of two templates.

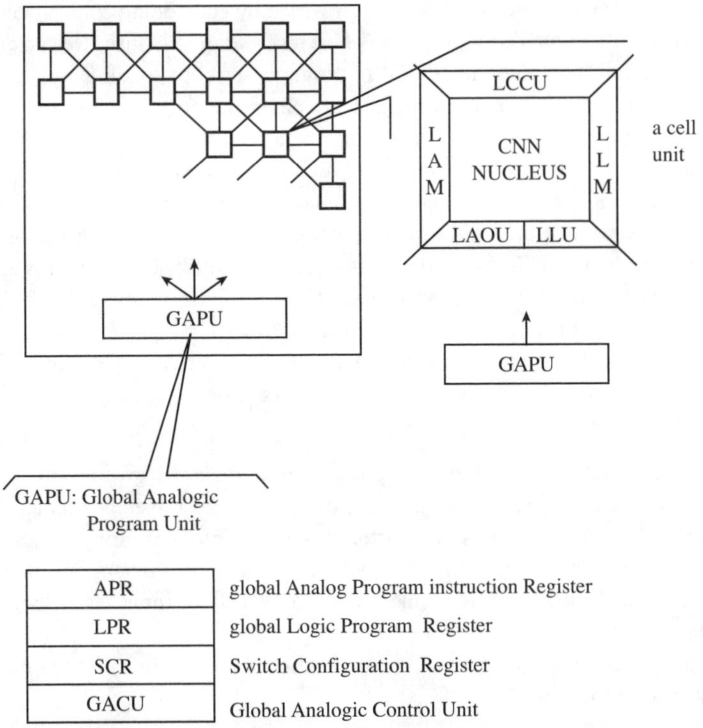

FIGURE 7.9 The global architecture of the CNN Universal Machine. *Source*: T. Roska and L. O. Chua, "The CNN Universal Machine: An analogic array computer," *IEEE Trans. Circuits Syst. I*, vol. 40, pp. 163–173, 1993. ® 1993 IEEE.

is that, in order to exploit the high speed of the CNN chips, we have to store the intermediate results cell by cell (pixel by pixel). Therefore, we need a local analog memory. By combining several template actions we can write more complex flowcharts for implementing almost any *analogic algorithms*. The name analogic is an acronym for "analog and logic." It is important to realize that analogic computation is completely different from hybrid computing. To cite just one point, among others, no A/D or D/A conversions occur during the computation of an analogic program. As with digital microprocessors, to control the execution of an analogic algorithm, we need a global programming unit. The global architecture of the CNN universal machine is shown in Figure 7.9.

As we can see from this figure, the CNN nucleus described in the previous section has been generalized to include several crucial functions depicted in the periphery. We have already discussed the role of the local analog memory (LAM) that provides the local (on-chip) storage of intermediate analog results. Because the results of many detection tasks in applications involve only black-and-white logic values, adding a local logic memory (LLM) in each cell is crucial. After applying several templates in a sequence, it is often necessary to combine their results. For example, to analyze motion, consecutive snapshots processed by CNN templates are compared. The local analog output unit (LAOU) and the local logic unit (LLU) perform these tasks, both on the local analog (gray scale) and the logical (black-and-white) values. The local communication and control unit (LCCU) of each cell decodes the various instructions coming from the global analogic program unit (GAPU).

The global control of each cell is provided by the GAPU. It consists of four parts:

1. The analog program (instruction) register (APR) stores the CNN template values (19 values for each CNN template instruction in the case of nearest interconnection). The templates stored here will be used during the run of the prescribed analogical algorithm.
2. The global logic program register (LPR) stores the code for the local logic units.

3. The flexibility of the extended CNN cells is provided by embedding controllable switches in each cell. By changing the switch configurations of each cell simultaneously, we can execute many tasks using the same cell. For example, the CNN program starts by loading a given template, storing the results of this template action in the local analog memory, placing this intermediate result back on the input to prepare the cell, starting the action with another template, etc. The switch configurations of the cells are coded in the switch configuration register (SCR).
4. Finally, the heart of the GAPU is the global analogic control unit (GACU), which contains the physical machine code of the logic sequence of analogical algorithm. It is important to emphasize that here the control code is *digital*; hence, although its internal operation is analog and logical, a CNN universal chip can be programmed with the same flexibility and ease as a digital microprocessor — except the language is much simpler. Indeed, a high-level language, a compiler, an operating system, and an algorithm development system are available for CNN universal chip architectures. Moreover, by fabricating optical sensors cell-by-cell on the chip [5], the image input is directly interfaced.

The CNN universal chip is called supercomputer chip because the execution speed of an analogic algorithm falls in the same range as the computing power of today's average digital supercomputers (a trillion operations per second). Another reason for this enormous computing power is that the reprogramming time of a new analog instruction (template) is of the same order, or less, than the analog array execution time (less than a microsecond). This is about one million times faster than some fully interconnected analog chips.

Based on the previously mentioned novel characteristics, the CNN universal chip can be considered to be an analogic microprocessor.

7.4 Applications

In view of its flexibility and its very high speed in image-processing tasks, the CNN Universal Machine is ideal for many applications. In the following, we briefly describe three areas. For more applications, the reader should consult the references at the end of this chapter.

Image Processing — Form, Motion, Color, and Depth

Image processing is currently the most popular application of CNN. Of the more than 100 different templates currently available, the vast majority are for image-processing tasks. Eventually, we will have templates for almost all conceivable local image-processing operations. Form (shape), motion, color, and depth can all be ideally processed via CNN. The interested reader can find many examples and applications in the references. CNN handles analog pixel values, so gray-scale images are processed directly.

Many templates detect simple features like different types of edges, convex or concave corners, lines with a prescribed orientation, etc. Other templates detect semiglobal features like holes, groups of objects within a given size of area, or delete objects smaller than a given size. There are also many CNN global operations like calculating the shadow, histogram, etc. Halftoning is commonly used in fax machines, laser printers, and newspapers. In this case, the local gray level is represented by black dots of identical size, whose density varies in accordance with the gray level. CNN templates can do this job as well. A simple example is shown in Fig. 7 .10. The original gray-scale image is shown on the left-hand side, the halftoned image is shown on the right-hand side. The "smoothing" function of our eye completes the image processing task.

More complex templates detect patterns defined within the neighborhood of interaction. In this case, the patterns of the ***A*** and ***B*** templates somehow reflect the pattern of the object to be detected.

Because the simplest templates are translation invariant, the detection or pattern recognition is translation invariant as well. By clever design, however, some rotationally invariant detection procedures have been developed.

FIGURE 7.10 Halftoning: an original gray-scale image (LHS) and its halftoned version (RHS). A low resolution is deliberately chosen in (b) in order to reveal the differing dot densities at various regions of the image.

Combining several templates according to some prescribed logic sequence, more complex pattern detection tasks can be performed, e.g., halftoning.

Color-processing CNN arrays represent the three basis colors by single layers via a multilayer CNN. For example, using the red-green-blue (RGB) representation in a three-layer CNN, simple color-processing operations can be performed. Combining them with logic, conversions between various color representations are possible.

One of the most complex tasks that has been undertaken by an analogic CNN algorithm is the recognition of bank notes. Recognition of bank notes in a few milliseconds is becoming more and more important. Recent advances in the copy machine industry have made currency counterfeiting easier. Therefore, automatic bank note detection is a pressing need. Figure 7.11 shows a part of this process (which involves color processing as well). The dollar bill shown in the foreground is analyzed and the circles of a given size are detected (colors are not shown). The "color cube" means that each color intensity is within prescribed lower and upper limit values.

Motion detection can be achieved by CNN in many ways. One approach to process motion is to apply two consecutive snapshots to the input and the initial state of the CNN cell. The CNN array calculates the various combinations between the two snapshots. The simplest case is just taking the difference to detect motion. Detecting direction, shape, etc. of moving objects are only the simplest problems that can be solved via CNN. In fact, even depth detection can be included as well.

Partial Differential Equations

As noted in the original paper [1], even the simple-cell CNN with the linear template

$$A = \begin{bmatrix} 0 & 1 & 0 \\ 1 & -3 & 1 \\ 0 & 1 & 0 \end{bmatrix} \quad B = 0 \quad I = 0$$

can approximate the solution of a *diffusion-type* partial differential equation on a discrete spatial grid. This solution maintains continuity in time, a nice property not possible in digital computers.

By adding just a simple capacitor to the output, i.e., by placing a parallel RC circuit across the output port of the cell of Figure 7.3, the following *wave equation* will be represented in a discrete space grid:

$$d^2p(t)/dt^2 = \Delta p$$

where $p(t) = P(x, y, t)$ is the state (intensity) variable on a two-dimensional plane (x, y), and Δ is the Laplacian operator (the sum of the second derivatives related to x and y).

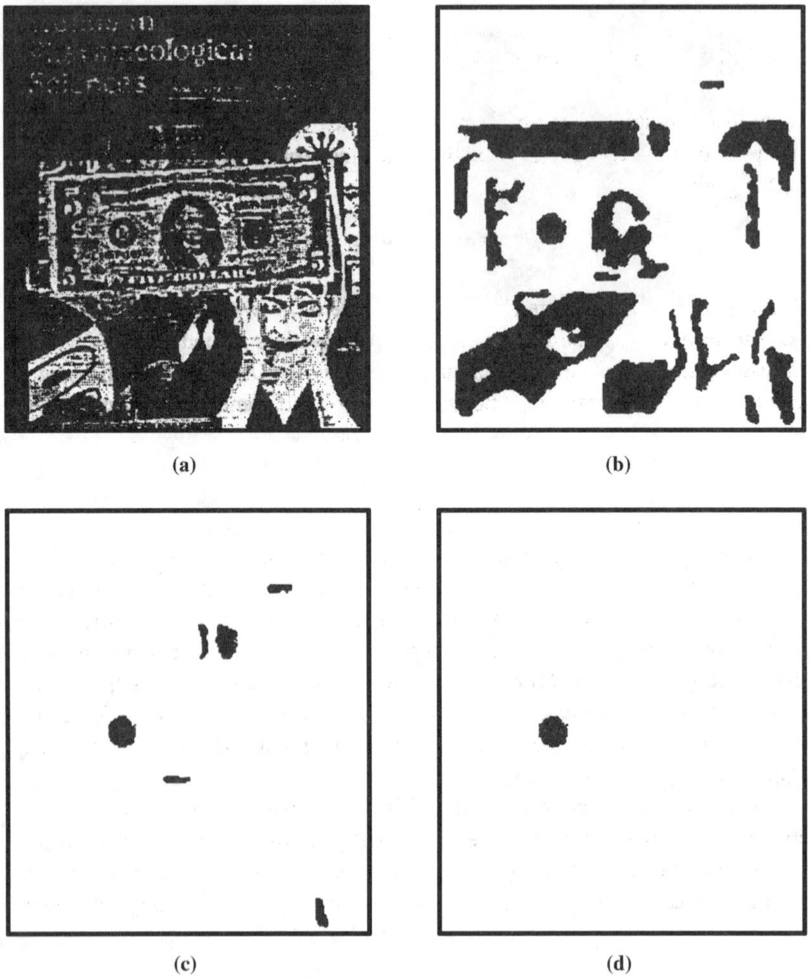

FIGURE 7.11 Some intermediate steps in the dollar bill recognition process. An input image (a) shown here in single color, results in the "color cube" (b), the convex objects (c), and the size classification (d). *Source:* A. Zarándy, F. Werblin, T. Roska, L. O. Chua, and "Novel type of analogical CNN algorithms for recognizing bank notes," Memorandum UCB/ERL, M94/29 1994, Electron. Res. Lab., Univ. California, Berkeley, 1994.

In some cases, it is useful to use a cell circuit that is chaotic. Using the canonical Chua's circuit, other types of partial differential equations can be modeled, generating effects like auto-waves, spiral waves, Turing patterns, and so on, e.g., Perez-Munuzuri et al. in [7].

Relation to Biology

Many topographical sensory organs have processing neural-cell structures very similar to the CNN model. Local connectivity in a few sheets of regularly situated neurons is very typical. Vision, especially the retina, reflects these properties strikingly. It is not surprising that, based on standard neurobiological models, CNN models have been applied to the modeling of the subcortical visual pathway [9]. Moreover, a new method has been devised to use the CNN universal machine for combining retina models of different species in a programmed way. Modalities from other sensory organs can be modeled similarly and combined with the retina models [12]. This has been called Bionic Eye.

Many of these models are neuromorphic. This means that there is a one-to-one correspondence between the neuroanatomy and the CNN structure. Moreover, the CNN template reflects the interconnection

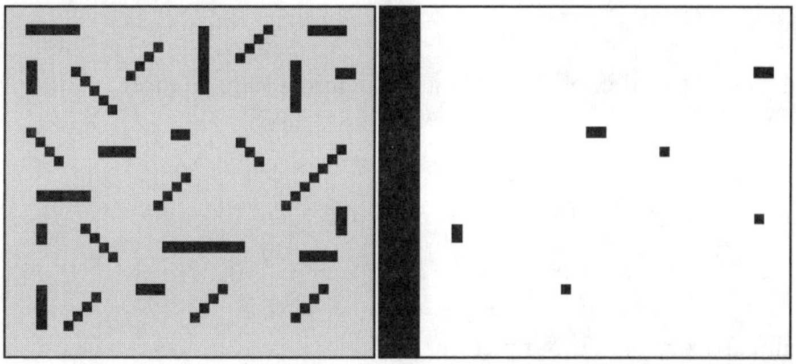

FIGURE 7.12 The length tuning effect. The input image on the LHS contains bars of different lengths. The out image on the RHS contains only those that are smaller than a given length. *Source:* T. Roska, J. Hámori, E. Lábos, K. Lotz. J. Takács, p. Venetianer, Z. Vidnyánszki and A. Zarándy, "The use of CNN models in the subcortical visual pathway," *IEEE Trans. Circuits Syst. I*, vol. pp. 182–195, 1993. © 1993 IEEE.

pattern of the neurons (called receptive field organizations). Length tuning is such an example. A corresponding input and output picture of the neuromorphic length tuning model is shown in Figure 7.12. Those bars are detected that have lengths smaller than or equal to 3 pixels.

7.5 Template Library: Analogical CNN Algorithms

During the last few years, after the invention of the cellular neural network paradigm and the CNN universal machine, many new cloning templates have been discovered. In addition, the number of innovative analogical CNN algorithms, combining both analog cloning templates and local as well as global logic, is presently steadily increasing at a rapid rate.

As an illustration, let us choose a couple of cloning templates from the CNN library [1], [11]. In each case, a name, a short description of the function, the cloning templates, and a representative input–output image pair are shown. With regard to the inputs, the default case means that the input and initial state are the same. If $B = 0$, then the input picture is chosen as the initial state.

Name: AVERAGE

Function. Spatial averaging of pixel intensities over the $r = 1$ convolutional window.

$$\boldsymbol{A} = \begin{bmatrix} 0 & 1 & 0 \\ 1 & 2 & 1 \\ 0 & 1 & 0 \end{bmatrix} \quad \boldsymbol{B} = \begin{bmatrix} 0 & 0 & 0 \\ 0 & 0 & 0 \\ 0 & 0 & 0 \end{bmatrix} \quad \boldsymbol{I} = 0$$

Example. Input and output picture.

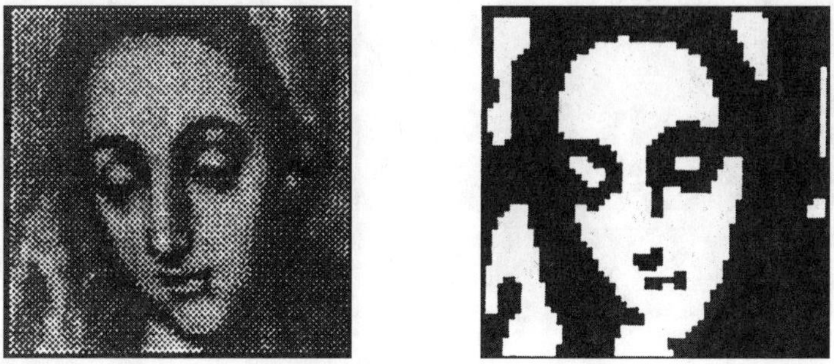

Name: AND

Function. Logical "AND" function of the input and the initial state pictures.

$$A = \begin{bmatrix} 0 & 0 & 0 \\ 0 & 1.5 & 0 \\ 0 & 0 & 0 \end{bmatrix} \quad B = \begin{bmatrix} 0 & 0 & 0 \\ 0 & 1.5 & 0 \\ 0 & 0 & 0 \end{bmatrix} \quad I = -1$$

Example. Input, initial state, and output picture.

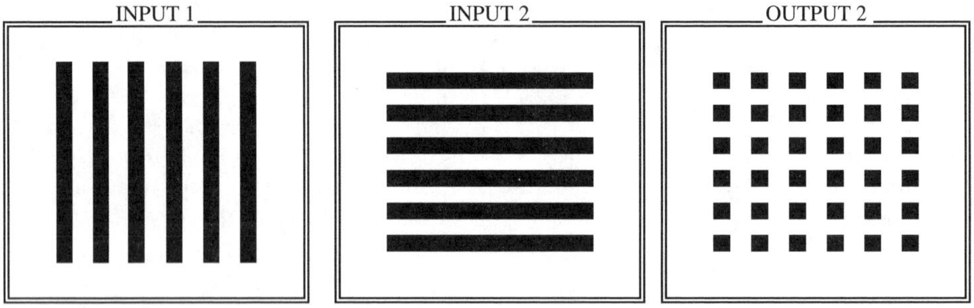

Name: CONTOUR

Function. Gray-scale contour detector.

$$A = \begin{bmatrix} 0 & 0 & 0 \\ 0 & 2 & 0 \\ 0 & 0 & 0 \end{bmatrix} \quad B = \begin{bmatrix} a & a & a \\ a & a & a \\ a & a & a \end{bmatrix} \quad I = 0.7$$

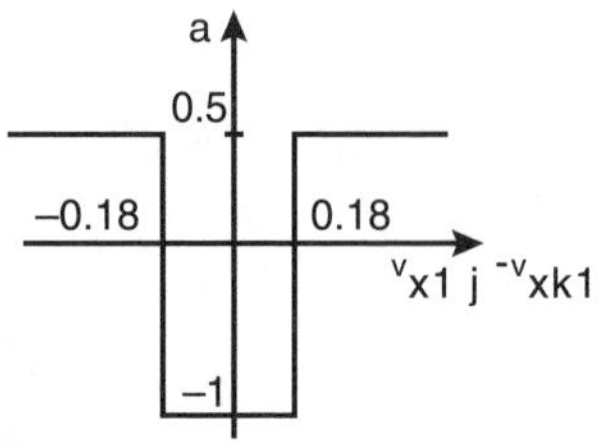

Example. Input and output picture.

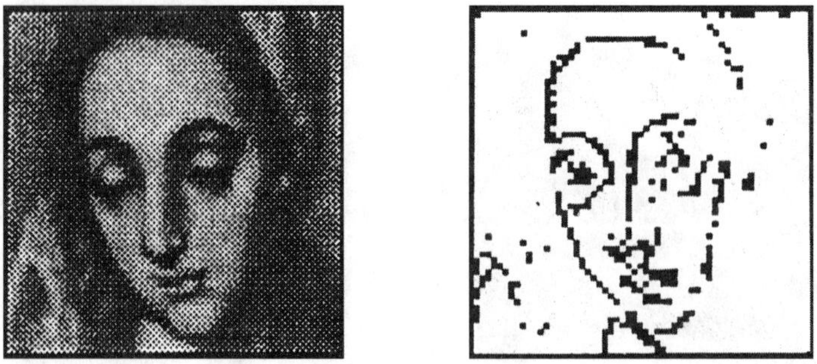

Name: CORNER

Function. Convex corner detector.

$$A = \begin{bmatrix} 0 & 0 & 0 \\ 0 & 2 & 0 \\ 0 & 0 & 0 \end{bmatrix} \quad B = \begin{bmatrix} -0.25 & -0.25 & -0.25 \\ -0.25 & 2 & -0.25 \\ -0.25 & -0.25 & -0.25 \end{bmatrix} \quad I = -3$$

Example. Input and output picture.

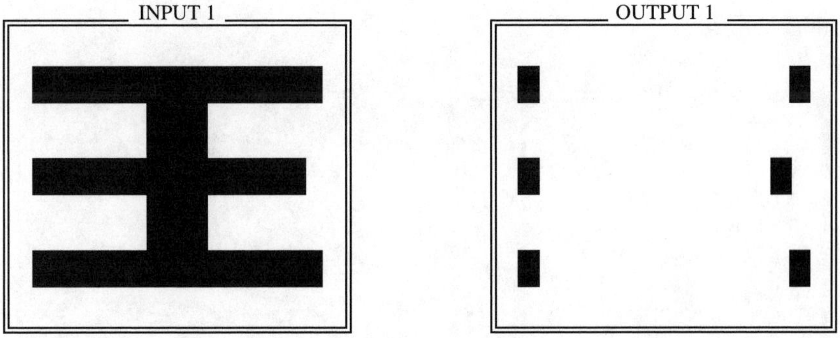

Name: DELDIAG1

Function. Deletes one pixel wide diagonal lines (5).

$$A = \begin{bmatrix} 0 & 0 & 0 \\ 0 & 2 & 0 \\ 0 & 0 & 0 \end{bmatrix} \quad B = \begin{bmatrix} -0.25 & 0 & -0.25 \\ 0 & 0 & 0 \\ -0.25 & 0 & -0.25 \end{bmatrix} \quad I = -2$$

Example. Input and output picture.

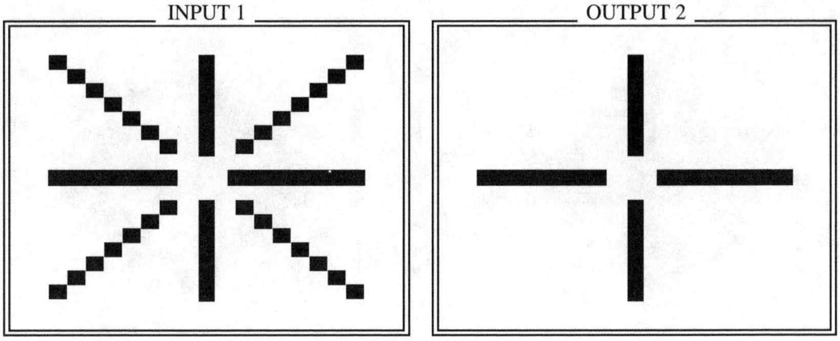

Name: DIAG

Function. Detects approximately diagonal lines situated in the SW-NE direction.

$$A = \begin{bmatrix} 0 & 0 & 0 & 0 & 0 \\ 0 & 0 & 0 & 0 & 0 \\ 0 & 0 & 2 & 0 & 0 \\ 0 & 0 & 0 & 0 & 0 \\ 0 & 0 & 0 & 0 & 0 \end{bmatrix} \quad B = \begin{bmatrix} -1 & -1 & -0.5 & 0.5 & 1 \\ -1 & -0.5 & 1 & 1 & 0.5 \\ -0.5 & 1 & 5 & 1 & -0.5 \\ 0.5 & 1 & 1 & -0.5 & -1 \\ 1 & 0.5 & -0.5 & -1 & -1 \end{bmatrix}$$

$$I = -9$$

Example. Input and output picture.

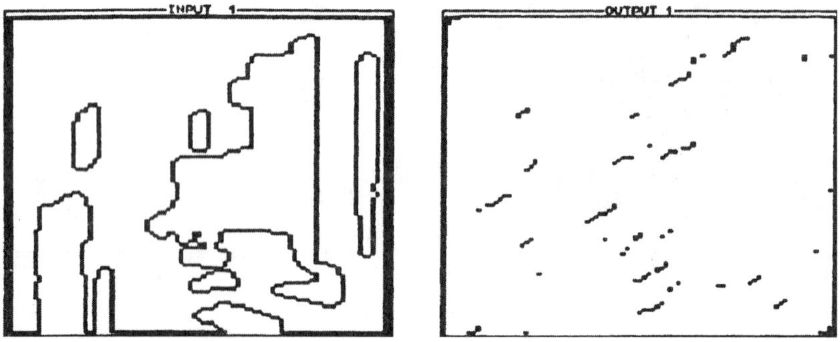

Name: EDGE

Function. Black and white edge detector.

$$A = \begin{bmatrix} 0 & 0 & 0 \\ 0 & 2 & 0 \\ 0 & 0 & 0 \end{bmatrix} \quad B = \begin{bmatrix} -0.25 & -0.25 & -0.25 \\ -0.25 & 2 & -0.25 \\ -0.25 & -0.25 & -0.25 \end{bmatrix} \quad I = -1.5$$

Example. Input and output picture.

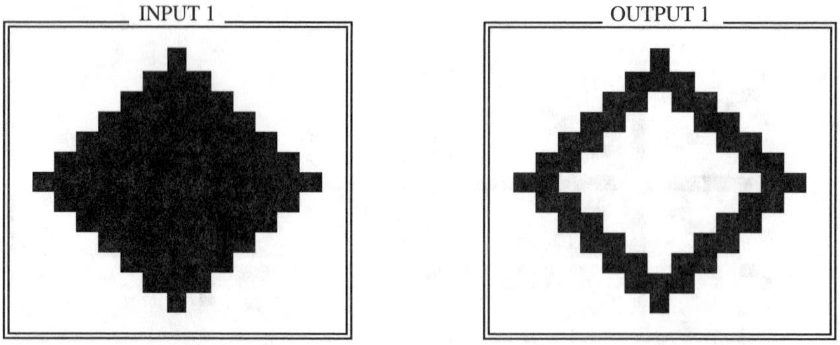

Name: MATCH

Function. Detects 3 × 3 patterns matching exactly the one prescribed by the template B, namely, having a black/white pixel where the template value is +1/–1, respectively.

$$A = \begin{bmatrix} 0 & 0 & 0 \\ 0 & 1 & 0 \\ 0 & 0 & 0 \end{bmatrix} \quad B = \begin{bmatrix} v & v & v \\ v & v & v \\ v & v & v \end{bmatrix} \quad I = -N + 0.5$$

where $v = +1$, if corresponding pixel is required to be black; $v = 0$, if corresponding pixel is don't care; $v = -1$, if corresponding pixel is required to be white; N = number of pixels required to be either black or white, i.e., the number of nonzero values in the $\boldsymbol{B}$ template.

Example. Input and output picture, using the following values:

$$A = \begin{bmatrix} 0 & 0 & 0 \\ 0 & 1 & 0 \\ 0 & 0 & 0 \end{bmatrix} \quad B = \begin{bmatrix} 1 & -1 & 1 \\ 0 & 1 & 0 \\ 1 & -1 & 1 \end{bmatrix} \quad I = -6.5$$

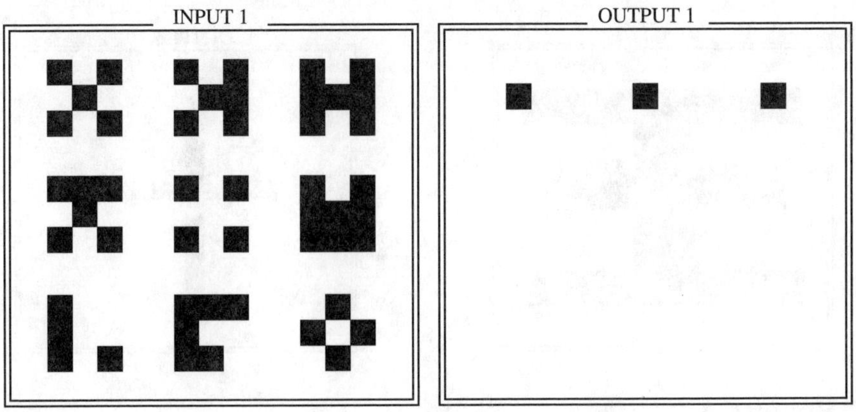

Name: OR

Function. Logical "OR" function of the input and the initial state.

$$A = \begin{bmatrix} 0 & 0 & 0 \\ 0 & 3 & 0 \\ 0 & 0 & 0 \end{bmatrix} \quad B = \begin{bmatrix} 0 & 0 & 0 \\ 0 & 3 & 0 \\ 0 & 0 & 0 \end{bmatrix} \quad I = 2$$

Example. Input, initial state, and output picture.

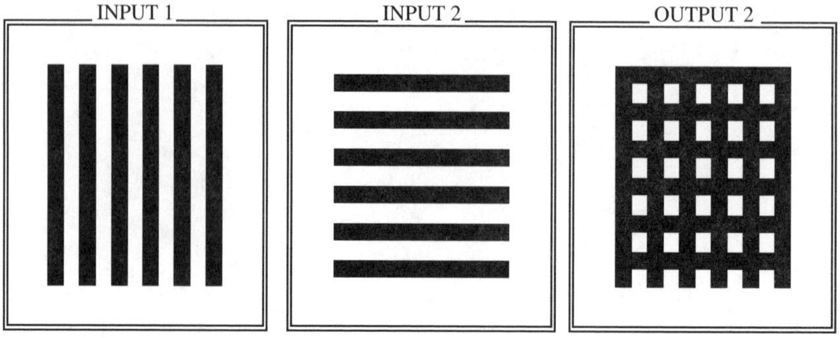

Name: PEELIPIX

Function. Peels one pixel from all directions.

$$A = \begin{bmatrix} 0 & 0.4 & 0 \\ 0.4 & 1.4 & 0.4 \\ 0 & 0.4 & 0 \end{bmatrix} \quad B = \begin{bmatrix} 4.6 & -2.8 & 4.6 \\ -2.8 & 1 & -2.8 \\ 4.6 & -2.8 & 4.6 \end{bmatrix} \quad I = -7.2$$

Example. Input and output picture.

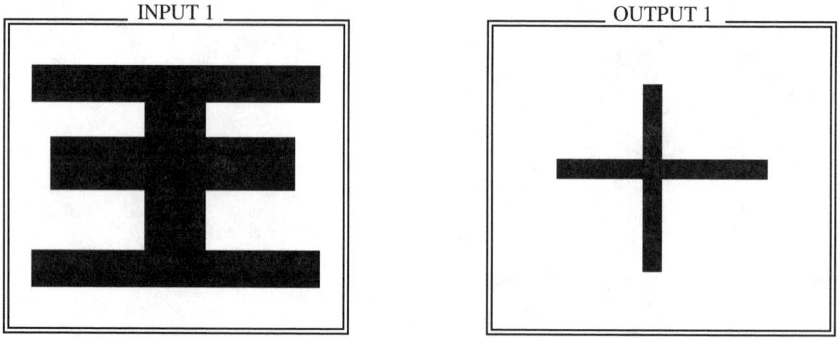

7.6 Recent Advances

After the first few integrated circuit implementations of the basic CNN circuits, stored programmable analogic CNN Universal Machine chips have been fabricated. Indeed, a full-fledged version of them [13] is the first visual microprocessor. All the 4096 cell processors of it contain an optical sensor right on the surface of the chip (a focal plane). This implementation represents, at the same time, the most complex operational, stored programmable analog CMOS integrated circuit ever reported, in terms of the number of transistors operating in analog mode (about 1 million). The equivalent digital computing power of this visual microprocessor is about a few TeraOPS (trillion operations per second). It processes gray-scale input images and has a gray-scale output. A 128×128 processor version has recently been fabricated. A binary input/output CNN Universal Machine Chip with 48×48 cell processors has a higher cell density [14], and another circuit design strategy [18] is aiming to implement 5×5 or even higher neighborhood templates.

These chips are the first examples of a new, analogic, topographic (spatial-temporal) computing technology. Its computational infrastructure (high level language, called Alpha, compiler, operating system, etc.) has also been developed [15], and the industrial applications have been started in a couple of companies worldwide. Moreover, a key application area of this technology is sensor-computing [17]. Integrating 2-D topographic sensor arrays with the CNN universal machine on a single chip, providing a direct, dynamic interaction with tuning of the sensors is a capability no other technology offers with comparable computational power.

Recently, it has been shown that partial differential equation (PDE)-based techniques, the most advanced methods for complex image processing problems, could solve tasks intractable with other methods. Their only drawback is the excessive digital computing power they need. In our cellular computing technology, however, the elementary instruction could be a solution of a PDE. It has been shown that, in addition to the simple diffusion PDE implementation described previously, almost all PDEs can be implemented by CNN [16]. Indeed, active waves [23] have been successfully applied using operational analogic CNN Universal Machine Chips with 4096 cell processors, manifesting at least 3 orders of magnitude speed advantage compared to fully digital chips of comparable IC technology feature size.

Following the first steps in modeling living sensory modalities, especially vision, motivated especially by a breakthrough in understanding the neurobiological constructs of the mammalian retina [19], new models and a modeling framework [20] have been developed based on CNNs. Their implementation in Complex Cell CNN Universal Machines [24] are under construction.

Studies in complexity related to CNN models and implementations have been emerging recently. Following the groundbreaking theoretical studies of Turing on the morphogenesis of CNN-like coupled nonlinear units [21] and a few experimental case studies of the well-publicized "complex systems", as well as many exotic waves generated by coupled ***A*** template CNNs, the root of complexity in pattern formation at the edge of chaos has been discovered [22]. As far as the computational complexity is concerned, the study of a new quality of computational complexity has been explored [25], showing qualitatively different properties compared to the classical digital complexity theory as well as the complexity on reals [30].

To further explore the vast amount of literature on CNN technology and analogic cellular computing, the interested reader could consult the bibliography at the Website of the Technical Committee on Cellular Neural Networks and Array Computing of the IEEE Circuits and Systems Society (http://www.ieee-cas.org/~cnnactc), some recent monographs [26, 27, 28], and an undergraduate textbook [29].

References

[1] L. O. Chua and L. Yang, "Cellular neural networks: Theory," *IEEE Trans. Circuits Syst.*, vol. 35, pp. 1257–1272, 1988;.

[2] L. O. Chua and L. Yang, "Cellular neural networks: Applications," *IEEE Trans. Circuits Syst.*, vol. 35, pp. 1273–1290, 1988.

[3] L. O. Chua and T. Roska, "The CNN paradigm," *IEEE Trans. Circuits Syst.*, I vol. 40, pp. 147–156, 1993.

[4] J. Cruz and L. O. Chua, "A CNN chip for connected component detection," *IEEE Trans. Circuits Syst.*, vol. 38, pp. 812–817, 1991.

[5] R. Dominguez-Castro, S. Espejo, A. Rodriguez-Vazquez, and R. Carmona, "A CNN Universal Chip in CMOS Technology," Proc. IEEE 3rd Int. Workshop on CNN and Applications, (CNNA-94), Rome, pp. 91–96, 1994.

[6] H. Harrer, J.A. Nossek, and R. Stelzl, "An analog implementation of discrete-time cellular neural networks," *IEEE Trans. Neural Networks*, vol. 3, pp. 466–476, 1992.

[7] J. A. Nossek and T. Roska, Eds. Special Issues on Cellular Neural Networks, *IEEE Trans. Circuits Syst.* I, vol. 40, Mar. 1993; Special Issue on Cellular Neural Networks, *IEEE Trans. Circuits Syst. II*, vol. 40, Mar. 1993.

[8] T. Roska, and L. O. Chua, "The CNN universal machine: An analogic array computer," *IEEE Trans. Circuits Syst. II*, vol. 40, pp. 163–173, 1993.

[9] T. Roska, J. Hamori, E. Labos, K. Lotz, K. Takacs, P. Venetianer, Z. Vidnyanszki, and A. Zarandy. "The use of CNN models in the subcortical visual pathway," *IEEE Trans. Circuits Syst.*, I, vol. 40, pp. 182–195, 1993.

[10] T. Roska and J. Vandewalle, Eds., *Cellular Neural Networks.* Chischester: Wiley, 1993.

[11] CANDY (CNN Analogic Nonlinear Dynamics) Simulator, guide and program (student Version) http://lab.analogic.sztaki.hu.

[12] F. Werblin, T. Roska, L.O. Chua, "The analogic cellular neural network as a bionic eye", *Int. J. CircuitTheory and Applications*, (CTA), Vol.23. N.6. pp. 541-569, 1995.

[13] G. Linán, S. Espejo, R. Dominguez-Castro, E. Roca, and A.Rodriguez-Vázquez, "CNNUC3: A mixed signal 64x64 CNN Universal Chip", *Proceedings of MicroNeuro*, pp. 61-68, 1999.

[14] A. Paasio, A. Davidzuk, K. Halonen, and V. Porra, "Minimum-size 0.5 micron CMOS Programmable 48 by 48 Test Chip", *Proceedings of ECCTD* 97, pp.154-156, 1997.

[15] T. Roska, Á. Zarándy, S. Zöld, P. Földesy and P. Szolgay, "The Computational Infrastructure of Analogic CNN Computing — Part I: The CNN-UM Chip Prototyping System", *IEEE Trans. on Circuits and Systems I*: Special Issue on Bio-Inspired Processors and Cellular Neural Networks for Vision, (CAS-I Special Issue), vol. 46, No.2, pp. 261-268, 1999.

[16] T. Roska, L. O. Chua, D. Wolf, T. Kozek, R. Tetzlaff, and F. Puffer, " Simulating Nonlinear Waves and Partial Differential Equations via CNN", *IEEE Trans. on Circuits and Systems I*,vol.42, pp. 807-815, 1995.

[17] T. Roska, "Computer-Sensors: spatial-temporal computers for analog array signals, dynamically integrated with sensors", *J. VLSI Signal Processing Systems*, vol. 23, pp. 221-238, 1999.

[18] W. C. Yen, C. Y. Wu, "The Design of Neuron-Bipolar Junction Transistor (vBJT) Cellular Neural Network (CNN) Structure with Multi-Neighborhood-Layer Templates", *Proceedings of IEEE Int. Workshop on Cellular Neural Networks and Their Applications*, (CNNA'2000), pp. 195-200, 2000.

[19] B. Roska and F. S. Werblin, "Vertical interactions across ten parallel, stacked representations in the mammalian retina", *Nature*, vol. 410, pp.583-587, March 29, 2001.

[20] F. Werblin, B. Roska, D. Bálya, Cs. Rekeczky, T. Roska, "Implementing a Retinal Visual Language in CNN: a Neuromorphic Case Study", *Proceedings of IEEE ISCAS* 2001, Vol. III, pp.333-336, 2001.

[21] A. M. Turing, "The chemical basis of morphogenesis", *Philos.Trans. R. Soc.*, London, vol. B237, pp. 37-72, 1952.

[22] L. O. Chua, *CNN: A Paradigm for Complexity*, World Scientific, Singapore, 1998.

[23] Cs. Rekeczky and L. O. Chua, "Computing with Front Propagation- Active Contours and Skeleton Models in Continuous-time CNN", *J. VLSI Signal Processing Systems*, vol. 23, pp. 373-402, 1999.

[24] Cs. Rekeczky, T. Serrano, T. Roska, and Á. Rodríguez-Vázquez, "A stored program 2nd order/3-layer Complex Cell CNN-UM", *Proceedings of IEEE Int.Workshop on Cellular Neural Networks and Their Applications* (CNNA-2000), pp.15-20, 2000.

[25] T. Roska, "AnaLogic Wave Computers – Wave-type Algorithms: Canonical Description, Computer Classes, and Computational Complexity", *Proceedings of IEEE ISCAS* 2001, vol. III, pp.41-44, 2001.

[26] G. Manganaro, P. Arena, and L. Fortuna, *Cellular Neural Networks – Chaos, Complexity and VLSI Processing*, Springer, Berlin, 1999.

[27] M. Hänggi and G. Moschitz, *Cellular Neural Networks – Analysis, Design and Optimization*, Kluwer Academic Publishers, Boston, 2000.

[28] T. Roska and Á. Rodríguez-Vázquez (eds), *Towards the Visual Microprocessor – VLSI Design and the Use of Cellular Neural Network Universal Machine*, J. Wiley, New York, 2001.

[29] L. O. Chua and T. Roska, *Cellular Neural Network and Visual Computing — Foundations and Applications*, Cambridge University Press, New York, 2002.

[30] L. Blum, F. Cucker, M. Shub, and S. Smale, *Complexity and Real Computation*, Springer, New York, 1998.

8

Bifurcation and Chaos

Michael Peter Kennedy
University College, Dublin Ireland

8.1 Introduction to Chaos

Electrical and Electronic Circuits as Dynamical Systems

A **system** is something having parts that may be perceived as a single entity. A dynamical system is one that changes with time; what changes is the state of the system. Mathematically, a dynamical system consists of a space of states (called the state space or phase space) and a rule, called the **dynamic**, for determining which state corresponds at a given future time to a given present state [8]. A deterministic dynamical system is one where the state, at any time, is completely determined by its initial state and dynamic. In this section, we consider only deterministic dynamical systems.

A deterministic dynamical system may have a continuous or discrete state space and a continuous-time or discrete-time dynamic.

A lumped[1] circuit containing resistive elements (resistors, voltage and current sources) and energy-storage elements (capacitors and/or inductors) may be modeled as a continuous-time deterministic dynamical system in $\mathbb{R}^n$. The evolution of the state of the circuit is described by a system of ordinary differential equations called state equations.

Discrete-time deterministic dynamical systems occur in electrical engineering as models of switched-capacitor and digital filters, sampled phase-locked loops, and sigma–delta modulators. Discrete-time dynamical systems also arise when analyzing the stability of steady-state solutions of continuous-time systems. The evolution of a discrete-time dynamical system is described by a system of *difference equations.*

Continuous-Time Dynamical Systems

Theorem 1: (Existence and Uniqueness of Solution for a Differential Equation) *Consider a continuous-time deterministic dynamical system defined by a system of ordinary differential equations of the form*

$$\dot{\mathbf{X}}(t) = \mathbf{F}(\mathbf{X}(t), t) \tag{8.1}$$

where $\mathbf{X}(t) \in \mathbb{R}^n$ is called the state, $\dot{\mathbf{X}}(t)$ denotes the derivative of $\mathbf{X}(t)$ with respect to time, $\mathbf{X}(t_0) = \mathbf{X}_0$ is called the initial condition, and the map $\mathbf{F}(\cdot,\cdot):\mathbb{R}^n \times \mathbb{R}_+ \to \mathbb{R}^n$ is (i) continuous almost everywhere[2] on $\mathbb{R}^n \times \mathbb{R}_+$ and (ii) globally Lipschitz[3] in $\mathbf{X}$. Then, for each $(\mathbf{X}_0, t_0) \in \mathbb{R}^n \times \mathbb{R}_+$, there exists a continuous function $\phi(\cdot; \mathbf{X}_0, t_0)$: $\mathbb{R}_+ \to \mathbb{R}^n$ such that

$$\phi(t_0; \mathbf{X}_0, t_0) = \mathbf{X}_0$$

and

$$\dot{\phi}(t; \mathbf{X}_0, t_0) = \mathbf{F}(\phi(t; \mathbf{X}_0, t_0), t) \tag{8.2}$$

Furthermore, this function is unique.

The function $\phi(\cdot; \mathbf{X}_0, t_0)$ is called the solution or trajectory $(\mathbf{X}_0, t_0)$ of the differential equation (8.1).

The image $\{\phi(t; \mathbf{X}_0, t_0) \in \mathbb{R}^n | t \in \mathbb{R}_+\}$ of the trajectory through $(\mathbf{X}_0, t_0)$ is a continuous curve in $\mathbb{R}^n$ called the orbit through $(\mathbf{X}_0, t_0)$.

$\mathbf{F}(\cdot, \cdot)$ is called the vector field of (8.1) because its image $\mathbf{F}(\mathbf{X}, t)$ is a vector that defines the direction and speed of the trajectory through $\mathbf{X}$ at time t.

The vector field $\mathbf{F}$ generates the *flow* ϕ, where $\phi(\cdot;\cdot,\cdot):\mathbb{R}_+ \times \mathbb{R}^n \times \mathbb{R}_+ \to \mathbb{R}^n$ is a collection of continuous maps $\{\phi(t;\cdot,\cdot):\mathbb{R}^n \times \mathbb{R}_+ \to \mathbb{R}^n | t \in \mathbb{R}_+\}$.

In particular, a point $\mathbf{X}_0 \in \mathbb{R}^n$ at t_0 is mapped by the flow into $\mathbf{X}(t) = \phi_t(t; \mathbf{X}_0, t_0)$ at time t.

Autonomous Continuous-Time Dynamical Systems

If the vector field of a continuous-time deterministic dynamical system depends only on the state and is *independent* of time t, then the system is said to be *autonomous* and may be written as

$$\dot{\mathbf{X}}(t) = \mathbf{F}[\mathbf{X}(t)]$$

---

[1]A *lumped* circuit is one with physical dimensions that are small compared with the wavelengths of its voltage and current waveforms [2].

[2]By *continuous almost everywhere*, we mean the following: let D be a set in $\mathbb{R}_+$ that contains a countable number of discontinuities and for each $\mathbf{X} \in \mathbb{R}^n$, assume that the function $t \in \mathbb{R}_+ \backslash D \to \mathbf{F}(\mathbf{X}, t) \in \mathbb{R}^n$ *is continuous* and for any $\tau \in D$ the left-hand and right-hand limits $\mathbf{F}(\mathbf{X}, \tau_-)$ and $\mathbf{F}(\mathbf{X}, \tau_+)$, respectively, are *finite* in $\mathbb{R}^n$ [1]. This condition includes circuits that contain switches and/or squarewave voltage and current sources.

[3]There is a piecewise continuous function $k(\cdot)$: $\mathbb{R}_+ \to \mathbb{R}_+$ such that $\|\mathbf{F}(\mathbf{X}, t) - \mathbf{F}(\mathbf{X}', t)\| \leq k(t) \|\mathbf{X} - \mathbf{X}'\|$, $\forall t \in \mathbb{R}_+$, $\forall \mathbf{X}, \mathbf{X}' \in \mathbb{R}^n$.

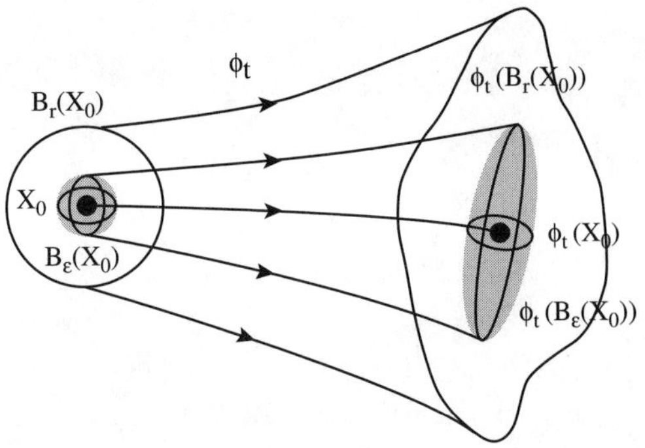

FIGURE 8.1 The vector field **F** of an autonomous continuous-time dynamical system generates a flow ϕ that maps a point $\mathbf{X}_0$ in the state space to its image $\phi_t(\mathbf{X}_0)$ t seconds later. A volume of state space $B_r(\mathbf{X}_0)$ evolves under the flow into a region $\phi_t[B_r(\mathbf{X}_0)]$. Sufficiently close to the trajectory $\phi_t(\mathbf{X}_0)$, the linearized flow maps a sphere of radius ε into an ellipsoid.

or simply

$$\dot{\mathbf{X}} = \mathbf{F}(\mathbf{X}) \tag{8.3}$$

If, in addition, the vector field $\mathbf{F}(\cdot): \mathbb{R}^n \to \mathbb{R}^n$ is *Lipschitz*[4], then there is a *unique* continuous function $\phi(\cdot, \mathbf{X}_0): \mathbb{R}_+ \to \mathbb{R}^n$ (called the trajectory through $\mathbf{X}_0$), which satisfies,

$$\dot{\phi}(t, \mathbf{X}_0) = \mathbf{F}[\phi(t, \mathbf{X}_0)], \quad \phi(t_0, \mathbf{X}_0) = \mathbf{X}_0 \tag{8.4}$$

Because the vector field is independent of time, we choose $t_0 \equiv 0$. For shorthand, we denote the flow by ϕ and the map $\phi(t,\cdot): \mathbb{R}^n \to \mathbb{R}^n$ by ϕ_t.

The t-advance map ϕ_t takes a state $\mathbf{X}_0 \in \mathbb{R}^n$ to state $\mathbf{X}(t) = \phi_t(\mathbf{X}_0)$ t seconds later. In particular, ϕ_0 is the identity mapping. Furthermore, $\phi_{t+s} = \phi_t\phi_s$, because the state $\mathbf{Y} = \phi_s(\mathbf{X})$ to which $\mathbf{X}$ evolves after time s evolves after an additional time t into the same state $\mathbf{Z}$ as that to which $\mathbf{X}$ evolves after time $t + s$:

$$\mathbf{Z} = \phi_t(\mathbf{Y}) = \phi_t[\phi_s(\mathbf{X})] = \phi_{t+s}(\mathbf{X})$$

A bundle of trajectories emanating from a ball $B_r(\mathbf{X}_0)$ of radius r centered at $\mathbf{X}_0$ is mapped by the flow into some region $\phi_t[B_r(\mathbf{X}_0)]$ after t seconds (see Figure 8.1). Consider a short segment of the trajectory $\phi_t(\mathbf{X}_0)$ along which the flow is differentiable with respect to $\mathbf{X}$: in a sufficiently small neighborhood of this trajectory, the flow is almost linear, so the ball $B_\varepsilon(\mathbf{X}_0)$ of radius ε about $\mathbf{X}_0$ evolves into an ellipsoid $\phi_t[B_\varepsilon(\mathbf{X}_0)]$, as shown.

An important consequence of Lipschitz continuity in an autonomous vector field and the resulting uniqueness of solution of (8.3) is that a trajectory of the dynamical system cannot go through the same point twice in two different directions. In particular, no two trajectories may cross each other; this is called the **noncrossing property** [18].

Nonautonomous Dynamical Systems

A nonautonomous, n-dimensional, continuous-time dynamical system may be transformed to an $(n + 1)$-dimensional autonomous system by appending time as an additional state variable and writing

$$\begin{aligned} \dot{\mathbf{X}}(t) &= \mathbf{F}[\mathbf{X}(t), X_{n+1}(t)] \\ \dot{X}_{n+1}(t) &= 1 \end{aligned} \tag{8.5}$$

[4]There exists a finite $k \in \mathbb{R}^n$ such that $\|\mathbf{F}(\mathbf{X}) - \mathbf{F}(\mathbf{X}')\| \le k\,\|\mathbf{X} - \mathbf{X}'\|$, $\forall \mathbf{X}, \mathbf{X}' \in \mathbb{R}^n$.

In the special case where the vector field is periodic with period T, as for example in the case of an oscillator with sinusoidal forcing, the periodically forced system (8.5) is equivalent to the $(n + 1)$st order autonomous system

$$\dot{\mathbf{X}}(t) = \mathbf{F}\big(\mathbf{X}(t), \theta(t)T\big)$$
$$\dot{\theta}(t) = \frac{1}{T} \tag{8.6}$$

where $\theta(t) = X_{n+1}/T$.

By identifying the n-dimensional hyperplanes corresponding to $\theta = 0$ and $\theta = 1$, the state space may be transformed from $\mathbb{R}^n \times \mathbb{R}_+$ into an equivalent cylindrical state space $\mathbb{R}^n \times S^1$, where S^1 denotes the circle. In the new coordinate system, the solution through $(\mathbf{X}_0, t_0)$ of (8.6) is

$$\begin{pmatrix} \mathbf{X}(t) \\ \theta_{S^1}(t) \end{pmatrix} = \begin{pmatrix} \phi_t(\mathbf{X}_0, t_0) \\ t/T \bmod 1 \end{pmatrix}$$

where $\theta(t) \in \mathbb{R}_+$ is identified with a point on S^1 (which has normalized angular coordinate $\theta_{S^1}(t) \in [0, 1)$) via the transformation $\theta_{S_1}(t) = \theta(t) \bmod 1$. Using this technique, periodically forced nonautonomous systems can be treated like autonomous systems.

Discrete-Time Dynamical Systems

Consider a discrete-time deterministic dynamical system defined by a system of difference equations of the form

$$\mathbf{X}(k+1) = \mathbf{G}\big(\mathbf{X}(k), k\big) \tag{8.7}$$

where $\mathbf{X}(k) \in \mathbb{R}^n$ is called the state, $\mathbf{X}(k_0) = \mathbf{X}_0$ is the initial condition, and $\mathbf{G}(\cdot,\cdot)$: $\mathbb{R}^n \times \mathbb{Z}_+ \to \mathbb{R}^n$ maps the current state $\mathbf{X}(k)$ into the next state $\mathbf{X}(k + 1)$, where $k_0 \in \mathbb{Z}_+$.

By analogy with the continuous-time case, there exists a function $\phi(\cdot, \mathbf{X}_0, k_0)$:$\mathbb{Z}_+ \to \mathbb{R}^n$ such that

$$\phi(k_0; \mathbf{X}_0, k_0) = \mathbf{X}_0$$

and

$$\phi(k+1; \mathbf{X}_0, k_0) = \mathbf{G}\big(\phi(k; \mathbf{X}_0, k_0), k\big)$$

The function $\phi(\cdot; \mathbf{X}_0, k_0)$: $\mathbb{Z}_+ \to \mathbb{R}^n$ is called the solution or trajectory through $(\mathbf{X}_0, k_0)$ of the difference equation (8.7).

The image $\{\phi(k; \mathbf{X}_0, k_0) \in \mathbb{R}^n \,|\, k \in \mathbb{Z}_+\}$ in $\mathbb{R}^n$ of the trajectory through $(\mathbf{X}_0, k_0)$ is called an orbit through $(\mathbf{X}_0, k_0)$.

If the map $\mathbf{G}(\cdot\,, \cdot)$ of a discrete-time dynamical system depends only on the state $\mathbf{X}(k)$ and is independent of k then the system is said to be autonomous and may be written more simply as

$$\mathbf{X}_{k+1} = \mathbf{G}(\mathbf{X}_k) \tag{8.8}$$

where $\mathbf{X}_k$ is shorthand for $\mathbf{X}(k)$ and the initial iterate k_0 is chosen, without loss of generality, to be zero. Using this notation, $\mathbf{X}_k$ is the image $\mathbf{X}_0$ after k iterations of the map $\mathbf{G}(\cdot)$: $\mathbb{R}^n \to \mathbb{R}^n$.

Example: Nonlinear Parallel *RLC* Circuit. Consider the parallel RLC circuit in Figure 8.2. This circuit contains a linear inductor L, a linear capacitor C_2, and a nonlinear resistor N'_R, where the continuous piecewise-linear driving-point (DP) characteristic (see Figure 8.3) has slope G'_a for $|V'_R| \le E$ and slope G'_b for $|V'_R| > E$. The DP characteristic of N'_R may be written explicitly

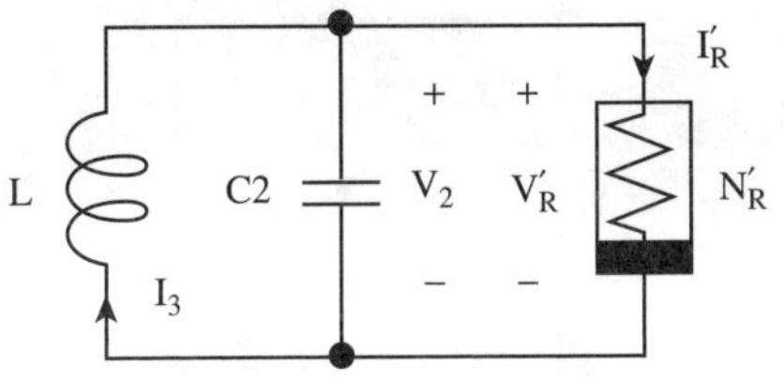

FIGURE 8.2 Parallel *RLC* circuit where the nonlinear resistor N'_R has a DP characteristic as illustrated in Figure 8.3. By Kirchhoff's voltage law, $V'_R = V_2$.

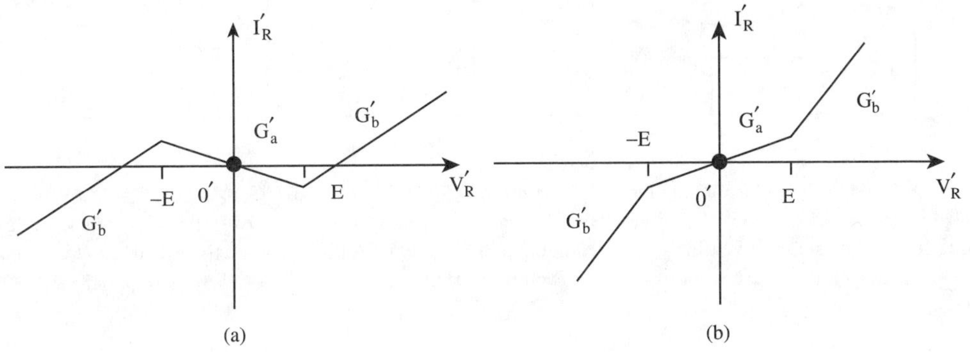

FIGURE 8.3 DP characteristic of N'_R in Figure 8.2 when (a) G'_a and (b) $G'_a > 0$.

$$I'_R(V'_R) = G'_b V'_R + \tfrac{1}{2}(G'_a - G'_b)\left(|V'_R + E| - |V'_R - E|\right)$$

This circuit may be described by a pair of ordinary differential equations and is therefore a second-order, continuous-time dynamical system. Choosing I_3 and V_2 as state variables, we write

$$\frac{dI_3}{dt} = -\frac{1}{L}V_2$$

$$\frac{dV_2}{dt} = \frac{1}{C_2}I_3 - \frac{1}{C_2}I'_R(V_2)$$

with $I_3(0) = I_{3_0}$ and $V_2(0) = V_{2_0}$.

We illustrate the vector field by drawing vectors at uniformly-spaced points in the two-dimensional state defined by (I_3, V_2). Starting from a given initial condition (I_{3_0}, V_{2_0}), a solution curve in state space is the locus of points plotted out by the state as it moves through the vector field, following the direction of the arrow at every point. Figure 8.4 illustrates typical vector fields and trajectories of the circuit.

If L, C_2, and G'_b are positive, the steady-state behavior of the circuit depends on the sign of G'_a. When $G'_a > 0$, the circuit is dissipative everywhere and all trajectories collapse toward the origin. The unique steady-state solution of the circuit is the stable dc equilibrium condition $I_3 = V_2 = 0$.

If $G'_a > 0$, N'_R looks like a negative resistor close to the origin and injects energy into the circuit, pushing trajectories away. Further out, where the characteristic has positive slope, trajectories are pulled in by the dissipative vector field. The resulting balance of forces produces a steady-state orbit called a *limit cycle*, which is approached asymptotically by all initial conditions of this circuit.

This limit cycle is said to be *attracting* because nearby trajectories move toward it and is structurally stable in the sense that, for almost all values of G'_a, a small change in the parameters of the circuit has little effect on it. In the special case when $G'_a \equiv 0$, a perturbation of G'_a causes the steady-state behavior to change from an equilibrium point to a limit cycle; this is called a *bifurcation*.

In the following subsections, we consider in detail steady-state behaviors, stability, structural stability, and bifurcations.

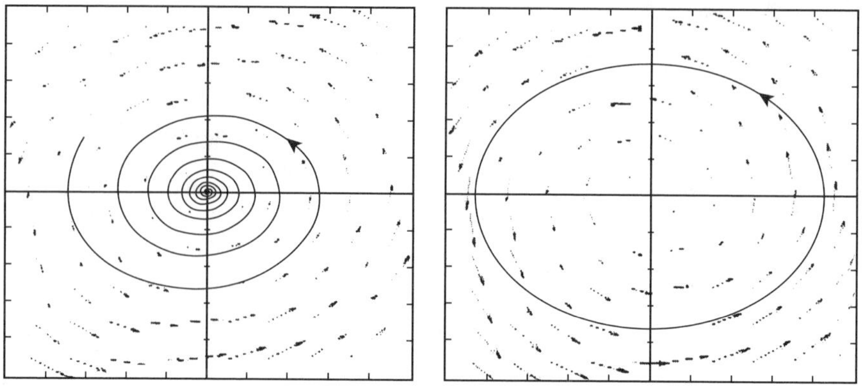

(a) (b)

FIGURE 8.4 Vector fields for the nonlinear *RLC* circuit in Figure 8.2. $L = 18$ mH, $C_2 = 100$ nF, $E = 0.47$ V. (a) $G'_a = 242.424$ μS, $G'_b = 1045.455$ μS: all trajectories converge to the origin (b) $G'_a = -257.576$ μS, $G'_b = 545.455$ μS: the unique steady-state solution is a limit cycle. Horizontal axis: I_3, 400 μA/div; vertical axis: V_2, 200 mV/div. *Source:* M. P. Kennedy, "Three steps to chaos — Part I: Evolution," *IEEE Trans. Circuits Syst. I*, vol. 40, p. 647, Oct. 1993. © 1993 IEEE.

Classification and Uniqueness of Steady-State Behaviors

A trajectory of a dynamical system from an initial state $\mathbf{X}_0$ settles, possibly after some transient, onto a set of points called a *limit set*. The **ω**-limit set corresponds to the asymptotic behavior of the system as $t \to +\infty$ and is called the steady-state response. We use the idea of recurrent states to determine when the system has reached steady-state.

A state **X** of a dynamical system is called recurrent under the flow ϕ if, for every neighborhood $B_\varepsilon(\mathbf{X})$ of **X** and for every $T > 0$, there is a time $t > T$ such that $\phi_t(\mathbf{X}) \cap B_\varepsilon(\mathbf{X}) \neq \varnothing$. Thus, a state **X** is recurrent if, by waiting long enough, the trajectory through **X** repeatedly returns arbitrarily close to **X** [7].

Wandering points correspond to transient behavior, while steady-state or asymptotic behavior corresponds to orbits of recurrent states.

A point $\mathbf{X}_\omega$ is an ω-limit point of $\mathbf{X}_0$ if and only if $\lim_{k\to+\infty} \phi_{t_k}(\mathbf{X}_0) = \mathbf{X}_\omega$ for some sequence $\{t_k | k \in \mathbb{Z}_+$ such that $t_k \to +\infty$. The set $L(\mathbf{X}_0)$ of ω-limit points of $\mathbf{X}_0$ is called ω-limit set of $\mathbf{X}_0$.[5]

A limit set L is called attracting if there exists a neighborhood U of L such that $L(\mathbf{X})_0 = L$ for all $\mathbf{X}_0 \in U$. Thus, nearby trajectories converge toward an attracting limit set as $t \to \infty$.

An attracting set $\mathcal{A}$ that contains at least one orbit that comes arbitrarily close to every point in $\mathcal{A}$ is called an attractor [7].

In an asymptotically stable linear system the limit set is independent of the initial condition and unique so it makes sense to talk of the steady-state behavior. By contrast, a nonlinear system may possess several different limit sets and therefore may exhibit a variety of steady-state behaviors, depending on the initial condition.

The set of all points in the state space that converge to a particular limit set L is called the **basin of attraction** of L.

Because nonattracting limit sets cannot be observed in physical systems, the asymptotic or steady-state behavior of a real electronic circuit corresponds to motion on an attracting limit set.

[5] The set of points to which trajectories converge from $\mathbf{X}_0$ as $t \to -\infty$ is called the α-limit set of $\mathbf{X}_0$. We consider only positive time, therefore, by *limit set*, we mean the ω-limit set.

Equilibrium Point

The simplest steady-state behavior of a dynamical system is an equilibrium point. An equilibrium point or stationary point of (8.3) is a state $\mathbf{X}_Q$ at which the vector field is zero. Thus, $\mathbf{F}(\mathbf{X}_Q) = 0$ and $\phi_t(\mathbf{X}_Q) = \mathbf{X}_Q$; a trajectory starting from an equilibrium point remains indefinitely at that point.

In state space, the limit set consists of a single nonwandering point $\mathbf{X}_Q$. A point is a zero-dimensional object. Thus, an equilibrium point is said to have dimension zero.

In the time domain, an equilibrium point of an equilibrium circuit is simply a dc solution or operating point.

An equilibrium point or fixed point of a discrete-time dynamical system is a point $\mathbf{X}_Q$ that satisfies

$$\mathbf{G}(\mathbf{X}_Q) = \mathbf{X}_Q$$

Example: Nonlinear Parallel *RLC* Circuit. The nonlinear *RLC* circuit shown in Figure 8.2 has just one equilibrium point $(I_{3Q}, V_{2Q}) = (0,0)$. When G_a' is positive, a trajectory originating at any point in the state converges to this attracting dc steady-state [as shown in Figure 8.4(a)]. The basin of attraction of the origin is the entire state space.

All trajectories, and not just those that start close to it, converge to the origin, so this equilibrium point is said to be a **global** attractor.

When $G_a' < 0$, the circuit possesses two steady-state solutions: the equilibrium point at the origin, and the limit cycle Γ. The equilibrium point is unstable in this case. All trajectories, except that which starts at the origin, are attracted to Γ.

Periodic Steady-State

A state $\mathbf{X}$ is called periodic if there exists $T > 0$ such that $\phi_T(\mathbf{X}) = \mathbf{X}$. A periodic orbit which is not a stationary point is called a **cycle**.

A limit cycle Γ is an isolated periodic orbit of a dynamical system [see Figure 8.5(b)]. The limit cycle trajectory visits every point on the simple closed curve Γ with period T. Indeed, $\phi_t(\mathbf{X}) = \phi_{t+T}(\mathbf{X}) \forall \mathbf{X} \in \Gamma$. Thus, every point on the limit cycle Γ is a nonwandering point.

A limit cycle is said to have dimension one because a small piece of it looks like a one-dimensional object: a line. Then, n components $X_i(t)$ of a limit cycle trajectory $\mathbf{X}(t) = [X_1(t), X_2(t), \ldots, X_n(t)]^T$ in $\mathbb{R}^n$ are periodic time waveforms with period T.

Every periodic signal $X(t)$ may be decomposed into a Fourier series — a weighted sum of sinusoids at integer multiples of a fundamental frequency. Thus, a periodic signal appears in the frequency domain as a set of spikes at integer multiples (**harmonics**) of the fundamental frequency. The amplitudes of these spikes correspond to the coefficient in the Fourier series expansion of $X(t)$. The Fourier transform is an extension of these ideas to aperiodic signals; one considers the distribution of the signal's power over a continuum of frequencies rather than on a discrete set of harmonics.

The distribution of power in a signal $X(t)$ is most commonly quantified by means of the power density spectrum, often simply called the power spectrum. The simplest estimator of the power spectrum is the periodogram [17], which given N uniformly spaced samples $X(k/f_s)$, $k = 0, 1, \ldots, N-1$ of $X(t)$, yields $N/2 + 1$ numbers $P(nf_s/N)$, $n = 0, 1, \ldots, N/2$, where f_s is the sampling frequency.

If one considers the signal $X(t)$ as being composed of sinusoidal components at discrete frequencies, then $P(nf_s/N)$ is an estimate of the power in the component at frequency nf_s/N. By Parseval's theorem, the sum of the power in each of these components equals the mean squared amplitude of the N samples of $X(t)$ [17].

If $X(t)$ is periodic with period T, then its power will be concentrated in a dc component, a fundamental frequency component $1/T$, and harmonics. In practice, the discrete nature of the sampling process causes power to "leak" between adjacent frequency components; this leakage may be reduced by "windowing" the measured data before calculating the periodogram [17].

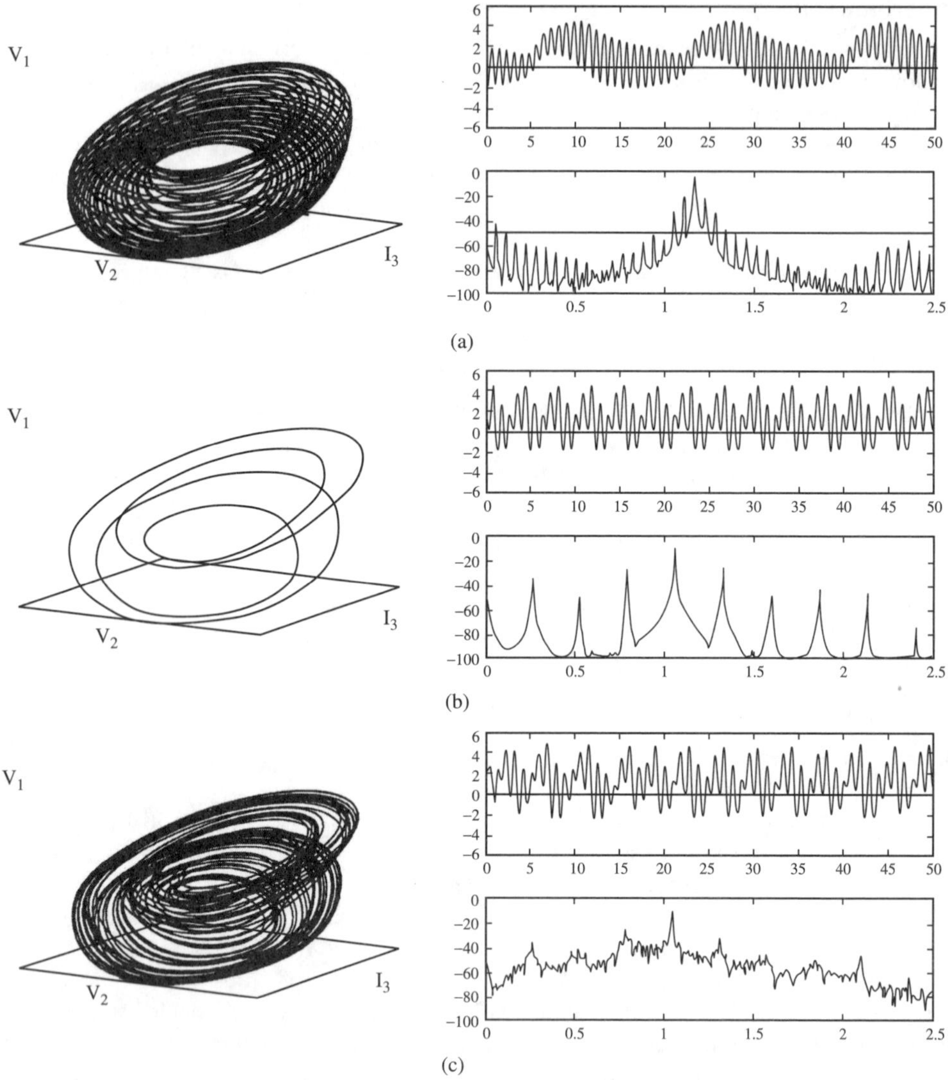

FIGURE 8.5 Quasiperiodicity (torus breakdown) route to chaos in Chua's oscillator. Simulated state space trajectories, time waveforms $V_1(t)$, and power spectra of $V_2(t)$. (a) Quasiperiodic steady-state — the signal is characterized by a discrete power spectrum with incommensurate frequency components; (b) periodic window — all spikes in the power spectrum are harmonically related to the fundamental frequency; (c) chaotic steady-state following breakdown of the torus — the waveform has a broadband power spectrum. Time plots; horizontal axis — t (ms); vertical axis — $V_1(V)$. Power spectra: horizontal axis — frequency (kHz); vertical axis — power (mean squared amplitude) of $V_2(t)$ (dB).

Example: Periodic Steady-State Solution. Figure 8.5(b) depicts a state-space orbit, time waveform, and power spectrum of a periodic steady-state solution of a third-order, autonomous, continuous-time dynamical system.

The orbit in state space is an asymmetric closed curve consisting of four loops. In the time domain, the waveform has four crests per period and a dc offset. In the power spectrum, the dc offset manifests itself as a spike at zero frequency. The period of approximately 270 Hz produces a fundamental component at that frequency. Notice that the fourth harmonic (arising from "four crests per period") has the largest magnitude. This power spectrum is reminiscent of subharmonic mode locking in a forced oscillator.

Subharmonic Periodic Steady-State

A subharmonic periodic solution or period-K orbit of a discrete-time dynamical system is a set of K points $\{\mathbf{X}_1, \mathbf{X}_2, \ldots, \mathbf{X}_k\}$ that satisfy

$$\mathbf{X}_2 = \mathbf{G}(\mathbf{X}_1) \qquad \mathbf{X}_3 = \mathbf{G}(\mathbf{X}_2) \ldots \mathbf{X}_K = \mathbf{G}(\mathbf{X}_{K-1}) \qquad \mathbf{X}_1 = \mathbf{G}(\mathbf{X}_K)$$

More succinctly, we may write $\mathbf{X}_i = \mathbf{G}^{(K)}(\mathbf{X}_i)$, where $\mathbf{G}^{(K)} = \mathbf{G}[\mathbf{G}(\cdots[\mathbf{G}(\cdot)]\cdots)]$ denotes $\mathbf{G}$ applied K times to the argument of the map; this is called Kth iterate of $\mathbf{G}$.

Subharmonic periodic solutions occur in systems that contain two or more competing frequencies, such as forced oscillators or sampled-data circuits. Subharmonic solutions also arise following period-doubling bifurcations (see the section on structural stability and bifurcations).

Quasiperiodic Steady-State

The next most complicated form of steady-state behavior is called **quasiperiodicity.** In state space, this corresponds to a *torus* [see Figure 8.5(a)]. Although a small piece of a limit cycle in $\mathbb{R}^3$ looks like a line, a small section of two-torus looks like a plane; a two-torus has dimension two.

A quasiperiodic function is one that may be expressed as a countable sum of periodic functions with incommensurate frequencies, i.e., frequencies that are not rationally related. For example, $X(t) = \sin(t) + \sin(2\pi t)$ is a quasiperiodic signal. In the time domain, a quasiperiodic signal may look like an amplitude- or phase-modulated waveform.

Although the Fourier spectrum of a periodic signal consists of a discrete set of spikes at integer multiples of a fundamental frequency, that of a quasiperiodic solution comprises a discrete set of spikes at incommensurate frequencies, as presented in Figure 8.5(a).

In principle, a quasiperiodic signal may be distinguished from a periodic one by determining whether the frequency spikes in the Fourier spectrum are harmonically related. In practice, it is impossible to determine whether a measured number is rational or irrational; therefore, any spectrum that appears to be quasiperiodic may simply be periodic with an extremely long period.

A two-torus in a three-dimensional state space looks like a doughnut. Quasiperiodic behavior on a higher dimensional torus is more difficult to visualize in state space but appears in the power spectrum as a set of discrete components at incommensurate frequencies. A K-torus has dimension K.

Quasiperiodic behavior occurs in discrete-time systems where two incommensurate frequencies are present. A periodically-forced or discrete-time dynamical system has a frequency associated with the period of the forcing or sampling interval of the system; if a second frequency is introduced that is not rationally related to the period of the forcing or the sampling interval, then quasiperiodicity may occur.

Example: Discrete Torus. Consider a map from the circle S^1 onto itself. In polar coordinates, a point on the circle is parameterized by an angle θ. Assume that θ has been normalized so that one complete revolution of the circle corresponds to a change in θ of 1. The state of this system is determined by the normalized angle θ and the dynamics by

$$\theta_{k+1} = (\theta_k + \Omega) \bmod 1$$

If Ω is a rational number (of the form J/K where $J, K \in \mathbb{Z}_+$), then the steady-state solution is a period-**K** (subharmonic) orbit. If Ω is irrational, we obtain quasiperiodic behavior.

Chaotic Steady-State

DC equilibrium periodic as well as quasiperiodic steady-state behaviors have been correctly identified and classified since the pioneering days of electronics in the 1920s. By contrast, the existence of more exotic steady-state behaviors in electronic circuits has been acknowledged only in the past 30 years. Although the notion of chaotic behavior in dynamical systems has existed in the mathematics literature since the turn of the century, unusual behaviors in the physical sciences as recently as the 1960s were described as "strange." Today, we classify as *chaos* the recurrent[6] motion in a deterministic dynamical system, which is characterized by a positive Lyapunov exponent.

From an experimentalist's point of view, chaos may be defined as bounded steady-state behavior in a deterministic dynamical system that is not an equilibrium point, nor periodic, and not quasiperiodic [15].

Chaos is characterized by repeated stretching and folding of bundles of trajectories in state space. Two trajectories started from almost identical initial conditions diverge and soon become uncorrelated; this is called sensitive dependence on initial conditions and gives rise to long-term unpredictability.

In the time domain, a chaotic trajectory is neither periodic nor quasiperiodic, but looks "random." This "randomness" manifests itself in the frequency domain as a broad "noise-like" Fourier spectrum, as presented in Figure 8.5(c).

Although an equilibrium point, a limit cycle, and a K-torus each have integer dimension, the repeated stretching and folding of trajectories in a chaotic steady state gives the limit set a more complicated structure that, for three-dimensional continuous-time circuits, is something more than a surface but not quite a volume.

Dimension

The structure of a limit set $L \subset \mathbb{R}^n$ of a dynamical system may be quantified using a generalized notion of dimension that considers not just the geometrical structure of the set, but also the time evolution of trajectories on L.

Capacity (D_0 Dimension). The simplest notion of dimension, called capacity (or D_0 dimension) considers a limit set simply as set of points, without reference to the dynamical system that produced it.

To estimate the capacity of L, cover the set with n-dimensional cubes having side length ε. If L is a D_0-dimensional object, then the minimum number $N(\varepsilon)$ of cubes required to cover L is proportional to ε^{-D_0}. Thus, $N(\varepsilon)\ \alpha\, \varepsilon^{-D_0}$.

The D_0 dimension is given by

$$D_0 = \lim_{\varepsilon \to 0} - \frac{\ln N(\varepsilon)}{\ln \varepsilon}$$

When this definition is applied to a point, a limit cycle (or line), or a two-torus (or surface) $\mathbb{R}^3$, the calculated dimensions are 0, 1, and 2, respectively, as expected. When applied to the set of nonwandering points that comprise a chaotic steady state, the D_0 dimension is typically noninteger. An object that has noninteger dimension is called a **fractal.**

Example: The Middle-Third Cantor Set. Consider the set of points that is obtained by repeatedly deleting the middle third of an interval, as indicated in Figure 8.6(a). At the first iteration, the unit interval is divided into 2^1 pieces of length 1/3 each; after k iterations, the set is covered by 2^k pieces of length $1/3^k$. By contrast, the set that is obtained by dividing the intervals into thirds but *not* throwing away the middle third each time [Figure 8.6(b)] is covered at the kth step by 3^k pieces of length $1/3^k$.

Applying the definition of capacity, the dimension of the unit interval is

$$\lim_{k \to \infty} \frac{k \ln 3}{k \ln 3} = 1.00$$

By contrast, the middle-third Cantor set has dimension

$$\lim_{k \to \infty} \frac{k \ln 2}{k \ln 3} \approx 0.63$$

[6]Because a chaotic steady-state does not settle down onto a single well-defined trajectory, the definition of recurrent states must be used to identify posttransient behavior.

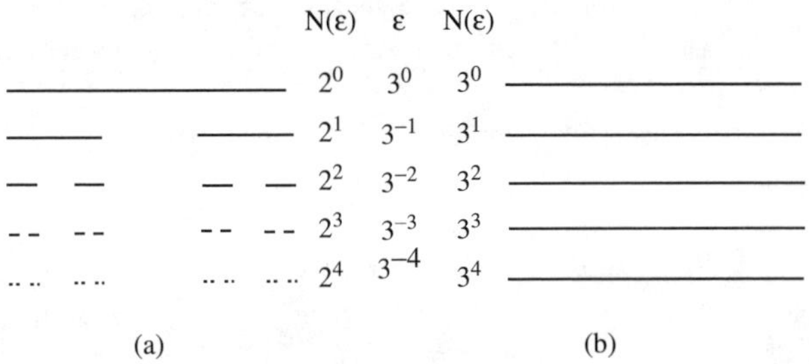

FIGURE 8.6 (a) The middle-third Cantor set is obtained by recursively removing the central portion of an interval. At the kth step, the set consists of $N(\varepsilon) = 2^k$ pieces of length $\varepsilon = 3^{-k}$. The limit set has capacity 0.63. (b) By contrast, the unit interval is covered by 3^k pieces of length 3^{-k}. The unit interval has dimension 1.00.

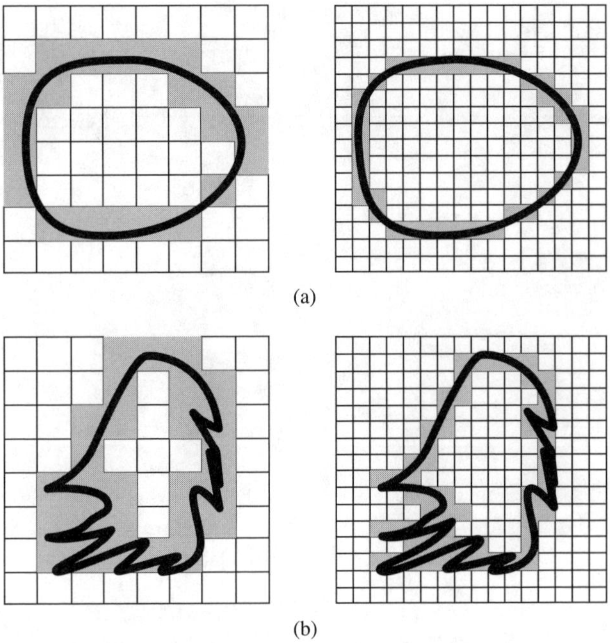

FIGURE 8.7 Coverings of two limit sets L_a (a) and L_b (b) with squares of sidelength ε_0 and $\varepsilon_0/2$, respectively.

The set is something more than a zero-dimensional object (a point) but not quite one-dimensional (like a line segment); it is a fractal.

Correlation (D_2) Dimension. The D_2 dimension considers not just the geometry of a limit set, but also the time evolution of trajectories on the set.

Consider the two limit sets L_a and L_b in $\mathbb{R}^2$ shown in Figure 8.7(a) and (b), respectively. The D_0 dimension of these sets may be determined by iteratively covering them with squares (two-dimensional "cubes") of sidelength $\varepsilon = \varepsilon_0/2^k$, $k = 0, 1, 2, \ldots$, counting the required number of squares $N(\varepsilon)$ for each ε, and evaluating the limit

$$D_0 = \lim_{k \to \infty} -\frac{\ln N(\varepsilon)}{\ln(\varepsilon)}$$

For the smooth curve L_a, the number of squares required to cover the set grows linearly with $1/\varepsilon$; hence $D_0 = 1.0$. By contrast, if the kinks and folds in set L_b are present at all scales, then the growth of $N(\varepsilon)$ versus $1/\varepsilon$ is superlinear and the object has a noninteger D_0 dimension between 1.0 and 2.0.

Imagine now that L_a and L_b are not simply static geometrical objects but are orbits of discrete-time dynamical systems. In this case, a steady-state trajectory corresponds to a sequence of points moving around the limit set.

Cover the limit set with the minimum number $N(\varepsilon)$ of "cubes" with sidelength ε, and label the boxes 1, 2..., i, ..., $N(\varepsilon)$. Count the number of times $n_i(N, \varepsilon)$ that a typical steady-state trajectory of length N visits box i and define

$$p_i = \lim_{N\to\infty} \frac{\mathrm{n_i}(N, \varepsilon)}{N}$$

where p_i is the relative frequency with which a trajectory visits the ith cube. The D_2 dimension is defined as

$$D_2 = \lim_{\varepsilon\to 0} \frac{\ln \sum_{i=1}^{N(\varepsilon)} p_i^2}{\ln \varepsilon}$$

In general, $D_2 \leq D_0$ with equality when a typical trajectory visits all $N(\varepsilon)$ cubes with the same relative frequency $p = 1/N(\varepsilon)$. In this special case,

$$\begin{aligned} D_2 &= \lim_{\varepsilon\to 0} \frac{\ln \sum_{i=1}^{N(\varepsilon)} \frac{1}{N(\varepsilon)^2}}{\ln \varepsilon} \\ &= \lim_{\varepsilon\to 0} - \frac{\ln N(\varepsilon)}{\ln \varepsilon} \\ &= D_0 \end{aligned}$$

An efficient algorithm (developed by Grassberger and Procaccia) for estimating D_2 is based on the approximation $\sum_{i=1}^{N(\varepsilon)} p_i^2 \approx C(\varepsilon)$ [15], where $C(\varepsilon) = \lim_{N\to 0} \frac{1}{N^2}$ (the number of pairs of points (X_i, X_j) such that $\|X_i - X_j\| < \varepsilon$) is called the correlation. The D_2 or correlation dimension is given by

$$D_2 = \lim_{\varepsilon\to 0} \frac{\ln C(\varepsilon)}{\ln \varepsilon}$$

Example: Correlation (D_2) Dimension. The correlation dimension of the chaotic attractor in Figure 8.5(c), estimated using INSITE, is 2.1, 2 while D_2 for the uniformly covered torus in Figure 8.5(a) is 2.0.

Stability of Steady-State Trajectories

Consider once more the nonlinear *RLC* circuit in Figure 8.2. IF G'_a is negative, this circuit settles to a periodic steady state from almost every initial condition. However, a trajectory started from the origin will, in principle, remain indefinitely at the origin since this is an equilibrium point. The circuit has two possible steady-state solutions. Experimentally, only the limit cycle will be observed. Why?

If trajectories starting from states close to a limit set converge to that steady-state, the limit set is called an attracting limit set. If, in addition, the attracting limit set contains at least one trajectory that comes arbitrarily close to every point in the set, then it is an attractor. If nearby points diverge from the limit set, it is called a repellor.

In the nonlinear *RLC* circuit with $G'_a < 0$, the equilibrium point is a repellor and the limit cycle is an attractor.

Stability of Equilibrium Points

Qualitatively, an equilibrium point is said to be stable if trajectories starting close to it remain nearby for all future time and unstable otherwise. Stability is a local concept, dealing with trajectories in a small neighborhood of the equilibrium point.

To analyze the behavior of the vector field in the vicinity of an equilibrium point $\mathbf{X}_Q$, we write $\mathbf{X} = \mathbf{X}_Q + \mathbf{x}$ and substitute into (8.3) to obtain

$$\dot{\mathbf{X}}_Q + \dot{\mathbf{x}} = \mathbf{F}(\mathbf{X}_Q + \mathbf{x})$$

$$\mathbf{F}(\mathbf{X}_Q) + \dot{\mathbf{x}} \approx \mathbf{F}(\mathbf{X}_Q) + \mathbf{D}_x\mathbf{F}(\mathbf{X}_Q)\mathbf{x}$$

where we have kept just the first two terms of the Taylor series expansion of $\mathbf{F}(\mathbf{X})$ about $\mathbf{X}_Q$. The Jacobian matrix $\mathbf{D}_X\mathbf{F}(\mathbf{X})$ is the matrix of partial derivatives of $\mathbf{F}(\mathbf{X})$ with respect to $\mathbf{X}$:

$$\mathbf{D}_X\mathbf{F}(\mathbf{X}) = \begin{bmatrix} \frac{\partial F_i(\mathbf{X})}{\partial X_1} & \frac{\partial F_1(\mathbf{X})}{\partial X_2} & \cdots & \frac{\partial F_1(\mathbf{X})}{\partial X_n} \\ \frac{\partial F_2(\mathbf{X})}{\partial X_1} & \frac{\partial F_2(\mathbf{X})}{\partial X_2} & \cdots & \frac{\partial F_2(\mathbf{X})}{\partial X_n} \\ \vdots & \vdots & \ddots & \vdots \\ \frac{\partial F_n(\mathbf{X})}{\partial X_1} & \frac{\partial F_n(\mathbf{X})}{\partial X_2} & \cdots & \frac{\partial F_n(\mathbf{X})}{\partial X_n} \end{bmatrix}$$

Subtracting $\mathbf{F}(\mathbf{X}_Q)$ from both sides of (8.9) we obtain the linear system

$$\dot{\mathbf{x}} = \mathbf{D}_x\mathbf{F}(\mathbf{X}_Q)\mathbf{x} \tag{8.9}$$

where the Jacobian matrix is evaluated at $\mathbf{X}_Q$. This linearization describes the behavior of the circuit in the vicinity of $\mathbf{X}_Q$; we call this the local behavior.

Note that the linearization is simply the small-signal equivalent circuit at the operating point $\mathbf{X}_Q$. In general, the local behavior of a circuit depends explicitly on the operating point $\mathbf{X}_Q$. For example, a *pn*-junction diode exhibits a small incremental resistance under forward bias, but a large small-signal resistance under reverse bias.

Eigenvalues

If $\mathbf{X}_Q$ is an equilibrium point of (8.3), a complete description of its stability is contained in the eigenvalues of the linearization of (8.3) about $\mathbf{X}_Q$. These are defined as the roots λ of the characteristic equation

$$\det[\lambda\mathbf{I} - \mathbf{D}_X\mathbf{F}(\mathbf{X}_Q)] = 0 \tag{8.10}$$

where $\mathbf{I}$ is the identity matrix.

If the real parts of all of the eigenvalues $\mathbf{D}_x\mathbf{F}(\mathbf{X}_Q)$ are strictly negative, then the equilibrium point $\mathbf{X}_Q$ is asymptotically stable and is called a sink because all nearby trajectories converge toward it.

If any of the eigenvalues has a positive real part, the equilibrium point is unstable; if all of the eigenvalues have positive real parts, the equilibrium point is called a source. An equilibrium point that has eigenvalues with both negative and positive real parts is called a saddle. A saddle is unstable.

An equilibrium point is said to be hyperbolic if all the eigenvalues of $\mathbf{D}_x\mathbf{F}(\mathbf{X}_Q)$ have nonzero real parts. All hyperbolic equilibrium points are either unstable or asymptotically stable.

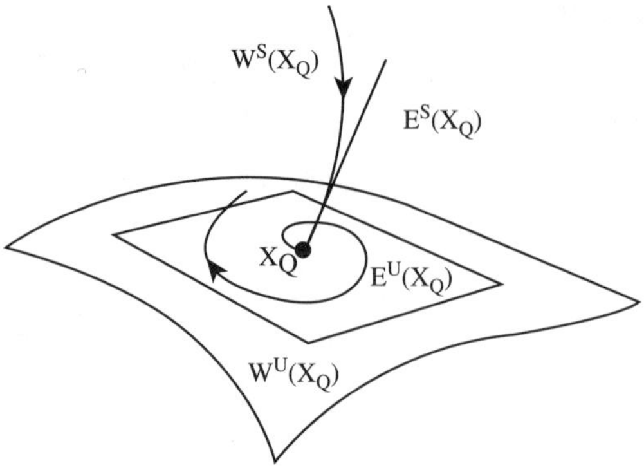

FIGURE 8.8 Stable and unstable manifolds $W^s(\mathbf{X}_Q)$ and $W^u(\mathbf{X}_Q)$ of an equilibrium point $\mathbf{X}_Q$. The stable and unstable eigenspaces $E^s(\mathbf{X}_Q)$ and $E^u(\mathbf{X}_Q)$ derived from the linearization of the vector field at $\mathbf{X}_Q$ are tangent to the corresponding manifolds W^s and W^u at $\mathbf{X}_Q$. A trajectory approaching the equilibrium point along the stable manifold is tangential to $E^s(\mathbf{X}_Q)$ at $\mathbf{X}_Q$; a trajectory leaving $\mathbf{X}_Q$ along the unstable manifold is tangential to $E^u(\mathbf{X}_Q)$ at $\mathbf{X}_Q$.

Discrete-Time Systems

The stability of a fixed point $\mathbf{X}_Q$ of a discrete-time dynamical system

$$\mathbf{X}_{k+1} = \mathbf{G}(\mathbf{X}_k)$$

is determined by the eigenvalues of the linearization $\mathbf{D_XG}(\mathbf{X}_Q)$ of the vector field $\mathbf{G}$, evaluated at $\mathbf{X}_Q$.

The equilibrium point is classified as stable if all of the eigenvalues of $\mathbf{D_XG}(\mathbf{X}_Q)$ are strictly less than unity in modulus, and unstable if any has modulus greater than unity.

Eigenvectors, Eigenspaces, Stable and Unstable Manifolds

Associated with each distinct eigenvalue λ of the Jacobian matrix $\mathbf{D_XF}(\mathbf{X}_Q)$ is an eigenvector $\vec{v}$ defined by

$$\mathbf{D_XF}(\mathbf{X}_Q)\vec{v} = \lambda\vec{v}$$

A real eigenvalue γ has a real eigenvector $\vec{\eta}$. Complex eigenvalues of a real matrix occur in pairs of the form $\sigma \pm j\omega$. The real and imaginary parts of the associated eigenvectors $\vec{\eta}_r \pm j\vec{\eta}_c$ span a plane called a complex eigenplane.

The n_s-dimensional subspace of $\mathbb{R}^n$ associated with the stable eigenvalues of the Jacobian matrix is called the stable eigenspace, denoted $E^s(\mathbf{X}_Q)$. The n_u-dimensional subspace corresponding to the unstable eigenvalues is called the unstable eigenspace, denoted $E^u(\mathbf{X}_Q)$.

The analogs of the stable and unstable eigenspaces for a general nonlinear system are called the local stable and unstable *manifolds*[7] $W^s(\mathbf{X}_Q)$ and $W^u(\mathbf{X}_Q)$.

The stable manifold $W^s(\mathbf{X}_Q)$ is defined as the set of all states from which trajectories remain in the manifold and converge under the flow to $\mathbf{X}_Q$. The unstable manifold $W^u(\mathbf{X}_Q)$ is defined as the set of all states from which trajectories remain in the manifold and diverge under the flow from $\mathbf{X}_Q$.

By definition, the stable and unstable manifolds are invariant under the flow (if $\mathbf{X} \in W^s$, then $\phi_t(\mathbf{X}) \in W^s$). Furthermore, the n_s- and n_u-dimensional tangent spaces to W^s and W^u at $\mathbf{X}_Q$ are E^s and E^u (as shown in Figure 8.8). In the special case of a linear or affine vector field **F**, the stable and unstable manifolds are simply the eigenspaces E^s and E^u.

[7] An m-dimensional manifold is a geometrical object every small section of which looks like $\mathbb{R}^m$. More precisely, M is an m-dimensional manifold if, for every $x \in M$, there exists an open neighborhood U of x and a smooth invertible map which takes U to some open neighborhood of $\mathbb{R}^m$. For example, a limit cycle of a continuous-time dynamical system is a one-dimensional manifold.

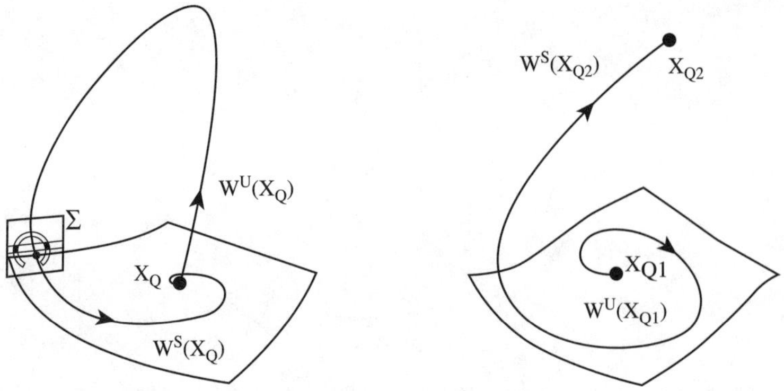

FIGURE 8.9 (a) A homoclinic orbit joins an isolated equilibrium point $\mathbf{X}_Q$ to itself along its stable and unstable manifolds. (b) A heteroclinic orbit joins two distinct equilibrium points, $\mathbf{X}_{Q1}$ and $\mathbf{X}_{Q2}$, along the unstable manifolds of one and the stable manifold of the other.

Chaos is associated with two characteristic connections of the stable and unstable manifolds. A homoclinic orbit [see Figure 8.9(a)] joins an isolated equilibrium point $\mathbf{X}_Q$ to itself along its stable and unstable manifolds. A heteroclinic orbit [Figure 8.9(b)] joins two distinct equilibrium points, $\mathbf{X}_{Q1}$ and $\mathbf{X}_{Q2}$, along the unstable manifold of one and the stable manifold of the other.

Stability of Limit Cycles

Although the stability of an equilibrium point may be determined by considering the eigenvalues of the linearization of the vector field near the point, how does one study the stability of a limit cycle, torus, or chaotic steady-state trajectory?

The idea introduced by Poincaré is to convert a continuous-time dynamical system into an equivalent discrete-time dynamical system by taking a transverse slice through the flow. Intersections of trajectories with this so-called Poincaré section define a Poincaré map from the section to themselves. Since the limit cycle is a fixed point $\mathbf{X}_Q$ of the associated discrete-time dynamical system, its stability may be determined by examining the eigenvalues of the linearization of the Pioncaré map at $\mathbf{X}_Q$.

Poincaré Sections

A Pioncaré section of an n-dimensional autonomus continuous-time dynamical system is an $(n - 1)$-dimensional hyperplane Σ in the state space that is intersected transversally[8] by the flow.

Let Γ be a closed orbit of the flow of a smooth vector field **F**, and let $\mathbf{X}_Q$ be a point of intersection of Γ with Σ. If T is the period of Γ and $\mathbf{X} \in \Sigma$ is sufficiently close to $\mathbf{X}_Q$, then the trajectory $\phi_t(\mathbf{X})$ through $\mathbf{X}$ will return to Σ after a time $\tau(\mathbf{X}) \approx T$ and intersect the hyperplane at a point $\phi_{\tau(\mathbf{X})}(\mathbf{X})$, as illustrated in Figure 8.10.

This construction implicitly defines a function (called a Poincaré map or first return map) $\mathbf{G}: U \to \Sigma$

$$\mathbf{G}(\mathbf{X}) = \phi_{\tau(\mathbf{X})}(\mathbf{X})$$

where U is a small region of Σ close to $\mathbf{X}_Q$. The corresponding discrete-time dynamical system

$$\mathbf{X}_{k+1} = \mathbf{G}(\mathbf{X}_k)$$

has a fixed point at $\mathbf{X}_Q$.

The stability of the limit cycle is determined by the eigenvalues of the linearization $\mathbf{D_X G}(\mathbf{X}_Q)$ of $\mathbf{G}$ at $\mathbf{X}_Q$. If all of the eigenvalues of $\mathbf{D_X G}(\mathbf{X}_Q)$ have modulus less than unity, the limit cycle is asymptotically stable; if any has modulus greater than unity, the limit cycle is unstable.

[8]A *transverse* intersection of manifolds in $\mathbb{R}^n$ is an intersection of manifolds such that, from any point in the intersection, all directions in $\mathbb{R}^n$ can be generated by linear combinations of vectors tangent to the manifolds.

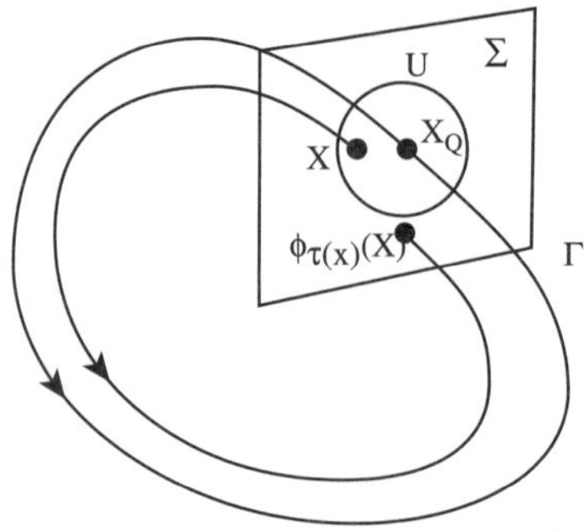

FIGURE 8.10 A transverse Poincaré section Σ through the flow of a dynamical system induces a discrete Poincaré map from a neighborhood U of the point of intersection X_Q to Σ.

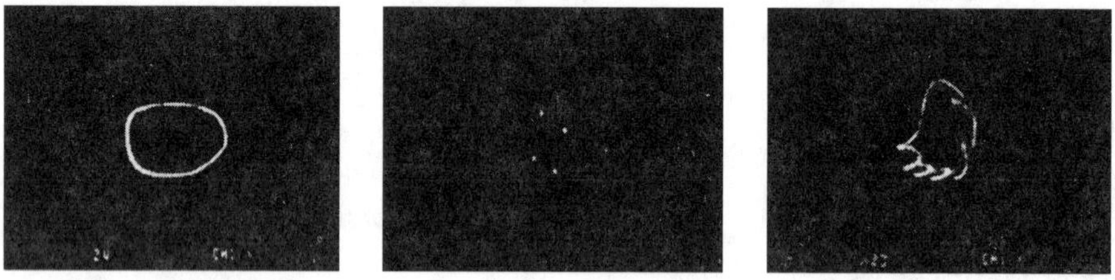

(a) (b) (c)

FIGURE 8.11 Experimental Poincaré sections corresponding to a torus breakdown sequence in Chua's oscillator. (a) Torus, (b) period-four orbit, (c) chaotic attractor resulting from torus breakdown. *Source*: L. O. Chua, C. W. Wu, A. Hung, and G.-Q. Zhong, "A universal circuit for studying and generating chaos — Part I: Routes to chaos," *IEEE Trans. Circuits Syst.*, vol. 40, pp. 738, 739, Oct. 1993.© 1993 IEEE.

Note that the stability of the limit cycle is independent of the position and orientation of the Poincaré plane, provided that the intersection is chosen *transverse* to the flow. For a nonautonomous system with periodic forcing, a natural choice for the hyperplane is at a fixed phase θ_o of the forcing.

In the Poincaré section, a limit cycle looks like a fixed point. A period-K subharmonic of a nonautonomous system with periodic forcing appears as a period-K orbit of the corresponding map [see Figure 8.11(b)].

The Poincaré section of a quasiperiodic attractor consisting of two incommensurate frequencies looks like a closed curve — a transverse cut through a two-torus [Figure 8.11(a)].

The Poincaré section of chaotic attractor has fractal structure, as depicted in Figure 8.11(c).

Horseshoes and Chaos

Chaotic behavior is characterized by sensitive dependence on initial conditions. This phrase emphasizes the fact that small differences in initial conditions are persistently magnified by the dynamics of the system so that trajectories starting from nearby initial conditions reach totally different states in a finite time.

Trajectories of the nonlinear *RLC* circuit in Figure 8.2 that originate near the equilibrium point are initially stretched apart exponentially by the locally negative resistance in the case $G'_a < 0$. Eventually, however, they are squeezed together onto a limit cycle, so the stretching is not persistent. This is a consequence of the noncrossing property and eventual passivity.

Although perhaps locally active, every physical resistor is eventually passive meaning that, for a large enough voltage across its terminals, it dissipates power. This in turn limits the maximum values of the voltages and currents in the circuit giving a bounded steady-state solution. All physical systems are bounded, so how can small differences be magnified persistently in a real circuit?

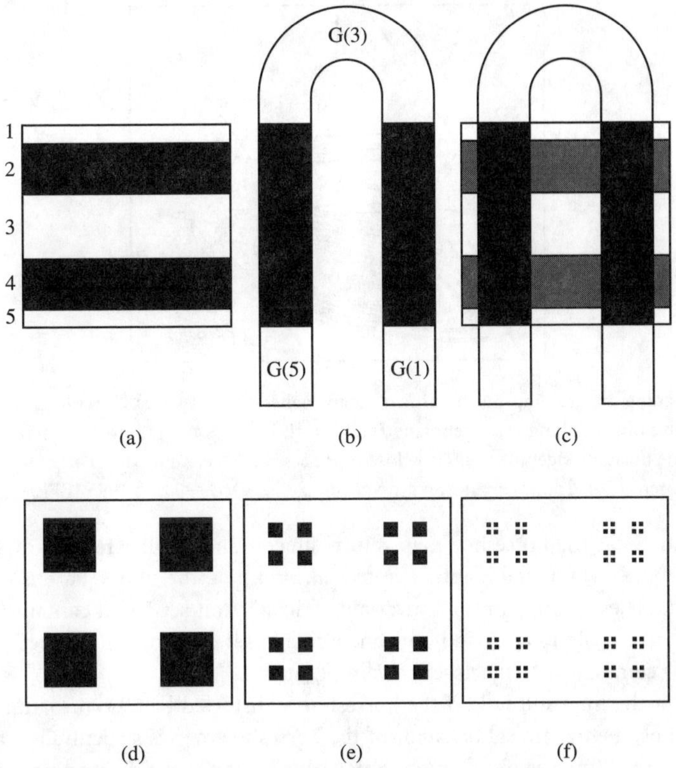

FIGURE 8.12 The Smale horseshoe map stretches the unit square (a), folds it into a horseshoe (b), and lays it back on itself (c), so that only points lying in bands 2 and 4 of (a) are mapped into the square. At the next iteration, only those points in (**G**(2) ∪ **G**(4)) ∩ (2 ∪ 4) (d) are mapped back to the square. Repeated iterations of the map (d)–(f) remove all points from the square except an invariant (fractal) set of fixed points.

Chaos in the Sense of Shil'nikov

Consider a flow ϕ in $\mathbb{R}^3$ that has an equilibrium point at the origin with a real eigenvalue $\gamma > 0$ and a pair of complex conjugate eigenvalues $\sigma \pm j\omega$ with $\sigma < 0$ and $\omega \neq 0$. Assume that the flow has a homoclinic orbit Γ through the origin.

One may define a Poincaré map for this system by taking a transverse section through the homoclinic orbit, as illustrated in Figure 8.9(a).

Theorem 2 (Shil'nikov): *If* $|\sigma/\gamma| < 1$, *the flow* ϕ *can be perturbed to* ϕ' *such that* ϕ' *has a homoclinic orbit* Γ' *near* Γ *and the Poincaré map of* ϕ' *defined in a neighborhood of* Γ' *has a countable number of horseshoes in its discrete dynamics.*

The characteristic horseshoe shape in the Poincaré map stretches and folds trajectories repeatedly (see Figure 8.12). The resulting dynamics exhibit extreme sensitivity to initial conditions [7].

The presence of horseshoes in the flow of a continuous-time system that satisfies the assumptions of Shil'nikov's theorem implies the existence of countable numbers of unstable periodic orbits of arbitrarily long period as well as an uncountable number of complicated bounded nonperiodic chaotic solutions [7].

Horseshoes. The action of the Smale horseshoe map is to take the unit square [Figure 8.12(a)], stretch it, fold it into a horseshoe shape [Figure 8.12(b)], and lay it down on itself [Figure 8.12(c)]. Under the action of this map, only four regions of the unit square are returned to the square.

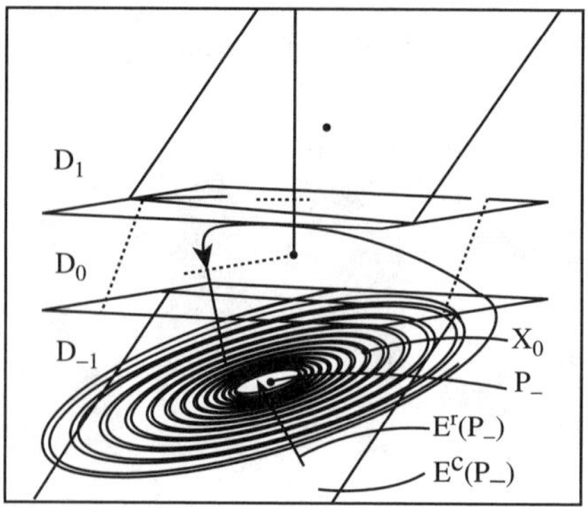

FIGURE 8.13 Stretching and folding mechanism of chaos generation in Chua's circuit. A trajectory spirals away from the equilibrium point P_- along the eigenplane $E^c(P_-)$ until it enters the D_0 region, where it is folded back into D_{-1} and returns to the unstable eigenplane $E^c(P_-)$ close to P_-. *Source*: M. P. Kennedy, "Three steps to chaos — Part II: A Chua's circuit primer," *IEEE Trans. Circuits Syst.* I, vol. 40, p. 657, Oct. 1993. ® 1993 IEEE.

Successive iterations of the horseshoe map return smaller and smaller regions of square to itself, as shown in Figure 8.12(d)–(f). If the map is iterated *ad infinitum*, the unit square is ultimately mapped onto a set of points. These points form an invariant (fractal) limit set L that contains a countable set of periodic orbits of arbitrarily long periods, an uncountable set of bounded nonperiodic orbits, and at least one orbit that comes arbitrarily close to every point in L.

The properties of the map still hold if the horseshoe is distorted by a perturbation of small size but arbitrary shape. Thus, the dynamical behavior of the horseshoe map is structurally stable.[9]

Although the invariant limit set of a horseshoe map consists of nonwandering points, it is not attracting. Therefore, the existence of a horseshoe in the flow of a third-order system does not imply that the system will exhibit chaotic steady-state behavior. However if a typical trajectory in the Poincaré map remains in a neighborhood of the invariant set, then the system may exhibit chaos. Thus, although Shil'nikov's theorem is a strong indicator of chaos, it does not provide definitive proof that a system is chaotic.

Example: Chaos in a Piecewise-Linear System. Although we have stated it for the case $\sigma < 0, \gamma > 0$, Shil'nikov's theorem also applies when the equilibrium point at the origin has an unstable pair of complex conjugate eigenvalues and a stable real eigenvalue. In that case, it is somewhat easier to visualize the stretching and folding of bundles of trajectories close to a homoclinic orbit.

Consider the trajectory in a three-region piecewise-linear vector field in Figure 8.13. We assume that the equilibrium point P_- has a stable real eigenvalue γ_1 [where the eigenvector is $E^r(P_-)$] and an unstable complex conjugate pair of eigenvalues $\sigma_1 \pm j\omega_1$, the real and imaginary parts of whose eigenvectors span the plane $E^c(P_-)$ [2], as illustrated. A trajectory originating from a point $\mathbf{X}_0$ on $E^c(P_-)$ spirals away from the equilibrium point along $E^c(P_-)$ until it enters the D_0 region, where it is folded back into D_{-1}. Upon reentering D_{-1}, the trajectory is pulled toward P_- roughly in the direction of the real eigenvector $E^r(P_-)$, as illustrated.

Now imagine what would happen if the trajectory entering D_{-1} from D_0 were in precisely the direction $E^r(P_-)$. Such a trajectory would follow $E^r(P_-)$ toward P_-, reaching the equilibrium point asymptotically as $t \to \infty$. Similarly, if we were to follow this trajectory backward in time through D_0 and back onto $E^c(P_-)$ in D_{-1}, it would then spiral toward P_-, reaching it asymptotically as $t \to -\infty$. The closed curve thus formed would be a homoclinic orbit, reaching the same equilibrium point P_- asymptotically in forward and reverse time.

Although the homoclinic orbit itself is not structurally stable, and therefore cannot be observed experimentally, horseshoes are structurally stable. A flow ϕ that satisfies the assumptions of Shil'nikov's

[9] *Structural stability* is discussed in more detail in the section on structural stability and bifurcations.

theorem contains a countable infinity of horseshoes; for sufficiently small perturbations ϕ' of the flow, finitely many of the horseshoes will persist. Thus, both the original flow and the perturbed flow exhibit chaos in the sense of Shil'nikov.

In Figure 8.13, we see that a trajectory lying close to a homoclinic orbit exhibits similar qualitative behavior: it spirals away from P_- along the unstable complex plane $E^c(P_-)$, is folded in D_0, reenters D_{-1} above $E^c(P_-)$, and is pulled back toward $E^c(P_-)$, only to be spun away from P_- once more.

Thus, two trajectories starting from distinct initial states close to P_- on $E^c(P_-)$ are stretched apart exponentially along the unstable eigenplane before being folded in D_1 and reinjected close to P_-; this gives rise to sensitive dependence on initial conditions. The recurrent stretching and folding continues *ad infinitum*, producing a chaotic steady-state solution.

Lyapunov Exponents

The notion of sensitive dependence on initial conditions may be made more precise through the introduction of Lyapunov exponents (LEs). Lyapunov exponents quantify the average exponential rates of separation of trajectories along the flow.

The flow in a neighborhood of asympototically stable trajectory is contracting so the LEs are zero or negative.[10] Sensitive dependence on initial conditions results from a positive LE.

To determine the stability of an equilibrium point, we considered the eigenvalues of the linearization of the vector field in the vicinity of equilibrium trajectory. This idea can be generalized to any trajectory of the flow.

The local behavior of the vector field along a trajectory $\phi_t(\mathbf{X}_0)$ of an autonomous continuous time dynamical system (8.3) is governed by the linearized dynamics

$$\begin{aligned}\dot{\mathbf{X}} &= \mathbf{D_X F(X)_X}, \qquad \mathbf{x}(0) = \mathbf{x}_0 \\ &= \mathbf{D_X F}[\phi_t(\mathbf{X}_0)]\mathbf{x}\end{aligned}$$

This is a linear time-varying system where the state transition matrix, $\mathbf{\Phi}_t(\mathbf{X}_0)$, maps a point $\mathbf{x}_0$ into $\mathbf{x}(t)$. Thus,

$$\mathbf{x}(t) = \mathbf{\Phi}_t(\mathbf{X}_0)\mathbf{x}_0$$

Note that $\mathbf{\Phi}_t$ is a linear operator. Therefore, a ball $B_\varepsilon(\mathbf{X}_0)$ of radius ε about $\mathbf{X}_0$ is mapped into an ellipsoid as presented in Figure 8.1. The principal axes of the ellipsoid are determined by the singular values of $\mathbf{\Phi}_t$.

The singular values $\sigma_1(t), \sigma_2(t), \ldots, \sigma_n(t)$ of $\mathbf{\Phi}_t$ are defined as the square roots of the eigenvalues of $\mathbf{\Phi}_t^H\mathbf{\Phi}_t$, where $\mathbf{\Phi}_t^H$ is the complex conjugate transpose of $\mathbf{\Phi}_t$. The singular values are ordered so that $\sigma_1(t) > \sigma_2(t) > \ldots > \sigma_n(t)$.

In particular, a ball of radius ε is mapped by the linearized flow into an ellipsoid (see Figure 8.1) the maximum and minimum radii of which are bounded by $\sigma_1(t)\varepsilon$ and $\sigma_n(t)\varepsilon$, respectively.

The stability of a steady-state orbit is governed by the average local rates of expansion and contraction of volumes of state space close to the orbit. The Lyapunov exponents (LEs) λ_i, are defined by

$$\lambda_i = \lim_{t\to\infty} \frac{1}{t} \ln \sigma_i(t)$$

whenever this limit exists. The LEs quantify the average exponential rates of separation of trajectories along the flow.

---

[10]A continuous flow that has a bounded trajectory not tending to an equlibrium point has a zero Lyapunov exponent (in the direction of flow).

The LEs are a property of a steady-state trajectory. Any transient effect is averaged out by taking the limit as $t \to \infty$. Furthermore, the LEs are global quantities of an attracting set that depend on the local stability properties of a trajectory within the set.

The set $\{\lambda_i, i, = 1, 2, \ldots, n\}$ is called the Lyapunov spectrum. An attractor has the property that the sum of its LEs is negative.

Lyapunov Exponents of Discrete-Time Systems. The local behavior along an orbit of the autonomous discrete-time dynamical system (8.8) is governed by the linearized dynamics

$$\mathbf{x}_{k+1} = \mathbf{D_x G}(\mathbf{X}_k)\mathbf{x}_k, \qquad k = 0, 1, 2, \ldots$$

where the state transition matrix, $\mathbf{\Phi}_k(\mathbf{X}_0)$, maps a point $\mathbf{x}_0$ into $\mathbf{x}_k$. Thus,

$$\mathbf{x}_k = \mathbf{\Phi}_k(\mathbf{X}_0)\mathbf{x}_0$$

The Lyapunov exponents λ_i for the discrete-time dynamical system (8.8) are defined by

$$\lambda_i = \lim_{t \to \infty} \frac{1}{k} \ln \sigma_i(k)$$

whenever this limit exists. $\sigma_i(k)$ denotes the ith singular value of $\mathbf{\Phi}_k^H \mathbf{\Phi}_k$.

Lyapunov Exponents of Steady-State Solutions. Consider once more the continuous-time dynamical system (8.3). If $\mathbf{D_x F}$ were constant along the flow, with n distinct eigenvalues $\tilde{\lambda}_i$

$i = 1, 2, \ldots, n$, then

$$\mathbf{\Phi}_t = \begin{pmatrix} \exp(\tilde{\lambda}_1 t) & 0 & \cdots & 0 \\ 0 & \exp(\tilde{\lambda}_2 t) & \cdots & 0 \\ \vdots & \vdots & \ddots & \vdots \\ 0 & 0 & \cdots & \exp(\tilde{\lambda}_n t) \end{pmatrix}$$

and

$$\mathbf{\Phi}_t^H \mathbf{\Phi}_t = \begin{pmatrix} \exp(2\,\mathrm{Re}(\tilde{\lambda}_1)t) & 0 & \cdots & 0 \\ 0 & \exp(2\,\mathrm{Re}(\tilde{\lambda}_2)t) & \cdots & 0 \\ \vdots & \vdots & \ddots & \vdots \\ 0 & 0 & \cdots & \exp(2\,\mathrm{Re}(\tilde{\lambda}_n)t) \end{pmatrix}$$

giving $\sigma_i(t) = \exp(\mathrm{Re}(\tilde{\lambda}_i)t)$ and

$$\begin{aligned} \lambda_i &= \lim_{t \to \infty} \frac{1}{t} \ln\left(\exp\left[\mathrm{Re}(\tilde{\lambda}_i)t\right]\right) \\ &= \mathrm{Re}(\tilde{\lambda}_i) \end{aligned}$$

In this case, the LEs are simply the real parts of the eigenvalues of $\mathbf{D_X F}$.

TABLE 8.1 Classification of Steady-State Behaviors According to their Limit Sets, Power Spectra, LEs, and Dimension

Steady State	Limit Set	Spectrum	LEs	Dimension
DC	Fixed Point	Spike at DC	$0 > \lambda_1 \geq \ldots \geq \lambda_n$	0
Periodic	Closed Curve	Fundamental Plus Integer Harmonics	$\lambda_1 = 0$ $0 > \lambda_2 \geq \ldots \geq \lambda_n$	1
Quasiperiodic	*K*-Torus	Incommensurate Frequencies	$\lambda_1 = \ldots = \lambda_K = 0$ $0 > \lambda_{K+1} \geq \ldots \geq \lambda_n$	K
Chaotic	Fractal	Broad Spectrum	$\lambda_1 > 0$ $\sum_{i=1}^{n} \lambda_i < 0$	Noninteger

All the eigenvalues of a stable equilibrium point have negative real parts and therefore the largest Lyapunov exponent of an attracting equilibrium point is negative.

Trajectories close to a stable limit cycle converge onto the limit cycle. Therefore, the largest LE of a periodic steady-state is zero (corresponding to motion along the limit cycle [15]), and all its other LEs are negative.

A quasiperiodic *K*-torus has *K* zero LEs because the flow is locally neither contracting nor expanding along the surface of the *K*-torus.

S chaotic trajectory is locally unstable and therefore has a positive LE; this produces sensitive dependence on initial conditions. Nevertheless, in the case of a chaotic attractor, this locally unstable chaotic trajectory belongs to an attracting limit set to which nearby trajectories converge.

The steady-state behavior of a four-dimensional continuous-time dynamical system which has two positive, one zero, and one negative LE is called hyperchaos.

The Lyapunov spectrum may be used to identify attractors, as summarized in Table 8.1.

Structural Stability and Bifurcations

Structural stability refers to the sensitivity of a phenomenon to small changes in the parameter of a system. A structurally stable vector field **F** is one for which sufficiently close vector fields **F**′ have equivalent[11] dynamics [18].

The behavior of a typical circuit depends on a set of parameters one or more of which may be varied in order to optimize some performance criteria. In particular, one may think of a one-parameter family of systems

$$\dot{\mathbf{X}} = \mathbf{F}_{\mu}(\mathbf{X}) \tag{8.11}$$

where the vector field is parametrized by a control parameter μ. A value μ_0 of (8.11) for which the flow of (8.11) is not structurally stable is a bifurcation value of μ [7].

The dynamics in the state space may be qualitatively very different from one value of μ to another. In the nonlinear *RLC* circuit example, the steady-state solution is a limit cycle if the control parameter G_a' is negative and an equilibrium point if G_a' is positive. If G_a' is identically equal to zero, trajectories starting from $I_{30} = 0$, $V_{20} < E$ yield sinusoidal solutions. These sinusoidal solutions are not structurally stable because the slightest perturbation on G_a' will cause the oscillation to decay to zero or converge to the limit cycle, depending on whether G_a' is made slightly larger or smaller than zero.

If we think of this circuit as being parametrized by G_a', then its vector field is not structurally stable at $G_a' \equiv 0$. We say that the equilibrium point undergoes a bifurcation (from stability to instability) as the value of the bifurcation parameter G_a' is reduced through the bifurcation point $G_a' = 0$.

[11]Equivalent means that there exists a continuous invertible function h that transforms **F** into **F**′.

Bifurcation Types

In this section, we consider three types of local bifurcation: the Hopf bifurcation, the saddle-node bifurcation, and the period-doubling bifurcation [18]. These bifurcations are called local because they may be understood by linearizing the system close to an equilibrium point or limit cycle.

Hopf Bifurcation. A Hopf bifurcation occurs in a continuous-time dynamical system (8.3) when a simple pair of complex conjugate eigenvalues of the linearization $\mathbf{D_x F}(\mathbf{X}_Q)$ of the vector field at an equilibrium point $\mathbf{X}_Q$ crosses the imaginary axis.

Typically, the equilibrium point changes stability from stable to unstable and a stable limit cycle is born. The bifurcation at $G_a' \equiv 0$ in the nonlinear *RLC* circuit is Hopf-like.[12]

When an equilibrium point undergoes a Hopf bifurcation, a limit cycle is born. When a limit cycle undergoes a Hopf bifurcation, motion on a two-torus results.

Saddle-Node Bifurcation. A saddle-node bifurcation occurs when a stable and an unstable equilibrium point merge and disappear; this typically manifests itself as the abrupt disappearance of an attractor.

A common example of a saddle-node bifurcation in electronic circuits is switching between equilibrium states in a Schmitt trigger. At the threshold for switching, a stable equilibrium point corresponding to the "high" saturated state merges with the high-gain region's unstable saddle-type equilibrium point and disappears. After a switching transient, the trajectory settles to the other stable equilibrium point, which corresponds to the "low" state.

A saddle-node bifurcation may also manifest itself as a switch between periodic attractors of different size, between a periodic attractor and a chaotic attractor, or between a limit cycle at one frequency and a limit cycle at another frequency.

Period-Doubling Bifurcation. A period-doubling bifurcation occurs in a discrete-time dynamical system (8.8) when a real eigenvalue of the linearization $\mathbf{D_X G}(\mathbf{X}_Q)$ of the map $\mathbf{G}$ at an equilibrium point crosses the unit circle at -1 [7].

In a continuous-time system, a period-doubling bifurcation occurs only from a periodic solution (an equilibrium point of the Poincaré map). At the bifurcation point, a periodic orbit with period T changes smoothly into one with period $2T$, as illustrated in Figure 8.14(a) and (b).

Blue Sky Catastrophe. A blue sky catastrophe is a global bifurcation that occurs when an attractor disappears "into the blue," usually because of a collision with a saddle-type limit set. Hysteresis involving a chaotic attractor is often caused by a blue sky catastrophe [18].

Routes to Chaos

Each of the three local bifurcations may give rise to a distinct route to chaos, and all three have been reported in electronic circuits. These routes are important because it is often difficult to conclude from experimental data alone whether irregular behavior is due to measurement noise or to underlying chaotic dynamics. If, upon adjusting a control parameter, one of the three prototype routes is observed, this indicates that the dynamics might be chaotic.

Periodic-Doubling Route to Chaos. The period-doubling route to chaos is characterized by a cascade of period-doubling bifurcations. Each period-doubling transforms a limit cycle into one at half the frequency, spreading the energy of the system over a wider range of frequencies. An infinite cascade of such doublings results in a chaotic trajectory of infinite period and a broad frequency spectrum that contains

[12]Note that the Hopf bifurcation theorem is proven for sufficiently smooth systems and does not strictly apply to piecewise-linear systems. However, a physical implementation of a piecewise-linear characteristic, such as that of N_R, is always smooth.

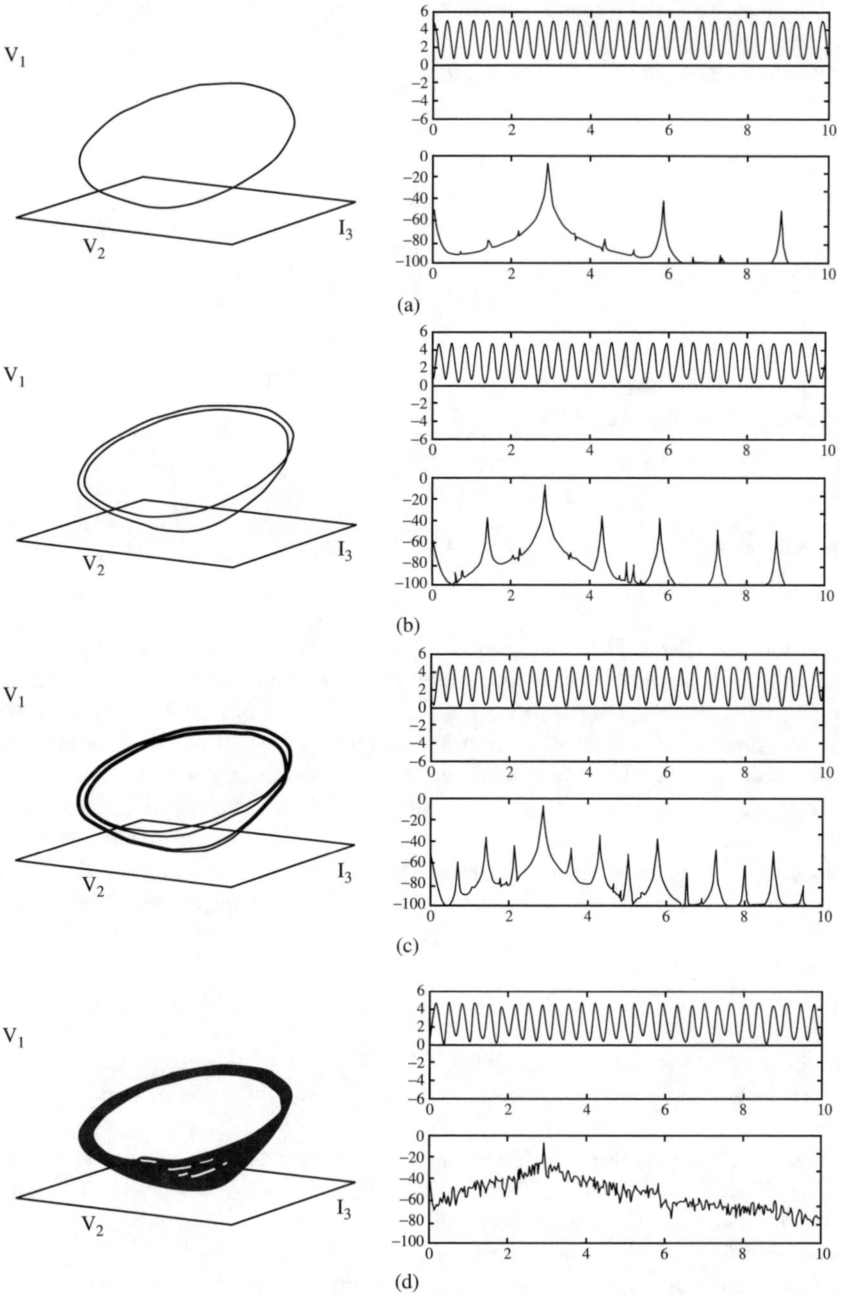

FIGURE 8.14 Period-doubling route to chaos in Chua's oscillator. Simulated state space trajectories, time waveforms $V_1(t)$, and power spectra of $V_2(t)$ (a) G = 530 μS: periodic steady-state — the signal is characterized by a discrete power spectrum with energy at integer multiples of the fundamental frequency f_0; (b) G = 537 μS; period-two — after a period-doubling bifurcation, the period of the signal is approximately twice that of (a). In the power spectrum, a spike appears at the new fundamental frequency $\approx f_0/2$. (c) G = 539 μS: period-four — a second period-doubling bifurcation gives rise to a fundamental frequency of $\approx f_0/4$; (d) G = 541 μS: spiral Chua's attractor — a cascade of period doublings results in a chaotic attractor that has a broadband power spectrum. Time plots: horizontal axis — t (ms); vertical axis — $V_1(V)$. Power spectra: horizontal axis — frequency (kHz); vertical axis — power [mean squared amplitude of $V_2\ (t)$] (db).

energy at all frequencies. Figure 8.14 is a set of snapshots of the period-doubling route to chaos in Chua's oscillator.

An infinite number of period-doubling bifurcations to chaos can occur over a finite range of the bifurcation parameter because of a geometric relationship between the intervals over which the control parameter must be moved to cause successive bifurcations. Period-doubling is governed by a universal scaling law that holds in the vicinity of the bifurcation point to chaos μ_∞.

Define the ratio δ_k of successive interval μ, in each of which there is a constant period of oscillation, as follows,

$$\delta_k = \frac{\mu_{2^k} - \mu_{2^{k-1}}}{\mu_{2^{k+1}} - \mu_{2^k}}$$

where μ_{2k} is the bifurcation point for the period from $2^k T$ to $2^{k+1}T$. In the limit as $k \to \infty$, a universal constant called the Feigenbaum number δ is obtained:

$$\lim_{k\to\infty} \delta_k = \delta = 4.6692\ldots$$

The period-doubling route to chaos is readily identified from a state-space plot, time series, power spectrum, or a Poincaré map.

Intermittency Route to Chaos. The route to chaos caused by saddle-node bifurcations comes in different forms, the common feature of which is direct transition from regular motion to chaos. The most common type is the intermittency route and results from a single saddle-node bifurcation. This is a route and not just a jump because straight after the bifurcation, the trajectory is characterized by long intervals of almost regular motion (called laminar phases) and short bursts of irregular motion. The period of the oscillations is approximately equal to that of the system just before the bifurcation. This is illustrated in Figure 8.15.

As the parameter passes through the critical value μ_c at the bifurcation point into the chaotic region, the laminar phases become shorter and the bursts become more frequent, until the regular intervals disappear altogether. The scaling law for the average interval of the laminar phases depends on $|\mu - \mu_c|$, so chaos is not fully developed until some distance from the bifurcation point [13].

Intermittency is best characterized in the time domain because its scaling law governs on the length of laminar phases.

Another type of bifurcation to chaos associated with saddle-nodes is the direct transition from a regular attractor (fixed point or limit cycle) to a coexisting chaotic one, without the phenomenon of intermittency.

Quasiperiodic (Torus Breakdown) Route to Chaos. The quasiperiodic route to chaos results from a sequence of Hopf bifurcations. Starting from a fixed point, the three-torus generated after three Hopf bifurcations is not stable in the sense that there exists an arbitrarily small perturbation of the system (in terms of parameters) for which the three-torus gives way to chaos.

A quasiperiodic–periodic–chaotic sequence corresponding to torus breakdown in Chua's oscillator is given in Figure 8.5.

Quasiperiodicity is difficult to detect from a time series; it is more readily identified by means of a power spectrum or Poincaré map (see Figure 8.11, for example).

Bifurcation Diagrams and Parameter Space Diagrams

Although state space, time- and frequency-domain measurements are useful for characterizing steady-state behaviors, nonlinear dynamics offers several other tools for summarizing qualitative information concerning bifurcations.

A bifurcation diagram is a plot of the attracting sets of a system versus a control parameter. Typically, one chooses a state variable and plots this against a single control parameter. In discrete systems, one

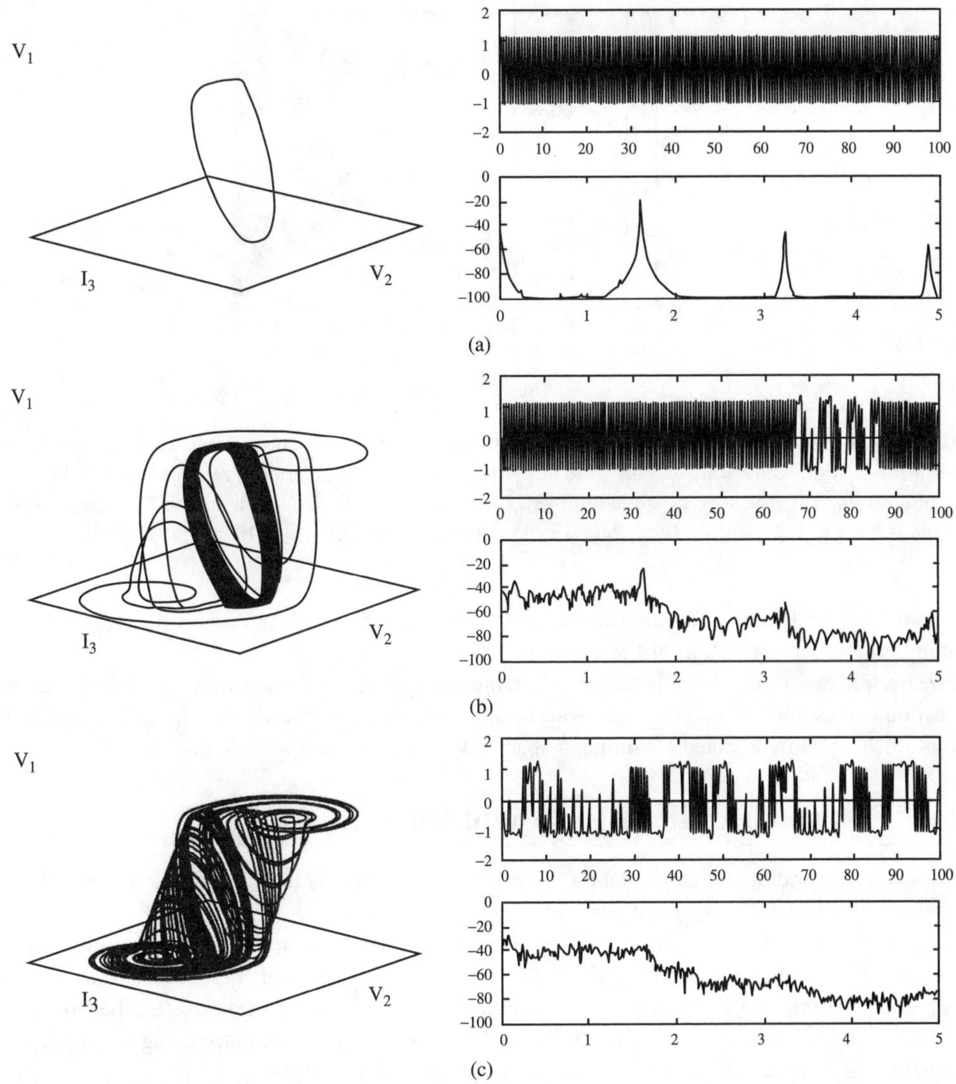

FIGURE 8.15 Intermittency route to chaos in Chua's oscillator. Simulated state space trajectories, time waveforms $V_1(t)$ and power spectra of $V_2(t)$ (a) Periodic steady-state — the signal is characterized by a discrete power spectrum with energy at integer multiples of the fundamental frequency; (b) onset of intermittency — the time signal contains long regular **"laminar" phases** and occasional **"bursts"** of irregular motion — in the frequency domain, intermittency manifests itself as a raising of the noise floor; (c) fully developed chaos-laminar phases are infrequent and the power spectrum is broad. Time plots: horizontal axis — t (ms); vertical axis — V_1 (V). Power spectra: horizontal axis — frequency (kHz); vertical axis — power [mean squared amplitude of $V_2(t)$] (dB).

simply plots successive values of a state variable. In the continuous-time case, some type of discretization is needed, typically by means of a Poincaré section.

Figure 8.16 is a bifurcation diagram of the logistic map $X_{k+1} = \mu X_k(1 - X_k)$ for $\mu \in [2.5, 4]$ and $X_k \in [0, 1]$. Period doubling from period-one to period-two occurs at μ_2; the next two doublings in the period-doubling cascade occur at μ_2 and μ_4, respectively. A periodic window in the chaotic region is indicated by μ_3. The map becomes chaotic by the period-doubling route if μ is increasing from μ_3 and by the intermittency route if μ is reduced out of the window.

When more than one control parameter is present in a system, the steady-state behavior may be summarized in a series of bifurcation diagrams, where one parameter is chosen as the control parameter,

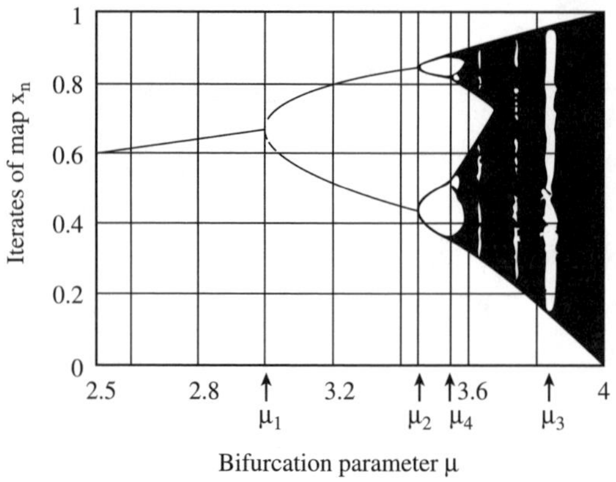

FIGURE 8.16 Bifurcation diagram for the logistic map: $X_{k+1} = \mu X_k(1 - X_k)$. The first period-doubling bifurcation occurs at $\mu = \mu_1$, the second at μ_2, and the third at $\mu_4 \cdot \mu_3$ corresponds to a period-three window. When $\mu = 4$, the entire interval (0, 1) is visited by a chaotic orbit $\{X_k, k = 0, 1, \ldots\}$. *Source*: C. W. Wu and N. R. Rul'kov, "Studying chaos via 1-D maps — A tutorial," *IEEE Trans. Circuits Syst. I*, vol. 40, p. 708, Oct. 1993. ® 1993 IEEE.

with the others held fixed, and only changed from one diagram to the next. This provides a complete but cumbersome representation of the dynamics [13].

A clearer picture of the global behavior is obtained by partitioning the parameter space by means of bifurcation curves, and labeling the regions according to the observed steady-state behaviors within these regions. Such a picture is called a parameter space diagram.

8.2 Chua's Circuit: A Paradigm for Chaos

Chaos is characterized by a stretching and folding mechanism; nearby trajectories of a dynamical system are repeatedly pulled apart exponentially and folded back together.

In order to exhibit chaos, as autonomous circuit consisting of resistors, capacitors, and inductors must contain (i) at least one locally active resistor, (ii) at least one nonlinear element, and (iii) at least three energy-storage elements. The active resistor supplies energy to separate trajectories, the nonlinearity provides folding, and the three-dimensional state space permits persistent stretching and folding in a bounded region without violating the noncrossing property of trajectories.

Chua's circuit (see Figure 8.17) is the simplest electronic circuit that satisfies these criteria. It consists of a linear inductor, a linear resistor, two linear capacitors, and a single nonlinear resistor N_R. The circuit is readily constructed at low cost using standard electronic components and exhibits a rich variety of bifurcation and chaos [10].

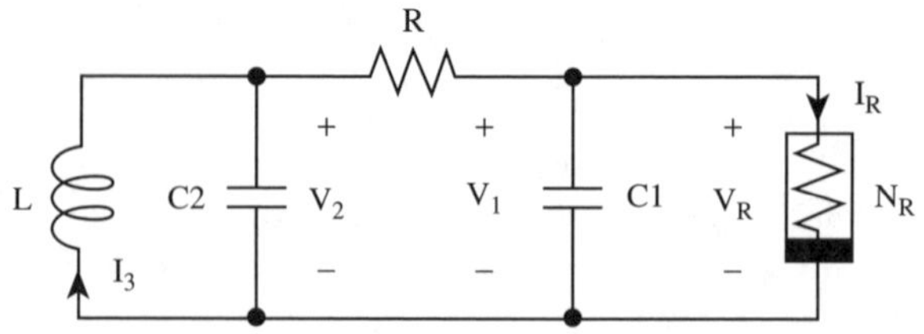

FIGURE 8.17 Chua's circuit consists of a linear inductor L, two linear capacitors (C_2, C_1), a linear resistor R, and a voltage-controlled nonlinear resistor N_R.

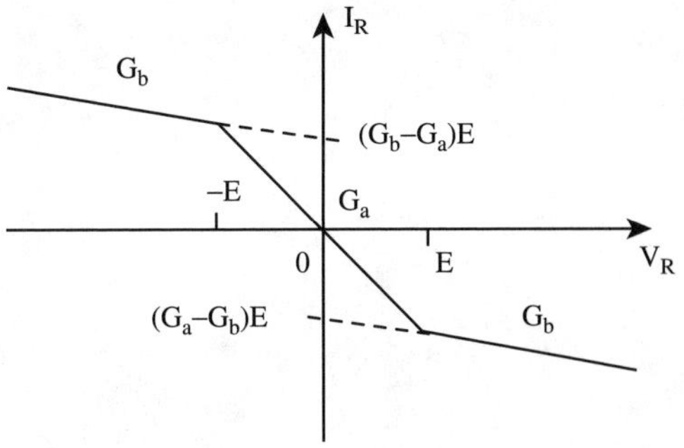

FIGURE 8.18 The driving-point characteristic of the nonlinear resistor N_R in Chua's circuit has breakpoints at $\pm E$ and slopes G_a and G_b in the inner and outer regions, respectively.

Dynamics of Chua's Circuit

State Equations

Chua's circuit may be described by three ordinary differential equations. Choosing V_1, V_2, and I_3 as state variables, we write

$$\begin{aligned}\frac{dV_1}{dt} &= \frac{G}{C_1}(V_2 - V_1) - \frac{1}{C_1} f(V_1) \\ \frac{dV_2}{dt} &= \frac{G}{C_2}(V_1 - V_2) + \frac{1}{C_2} I_3 \\ \frac{dI_3}{dt} &= -\frac{1}{L} V_2\end{aligned} \tag{8.12}$$

where $G = 1/R$ and $f(V_R) = G_b V_R + 1/2(G_a + G_b)\,(|V_R + E| - |V_R - E|)$, as depicted in Figure 8.18.

Because of the piecewise-linear nature of N_R, the vector field of Chua's circuit may be decomposed into three distinct affine regions: $V_1 < -E$, $|V_1| \le E$, and $V_1 > E$. We call these the D_{-1}, D_0 and D_1 regions, respectively. The global dynamics may be determined by considering separately the behavior in each of the three regions (D_{-1}, D_0 and D_1) and then gluing the pieces together along the boundary planes U_{-1} and U_1.

Piecewise-Linear Dynamics

In each region, the circuit is governed by a three-dimensional autonomous affine dynamical system of the form

$$\dot{\mathbf{X}} = \mathbf{AX} + \mathbf{b} \tag{8.13}$$

where $\mathbf{A}$ is the (constant) system matrix and $\mathbf{b}$ is a constant vector.

The equilibrium points of the circuit may be determined graphically by intersecting the load line $I_R = -GV_R$ with the DP characteristic $I_R = f(V_R)$ of the nonlinear resistor N_R, as presented in Figure 8.19 [2]. When $G > |G_a|$ or $G > |G_b|$, the circuit has a unique equilibrium point at the origin (and two virtual equilibria P_- and P_+); otherwise, it has three equilibrium points at P_-, 0, and P_+.

The dynamics close to an equilibrium point $\mathbf{X}_Q$ are governed locally by the linear system

$$\dot{\mathbf{x}} = \mathbf{Ax} \tag{8.14}$$

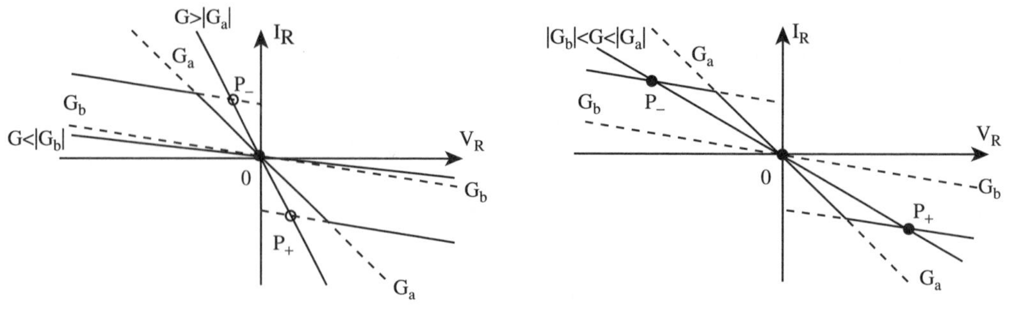

FIGURE 8.19 DC equilibrium points of Figure 8.17 may be determined graphically by intersecting the load line $I_R = -GV_R$ with the DP characteristic of N_R. (a) If $G > |G_a|$ or $G < |g_b|$, the circuit has a unique equilibrium point at the origin (P_- and P_+ are virtual equilibria in this case). (b) When $|G_b| < G < |G_a|$, the circuit has three equilibrium points at P_-, 0 and P_+.

If the eigenvalues λ_1, λ_2, and λ_3 of $\mathbf{A}$ are distinct, then every solution $\mathbf{x}(t)$ of (8.14) may be expressed in the form

$$\mathbf{x}(t) = c_1 \exp(\lambda_1 t)\vec{\xi}_1 + c_2 \exp(\lambda_2 t)\vec{\xi}_2 + c_3 \exp(\lambda_3 t)\vec{\xi}_3$$

where $\vec{\xi}_1$, $\vec{\xi}_2$, and $\vec{\xi}_3$ are the (possibly complex) eigenvectors associated with the eigenvalues λ_1, λ_2, and λ_3, respectively, and the c_k's are (possibly complex) constants that depend on the initial state $\mathbf{X}_0$.

In the special case when $\mathbf{A}$ has one real eigenvalue γ and a complex conjugate pair of eigenvalues $\sigma \pm j\omega$, the solution of (8.14) has the form

$$\mathbf{x}(t) = c_r \exp(\gamma t)\vec{\xi}_\gamma + 2c_c \exp(\sigma t)\left[\cos(\omega t + \phi_c)\vec{\eta}_r - \sin(\omega t + \phi_c)\vec{\eta}_i\right]$$

where $\vec{\eta}_r$ and $\vec{\eta}_i$ are the real and imaginary parts of the eigenvectors associated with the complex conjugate pair of eigenvalues, $\vec{\xi}_\gamma$ is the eigenvector defined by $\mathbf{A}\vec{\xi}_\gamma = \gamma\, \vec{\xi}_\gamma$ and c_r, c_c, and ϕ_c are real constants that are determined by the initial conditions.

Let us relabel the real eigenvector E^r, and define E^c as the complex eigenplane spanned by $\vec{\eta}_r$ and $\vec{\eta}_i$.

We can think of the solution $\mathbf{x}(t)$ of (8.14) as being the sum of two distinct components $\mathbf{x}_r(t) \in E^r$ and $\mathbf{x}_c(t) \in E^c$:

$$\mathbf{x}_r(t) = c_r \exp(\gamma t)\vec{\xi}_\gamma$$

$$\mathbf{x}_c(t) = 2c_c \exp(\sigma t)\left[\cos(\omega t + \phi_c)\vec{\eta}_r - \sin(\omega t + \phi_c)\vec{\eta}_i\right]$$

The complete solution $\mathbf{X}(t)$ of (8.13) may be found by translating the origin of the linearized coordinate system to the equilibrium point $\mathbf{X}_Q$. Thus,

$$\begin{aligned}\mathbf{X}(t) &= \mathbf{X}_Q + \mathbf{x}(t) \\ &= \mathbf{X}_Q + \mathbf{x}_r(t) + \mathbf{x}_c(t)\end{aligned}$$

We can determine the qualitative behavior of the complete solution $\mathbf{X}(t)$ by considering separately the components $\mathbf{x}_r(t)$ and $\mathbf{x}_c(t)$ along E^r and E^c, respectively.

If $\gamma > 0$, $\mathbf{x}_r(t)$ grows exponentially in the direction of E^r; if $\gamma < 0$ the component $\mathbf{x}_r(t)$ tends asymptotically to zero. When $\sigma > 0$ and $\omega \neq 0$, $\mathbf{x}_c(t)$ spirals away from $\mathbf{X}_Q$ along the complex eigenplane E^c, and if $\sigma < 0$, $\mathbf{x}_c(t)$ spirals toward $\mathbf{X}_Q$ and E^c.

We remark that the vector E^r and plane E^c are invariant under the flow of (8.13): if $\mathbf{X}(0) \in E^r$, then $\mathbf{X}(t) \in E^r$ for all t; if $\mathbf{X}(0) \in E^c$, then $\mathbf{X}(t) \in E^c$ for all t. An important consequence of this is that a trajectory $\mathbf{X}(t)$ cannot cross through the complex eigenspace E^c; suppose $\mathbf{X}(t_0) \in E^c$ at some time t_0, then $\mathbf{X}(t) \in E^c$, for all $t > t_0$.

Chaos in Chua's Circuit

In the following discussion, we consider a fixed set of component values: L = 18 mH, C_2 = 100 nF, C_1 = 10 nF, $G_a = -50/66$ mS = –757.576 μS, $G_b = -9/22$ mS = –409.091 μS, and E = 1 V. When G = 550 μS, three equilibrium points occur at P_+, 0, and P_-. The equilibrium point at the origin (0) has one unstable real eigenvalue γ_0 and a stable complex pair $\sigma_0 \pm j\omega_0$. The outer equilibria (P_- and P_+) each have a stable real eigenvalue γ_1 and an unstable complex pair $\sigma_0 \pm j\omega_1$.

Dynamics of D_0

A trajectory starting from some initial state $\mathbf{X}_0$ in the D_0 region may be decomposed into its components along the complex eigenplane $E^c(0)$ and along the eigenvector $E^r(0)$. When $\gamma_0 > 0$ and $\sigma_0 < 0$, the component along $E^c(0)$ spirals toward the origin along this plane while the component in the direction $E^r(0)$ grows exponentially. Adding the two components, we see that a trajectory starting slightly above the stable complex eigenplane $E^c(0)$ spirals toward the origin along the $E^c(0)$ direction, all the while being pushed away from $E^c(0)$ along the unstable direction $E^r(0)$. As the (stable) component along $E^c(0)$ shrinks in magnitude, the (unstable) component grows exponentially, and the trajectory follows a helix of exponentially decreasing radius whose axis lies in the direction of $E^r(0)$; this is illustrated in Figure 8.20.

Dynamics of D_{-1} and D_1

Associated with the stable real eigenvalue γ_1 in the D_1 region is the eigenvector $E^r(P_+)$. The real and imaginary parts of the complex eigenvectors associated with $\sigma_1 \pm j\omega_1$ define a complex eigenplane $E^c(P_+)$.

A trajectory starting from some initial state $\mathbf{X}_0$ in the D_1 region may be decomposed into its components along the complex eigenplane $E^c(P_+)$ and the eigenvector $E^r(P_+)$. When $\gamma_1 < 0$ and $\sigma_1 > 0$, the component on $E^c(P_+)$ spirals away from P_+ along this plane while the component in the direction of $E^r(0)$ tends asymptotically toward P_+. Adding the two components, we see that a trajectory starting close to the stable real eigenvector $E^r(P_+)$ above the complex eigenplane moves toward $E^c(P_+)$ along a helix of exponentially increasing radius. Because the component along $E^r(P_+)$ shrinks exponentially in magnitude and the component on $E^c(P_+)$ grows exponentially, the trajectory is quickly flattended onto $E^c(P_+)$, where it spirals away from P_+ along the complex eigenplane; this is illustrated in Figure 8.21.

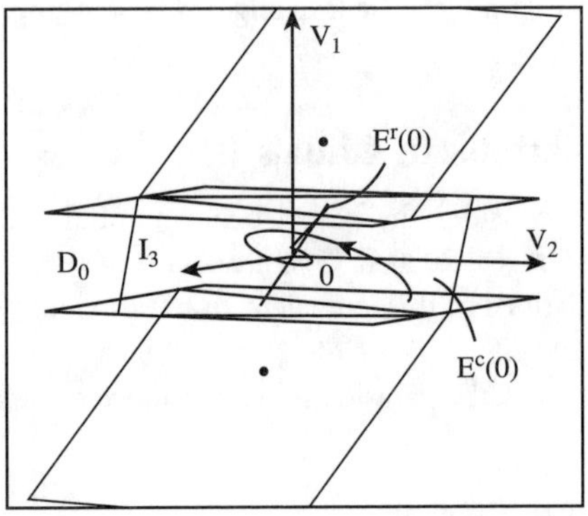

FIGURE 8.20 Dynamics of the D_0 region. A trajectory starting slightly above the stable complex eigenplane $E^c(0)$ spirals toward the origin along this plane and is repelled close to 0 in the direction of the unstabe eigenvector $E^r(0)$. *Source*: M. P. Kennedy, "Three steps to chaos — Part II: A Chua's circuit primer," *IEEE Trans. Circuits Syst. I*, vol. 40, p. 660, Oct. 1993. ® 1993 IEEE.

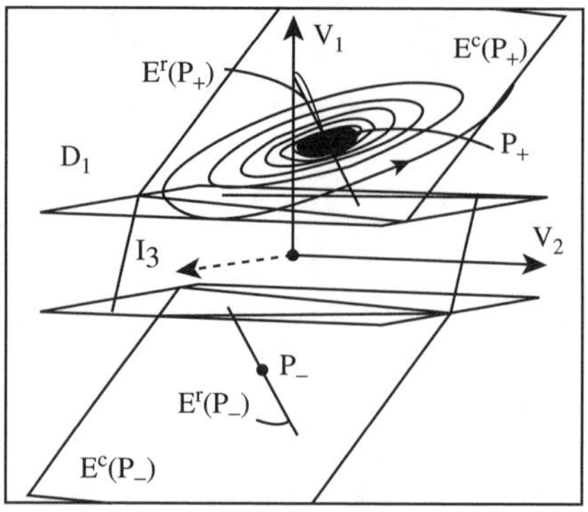

FIGURE 8.21 Dynamics of the D_1 region. A trajectory starting above the unstable complex eigenplane $E^c(P_+)$ close to the eigenvector $E^r(P_+)$ moves toward the plane and spirals away from P_+ along $E^c(P_+)$. By symmetry, the D_{-1} region has equivalent dynamics. *Source*: M. P. Kennedy, "Three steps to chaos — Part II: A Chua's circuit primer," *IEEE Trans. Circuits Syst.*, vol. 40, p. 662, Oct. 1993. © 1993 IEEE.

By symmetry, the equilibrium point P_- in the D_{-1} region has three eigenvalues: γ_1 and $\sigma_1 \pm j\omega_1$. The eigenvector $E^r(P_-)$ is associated with the stable real eigenvalue γ_1; the real and imaginary parts of the eigenvectors associated with the unstable complex pair $\sigma_1 \pm j\omega_1$ define an eigenplane $E^c(P_-)$, along which trajectories spiral away from P_-.

Global Dynamics

With the given set of parameter values, the equilibrium point at the orgin has an unstable real eigenvalue and a stable pair of complex conjugate eigenvalues; the outer equilibrium point P_- has a stable real eigenvalue and an unstable complex pair.

In particular, P_- has a pair of unstable complex conjugate eigenvalues $\sigma_1 \pm \omega_1$ ($\sigma_1 > 0$, $\omega_1 \neq 0$) and a stable real eigenvalue γ_1, where $|\sigma_1| < |\omega_1|$. In order to prove that the circuit is chaotic in the sense of Shil'nikov, it is necessary to show that it possesses a homoclinic orbit for this set of parameter values.

A trajectory starting on the eigenvector $E^r(0)$ close to 0 moves away from the equilibrium point until it crosses the boundary U_1 and enters D_1, as illustrated in Figure 8.20. If this trajectory is folded back into D_0 by the dynamics of the outer region, and reinjected toward 0 along the stable complex eigenplane $E^c(0)$ then a homoclinic orbit is produced.

That Chua's circuit is chaotic in the sense of Shil'nikov was first proven by Chua et al. [21] in 1985. Since then, there has been an intensive effort to understand every aspect of the dynamics of this circuit with a view to developing it as a paradigm for learning, understanding, and teaching about nonlinear dynamics and chaos [3].

Steady-States and Bifurcations in Chua's Circuit

In the following discussion, we consider the global behavior of the circuit using our chosen set of parameters with R in the range $0 \leq R \leq 2000\ \Omega$ ($500\ \mu$ s $\leq G < \infty$ s).

Figure 8.14 is a series of simulations of the equivalent circuit in Figure 8.26 with the following parameter values: $L = 18$ mH, $C_2 = 100$ nF, $C_1 = 10$ nF, $G_a = -50/66$ mS $= -757.576\ \mu$S, $G_b = -9/22$ mS $=$ $-409.091\ \mu$S, and $E = 1$ V. $R_0 = 12.5\ \Omega$, the parasitic series resistance of a real inductor. R is the bifurcation parameter.

Equilibrium Point and Hopf Bifurcation

When R is large (2000 Ω), the outer equilibrium points P_- and P_+ are stable ($\gamma_1 < 0$ and $\sigma_1 < 0$, $\omega_1 \neq 0$); the inner equilibrium point 0 is unstable ($\gamma_0 > 0$ and $\sigma_0 < 0, \omega_0 \neq 0$).

Depending on the initial state of the circuit, the system remains at one outer equilibrium point or the other. Let us assume that we start at P_+ in the D_1 region. This equilibrium point has one negative real eigenvalue and a complex pair with negative real parts. The action of the negative real eigenvalue γ_1 is to squeeze trajectories down onto the complex eigenplane $E^c(P_+)$, where they spiral toward the equilibrium point P_+.

As the resistance R is decreased, the real part of the complex pair of eigenvalues changes sign and becomes positive. Correspondingly, the outer equilibrium points become unstable as σ_1 passes through 0; this is a Hopf-like bifurcation.[13] The real eigenvalue of P_+ remains negative so trajectories in the D_1 region converge toward the complex eigenplane $E^c(P_+)$. However, they spiral away from the equilibrium point P_+ along $E^c(P_+)$ until they reach the dividing plane U_1 (defined by $V_1 \equiv E$) and enter the D_0 region.

The equilibrium point at the origin in the D_0 region has a stable complex pair of eigenvalues and an unstable real eigenvalue. Trajectories that enter the D_0 region on the complex eigenplane $E^c(0)$ are attracted to the origin along this plane. Trajectories that enter D_0 from D_1 below or above the eigenplane either cross over to D_{-1} or are turned back toward D_1, respectively. For R sufficiently large, trajectories that spiral away from P_+ along $E^c(P_+)$ and enter D_0 above $E^c(0)$ are returned to D_1, producing a stable period-one limit cycle. This is illustrated in Figure 8.14.

Period-Doubling Cascade

As the resistance R is decreased further, a period-doubling bifurcation occurs. The limit cycle now closes on itself after encircling P_+ twice; this is called a period-two cycle because a trajectory takes approximately twice the time to complete this closed orbit as to complete the preceding period-one orbit [see Figure 8.14(b)].

Decreasing the resistance R still further produces a cascade of period-doubling bifurcations to period-four [Figure 8.14(c)], period-eight, period-sixteen, and so on until an orbit of infinite period is reached, beyond which we have chaos [see Figure 8.14(d)]. This is a spiral Chua's **chaotic attractor**.

The spiral Chua's attractor in Figure 8.14(d) looks like a ribbon or band that is smoothly folded on itself; this folded band is the simplest type of chaotic attractor [18]. A trajectory from an initial condition $\mathbf{X}_0$ winds around the strip repeatedly, returning close to $\mathbf{X}_0$, but never closing on itself.

Periodic Windows

Between the chaotic regions in the parameter space of Chua's circuit, there exist ranges of the bifurcation parameter R over which stable periodic motion occurs. These regions of periodicity are called periodic windows and are similar to those that exist in the bifurcation diagram of the logistic map (see Figure 8.16).

Periodic windows of periods three and five are readily found in Chua's circuit. These limit cycles undergo period-doubling bifurcations to chaos as the resistance R is decreased.

For certain sets of parameters, Chua's circuit follows the intermittency route to chaos as R is increased out of the period-three window.

Spiral Chua's Attractor

Figure 8.22 outlines three views of another simulated spiral Chua's chaotic attractor. Figure 8.22(b) is a view along the edge of the outer complex eigenplanes $E^c(P_+)$ and $E^c(P_-)$; notice how trajectories in the D_1 region are compressed toward the complex eigenplane $E^c(P_+)$ along the direction of the stable real eigenvector $E^c(P_+)$ and they spiral away from the equilibrium point P_+ along $E^c(P_+)$.

When a trajectory enters the D_0 region through U_1 from D_1, it is twisted around the unstable real eigenvector $E^r(0)$ and returned to D_1.

Figure 8.22(c) illustrates clearly that when the trajectory enters D_0 from D_1, it crosses U_1 above the eigenplace $E^c(0)$. The trajectory cannot cross through this eigenplane and therefore it must return to the D_1 region.

Double-Scroll Chua's Attractor

Because we chose a nonlinear resistor with a symmetric nonlinearity, every attractor that exists in the D_1 and D_0 regions has a counterpart (mirror image) in the D_{-1} and D_0 regions. As the coupling resistance

[13]Recall that the Hopf bifurcation theorem strictly applies only for sufficiently smooth systems, but that physical implementations of piecewise-linear characteristics are typically smooth.

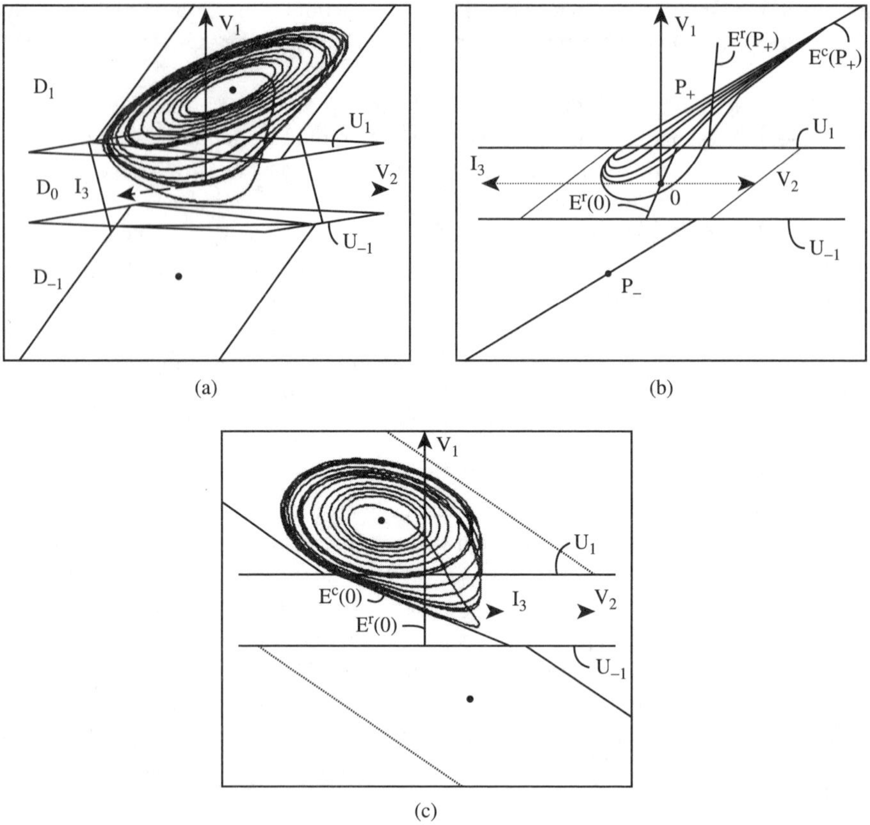

FIGURE 8.22 Three views of a simulated spiral Chua's attractor in Chua's oscillator with $G = 550\mu S$. (a) Reference view [compare with Figure 8.14(d)]. (b) View of the edge of the outer complex eigenplanes $E^c(P_+)$ and $E^c(P_-)$; note how trajectory in D_1 is flattened onto $E^c(P_+)$. (c) View along the edge of the complex eigenplane $E^c(0)$; trajectories cannot cross this plane. *Source*: M. P. Kennedy, "Three steps to chaos — Part II: A Chua's circuit primer," *IEEE Trans. Circuits Syst. I*, vol. 40, p. 664, Oct. 1993. © 1993 IEEE.

R is decreased further, the spiral Chua's attractor "collides" with its mirror image and the two merge to form a single compound attractor called a double-scroll Chua's chaotic attractor [10], as presented in Figure 8.23.

Once more, we show three views of this attractor in order to illustrate its geometrical structure. Figure 8.23(b) is a view of the attractor along the edge of the outer complex eigenplanes $E^c(P_+)$ and $E^c(P_-)$. Upon entering the D_1 region form D_0, the trajectory collapses onto $E^c(P_+)$ and spirals away from P_+ along this plane.

Figure 8.23(c) is a view of the attractor along the edge of the complex eigenplane $E^c(0)$ in the inner region. Notice once more that when the trajectory crosses U_1 into D_0 above $E^c(0)$, it must remain above $E^c(0)$ and so returns to D_1. Similarly, if the trajectory crosses U_1 below $E^c(0)$, it must remain below $E^c(0)$and therefore crosses over to the D_{-1} region. Thus, $E^c(0)$ presents a knife-edge to the trajectory as it crosses U_1 into the D_0 region, forcing it back toward D_1 or across D_0 to D_{-1}.

Boundary Crisis

Reducing the resistance R still further produces more regions of chaos, interspersed with periodic windows. Eventually, for a sufficiently small value of R, the unstable saddle trajectory that normally resides outside the stable steady-state solution collides with the double-scroll Chua's attractor and a blue sky catastrophe called a boundary crisis [10] occurs. After this, all trajectories become unbounded.

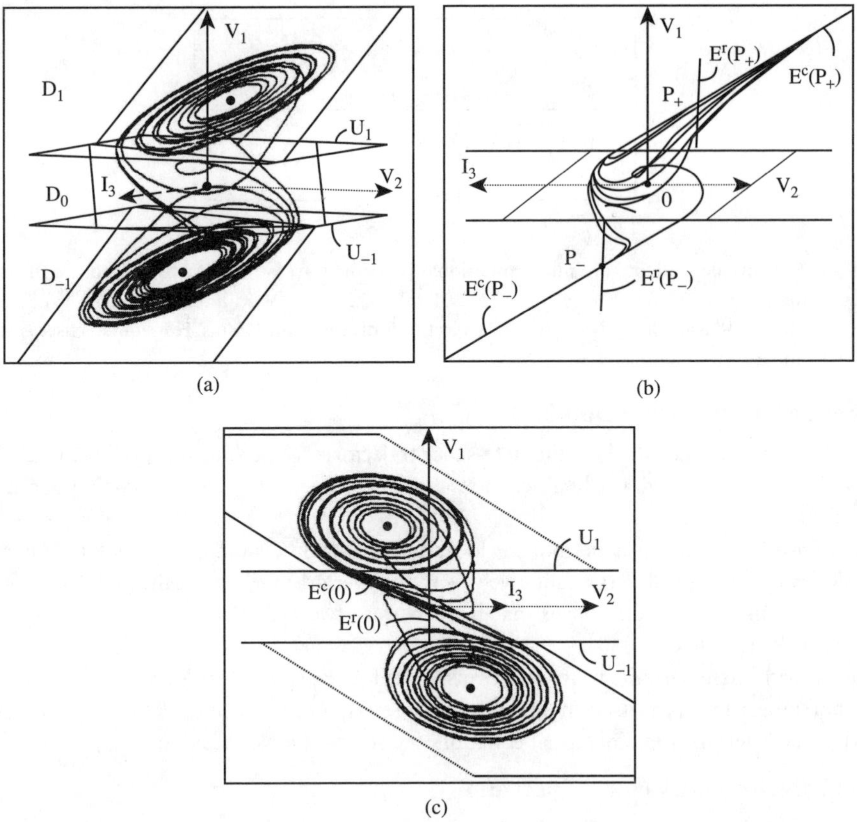

FIGURE 8.23 Three views of a simulated double-scroll Chua's attractor in Chua's oscillator with G = 565 μS. (a) Reference view [compare with Figure 8.14(d)]. (b) View along the edge of the outer complex eigenplanes $E^c(P_+)$ and $E^c(P_-)$; note how the trajectory in D_1 is flattened onto $E^c(P_+)$ and onto $E^c(P_-)$ in D_{-1}. (c) View along the edge of the complex eigenplane $E^c(0)$; a trajectory entering D_0 from D_1 above this plane returns to D_1 while one entering D_0 below $E^c(0)$ crosses to D_{-1}. *Source*: M. P. Kennedy, "Three steps to chaos — Part II: A Chua's circuit primer," *IEEE Trans. Circuits Syst. I*, vol. 40, p. 665, Oct. 1993. © 1993 IEEE.

Manifestations of Chaos

Sensitive Dependence on Initial Conditions

Consider once more the double-scroll Chau's attractor shown in Figure 8.23. Two trajectories starting from distinct but almost identical initial states in D_1 will remain "close together" until they reach the separating plane U_1. Imagine that the trajectories are still "close" at the knife-edge, but that one trajectory crosses into D_0 slightly above $E^c(0)$ and the other slightly below $E^c(0)$. The former trajectory returns to D_1 and the latter crosses over to D_1: their "closeness" is lost.

The time-domain waveforms $V_1(t)$ for two such trajectories are shown in Figure 8.24. These are solutions of Chua's oscillator with the same parameters as in Figure 8.23; the initial conditions are (I_3, V_2, V_1) = (1.810 mA, 222.014 mV, –2.286 V) [solid line] and (I_3, V_2, V_1) = (1.810 mA, 222.000 mV, –2.286 V) [dashed line]. Although the initial conditions differ by less than 0.01 percent in just one component (V_2), the trajectories diverge and become uncorrelated within 0.01 percent in just one component (V_2), the trajectories diverge and become uncorrelated within 5 ms because one crosses the knife-edge before the other.

This rapid decorrelation of trajectories that originate in nearby initial states, commonly called sensitive dependence on initial conditions, is a generic property of chaotic systems. It gives rise to an apparent randomness in the output of the system and long-term unpredictability of the state.

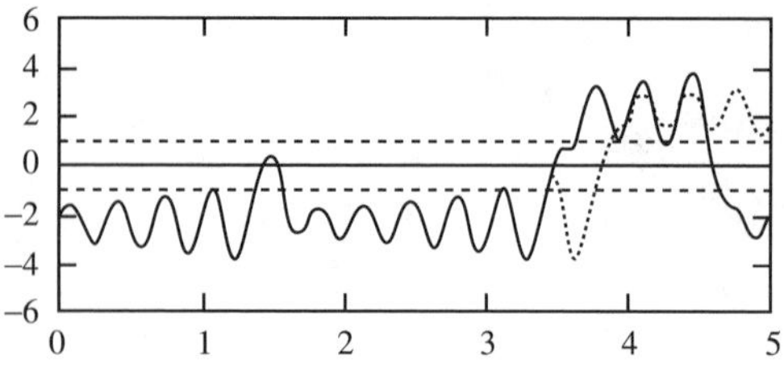

FIGURE 8.24 Sensitive dependence on initial conditions. Two time waveforms $V_1(t)$ from Chua's oscillator with $G = 550\ \mu S$, starting from $(I_3, V_2, V_1) = (1.810\ mA, 222.01\ mV, -2.286\ V)$ [solid line] and $(I_3, V_2, V_1) = (1.810\ mA, 222.000\ mV, -2.286\ V)$ [dashed line]. Note that the trajectories diverge within 5 ms. Horizontal axis: t (ms); vertical axis: V_1 (V). Compare with Figure 8.23.

"Randomness" in the Time Domian

Figures 8.14(a), (b), (c), and (d) show the state-space trajectories of period-one, period-two, and period-four periodic attractors, a spiral Chua's chaotic attractor, respectively, and the corresponding voltage waveforms $V_1(t)$.

The "period-one" waveform is periodic; it looks like a slightly distorted sinusoid. The "period-two" waveform is also periodic. It differs qualitatively from the "period-one" in that the pattern of a large peak followed by a small peak repeats approximately once every two cycles of the period-one signal; that is why it is called "period-two."

In contrast with these periodic time waveforms, $V_1(t)$ for the spiral Chua's attractor is quite irregular and does not appear to repeat itself in any observation period of finite length. Although it is produced by a third-order deterministic differential equation, the solution looks "random."

Broadband "Noise-Like" Power Spectrum

In the following discussion, we consider 8192 samples of $V_2(t)$ recorded at 200 kHz; leakage in the power spectrum is controlled by applying a Welch window [17] to the data.

We remarked earlier that the period-one time waveform corresponding to the attractor in Figure 8.14(a), is almost sinusoidal; we expect, therefore, that most of its power should be concentrated at the fundamental frequency. The power spectrum of the period-one waveform $V_2(t)$ shown in Figure 8.14(a) consists of a sharp spike at approximately 3 kHz and higher harmonic components that are over 30 dB below the fundamental.

Because the period-two waveform repeats roughly once every 0.67 ms, this periodic signal has a fundamental frequency component at approximately 1.5 kHz [see Figure 8.14(b)]. Notice, however, that most of the power in the signal is concentrated close to 3 kHz.

The period-four waveform repeats roughly once every 1.34 ms, corresponding to a fundamental frequency component at approximately 750 Hz [See Figure 8.14(c)]. Note once more that most of the power in the signal is still concentrated close to 3 kHz.

The spiral Chua's attractor is qualitatively different from these periodic signals. The aperiodic nature of its time-domain waveforms is reflected in the broadband noise-like power spectrum [Figure 8.14(d)]. No longer is the power of the signal concentrated in a small number of frequency components; rather, it is distributed over a broad range of frequencies. This broadband structure of the power spectrum persists even if the spectral resolution is increased by sampling at a higher frequency f_s. Notice that the spectrum still contains a peak at approximately 3 kHz that corresponds to the average frequency of rotation of the trajectory about the fixed point.

Practical Realization of Chua's Circuit

Chua's circuit can be realized in a variety of ways using standard or custom-made electronic components. All the linear elements (capacitor, resistor, and inductor) are readily available as two-terminal devices. A

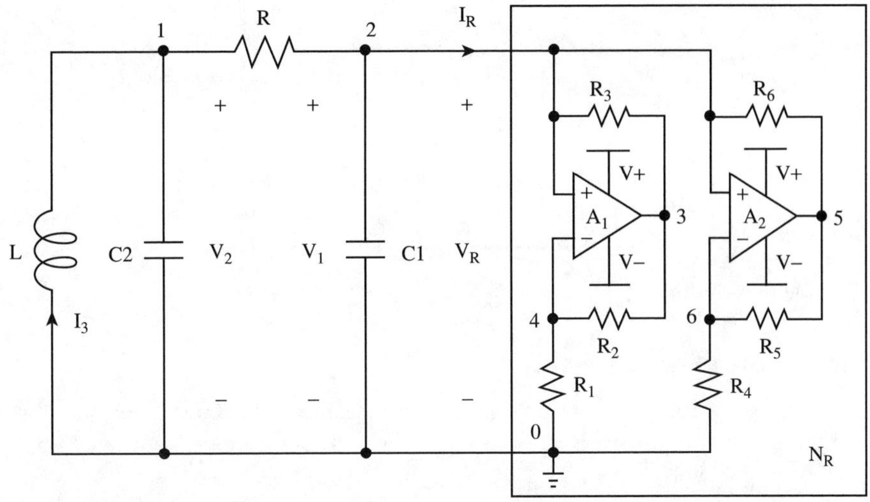

FIGURE 8.25 Practical implementation of Chua's circuit using two op amps and six resistors to realize the Chua diode [10]. Component values are listed in Table 8.2.

TABLE 8.2 Component List for the Practical Implementation of Chua's Circuit, depicted in Figure 8.25

Element	Description	Value	Tolerance
A_1	Op Amp ($\frac{1}{2}$AD712, TL082, or Equivalent)	—	—
A_2	Op Amp ($\frac{1}{2}$AD712, TL082, or Equivalent)	—	—
C_1	Capacitor	10 nF	±5%
C_2	Capacitor	100 nF	±5%
R	Potentiometer	2 kΩ	—
R_1	$\frac{1}{4}$ W Resistor	3.3 kΩ	±5%
R_2	$\frac{1}{4}$ W Resistor	22 kΩ	±5%
R_3	$\frac{1}{4}$ W Resistor	22 kΩ	±5%
R_4	$\frac{1}{4}$ W Resistor	2.2 kΩ	±5%
R_5	$\frac{1}{4}$ W Resistor	220 Ω	±5%
R_6	$\frac{1}{4}$ W Resistor	220 Ω	±5%
L	Inductor (TOKO-Type 10 RB, or Equivalent)	118 mH	±10%

nonlinear resistor N_R with the prescribed DP characteristic (called a Chua diode [10]) may be implemented by connecting two negative resistance converters in parallel as outlined in Figure 8.25. A complete list of components is given in Table 8.2.

The op amp subcircuit consisting of A_1, A_2 and R_1–R_6 functions as a **negative resistance converter** N_R with driving-point characteristic as shown in Figure 8.28(b). Using two 9-V batteries to power the op amps gives $V^+ = 9$ V and $V^- = -9$ V. From measurements of the saturation levels of the AD712 outputs, $E_{sat} \approx 8.3$ V, giving $E \approx 1$ V. With $R_2 = R_3$ and $R_5 = R_6$, the nonlinear characteristic is defined by $G_a = -1/R_1 - 1/R_4 = -50/66$ mS, $G_b = 1/R_3 - 1/R_4 = -9/22$ mS, and $E = R_1 E_{sat}/(R_1 + R_2) \approx 1$ V[10].

The equivalent circuit of Figure 8.25 is presented in Figure 8.26, where the real inductor is modeled as a series connection of an ideal linear inductor L and a linear resistor R_0. When the inductor's resistance is modeled explicitly in this way, the circuit is called **Chua's oscillator** [5].

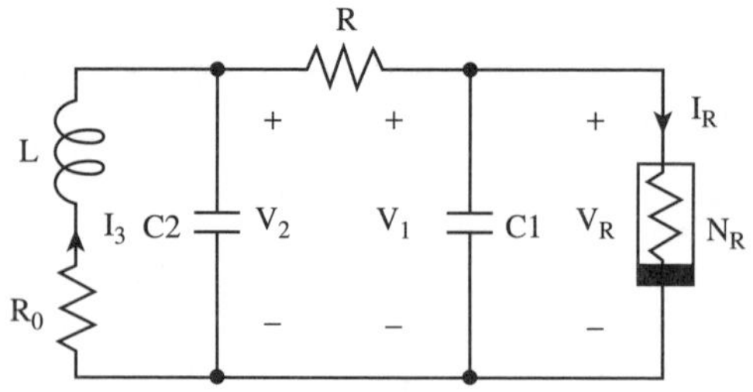

FIGURE 8.26 Chua's oscillator.

Experimental Steady-State Solutions

A two-dimensional projection of the steady-state attractor in Chua's circuit may be obtained by connecting V_2 and V_1 to the X and Y channels, respectively, of an oscilloscope in X–Y mode.

Bifurcation Sequence with *R* as Control Parameter

By reducing the variable resistor R in Figure 8.25 from 2000 Ω toward zero, Chua's circuit exhibits a Hopf bifurcation from dc equilibrium, a sequence of period-doubling bifurcations to a spiral Chua's attractor, periodic windows, a double-scroll Chua's chaotic attractor, and a boundary crisis, as illustrated in Figure 8.27.

Notice that varying R in this way causes the size of the attractors to change: the period-one orbit is large, period-two is smaller, the spiral Chua's attractor is smaller again, and the double-scroll Chua's attractor shrinks considerably before it dies. This shrinking is due to the equilibrium points P_+ and P_- moving closer towards the origin as R is decreased. Consider the load line in Figure 8.19(b): as R is decreased, the slope G increases, and the equilibrium points P_- and P_+ move toward the origin. Compare also the positions of P_+ in Figures 8.22(a) and 8.23(a)

The Outer Limit Cycle

No physical system can have unbounded trajectories. In particular, any physical realization of a Chua diode is **eventually passive**, meaning simply that for a large enough voltage across its terminals, the instantaneous power $P_R(t)$ $[= V_R(t)I_R(t)]$ consumed by the device is positive.

Hence, the DP characteristic of a real Chua diode must include at least two outer segments with positive slopes which return the characteristic to the first and third quadrants [see Figure 8.28(b)]. From a practical point of view, as long as the voltages and currents on the attractor are restricted to the negative resistance region of the characteristic, these outer segments will not affect the circuit's behavior.

The DP characteristic of the op-amp-based Chua diode differs from the desired piecewise linear characteristic depicted in Figure 8.28(a) in that it has five segments, the outer two of which have positive slopes $G_c = 1/R_5 = 1/220$ S.

The "unbounded" trajectories that follow the boundary crisis in the ideal three-region system are limited in amplitude by these dissipative outer segments and a large limit cycle results, as illustrated in Figure 8.27(i). This effect could, of course, be simulated by using a five-segment DP characteristic for N_R as illustrated in Figure 8.28(b).

The parameter value at which the double-scroll Chua's attractor disappears and the outer limit cycle appears is different from that at which the outer limit cycle disappears and the chaotic attractor reappears. This "hysteresis" in parameter space is characteristic of a blue sky catastrophe.

Simulation of Chua's Circuit

Our experimental observations and qualitative descriptive description of the global dynamics of Chua's circuit may be confirmed by simulation using a specialized nonlinear dynamics simulation package such as **INSITE** [15] or by employing a customized simulator such as "**ABC**" [10].

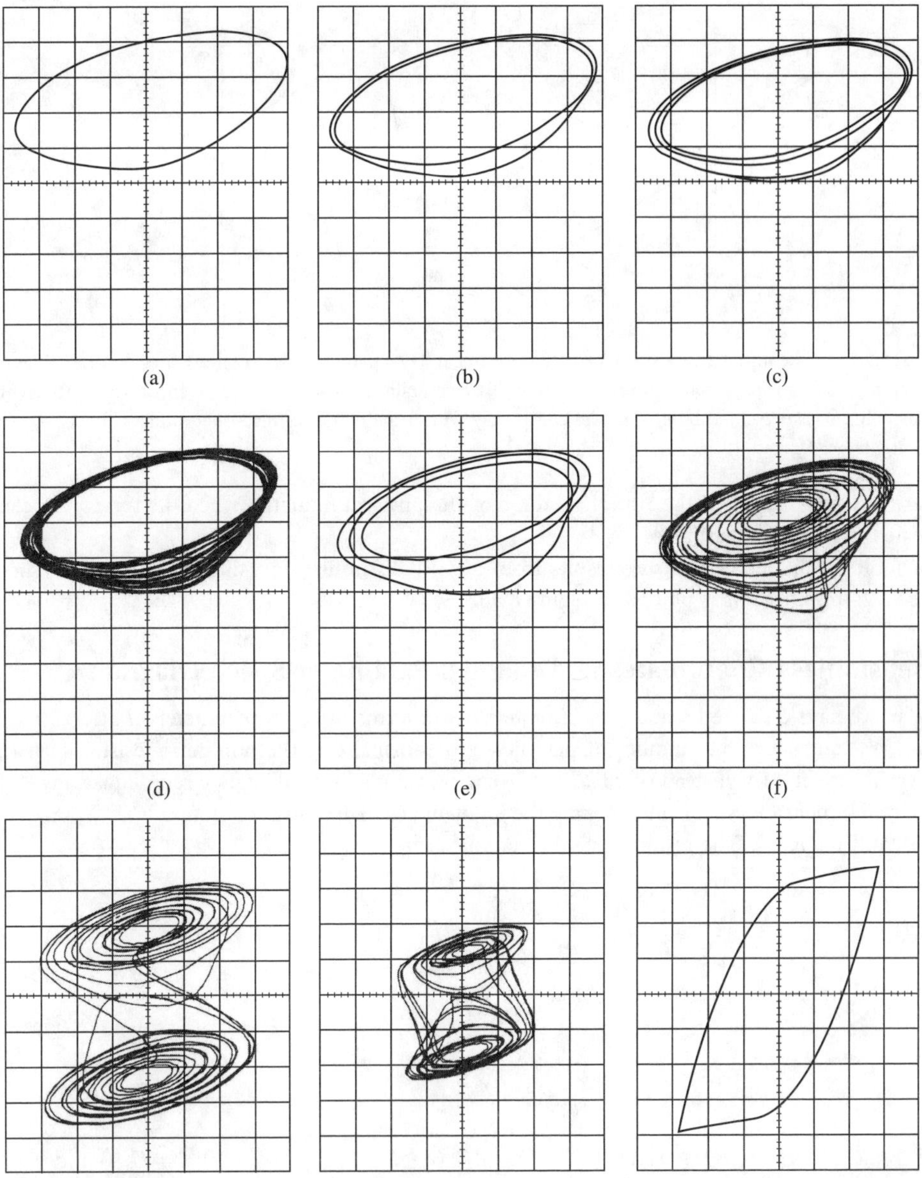

FIGURE 8.27 Typical experimental bifurcation sequence in Chua's circuit (component values as in Table 8.2) recorded using a digital storage oscilloscope. Horizontal axis V_2 (a)–(h) 200 mV/div, (i) 2 V/div; vertical axis V_1 (a)–(h) 1 V/div, (i) 2 V/div. (a) R = 1.83 kΩ, period–1; (b) R = 1.82 kΩ, period-2; (c) R = 1.81 kΩ, period-4; (d) R = 1.80 kΩ, spiral Chua's attractor; (e) R = 1.797 kΩ, period-3 window; (f) R = 1.76 kΩ, spiral Chua's attractor; (g) R = 1.73 kΩ, double-scroll Chua's attractor; (h) R = 1.52 kΩ, double-scroll Chua's attractor; (i) R = 1.42 kΩ, large limit cycle corresponding to the outer segments of the Chua diode's DP characteristic. *Source*: M. P. Kennedy, "Three steps to chaos — Part II: A Chua's circuit primer," *IEEE Trans. Circuit Syst. I*, vol. 40, pp. 669, 670, Oct. 1993. © 1993 IEEE.

For electrical engineers who are familiar with the **SPICE** circuit simulator but perhaps not with chaos, we present a net-list and simulation results for a robust op-amp-based implementation of Chua's circuit. The AD712 op amps in this realization of the circuit are modeled using Analog Devices' AD712 macro-model. The TOKO 10RB inductor has a nonzero series resistance that we have included in the SPICE net-list; a typical value of RO for this inductor is 12.5 Ω. Node numbers are as Figure 8.25: the power

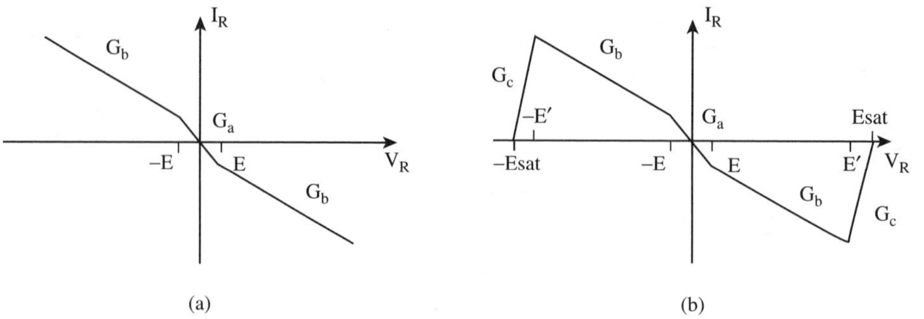

FIGURE 8.28 (a) Required three-segment piecewise-linear DP characteristic for the Chua diode in Figure 8.17. (b) Every physically realizable nonlinear resistor N_R is eventually passive — the outermost segments (while not necessarily linear as presented here) must lie completely within the first and third quadrants of the V_R–I_R place for sufficiently large $|V_R|$ and $|I_R|$.

rails are 111 and 222; 10 is the "internal" node of the physical inductor, where its series inductance is connected to its series resistance.

A double-scroll Chua's attractor results from a PSPICE simulation using the input deck shown in Figure 8.29; this attractor is plotted in Figure 8.30.

Dimensionless Coordinates and the α–β Parameter-Space Diagram

Thus far, we have discussed Chua's circuit equations in terms of seven parameters: L, C_2, G, C_1, E, G_a, and G_b. We can reduce the number of parameters by normalizing the nonlinear resistor such that its breakpoints are at ±1 V instead of ±E V. Furthermore, we may write Chua's circuit equations (8.12) in normalized dimensionless form by making the following change of variables: $X_1 = V_1/E$, $X_2 = V_2/E$, $X_3 = I_3/(EG)$, and $\tau = tG/C_2$. The resulting state equations are

$$\frac{dX_1}{d\tau} = \alpha\left[X_2 - X_1 - f(X_1)\right]$$

$$\frac{dX_2}{d\tau} = X_1 - X_2 + X_3 \tag{8.15}$$

$$\frac{dX_3}{d\tau} = -\beta X_2$$

where $\alpha = C_2/C_1$, $\beta = C_2/(LG^2)$, and $f(X) = bX + 1/2\,(a - b)\,(|X + 1| - |X - 1|)$; $a = G_a/G$ and $b = G_b/G$. Thus, each set of seven circuit parameters has an equivalent set of four normalized dimensionless parameters $\{\alpha, \beta, a, b\}$. If we fix the values of a and b (which correspond to the slopes G_a and G_b of the Chua diode), we can summarize the steady-state dynamical behavior of Chua's circuit by means of a two-dimensional parameter-space diagram.

Figure 8.31 presents the (α,β) parameter-space diagram with $a = -8/7$ and $b = -5/7$. In this diagram, each region denotes a particular type of steady-state behavior: for example, an equilibrium point, period-one orbit, period-two, spiral Chua's attractor, double-scroll Chua's attractor. Typical state-space behaviors are shown in the insets. For clarity, we show chaotic regions in a single shade; it should be noted that these chaotic regions are further partitioned by periodic windows and "islands" of periodic behavior.

To interpret the α–β diagram, imagine fixing the value of $\beta = C_2/(LG_2)$ and increasing $\alpha = C_2/C_1$ from a positive value to the left of the curve labeled "Hopf at $P^{\pm}$"; experimentally, this corresponds to fixing the parameters L, C_2, G, E, G_a, and G_b, and reducing the value of C_1 — this is called a "C_1 bifurcation sequence."

Initially, the steady-state solution is an equilibrium point. As the value of C_1 is reduced, the circuit undergoes a Hopf bifurcation when α crosses the "Hopf at $P\pm$" curve. Decreasing C_1 still further, the

```
ROBUST OP AMP REALIZATION OF CHUA'S CIRCUIT
V+   111 0      DC 9
V-   0   222    DC 9
L    1 10    0.018
R0   10 0     12.5
R    1 2     1770
C2   1 0       100.0N
C1   2 0       10.0N
XA1  2 4 111 222 3 AD712
R1   2 3  220
R2   3 4  220
R3   4 0  2200
XA2  2 6 111 222  5 AD712
R4   2 5  22000
R5   5 6  22000
R6   6 0  3300

* AD712 SPICE Macro-model                1/91, Rev. A
* Copyright 1991 by Analog Devices, Inc. (reproduced with permission)
*
.SUBCKT AD 712 13  15  21 16  14
*
VOS 15  8 DC  0
EC 9 0 14 0 1
C1 6 7 .5P
RP 16 12 12K
GB 11 0 3 0 1.67K
RD1 6 16 16k
RD2 7 16 16k
ISS 12 1 DC 100U
CCI 3 11 150P
GCM 0 3 0 1  1.76N
GA 3 0 7 6  2.3M
RE 1 0 2.5MEG
RGM 3 0 1.69K
VC  12 2 DC 2.8
VE  10 16 DC 2.8
RO1 11 14 25
CE 1 0 2P
RO2 0 11 30
RS1 1 4 5.77K
RS2 1 5 5.77K
J1 6 13 4 FET
J2 7 8 5 FET
DC 14 2 DIODE
DE 10 14 DIODE
DP 16 12 DIODE
D1 9 11 DIODE
D2 11 9 DIODE
IOS 15 13 5E-12
.MODEL DIODE D
.MODEL FET PJF(VTO = -1 BETA = 1M IS = 25E-12)
.ENDS

.IC V(2) = 0.1 V(1) = 0
.TRAN 0.01MS 100MS 50MS
.OPTIONS RELTOL = 1.0E-4 ABSTOL = 1.0E-4
.PRINT TRAN V(2) V(1)
.END
```

FIGURE 8.29 SPICE deck to simulate the transient response of the dual op amp implementation of Chua's circuit. Node numbers are as in Figure 8.25. The op amps are modeled using the Analog Devices AD712 macro-model. RO models the series resistance of the real inductor *L*.

steady-state behavior bifurcates from period-one to period-two to period-four and so on to chaos, periodic windows, and a double-scroll Chua's attractor. The right-hand side edge of the chaotic region is delimited by a curve corresponding to the boundary crisis and "death" of the attractor. Beyond this curve, trajectories diverge toward infinity. Because of eventual passivity in a real circuit, these divergent trajectories will of course converge to a limit cycle in any physical implementation of Chua's circuit.

8.3 Chua's Oscillator

Chua's oscillator [5] (see Figure 8.26) is derived from Chua's circuit by adding a resistor R_0 in series with the inductor L. The oscillator contains a linear inductor, two linear resistors, two linear capacitors, and

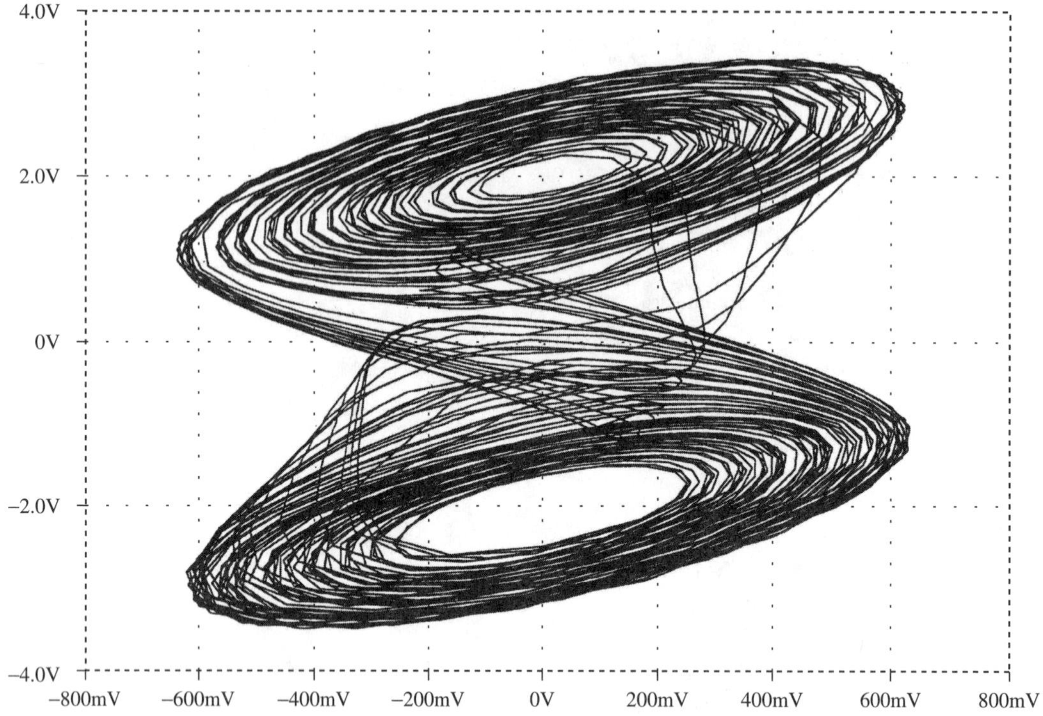

FIGURE 8.30 **PSpice** (evaluation version 5.4, July 1993) simulation of Figure 8.25 using the input deck from Figure 8.29 yields this double-scroll Chua's attractor. Horizontal axis V_2 (V); vertical axis V_1 (V).

a single Chua diode N_R. N_R is a voltage-controlled piecewise-linear resistor whose continuous odd-symmetric three-segment driving-point characteristic (see Figure 8.18) is described explicitly by the relationship

$$I_R = G_b V_R + \tfrac{1}{2}(G_a - G_b)(|V_R + E| - |V_R - E|)$$

The primary motivation for studying this circuit is that the vector field of Chua's oscillator is topologically conjugate to the vector field of a large class of three-dimensional, piecewise-linear vector fields. In particular, the oscillator can exhibit every dynamical behavior known to be possible in an autonomous three-dimensional, continuous-time dynamical system described by a continuous odd-symmetric three-region piecewise-linear vector field. With appropriate choices of component values, the circuit follows the period-doubling, intermittency, and quasiperiodic routes to chaos.

State Equations

Choosing V_1, V_2, V, and I_3 as state variables, Chua's oscillator may be described by three ordinary differential equations:

$$\frac{dV_1}{dt} = \frac{G}{C_1}(V_2 - V_1) - \frac{1}{C_1} f(V_1)$$

$$\frac{dV_2}{dt} = \frac{G}{C_2}(V_1 - V_2) + \frac{1}{C_2} I_3$$

$$\frac{dI_3}{dt} = -\frac{1}{L} V_2 - \frac{R_0}{L} I_3$$

where $G = 1/R$ and $f(V_R) = G_b V_R + 1/2\,(G_a - G_b)\,(|V_R + E| - |V_R - E|)$.

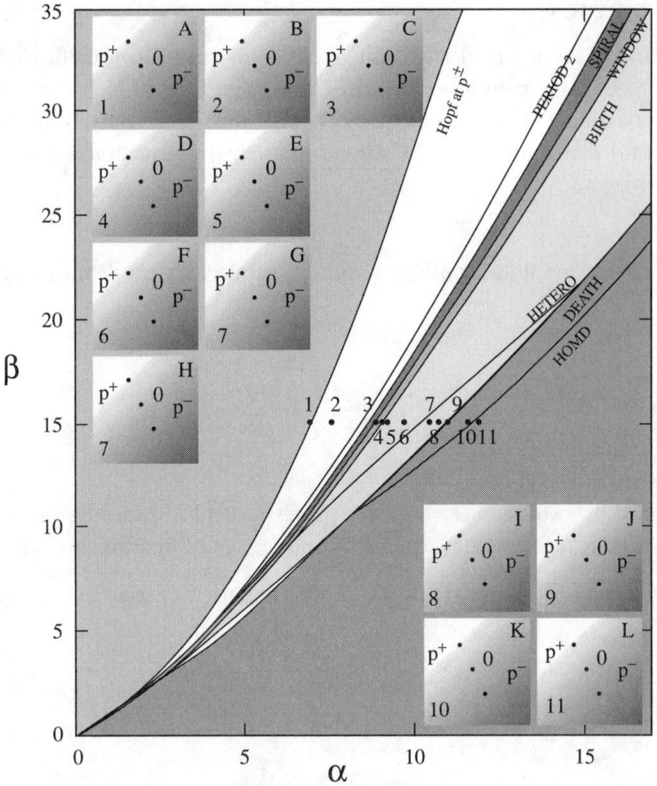

FIGURE 8.31 α–β parameter space diagram for the normalized dimensionless Chua's circuit equations (8.31) with $a = -8/7$ and $b = -5/7$. *Source*: M. P. Kennedy, "Three steps to chaos — Part II: A Chua's circuit primer," *IEEE Trans. Circuits Syst. I*, vol. 40, p. 673, Oct. 1993. ® 1993 IEEE.

The vector field is parameterized by eight constants: L, C_2, G, C_1, R_0, E, G_a, and G_b. We can reduce the number of parameters by normalizing the nonlinear resistor such that its breakpoints are at ±1 V instead of ±*E* V, scaling the state variables, and scaling time.

By making the following change of variables: $X_1 = V_1/E$, $X_2 = V_2/E$, $X_3 = I3/(EG)$, $\tau = t|G/C_2|$, and $k =$ *sgn* $(G/C2)$,[14] *we can rewrite the state equations* (8.16) *in normalized dimensionless form:*

$$\frac{dX_1}{d\tau} = k\alpha\left[X_2 - X_1 - f(X_1)\right]$$

$$\frac{dX_2}{d\tau} = k\left(X_1 - X_2 + X_3\right)$$

$$\frac{dX_3}{d\tau} = -k\left(\beta X_2 + \gamma X_3\right)$$

where $\alpha = C_2/C_1$, $\beta = C_2/(LG^2)$, $\gamma = R_0C_2/(LG)$, and $f(x) = bX + 1/2(a - b)(|X + 1| - |X - 1|)$ with $a = G_a/G$ and $b = G_b/G$. Thus, each set of eight circuit parameters has an equivalent set of six normalized dimensionless parameters $\{\alpha, \beta, \gamma, a, b, k\}$.

---

[14]The signum function is defined by sgn(x) = x if x > 0, sgn(x) = –x if x < 0, and sgn(0) = 0.

Topological Conjugacy

Two vector fields **F** and **F′** are topologically conjugate if there exists a continuous map h (which has a continuous inverse) such that h maps trajectories of **F** into trajectories of **F′**, preserving time orientation and parametrization of time. If ϕ_t and ϕ'_t are the flows of **F** and **F′**, respectively, then $\phi_t \circ h = h \circ \phi'_t$ for all t. This means that the dynamics of **F** and **F′** are qualitatively the same. If h is linear, then **F** and **F′** are said to be linearly conjugate.

Class $\mathcal{C}$

The three-dimensional, autonomous, continuous-time dynamical system defined by the state equation

$$\dot{\mathbf{X}} = \mathbf{F}(\mathbf{X}) \qquad \mathbf{X} \in \mathbb{R}^3$$

is said to belong to class $\mathcal{C}$ iff

1. $\mathbf{F}$: $\mathbb{R}^3 \rightarrow \mathbb{R}^3$ is continuous
2. F is odd-symmetric, i.e., $\mathbf{F}(-X) = -\mathbf{F}(X)$
3. $\mathbb{R}^3$ is partitioned by two parallel boundary planes U_1 and U_{-1} into an inner region D_0, which contains the origin, and two outer regions D_1 and D_{-1}, and **F** is affine in each region.

Without loss of generality, the boundary planes and the regions they separate can be chosen as follows:

$$D_{-1} = \{\mathbf{X}: \mathbf{X}_1 \leq -1\}$$

$$U_{-1} = \{\mathbf{X}: \mathbf{X}_1 = -1\}$$

$$D_0 = \{\mathbf{X}: |\mathbf{X}_1| \leq 1\}$$

$$U_{-1} = \{\mathbf{X}: \mathbf{X}_1 = 1\}$$

$$D_1 = \{\mathbf{X}: \mathbf{X}_1 \geq 1\}$$

Any vector field in the family $\mathcal{C}$ can then be written in the form

$$\dot{\mathbf{X}} = \begin{cases} \mathbf{A}_{-1}\mathbf{X} - \mathbf{b} & \mathbf{X}_1 \leq -1 \\ \mathbf{A}_0\mathbf{X} & -1 \leq \mathbf{X}_1 \leq 1 \\ \mathbf{A}_1\mathbf{X} + \mathbf{b} & \mathbf{X}_1 \geq 1 \end{cases}$$

where

$$\mathbf{A}_{-1} = \mathbf{A}_1 = \begin{bmatrix} a_{11} & a_{12} & a_{13} \\ a_{21} & a_{22} & a_{23} \\ a_{31} & a_{32} & a_{33} \end{bmatrix} \quad \text{and} \quad \mathbf{b} = \begin{bmatrix} b_1 \\ b_2 \\ b_3 \end{bmatrix}$$

By continuity of the vector field across the boundary planes,

$$\mathbf{A}_0 = \begin{bmatrix} (a_{11} + b_1) & a_{12} & a_{13} \\ (a_{21} + b_2) & a_{22} & a_{23} \\ (a_{31} + b_3) & a_{32} & a_{33} \end{bmatrix}$$

Equivalent Eigenvalue Parameters

Let (μ_1, μ_2, μ_3) denote the eigenvalues associated with the linear vector field in the D_0 region and let (ν_1, ν_2, ν_3) denote the eigenvalues associated with the affine vector fields in the outer regions D_1 and D_{-1}. Define

$$\left.\begin{aligned} p_1 &= \mu_1 + \mu_2 + \mu_3 \\ p_2 &= \mu_1\mu_2 + \mu_2\mu_3 + \mu_3\mu_1 \\ p_3 &= \mu_1\mu_2\mu_3 \\ \\ q_1 &= \nu_1 + \nu_2 + \nu_3 \\ q_2 &= \nu_1\nu_2 + \nu_2\nu_3 + \nu_3\nu_1 \\ q_3 &= \nu_1\nu_2\nu_3 \end{aligned}\right\} \tag{8.16}$$

Because the six parameters $\{p_1, p_2, p_3, q_1, q_2, q_3\}$ are uniquely determined by the eigenvalues $\{\mu_1, \mu_2, \mu_3, \nu_1, \nu_2, \nu_3\}$ and vice versa, the former are called the equivalent eigenvalue parameters. Note that the equivalent eigenvalues are real; they are simply the coefficients of the characteristic polynomials:

$$(s-\mu_1)(s-\mu_2)(s-\mu_3) = s^3 - p_1 s^2 + p_2 s - p_3$$

$$(s-\nu_1)(s-\nu_2)(s-\nu_3) = s^3 - q_1 s^2 + q_2 s - q_3$$

Theorem 3 (Chua et al.) [5]: *Let* $\{\mu_1, \mu_2, \mu_3, \nu_1, \nu_2, \nu_3\}$ *be the eigenvalues associated with a vector field* $F(X) \in \mathscr{C}/\mathscr{E}_0$, *where* $\mathscr{E}_0$ *is the set of measure zero in the space of equivalent eigenvalue parameters where one of* (8.17) *is satisfied. Then, Chua's oscillator with parameters defined by* (8.18) *and* (8.19) *is linearly conjugate to this vector field.*

$$\left.\begin{aligned} p_1 - q_1 &= 0 \\ p_2 - \left(\frac{p_3 - q_3}{p_1 - q_1}\right) - \left(\frac{p_2 - q_2}{p_1 - q_1}\right)\left(p_1 - \frac{p_2 - q_2}{p_1 - q_1}\right) &= 0 \\ -\left(\frac{p_2 - q_2}{p_1 - q_1}\right) - \frac{k_1}{k_2} &= 0 \\ -k_1 k_3 + k_2\left(\frac{p_3 - q_3}{p_1 - q_1}\right) &= 0 \\ \det \mathbf{K} = \det\begin{bmatrix} 1 & 0 & 0 \\ a_{11} & a_{12} & a_{13} \\ K_{31} & K_{32} & K_{33} \end{bmatrix} = a_{12}k_{33} - a_{13}k_{32} &= 0 \end{aligned}\right\} \tag{8.17}$$

where

$$K_{3i} = \sum_{j=i}^{3} a_{1j} a_{ji} \qquad i = 1, 2, 3$$

We denote by $\tilde{\mathscr{C}}$ the set of vector fields $\mathscr{C}/\mathscr{E}_0$. Two vector fields in $\tilde{\mathscr{C}}$ are linearly conjugate if they have the same eigenvalues in each region.

Eigenvalues-to-Parameters Mapping Algorithm for Chua's Oscillator

Every continuous, third-order, odd-symmetric, three-region, piecewise-linear vector field $\mathbf{F}'$ in $\mathcal{C}$ may be mapped onto a Chua's oscillator (where the vector field $\mathbf{F}$ is topologically conjugate to $\mathbf{F}'$) by means of the following algorithm [5]:

1. Calculate the eigenvalues (μ_1', μ_2', μ_3') and (ν_1', ν_2', ν_3') associated with the linear and affine regions, respectively, of the vector field $\mathbf{F}'$ of the circuit or system whose attractor is to be reproduced (up to linear conjugacy) by Chua's oscillator.
2. Find a set of circuit parameters $\{C_1, C_2, L, R, R_0, G_a, G_b, E\}$ (or dimensionless parameters $\{\alpha, \beta, \gamma, a, b, k\}$) so that the resulting eigenvalues μ_j and ν_j for Chua's oscillator satisfy $\mu_j = \mu_j'$ and $\nu_j = \nu_j'$, $j = 1, 2, 3$.

Let $\{p_1, p_2, p_3, q_1, q_2, q_3\}$ be the equivalent eigenvalue parameters defined by (8.16). Furthermore, let

$$\left.\begin{aligned}
k_1 &= -p_3 + \left(\frac{p_3 - q_3}{p_1 - q_1}\right)\left(p_1 - \frac{p_2 - q_2}{p_1 - q_1}\right)\\
k_2 &= p_2 - \left(\frac{p_3 - q_3}{p_1 - q_1}\right) - \left(\frac{p_2 - q_2}{p_1 - q_1}\right)\left(p_1 - \frac{p_2 - q_2}{p_1 - q_1}\right)\\
k_3 &= -\left(\frac{p_2 - q_2}{p_1 - q_1}\right) - \frac{k_1}{k_2}\\
k_4 &= -k_1 k_3 + k_2\left(\frac{p_3 - q_3}{p_1 - q_1}\right)
\end{aligned}\right\} \tag{8.18}$$

The corresponding circuit parameters are given by

$$\left.\begin{aligned}
C_1 &= 1\\
C_2 &= -\frac{k_2}{k_3^2}\\
L &= -\frac{k_3^2}{k_4}\\
R &= -\frac{k_3}{k_2}\\
R_0 &= -\frac{k_1 k_3^2}{k_2 k_4}\\
G_a &= -p_1 + \left(\frac{p_2 - q_2}{p_1 - q_1}\right) + \frac{k_2}{k_3}\\
G_b &= -q_1 + \left(\frac{p_2 - q_2}{p_1 - q_1}\right) + \frac{k_2}{k_3}
\end{aligned}\right\} \tag{8.19}$$

The breakpoint E of the piecewise-linear Chua diode can be chosen arbitrarily because the choice of E does not affect either the eigenvalues or the dynamics; it simply scales the circuit variables. In a practical realization of the circuit, one should scale the voltages and currents so that they lie within the inner three segments of the nonlinear resistor N_R.

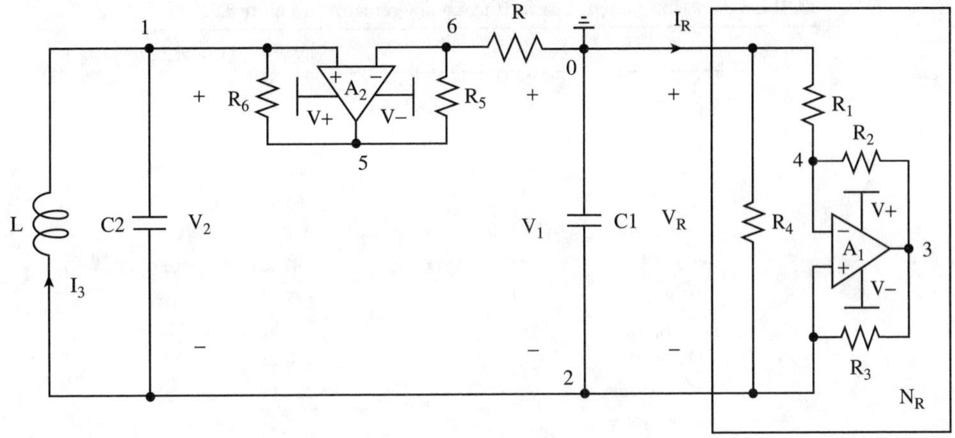

FIGURE 8.32 Practical implementation of Chua's oscillator using an op amp and four resistors to realize the Chua diode [10]. The negative resistor G is realized by means of a negative resistance converter (a_2, R_5, R_6, and positive resistor R). If $R_2 = R_3$ and $R_5 = R_6$, and $G_a = 1/R_4 - 1/R_1$, $G_b = 1/R_4 + 1/R_2$, and $G = -1/R$. Component values are listed in Table 8.3.

The dimensionless parameters can be calculated as follows:

$$\left.\begin{aligned} \alpha &= -\frac{k_2}{k_3^2} \\ \beta &= \frac{k_4}{k_2 k_3^2} \\ \gamma &= \frac{k_1}{k_2 k_3} \\ a &= -1 + \frac{k_3}{k_2}\left(p_1 - \frac{p_2 - q_2}{p_1 - q_1}\right) \\ b &= -1 + \frac{k_3}{k_2}\left(q_1 - \frac{p_2 - q_2}{p_1 - q_1}\right) \\ k &= \operatorname{sgn}(k_3) \end{aligned}\right\} \tag{8.20}$$

Example: Torus

Figure 8.32 shows a practical implementation of Chua's oscillator that exhibits a transition to chaos by torus breakdown. A complete list of components is given in Table 8.3.

A SPICE simulation of this circuit produces a quasiperiodic voltage $v(2)$ $(= -V_1)$, as expected (see Figure 8.33). The resistor RO is not explicitly added to the circuit, but models the dc resistance of the inductor.

8.4 Van der Pol Neon Bulb Oscillator

In a paper titled "Frequency Demultiplication," the eminent Dutch electrical engineer Balthazar van der Pol described an experiment in which, by tuning the capacitor in a neon bulb RC relaxation oscillator

TABLE 8.3 Component List for the Chua Oscillator in Figure 8.32

Element	Description	Value	Tolerance
A_1	Op Amp($\frac{1}{2}$AD712, TL082, or Equivalent)	—	—
A_2	Op Amp($\frac{1}{2}$AD712, TL082, or Equivalent)	—	—
C_1	Capacitor	47 nF	±5%
C_2	Capacitor	820 nF	±5%
R_1	$\frac{1}{4}$ W Resistor	6.8 kΩ	±5%
R_2	œmp W Resistor	47 kΩ	±5%
R_3	$\frac{1}{4}$ W Resistor	47 kΩ	±5%
R_4	Potentiometer	2 kΩ	—
R_5	W Resistor	220 Ω	±5%
R_6	$\frac{1}{4}$ W Resistor	220 Ω	±5%
R	Potentiometer	2 kΩ	—
L	Inductor(TOKO-Type10RB,orEquivalent)	18 mH	±10%

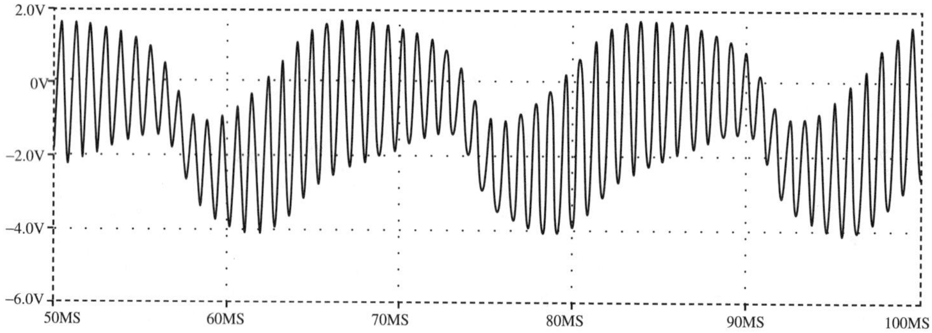

FIGURE 8.33 PSpice simulation (.TRAN 0.01MS 100 MS 50 MS) of Figure 8.32 with initial conditions .IC V(2) = –0.1 V(1) = –0.1 and tolerances .OPTIONS RELTOL = 1 E -4 ABSTOL = 1 E – 4 yields this quasiperiodic voltage waveform at node 2.

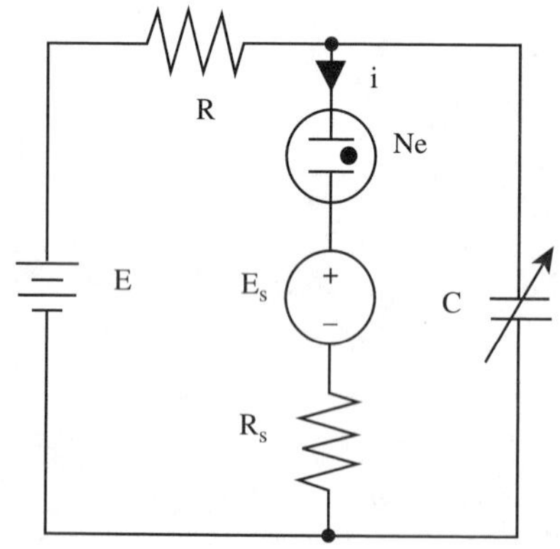

FIGURE 8.34 Sinusoidally driven **neon bulb relaxation oscillator**. Ne is the neon bulb.

driven by a sinusoidal voltage source (see Figure 8.34), "currents and voltages appear in the system which are whole submultiples of the driving frequency" [11].

The circuit consists of a high-voltage dc source E attached via a large series resistance R to a neon bulb and capacitor C that are connected in parallel; this forms the basic relaxation oscillator. Initially, the capacitor is discharged and the neon bulb is nonconducting. The dc source charges C with time constant

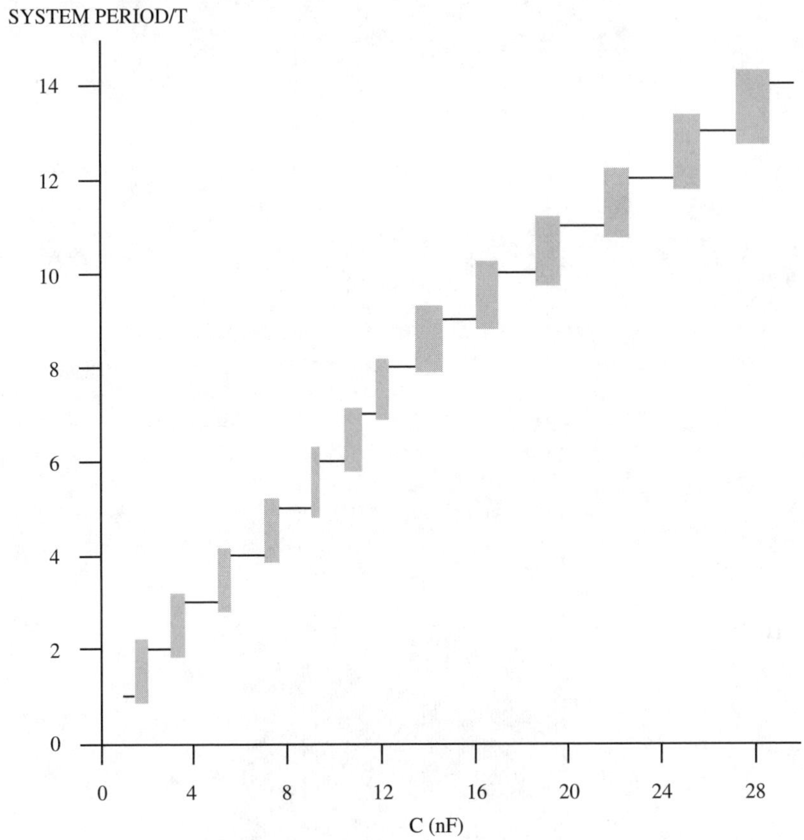

FIGURE 8.35 Normalized current pulse pattern repetition rate vs. C for the sinusoidally driven neon relaxation oscillator in Figure 8.34, showing a coarse staircase structure of mode-lockings. *Source*: M. P. Kennedy and L. O. Chua, "Van der Pol and Chaos," *IEEE Trans. Circuits Syst.*, vol. CAS-33, p. 975, Oct. 1986. ® 1986 IEEE.

RC until the voltage across the neon bulb is sufficient to turn it on. Once lit, he bulb presents shunt low resistance path to the capacitor. The voltage across the capacitor falls exponentially until the neon arc is quenched, the bulb is returned to its "off" state, and the cycle repeats.

In series with the neon bulb is inserted a sinusoidal voltage source $E_s = E_0 \sin(2\pi f_s t)$; its effect is to perturb the "on" and "off" switching thresholds of the capacitor voltage.

Experimental results for this circuit are summarized in Figure 8.35, where the ratio of the system period (time interval before the pattern of current pulses repeats itself) to the period T of the forcing is plotted versus the capacitance C.

Van der Pol noted that as the capacitance was increased from that values (C_0) for which the natural frequency f_0 of the undriven relaxation oscillator equaled that of the sinusoidal source (system period/$T = 1$), the system frequency made "discrete jumps from one whole submultiple of the driving frequency to the next" (detected by means of "a telephone coupled loosely in some way to the system"). Van der Pol noted that "often an irregular noise is heard in the telephone receiver before the frequency jumps to the next lower value"; van der Pol had observed chaos. Interested primarily in frequency demultiplication, he dismissed the "noise" as "a subsidiary phenomenon."

Typical current waveforms, detected by means of a small current-sensing resistor R_s placed in series with the bulb are shown in Figure 8.36. These consist of a series of sharp spikes, corresponding to the periodic firing of the bulb. Figure 8.36(c) shows a nonperiodic "noisy" signal of the type noticed by van der Pol.

The frequency locking behavior of the driven neon bulb oscillator circuit is characteristic of forced oscillators that contain two competing frequencies: the natural frequency f_0 of the undriven oscillator

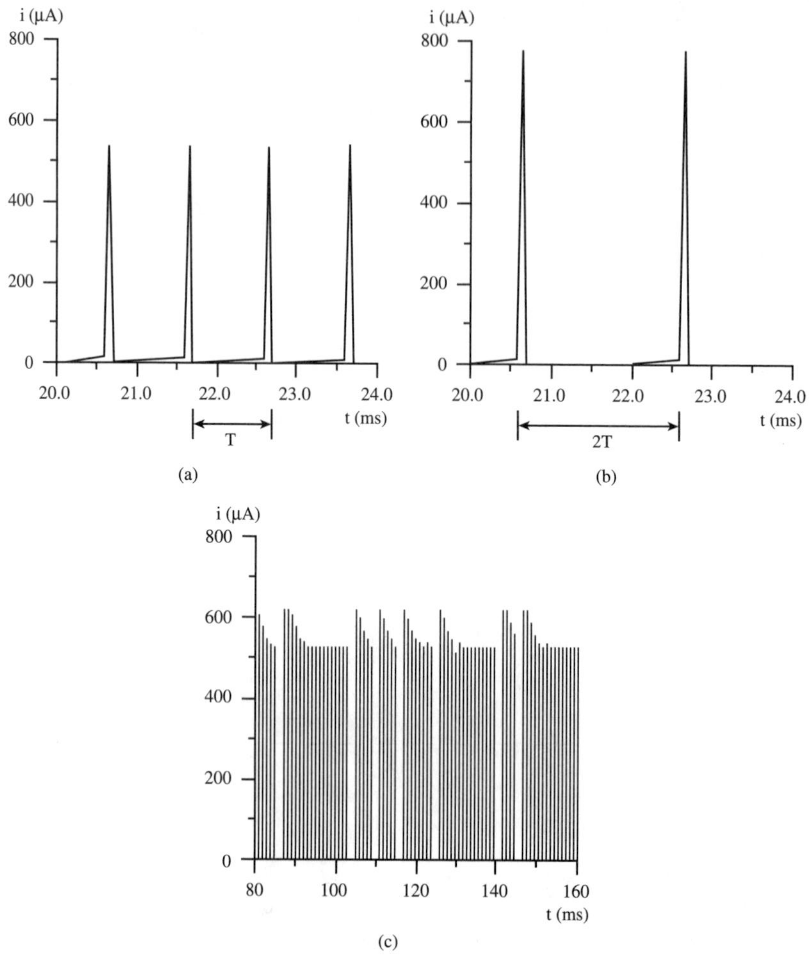

FIGURE 8.36 Periodic and chaotic neon bulb current waveforms. (a) One current pulse per cycle of E_s:$f_s/f_d = 1/1$; (b) one current pulse every two cycles of E_s: $f_s/f_d = 2/1$; (c) "noisy" current waveform.

and the driving frequency f_s. If the amplitude of the forcing is small, either quasiperiodicity or mode-locking occurs. For a sufficiently large amplitude of the forcing, the system may exhibit chaos.

Winding Numbers

Subharmonic frequency locking in a forced oscillator containing two competing frequencies f_1 and f_2 may be understood in terms of a **winding number**. The concept of a winding number was introduced by Poincaré to describe periodic and quasiperiodic trajectories on a torus.

A trajectory on a torus that winds around the minor axis of the torus with frequency f_1 revolutions per second, and completes revolution of the major axis with frequency f_2, may be parametrized by two angular coordinates $\theta_1 \equiv f_1 t$ and $\theta_2 \equiv f_2 t$, as illustrated in Figure 8.37. The angles of rotation θ_1 and θ_2 about the major and minor axes of the torus are normalized so that one revolution corresponds to a change in θ of 1.

A Poincaré map for this system can be defined by sampling the state θ_1 with period $\tau = 1/f_2$. Let $\theta_k = \theta_1(k\tau)$. The Poincaré map has the form

$$\theta_{k+1} = G(\theta_k), \quad k = 0, 1, 2, \ldots$$

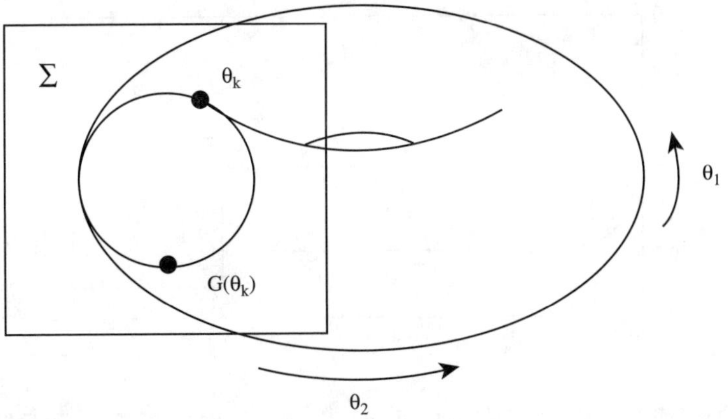

FIGURE 8.37 A trajectory on a torus is characterized by two normalized angular coordinates. $\theta_1 = f_1 t$ is the angle of rotation about the minor axis of the torus, while $\theta_2 = f_2 t$ is the angle of rotation along the major axis, where f_1 and f_2 are the frequencies of rotation about the corresponding axes. A Poincaré map $\theta_{k+1} = G(\theta_k)$ is defined by sampling the trajectory with frequency $1/f_2$. The winding number w counts the average number of revolutions in the Poincaré section per iteration of the map.

If $f_1/f_2 = p/q$ is rational, then the trajectory is periodic, closing on itself after completing q revolutions about the major axis of the torus. In this case, we say that the system is periodic with period q and completes p cycles per period. If the ratio p/q is irrational then the system is quasiperiodic; a trajectory covers the surface of the torus, coming arbitrarily close to every point on it, but does not close on itself.

The winding number w is defined by

$$w = \lim_{k\to\infty} \frac{G^{(k)}(\theta_0)}{k}$$

where $G^{(k)}$ denotes the k-fold iterate of G and θ_0 is the initial state.

The winding number counts the average number of revolutions in the Poincaré section per iteration. Equivalently, w equals the average number of turns about the minor axis per revolution about the major axis of the torus.[15] Periodic orbits possess rational winding numbers and are called resonant; quasiperiodic trajectories have irrational winding numbers.

The Circle Map

A popular paradigm for explaining the behavior of coupled nonlinear oscillators with two competing frequencies is the circle map:

$$\theta_{k+1} = \left(\theta_k + \frac{K}{2\pi}\sin(2\pi\theta_k) + \Omega\right) \bmod 1, \quad k = 0, 1, 2, \ldots \tag{8.21}$$

so-called because it maps the circle into itself. The sinusoidal term represents the amplitude of the forcing, and Ω is the ratio of the natural frequency of the unperturbed system and the forcing frequency [18].

When $K \equiv 0$, the steady state of the discrete-time dynamical system (8.22) is either periodic or quasiperiodic, depending on whether Ω is rational or irrational.

[15]Either frequency may be chosen to correspond to the major axis of the torus, so the winding number and its reciprocal are equivalent.

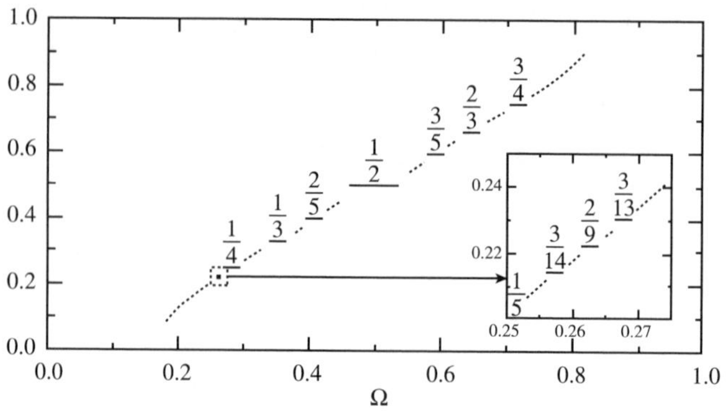

FIGURE 8.38 Devil's staircase for the circle map with $K = 1$. The steps indicate the regions in which w is constant. The staircase is self-similar in the sense that its structure is reproduced qualitatively at smaller scales (see inset). *Source*: J. A. Glazier and A. Libchaber, "Quasi-periodicity and dynamical systems: An experimentalist's view," *IEEE Trans. Circuits Syst.*, vol. 35, p. 793, July 1988. © 1988 IEEE.

If the amplitude K of the forcing is nonzero but less than unity, the steady-state is q-periodic when $\Omega = p/q$ is rational. In this case, a nonzero mode-locked window $[\Omega_{\min}(w), \Omega_{\max}(w)]$ occurs, over which $w = p/q$. A mode-locked region is delimited by saddle-node bifurcations at $\Omega_{\min}(w)$ and $\Omega_{\max}(w)$ [18].

The function $w(\Omega)$ in Figure 8.38 is monotone increasing and forms a **Devil's staircase** with plateaus at every rational value of w — for example, the step with winding number 1/2 is centered at $\Omega = 0.5$. An experimental Devil's staircase for the driven neon bulb circuit, with low-amplitude forcing, is shown in Figure 8.39.

As the amplitude k is increased, the width of each locked interval in the circle map increases so that mode-locking becomes more common and quasiperiodicity occurs over smaller ranges of driving frequencies. The corresponding (K, Ω) parameter space diagram (see Figure 8.40) consists of a series of distorted triangles, known as **Arnold Tongues**, with apexes that converge to rational values of Ω at $K = 0$.

Within a tongue, the winding number is constant, yielding one step of the Devil's staircase. The winding numbers of adjacent tongues are related by a *Farey tree* structure. Given two periodic windows with winding number $w_1 = p/q$ and $w_2 = r/s$, another periodic window with winding number $w = (\alpha p + \beta r)/(\alpha q + \beta s)$ can always be found, where p, q, r, and s are relatively prime and α and β are strictly positive integers. Furthermore, the widest mode-locked window between w_1 and w_2 has winding number $(p + r)/(q + s)$. For example, the widest step between those with winding numbers 1/2 and 2/3 in Figure 8.38 has $w = 3/5$.

The sum of the widths of the mode-locked states increases monotonically from zero at $K = 1$ to unity at $K = 1$. Below the critical line $K = 1$, the tongues bend away from each other and do not overlap. At $K = 1$, tongues begin to overlap, a kink appears in the Poincaré section and the Poincaré map develops a horseshoe; this produces coexisting attractors and chaos.

The transition to chaos as K is increased through $K = 1$ may be by a period-doubling cascade within a tongue, intermittency, or directly from a quasiperiodic trajectory by the abrupt disappearance of that trajectory (a blue sky catastrophe). This qualitative behavior is observed in van der Pol's neon bulb circuit.

Experimental Observations of Mode-Locking and Chaos in van der Pol's Neon Bulb Circuit

With the signal source E_s zeroed, the natural frequency of the undriven relaxation oscillator is set to 1 kHz by tuning capacitance C to C_0. A sinusoidal signal with frequency 1 kHz and amplitude E_0 is applied as shown in Figure 8.34. The resulting frequency of the current pulses (detected by measuring the voltage across R_s) is recorded with C as the bifurcation parameter.

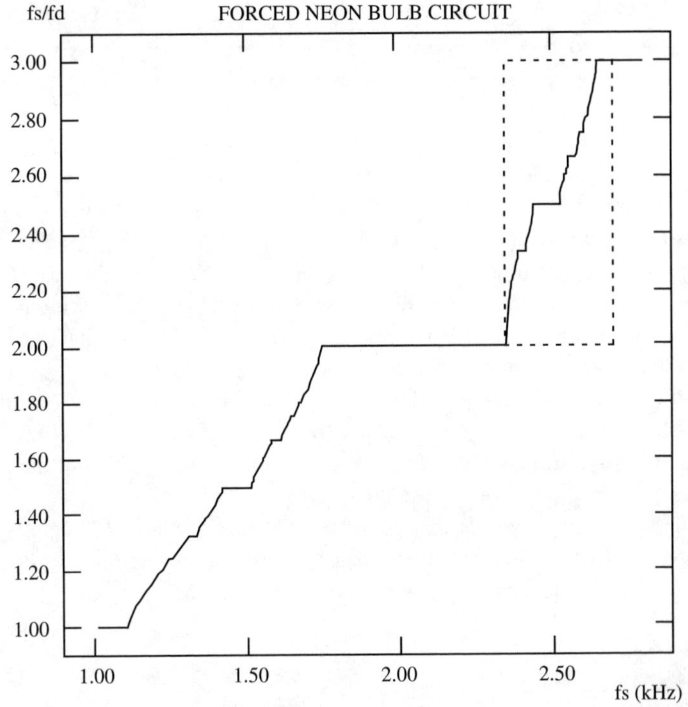

FIGURE 8.39 Experimentally measured staircase structure of lockings for a forced neon bulb relaxation oscillator. The winding number is given by f_s/f_d, the ratio of the frequency of the sinusoidal driving signal to the average frequency of current pulses through the bulb. *Source*: M. P. Kennedy, K. R. Krieg, and L. O. Chua, "The Devil's staircase: The electrical engineer's fractal," *IEEE Trans. Circuits Syst.*, vol. 36, p. 1137, Aug. 1989. © 1989 IEEE.

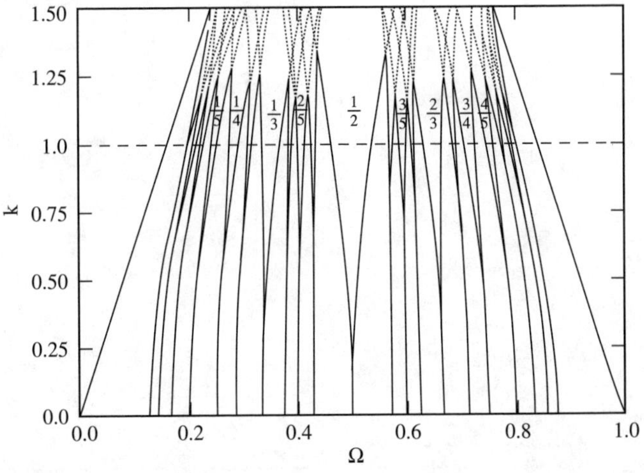

FIGURE 8.40 Parameter space diagram for the circle map showing Arnold tongue structure of lockings in the K–Ω plane. The relative widths of the tongues decrease as the denominator of the winding number increases. Below the critical line $K = 1$, the tongues bend away from each other and do not overlap; for $K > 1$, the Poincaré map develops a fold and chaos can occur. *Source*: J. A. Glazier and A. Libchaber, "Quasi-periodicity and dynamical systems: An experimentalist's view," *IEEE Trans. Circuits Syst.*, vol. 35, p. 793, July 1988. © IEEE.

C Bifurcation Sequence

If a fixed large-amplitude forcing E_s is applied and C is increased slowly, the system at first continues to oscillate and 1 kHz [Figure 8.36(a)] over a range of C, until the frequency "suddenly" drops to 1000/2 Hz [Figure 8.36(b)], stays at that value over an additional range of capacitance, drops to 1000/3 Hz, then 1000/4 Hz, 1000/5 Hz, and so on as far as 1000/20 Hz. These results are summarized in Figure 8.35.

Between each two submultiples of the oscillator driving frequency, a further rich structure of submultiples is found. At the macroscopic level (the coarse structure examined by van der Pol) increasing the value of C causes the system periods to step from T (1 ms) to $2T$, $3T$, $4T$, …, where the range of C for which the period is fixed is much greater than that over which the transitions occur (Figure 8.35).

Examining the shaded transition regions more closely, one finds that between any two "macroscopic" regions where the period is fixed at $(n - 1)T$ and nT $(n > 1)$, respectively, there lies a narrower region over which the system oscillates with stable period $(2n - 1)T$. Further, between $(n - 1)T$ and $(2n - 1)T$, one finds a region of C for which the period is $(3n - 2)T$, and between $(2n - 1)T$ and nT, a region with period $(3n - 1)T$. Indeed, between any two stable regions with periods $(n - 1)T$ and nT, respectively, a region with period $(2n - 1)T$ can be found. Figure 8.41 depicts an enlargement of the C axis in the region of the T to $2T$ macro-transition, showing the finer period-adding structure.

Between T and $2T$ is a region with stable period $3T$. Between this and $2T$, regions of periods $5T$, $7T$, $9T$, … up to $25T$ are detected. Current waveforms corresponding to period-3 and period-5 steps, with winding numbers 3/2 and 5/3, respectively, are shown in Figure 8.42.

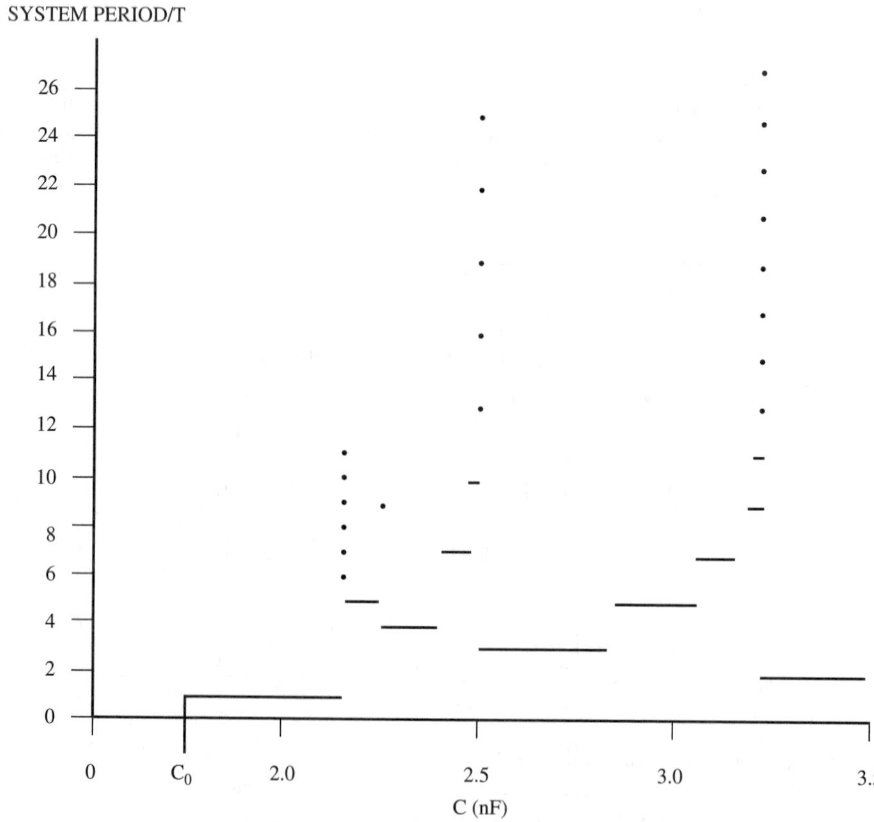

FIGURE 8.41 Experimental pulse pattern repetition rate versus C for van der Pol's forced neon bulb oscillator, showing fine period-adding structure. *Source*: M. P. Kennedy and L. O. Chua, "Van der Pol and chaos," *IEEE Trans, Circuits Syst.*, vol. CAS-33, p. 975, Oct. 1986 © 1986 IEEE.

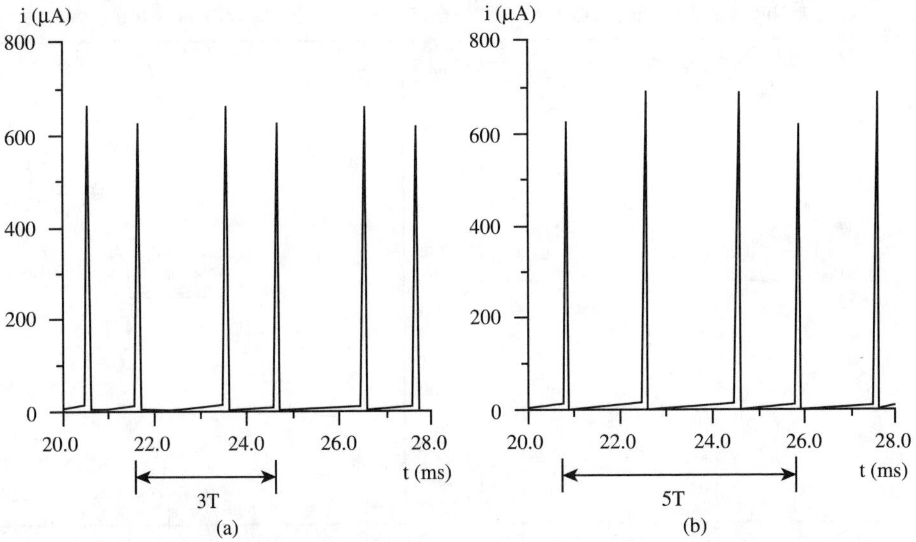

FIGURE 8.42 Neon bulb current waveforms. (a) Two pulses every three cycles of Ezi_s:$f_s/f_d = 3/2$; (b) three pulses every five cycles $f_s/f_d = 5/3$.

A region of period $4T$ lies between T and $3T$, with steps $7T$, $10T$, $13T$, ... up to $25T$ between that and $3T$. In practice, it becomes difficult to observe cycles with longer periods because stochastic noise in the experimental circuit can throw the solution out of the narrow window of existence of a high period orbit.

Experimental f_s Bifurcation Sequence with Low-Amplitude Forcing

An experimental Devil's staircase may be plotted for this circuit by fixing the parameters of the relaxation oscillator and the amplitude of the sinusoidal forcing signal, and choosing the forcing frequency f_s as the bifurcation parameter. The quantity f_s/f_d is the equivalent winding number in this case, where f_d is the average frequency of the current pulses through the neon bulb.

Experimental results for the neon bulb circuit with low-amplitude forcing are presented in Figures 8.39 and 8.43. The monotone staircase of lockings is consistent with a forcing signal of small amplitude. Note that the staircase is self-similar in the sense that its structure is reproduced qualitatively at smaller scales of the bifurcation parameter.

If the amplitude of the forcing is increased, the onset of chaos is indicated by a nonmonotonicity in the staircase.

Circuit Model

The experimental behavior of van der Pol's sinusoidally driven neon bulb circuit may be reproduced in simulation by an equivalent circuit (see Figure 8.44) in which the only nonlinear element (the neon bulb) is modeled by a nonmonotone current-controlled resistor with a series parasitic inductor L_p.

The corresponding state equations are

$$\frac{dV_C}{dt} = -\frac{1}{RC}V_C - \frac{1}{C}I_L + \frac{E}{RC}$$

$$\frac{dI_L}{dt} = \frac{1}{L_p}V_C - \frac{R_s}{L_p}I_L - \frac{f(I_L)}{L_p} - \frac{E_0\sin(2\pi f_s t)}{L_p}$$

where $V = f(I)$ is the driving-point characteristic of the current-controlled resistor (see Figure 8.45).

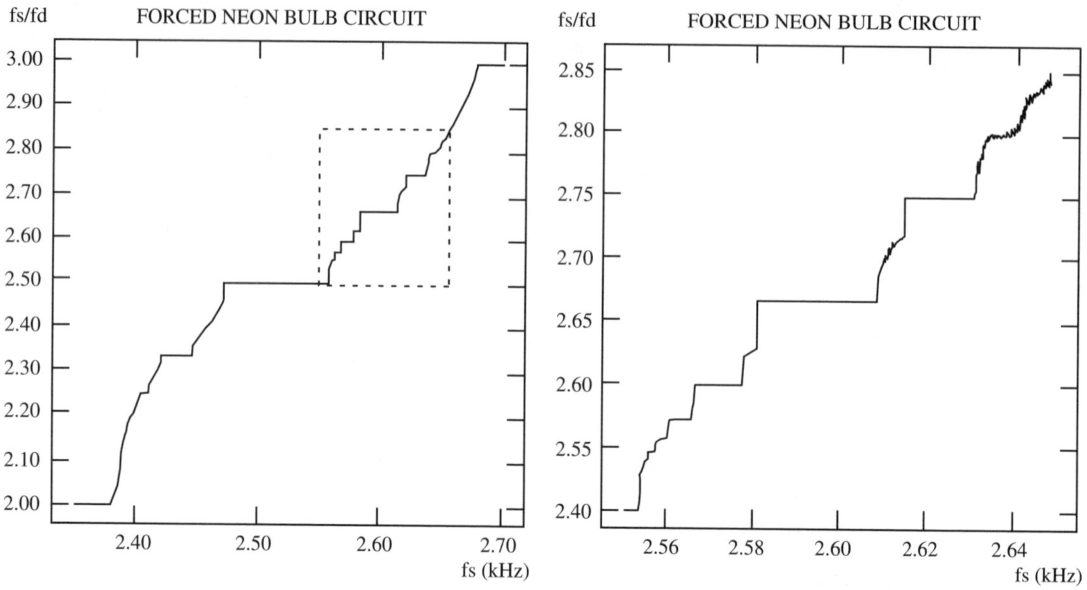

FIGURE 8.43 Magnification of Figure 8.39 showing self-similarity. *Source*: M. P. Kennedy, K. R. Krieg, and L. O. Chua, "The Devil's staircase: The electrical engineer's fractal," *IEEE Trans. Circuits Syst.*, vol. 36, p. 1137, Aug. 1989. © 1989 IEEE.

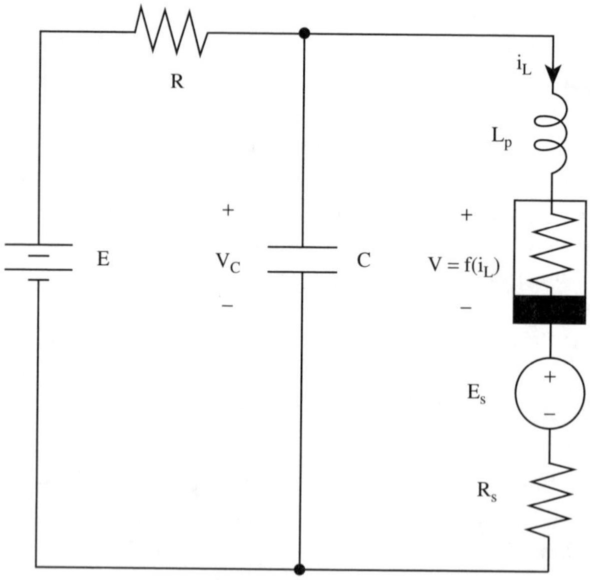

FIGURE 8.44 Van der Pol's neon bulb circuit — computer model. The bulb is modeled by a nonmonotonic current-controlled nonlinear resistor with parasitic transit inductance L_p.

8.5 Synchronization of Chaotic Circuits

Chaotic steady-state solutions are characterized by sensitive dependence on initial conditions; trajectories of two identical autonomous continuous-time dynamical systems started from slightly different initial

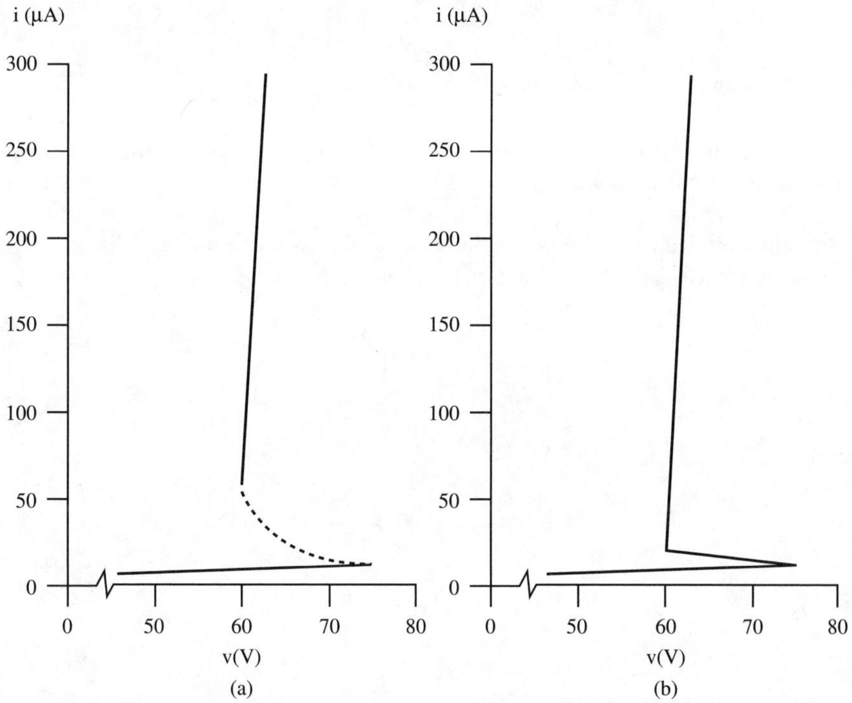

FIGURE 8.45 Neon bulb driving-point characteristics: (a) measured and (b) simulated. *Source*: M. P. Kennedy and L. O. Chua, "Van der Pol and chaos," *IEEE Trans. Circuit Syst.*, vol. CAS-33, p. 976, Oct. 1986. © 1986 IEEE.

conditions quickly become uncorrelated. Surprisingly perhaps, it is nevertheless possible to synchronize these systems in the sense that a trajectory of one asymptotically approaches that of the other. Two trajectories $X_1(t)$ and $X_2(t)$ are said to *synchronize* if

$$\lim_{t\to\infty} \|\mathbf{X}_1(t) - \mathbf{X}_2(t)\| = 0$$

In this section, we describe two techniques for synchronizing chaotic trajectories.

Linear Mutual Coupling

The simplest technique for synchronizing two dynamical systems

$$\begin{aligned} \dot{\mathbf{X}}_1 &= \mathbf{F}_1(\mathbf{X}_1) \qquad & \mathbf{X}_1(0) &= \mathbf{X}_{1_0} \\ \dot{\mathbf{X}}_2 &= \mathbf{F}_2(\mathbf{X}_2) \qquad & \mathbf{X}_2(0) &= \mathbf{X}_{2_0} \end{aligned}$$

is by linear mutual coupling of the form

$$\begin{aligned} \dot{\mathbf{X}}_1 &= \mathbf{F}_1(\mathbf{X}_1) + \mathbf{K}(\mathbf{X}_2 - \mathbf{X}_1) \qquad & \mathbf{X}_1(0) &= \mathbf{X}_{1_0} \\ \dot{\mathbf{X}}_2 &= \mathbf{F}_2(\mathbf{X}_2) + \mathbf{K}(\mathbf{X}_2 - \mathbf{X}_1) \qquad & \mathbf{X}_2(0) &= \mathbf{X}_{2_0} \end{aligned} \tag{8.22}$$

where $\mathbf{X}_1, \mathbf{X}_2 \in \mathbb{R}^n$ and K = diag$(K_{11}, K_{22}, \ldots, K_{nn})^T$.

Here, $\mathbf{X}_1(t)$ is called the goal dynamics. The synchronization problem may be stated as follows: find K such that

$$\lim_{t\to\infty}\left\|\mathbf{X}_1(t)-\mathbf{X}_2(t)\right\|=0$$

that is, that the solution $\mathbf{X}_2(t)$ synchronizes with the goal trajectory $\mathbf{X}_1(t)$.

In general, it is difficult to prove that synchronization occurs, unless an appropriate Lyapunov function[16] of the error system $\mathbf{E}(t) = \mathbf{X}_1(t) - \mathbf{X}_2(t)$ can be found. However, several examples exist in the literature where mutually coupled chaotic systems synchronize over particular ranges of parameters.

Example: Mutually Coupled Chua's Circuits. Consider a linear mutual coupling of two Chua's circuits. In dimensionless coordinates, the system under consideration is

$$\frac{dX_1}{dt}=\alpha\left[X_2-X_1-f(X_1)\right]+K_{11}(X_4-X_1)$$

$$\frac{dX_2}{dt}=X_1-X_2-X_3+K_{22}(X_5-X_2)$$

$$\frac{dX_3}{dt}=-\beta y+K_{33}(X_6-X_3)$$

$$\frac{dX_4}{dt}=\alpha\left(X_5-X_4-f(X_4)\right)+K_{11}(X_1-X_4)$$

$$\frac{dX_5}{dt}=X_4-X_5-X_6+K_{22}(X_2-X_5)$$

$$\frac{dX6}{dt}=-\beta X_5+K_{33}(X_3-X_6)$$

Two mutually coupled Chua's circuits characterized by $\alpha = 10.0$, $\beta = 14.87$, $a = -1.27$, and $b = -0.68$ will synchronize (the solutions of the two systems will approach each other asymptotically) for the following matrices K:

X_1 – **coupling** $K_{11} > 0.5$, $K_{22} = K_{33} = 0$
X_2 – **coupling** $K_{22} > 5.5$, $K_{11} = K_{33} = 0$
X_3 – **coupling** $0.7 < K_{33} < 2$, $K_{11} = K_{22} = 0$

Coupling between states X_1 and X_4 may be realized experimentally by connecting a resistor between the tops of the nonlinear resistors, as shown in Figure 8.46. States X_2 and X_5 may be coupled by connecting the tops of capacitors C_2 by means of a resistor.

An INSITE simulation of the system, which confirms synchronization of the two chaotic Chau's circuits in the case of linear mutual coupling between state V_{C_1} and V'_{C_1} is presented in Figure 8.47.

Pecora–Carroll Drive-Response Concept

The drive-response synchronization scheme proposed by Pecora and Carroll applies to systems that are drive-decomposable [16]. A dynamical system is called drive-decomposable if it can be partitioned into two subsystems that are coupled so that the behavior of the second (called the response subsystem) depends on that of the first, but the behavior of the first (called the drive subsystem) is independent of that of the second.

To construct a drive-decomposable system, an n-dimensional autonomous continuous-time dynamical system

[16]For a comprehensive exposition of Lyapunov stability theory, see [19].

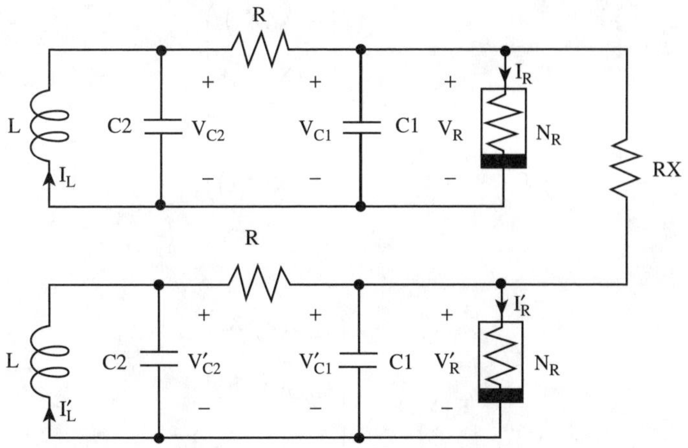

FIGURE 8.46 Synchronization of two Chua's circuits by means of resistive coupling between V_{C_1} and V'_{C_1}.

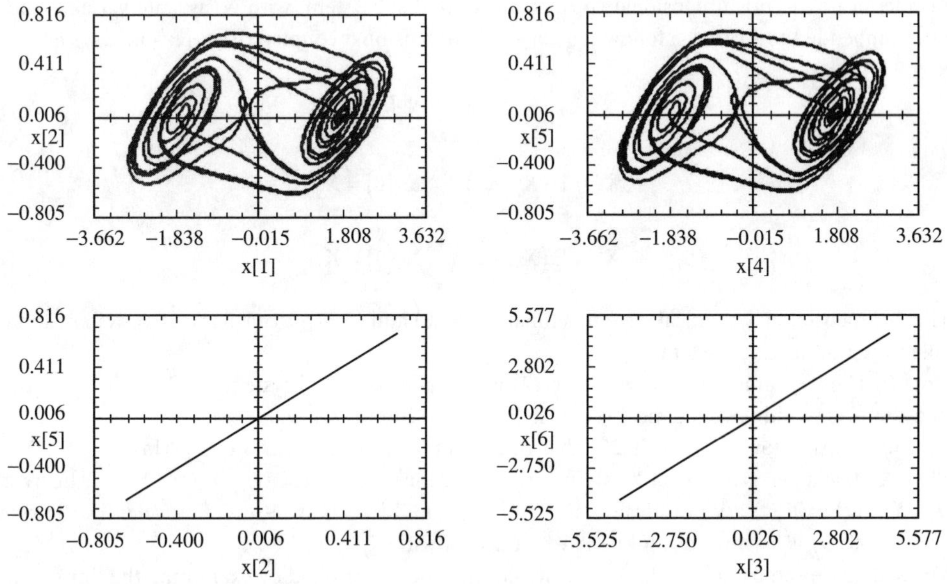

FIGURE 8.47 INSITE simulation of the normalized dimensionless form of Figure 8.46, illustrating synchronization by mutual coupling of state variables. Identify {$x[1]$, $x[2]$, $x[3]$} with {V_{C_1}, V_{C_2}, I_L} and {$x[4]$, $x[5]$, $x[6]$} with {V'_{C_1}, V'_{C_2}, I'_L} synchronizes with $V_{C_2}(t)$ and $I'_L(t)$ synchronizes with $I_L(t)$. *Source*: L. O. Chua, M. Itoh, L. Kočarev, and K. Eckert, "Chaos synchronization in Chua's Circuit," *J. Circuits Syst. Comput.*, vol. 3, no. 1, p. 99, Mar. 1993.

$$\dot{\mathbf{X}} = \mathbf{F}(\mathbf{X}) \qquad \mathbf{X}(0) = \mathbf{X}_0 \tag{8.23}$$

where $\mathbf{X} = (X_1, X_2, \ldots, X_n)^T$ and $\mathbf{F}(\mathbf{X}) = [F_1(\mathbf{X}), F_2(\mathbf{X}), \ldots, F_n(\mathbf{X})]^T$, is first partitioned into two subsystems

$$\dot{\mathbf{X}}_1 = \mathbf{F}_1(\mathbf{X}_1, \mathbf{X}_2) \qquad \mathbf{X}_1(0) = \mathbf{X}_{1_0} \tag{8.24}$$

$$\dot{\mathbf{X}}_2 = \mathbf{F}_2(\mathbf{X}_1, \mathbf{X}_2) \qquad \mathbf{X}_2(0) = \mathbf{X}_{2_0} \tag{8.25}$$

where $\mathbf{X}_1 = (X_1, X_2, \ldots, X_m)^T$, $\mathbf{X}_2 = (X_{m+1}, X_{m+2}, \ldots, X_n)^T$,

$$\mathbf{F}_1(\mathbf{X}_1,\mathbf{X}_2)=\begin{pmatrix} F_1(\mathbf{X}_1,\mathbf{X}_2) \\ F_2(\mathbf{X}_1,\mathbf{X}_2) \\ \vdots \\ F_m(\mathbf{X}_1,\mathbf{X}_2) \end{pmatrix}$$

and

$$\mathbf{F}_2(\mathbf{X}_1,\mathbf{X}_2)=\begin{pmatrix} F_{m+1}(\mathbf{X}_1,\mathbf{X}_2) \\ F_{m+2}(\mathbf{X}_1,\mathbf{X}_2) \\ \vdots \\ F_n(\mathbf{X}_1,\mathbf{X}_2) \end{pmatrix}$$

An identical $(n-m)$-dimensional copy of the second subsystem, with $\mathbf{X}_3$ as state variable and $\mathbf{X}_1$ as input, is appended to form the following $(2n-m)$-dimensional coupled drive-response system:

$$\dot{\mathbf{X}}_1=\mathbf{F}_1(\mathbf{X}_1,\mathbf{X}_2) \qquad \mathbf{X}_1(0)=\mathbf{X}_{1_0} \tag{8.26}$$

$$\dot{\mathbf{X}}_2=\mathbf{F}_2(\mathbf{X}_1,\mathbf{X}_2) \qquad \mathbf{X}_2(0)=\mathbf{X}_{2_0} \tag{8.27}$$

$$\dot{\mathbf{X}}_3=\mathbf{F}_2(\mathbf{X}_1,\mathbf{X}_3) \qquad \mathbf{X}_3(0)=\mathbf{X}_{3_0} \tag{8.28}$$

The n-dimensional dynamical system defined by (8.26) and (8.27) is called the drive system and (8.28) is called the response subsystem.

Note that the second drive subsystem (8.27) and the response subsystem (8.28) lie in state spaces of dimension $\mathbb{R}^{(n-m)}$ and have identical vector fields $\mathbf{F}_2$ and inputs $\mathbf{X}_1$.

Consider a trajectory $\mathbf{X}_3(t)$ of (8.29) that originates from an initial state $\mathbf{X}_{3_0}$ "close" to $\mathbf{X}_{2_0}$. We may think of $\mathbf{X}_2(t)$ as a perturbation of $\mathbf{X}_3(t)$. In particular, define the error $\mathbf{X}_3(t)=\mathbf{X}_2(t)-\mathbf{X}_3(t)$. The trajectory $\mathbf{X}_2(t)$ approaches $\mathbf{X}_3(t)$ asymptotically (synchronizes) if $\|\mathbf{X}_2\| \to 0$ as $t \to \infty$. Equivalently, the response subsystem (8.29) is asymptotically stable when driven with $\mathbf{X}_1(t)$.

The stability of an orbit of a dynamical system may be determined by examining the linearization of the vector field along the orbit. The linearized response subsystem is governed by

$$\dot{\mathbf{x}}_3=\mathbf{D}_{\mathbf{X}_3}\mathbf{F}_2(\mathbf{X}_1(t),\mathbf{X}_3)\mathbf{x}_3, \qquad \mathbf{x}_3(0)=\mathbf{x}_{3_0}$$

where $\mathbf{D}_{\mathbf{x}_3}\mathbf{F}_2(\mathbf{X}_1(t),\mathbf{X}_3)$ denotes the partial derivatives of the vector field $\mathbf{F}_2$ of the response subsystem with respect to $\mathbf{X}_3$. This is linear time-varying system whose state transition matrix $\Phi_t(\mathbf{X}_{1_0},\mathbf{X}_{3_0})$ maps a point $\mathbf{x}_{3_0}$ into $\mathbf{X}_3(t)$. Thus,

$$\mathbf{x}_3(t)=\Phi_t(\mathbf{X}_{1_0},\mathbf{X}_{3_0})\mathbf{x}_{3_0}$$

Note that Φ_t is a linear operator. Therefore, an $(n-m)$-dimensional ball $B_\varepsilon(\mathbf{X}_{3_0})$ of radius ε about $\mathbf{X}_{3_0}$ is mapped into an ellipsoid whose principal axes are determined by the singular values of Φ_t. In particular, a ball of radius ε is mapped by Φ_t into an ellipsoid, the maximum and minimum radii of which are bounded by the largest and smallest singular values, respectively, of Φ_t.

The conditional Lyapunov exponents λ_i ($\mathbf{X}_{1_0}$, $\mathbf{X}_{2_0}$) (hereafter denoted CLE) are defined by

$$\lambda_i\left(\mathbf{X}_{1_0},\mathbf{X}_{2_0}\right)=\lim_{t\to\infty}\frac{1}{t}\ln\sigma_i\left[\phi_t\left(\mathbf{X}_{1_0},\mathbf{X}_{2_0}\right)\right],\qquad i=1,2,\ldots,(n-m)$$

whenever the limit exists.

The term conditional refers to the fact that the exponents depend explicitly on the trajectory $\phi_t(\mathbf{X}_{1_0}, \mathbf{X}_{2_0})$ of the drive system.

Given that ε remains infinitesimally small, one of considering the local linearized dynamics along the flow determined by $\phi_t(\mathbf{X}_{1_0}, \mathbf{X}_{2_0})$ and determining the average local exponential rates of expansion and contraction along the principal axes of an ellipsoid. If all CLEs are negative, the response subsystem is asymptotically stable. A subsystem, where all the CLEs are negative, is called a stable subsystem.

A stable subsystem does not necessarily exhibit dc steady-state behavior. For example, while an asymptotically stable linear parallel *RLC* circuit has all negative LEs, the system settles to a periodic steady-state solution when driven with a sinusoidal current. Although the *RLC* subcircuit has negative CLEs in this case, the complete forced circuit has one nonnegative LE corresponding to motion along the direction of the flow.

Theorem 4 (Pecora and Carroll): *The trajectories* $\mathbf{X}_2(t)$ *and* $\mathbf{X}_3(t)$ *will synchronize only if the CLEs of the response system* (8.28) *are all negative.*

Note that this is a necessary but not sufficient condition for synchronization. If the response and second drive subsystems are identical and the initial conditions $\mathbf{X}_{2_0}$ and $\mathbf{X}_{3_0}$ are sufficiently close, and the CLEs of (8.28) are all negative, synchronization will occur. However, if the systems are not identical or the initial conditions are not sufficiently close, synchronization might not occur, even if all of the CLEs are negative.

Although we have described it only for an autonomous continuous-time system, the drive-response technique may also be applied for synchronizing nonautonomous and discrete-time circuits.

Cascaded Drive-Response Systems

The drive-response concept may be extended to the case where a dynamical system can be partitioned into more than two parts. A simple two-level drive-response cascade is constructed as follows. Divide the dynamical system

$$\dot{\mathbf{X}}=\mathbf{F}(\mathbf{X}),\quad \mathbf{X}(0)=\mathbf{X}_0 \tag{8.29}$$

into three parts:

$$\dot{\mathbf{X}}_1=\mathbf{F}_1(\mathbf{X}_1,\mathbf{X}_2,\mathbf{X}_3)\quad \mathbf{X}_1(0)=\mathbf{X}_{1_0} \tag{8.30}$$

$$\dot{\mathbf{X}}_2=\mathbf{F}_2(\mathbf{X}_1,\mathbf{X}_2,\mathbf{X}_3)\quad \mathbf{X}_2(0)=\mathbf{X}_{2_0} \tag{8.31}$$

$$\dot{\mathbf{X}}_3=\mathbf{F}_3(\mathbf{X}_1,\mathbf{X}_2,\mathbf{X}_3)\quad \mathbf{X}_3(0)=\mathbf{X}_{3_0} \tag{8.32}$$

Now, construct an identical copy of the subsystems corresponding to (8.31) and (8.32) with $\mathbf{X}_1(t)$ as input:

$$\dot{\mathbf{X}}_4=\mathbf{F}_2(\mathbf{X}_1,\mathbf{X}_4,\mathbf{X}_5)\quad \mathbf{X}_4(0)=\mathbf{X}_{4_0} \tag{8.33}$$

$$\dot{\mathbf{X}}_5=\mathbf{F}_3(\mathbf{X}_1,\mathbf{X}_4,\mathbf{X}_5)\quad \mathbf{X}_5(0)=\mathbf{X}_{5_0} \tag{8.34}$$

If all the CLEs of the driven subsystem composed of (8.33) and (8.34) are negative then, after the transient decays, $\mathbf{X}_4(t) = \mathbf{X}_2(t)$ and $\mathbf{X}_5(t) = \mathbf{X}_3(t)$.

Note that (8.30)–(8.34) together define one large coupled dynamical system. Hence, the response subsystem can exhibit chaos even if all of its CLEs are negative.

Proceeding one step further, we reproduce subsystem (8.30):

$$\dot{\mathbf{X}}_6 = \mathbf{F}_1(\mathbf{X}_6, \mathbf{X}_4, \mathbf{X}_5) \quad \mathbf{X}_6(0) = \mathbf{X}_{6_0} \tag{8.35}$$

As before, if all of the conditional Lyapunov exponents of (8.35) are negative, then $\|\mathbf{X}_6(t) - \mathbf{X}_1(t)\| \rightarrow 0$. If the original system could be partitioned so that (8.30) is one-dimensional, then using (8.33)–(8.35) as a driven response system, *all* of the variables in the drive system could be reproduced by driving with just *one* variable $X_1(t)$. This principle can be exploited for spread-spectrum communication using a chaotic carrier signal.

Example: Synchronization of Chua's Circuits Using the Drive-Response Concept. Chua's circuit may be partitioned in three distinct ways to form five-dimensional, drive-decomposable systems:

X_1-drive configuration

$$\frac{dX_1}{dt} = \alpha[X_2 - X_1 - f(X_1)]$$

$$\frac{dX_2}{dt} = X_1 - X_2 - X_3$$

$$\frac{dX_3}{dt} = -\beta X_2$$

$$\frac{dX_4}{dt} = X_1 - X_4 - X_5$$

$$\frac{dX_5}{dt} = -\beta X_5$$

With $\alpha = 10.0$, $\beta = 14.87$, $a = -1.27$, and $b = -0.68$, the CLEs for the (X_2, X_3) subsystem (calculated using INSITE) are (–0.5, –0.5).

X_2-drive configuration

$$\frac{dX_2}{dt} = X_1 - X_2 - X_3$$

$$\frac{dX_1}{dt} = \alpha[X_2 - X_1 - f(X_1)]$$

$$\frac{dX_2}{dt} = -\beta X_2$$

$$\frac{dX_4}{dt} = \alpha[X_2 - X_4 - f(X_4)]$$

$$\frac{dX_5}{dt} = -\beta X_5$$

This case is illustrated in Figure 8.48. The CLEs of the (X_1, X_3) subsystem are 0 and -2.5 ± 0.05. Because of the zero CLE, states $X_5(t)$ and $X_3(t)$ remain a constant distance $|X_{3_0} - X_{5_0}|$ apart, as depicted in Figure 8.49.

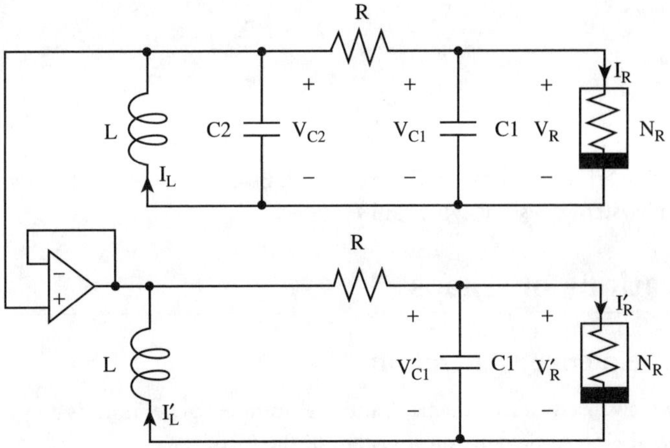

FIGURE 8.48 Synchronization of two Chua's circuits using the Pecora–Carroll drive-response method with V_{C_2} as drive variable.

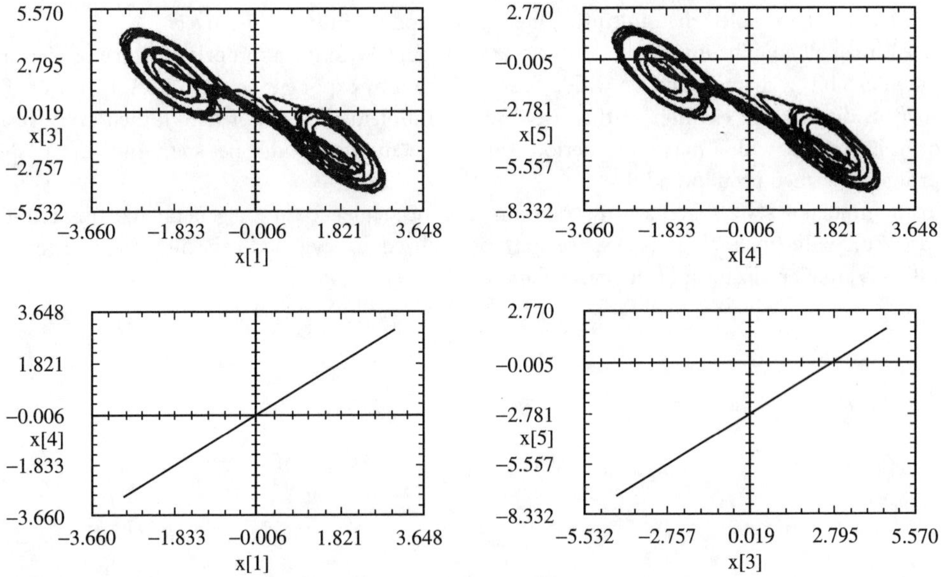

FIGURE 8.49 INSITE simulation of the normalized dimensionless form of Figure 8.48, illustrating synchronization of state variables. Identify {$x[1]$, $x[2]$, $x[3]$} with {V_{C_1}, V_{C_2}, I_L}, and {$x[4]$, $x[5]$} with {V'_{C_1}, I'_L} · $V'_{C_1}(t)$ synchronizes with $V'_{C_1}(t)$, and $I'_L(t)$ synchronizes with $I_L(t)$. Because one of the CLEs of the response subsystem is zero, the difference in the initial conditions $I_{L_0} - I_{L_0}$ does not decay to zero. *Source:* L. O. Chua, M. Itoh, L. Kočarev, and K. Eckert, "Chaos synchronization in Chua's circuits," *J. Circuits Syst. Comput.*, vol. 3, no. 1, p. 106, Mar. 1993.

X_3-drive configuration

$$\frac{dX_3}{dt} = -\beta X_2$$

$$\frac{dX_1}{dt} = \alpha\left[X_2 - X_1 - f(X_1)\right]$$

$$\frac{dX_2}{dt} = X_1 - X_2 - X_3$$

$$\frac{dX_4}{dt} = \alpha\left[X_5 - X_4 - f(X_4)\right]$$

$$\frac{dX_5}{dt} = X_4 - X_5 - X_3$$

The CLEs in this case are 1.23 ± 0.03 and –5.42 ± 0.02. Because the (X_1, X_2) subsystem has a positive CLE, the response subsystem does not synchronize.

8.6 Applications of Chaos

Pseudorandom Sequence Generation

One of the most widely used deterministic "random" number generators is the linear congruential generator, which is a discrete-time dynamical system of the form

$$X_{k+1} = (AX_k + B) \bmod M, \qquad k = 0, 1, \ldots \tag{8.36}$$

where A, B, and M are called the **multiplier**, **increment**, and **modulus**, respectively.

If $A > 1$, then all equilibrium points of (8.36) are unstable. With the appropriate choice of constants, this system exhibits a chaotic solution with a positive Lyapunov exponent equal to In A. However, if the state space is discrete, for example in the case of digital implementations of (8.37), then every steady-state orbit is periodic with a maximum period equal to the number of distinct states in the state space; such orbits are termed pseudorandom.

By using in analog state space, a truly "random" chaotic sequence can be generated. A discrete-time chaotic circuit with an analog state space may be realized in switched-capacitor (SC) technology. Figure 8.50 is an SC realization of the parabolic map

$$x_{k+1} = V - 0.5x_k^2 \tag{8.37}$$

which, by the change of variables

$$X_k = Ax_k + B$$

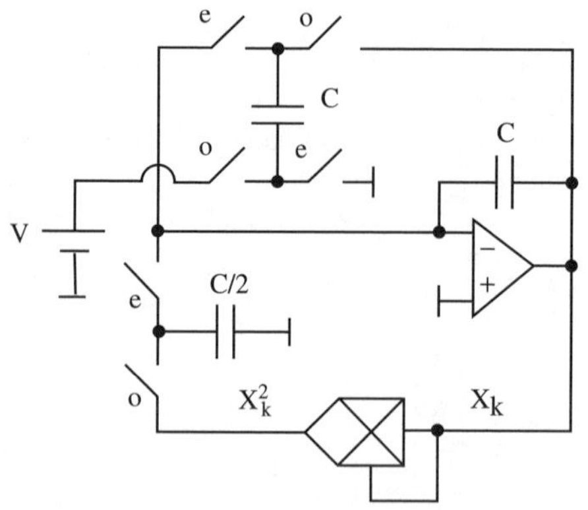

FIGURE 8.50 Switched-capacitor (SC) realization of the parabolic map $x_{k+1} = V - 0.5\,X_k^2$. The switches labeled o and e are driven by the odd and even phases, respectively, of a nonoverlapping two-phase clock.

with $\mu = 1/(2A)$, $B = 0.5$, and $A = (-1 \pm \sqrt{1+2V})/(4V)$, is equivalent to the logistic map

$$X_{k+1} = \mu X_k (1 - X_k) \tag{8.38}$$

The logistic map is chaotic for $\mu = 4$ with Lyapunov exponent ln 2 [18]. Figure 8.16 is a bifurcation diagram of (8.38) with $0 \le \mu \le 4$.

For $V < 1.5$, the steady-state solution of the SC parabolic map described by (8.37) is a fixed point. As the bifurcation parameter V is increased from 1.5 to 3 V, the circuit undergoes a series of period-doubling bifurcations to chaos. $V = 4$ corresponds to fully developed chaos on the open interval $(0 < X_k < 1)$ in the logistic map with $\mu = 4$.

Spread-Spectrum and Secure Communications

Modulation and coding techniques for mobile communication systems are driven by two fundamental requirements: that the communication channel should be secure and the modulation scheme should be tolerant of multipath effects. Security is ensured by coding and immunity from multipath degradation may be achieved by using a spread-spectrum transmission.

With appropriate modulation and demodulation techniques, the "random" nature and "noise-like" spectral properties of chaotic circuits can be exploited to provide simultaneous coding and spreading of a transmission.

Chaotic Switching

The simplest idea for data transmission using a chaotic circuit is use the data to modulate some parameter(s) of the transmitter. This technique is called **parameter modulation**, **chaotic switching**, or **chaos shift keying** (CSK).

In the case of binary data, the information signal is encoded as a pair of circuit parameter sets which produce distinct attractors in a dynamical system (the transmitter). In particular, a single control parameter μ may be switched between two values μ_0, corresponding to attractor $\mathcal{A}_0$ and μ_1, corresponding to $\mathcal{A}_1$. By analogy with FSK and PSK, this technique is known as chaos shift keying.

The binary sequence to be transmitted is mapped into the appropriate control signal $\mu(t)$ and the corresponding trajectory switches, as required, between $\mathcal{A}_1$ and $\mathcal{A}_0$.

One of the state variables of the transmitter is conveyed to the receiver, where the remaining state variables are recovered by drive-response synchronization. These states are then applied to the second stage of a drive-response cascade.

At the second level, two matched receiver subsystems are constructed, one of which synchronizes with the incoming signal if a "zero" was transmitted, the other of which synchronizes only if a "one" was transmitted. The use of two receiver circuits with mutually exclusive synchronization properties improves the reliability of the communication system.

Chaos shift keying has been demonstrated both theoretically and experimentally. Figure 8.51 depicts a CSK transmitter and receiver based on Chua's circuit. The control parameter is a resistor with conductance ΔG whose effect is to modulate the slopes G_a and G_b of the Chua diode. Switch S is opened and closed by the binary data sequence and V_{C_1} is transmitted. At the receiver, the first subsystem (a copy of the (V_{C_2}, I_L) subsystem of the transmitter) synchronizes with the incoming signal, recovering $V_{C_2}(t)$. Thus, $V_{C_{21}}(t) \to V_{C_2}(t)$.

The synchronized local copy of $V_{C_1}(t)$ is then used to synchronize two further subsystems corresponding to the V_{C_1} subsystem of the transmitter with and without the resistor ΔG.

If the switch is closed at the transmitter, V'_{C12} (but not V_{C12}) synchronizes with V_{C_1} and if the switch is open, V_{C12} (but not V'_{C12}) synchronizes with V_{C_1}.

Figure 8.52 presents simulated results for a similar system consisting of two Chua's circuits. At the receiver, a decision must be made as to which bit has been transmitted. In this case, b_{out} was derived using the rule

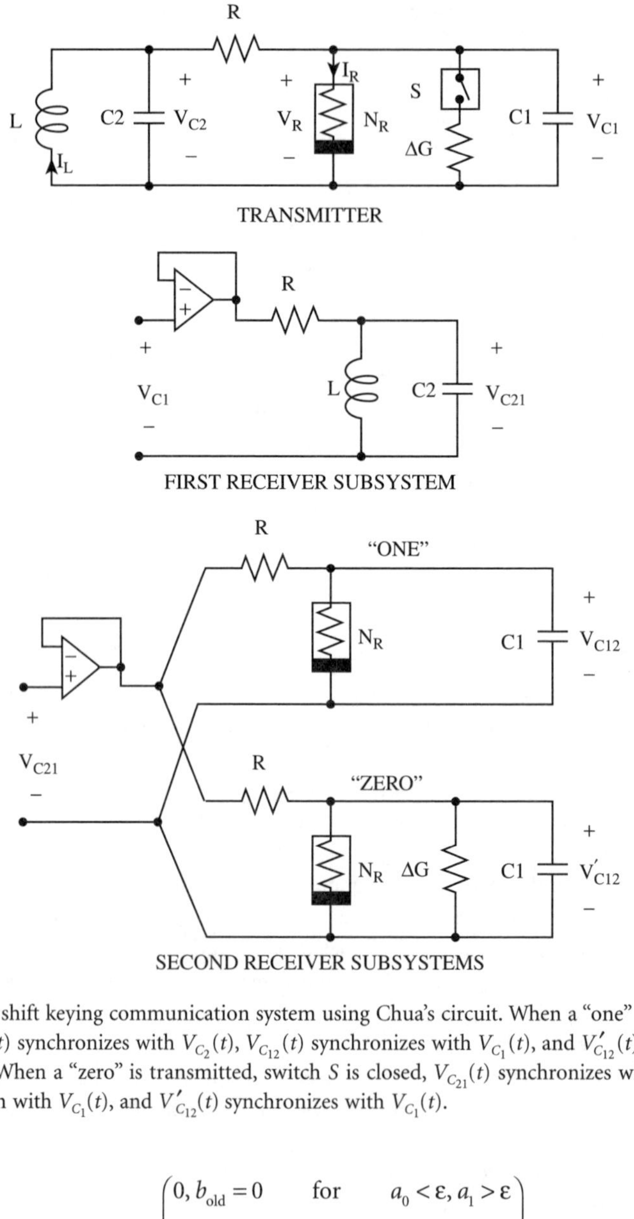

FIGURE 8.51 Chaos shift keying communication system using Chua's circuit. When a "one" is transmitted, switch S remains open, $V_{C_{21}}(t)$ synchronizes with $V_{C_2}(t)$, $V_{C_{12}}(t)$ synchronizes with $V_{C_1}(t)$, and $V'_{C_{12}}(t)$ falls out of synchronization with $V_{C_1}(t)$. When a "zero" is transmitted, switch S is closed, $V_{C_{21}}(t)$ synchronizes with $V_{C_2}(t)$, $V_{C_{12}}(t)$ falls out of synchronization with $V_{C_1}(t)$, and $V'_{C_{12}}(t)$ synchronizes with $V_{C_1}(t)$.

$$b_{\text{out}} = \begin{cases} 0, b_{\text{old}} = 0 & \text{for} \quad a_0 < \varepsilon, a_1 > \varepsilon \\ 1, b_{\text{old}} = 1 & \text{for} \quad a_0 > \varepsilon, a_1 < \varepsilon \\ b_{\text{old}} & \text{for} \quad a_0 < \varepsilon, a_1 < \varepsilon \\ 1 - b_{\text{old}} & \text{for} \quad a_0 > \varepsilon, a_1 > \varepsilon \end{cases}$$

where b_{old} is the last bit received and b_{out} is the current bit [14].

Chaotic Masking

Chaotic masking is a method of hiding an information signal by adding it to a chaotic carrier at the transmitter. The drive-response synchronization technique is used to recover the carrier at the receiver.

Figure 8.53 is a block diagram of a communication system using matched Chua's circuits. The receiver has the same two-layer structure as in the previous example. The first subcircuit, which has very negative

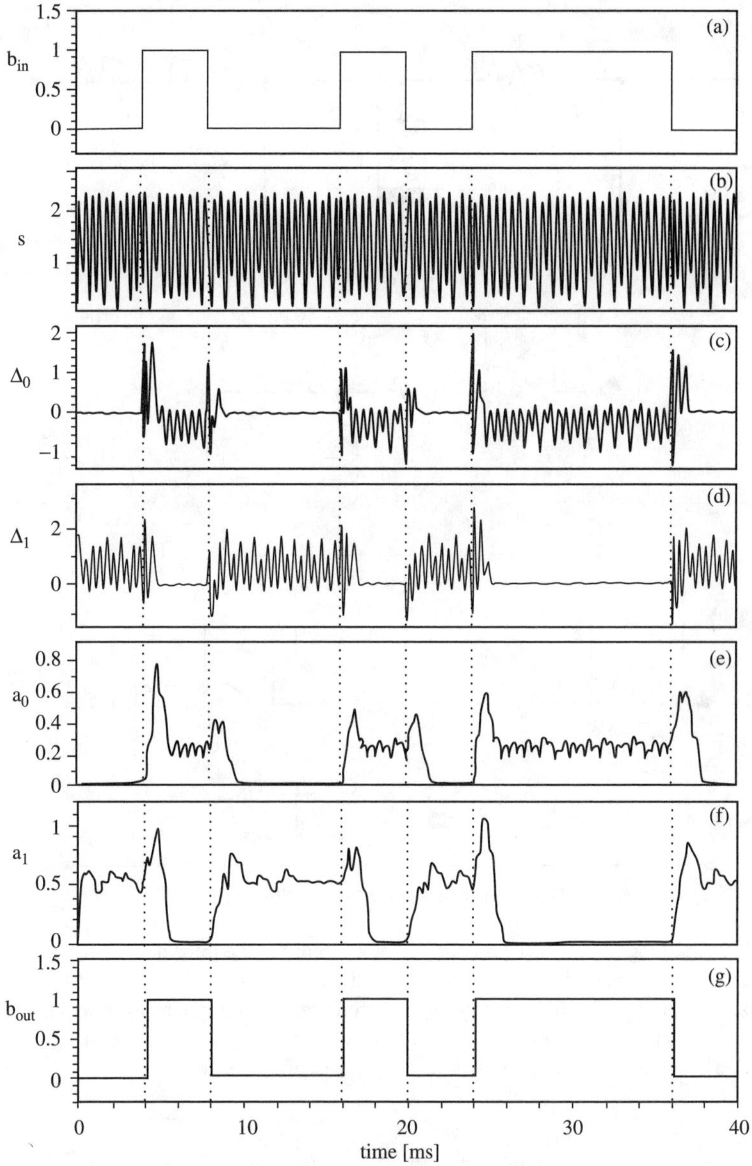

FIGURE 8.52 Chaos shift keying waveforms. (a) Binary input signal b_{in}; (b) transmitted signal $s(t)$;(c) response $\Delta_0 = V_{C_{12}} - V_{C_1}$; (d) response $\Delta_0 = V'_{C_{12}} - V_{C_1}$; (e) 40-point moving average of Δ_0; (f) 40-point moving average of Δ_1; (g) output binary signal b_{out} when $\varepsilon = 0.1$. *Source*: M. Ogorzalek, "Taming chaos — Part I: Synchronization," *IEEE Trans. Circuits Syst. I*, vol. 40, p. 696, Oct. 1993. © 1993 IEEE.

conditional Lyapunov exponents, synchronizes with the incoming signal, despite the perturbation $s(t)$, and recovers V_{C_2}. The second subcircuit, when driven by V_{C_2}, produces the receiver's copy of V_{C_1}. The information signal $r(t)$ is recovered by subtracting the local copy of V_{C_1} from the incoming signal $V_{C_1}+s(t)$.

Vector Field Modulation

With *vector field modulation*, an information-carrying signal is added to the vector field at the transmitter and recovered at the receiver [9].

A dynamical system is partitioned as follows:

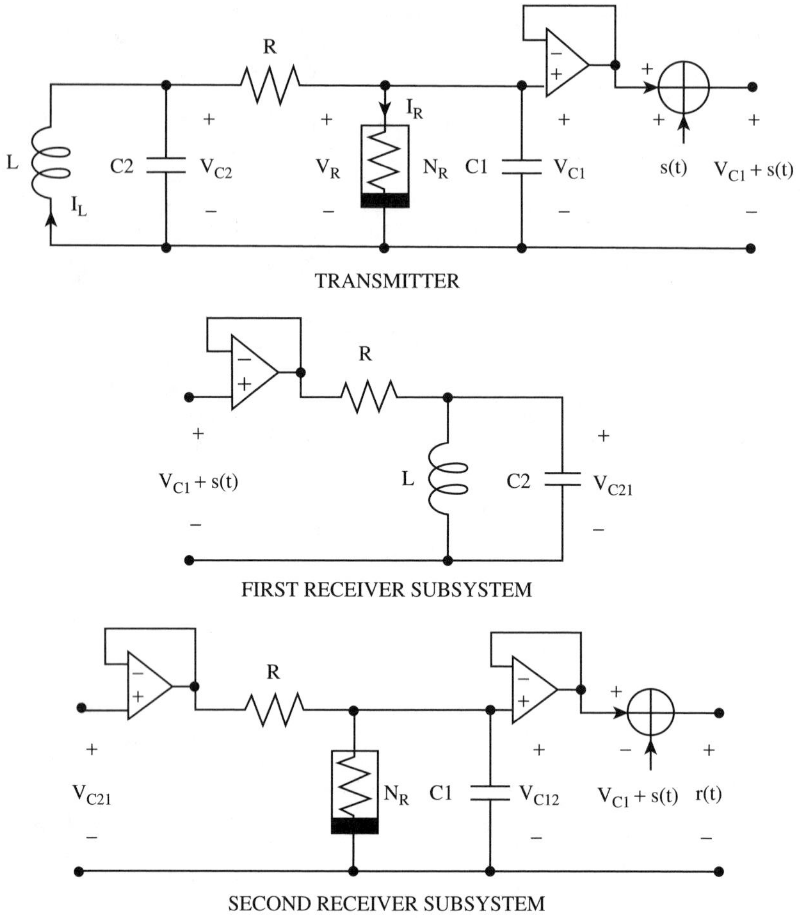

FIGURE 8.53 Chaos masking using Chua's circuits. At the transmitter, the information signal $s(t)$ is added to the chaotic carrier signal $V_{C_1}(t)$. Provided $s(t)$ is sufficiently small and the first receiver subsystem is sufficiently stable, $V_{C_{21}}(t)$ synchronizes with $V_{C_2}(t)$. This signal is applied to a second receiver subsystem, from which an estimate $V_{C_{12}}(t)$ of the unmodulated carrier $V_{C_1}(t)$ is derived. $V_{C_{12}}$ is subtracted from the incoming signal $V_{C_1} + s(t)$ to yield the received signal $r(t)$. The method works well only if $s(t)$ is much smaller than $V_{C_1}(t)$.

$$\dot{\mathbf{X}}_1 = \mathbf{F}_1(\mathbf{X}_1, \mathbf{X}_2) \qquad \mathbf{X}_1(0) = \mathbf{X}_{1_0}$$

$$\dot{\mathbf{X}}_2 = \mathbf{F}_2(\mathbf{X}_1, \mathbf{X}_2) \qquad \mathbf{X}_2(0) = \mathbf{X}_{2_0}$$

The information signal $\mathbf{S}(t)$ is added into the transmitter

$$\dot{\mathbf{X}}_1 = \mathbf{F}_1(\mathbf{X}_1, \mathbf{X}_2) + \mathbf{S}(t) \qquad \mathbf{X}_1(0) = \mathbf{X}_{1_0} \tag{8.39}$$

$$\dot{\mathbf{X}}_2 = \mathbf{F}_2(\mathbf{X}_1, \mathbf{X}_2) \qquad \mathbf{X}_2(0) = \mathbf{X}_{2_0} \tag{8.40}$$

and the state $\mathbf{X}_1(t)$ transmitted.

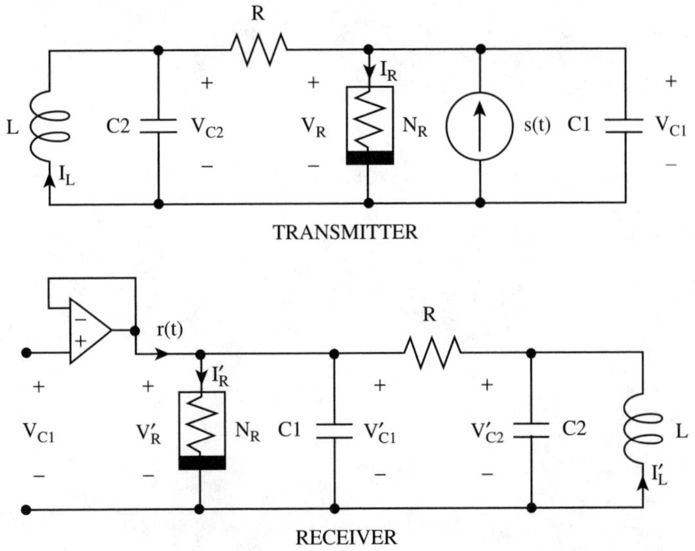

FIGURE 8.54 Vector field modulation using Chua's circuit. The signal s(t) to be transmitted is applied as perturbation of the vector field at the transmitter. The receiver's (V'_{C_2}, I'_L) subsystem synchronizes with the corresponding subsystem at the transmitter. This synchronized subsystem drives the V'_{C_1} subsystem. The transmitted signal $s(t)$ is recovered as an excess current $r(t)$ at the receiver.

The receiver contains a copy of the second drive subsystem (8.41)

$$\dot{\mathbf{X}}_3 = \mathbf{F}_2(\mathbf{X}_1, \mathbf{X}_3) \qquad \mathbf{X}_3(0) = \mathbf{X}_{3_0}$$

where the state $\mathbf{X}_3(t)$ synchronizes with $\mathbf{X}_2(t)$, and a demodulator,

$$\mathbf{R}(t) = \dot{\mathbf{X}}_1 - \mathbf{F}_1(\mathbf{X}_1, \mathbf{X}_3)$$

which is used to recover $\mathbf{S}(t)$.

If all of the CLEs of the response system are negative, and the initial conditions are sufficiently close, then $\mathbf{X}_3(t) \rightarrow \mathbf{X}_2(t)$ and the recovered signal $\mathbf{R}(t)$ equals $\mathbf{S}(t)$.

Example: Communication via Vector Field Modulation Using Chua's Circuits

A communication system based on Chua's circuit that uses the vector field modulation technique and a chaotic carrier is illustrated in Figure 8.54. In this case, the signal to be transmitted is the scalar current $s(t)$, which is recovered as a current $r(t)$ at the receiver.

The state equations of the coupled drive-response system are:

$$C_1 \frac{dV_{C_1}}{dt} = G\left(V_{C_2} - V_{C_1}\right) - f\left(V_{C_1}\right) + s(t)$$

$$C_2 \frac{dV_{C_2}}{dt} = G\left(V_{C_1} - V_{C_2}\right) + I_L$$

$$L \frac{dI_L}{dt} = -V_{C_2}$$

$$C_2 \frac{dV'_{C_2}}{dt} = G\left(V_{C_1} - V'_{C2}\right) + I'_L$$

$$L \frac{dI'_L}{dt} = -V'_{C_2}$$

Because the CLEs of the (V_{C_2}, I_L) subsystem are negative, $V'_{C_2}(t) \rightarrow V_{C_2}(t)$ and $I'_L(t) \rightarrow I_L(t)$. The current r(t) at the receiver is given by

$$\begin{aligned} r(t) &= C_1 \frac{dV_{C_1}}{dt} - G\left(V'_{C_2} - V_{C_1}\right) - f\left(V_{C_1}\right) \\ &= C_1 \frac{dV_{C_1}}{dt} - G\left(V_{C_2} - V_{C_1}\right) - f\left(V_{C_1}\right) \\ &= s(t) \end{aligned}$$

Miscellaneous

Chaotic circuits may also be used for suppressing spurious tones in ΣΔ modulators, for modeling musical instruments, fractal pattern generation, image-processing, and pattern recognition [3].

A chaotic attractor contains an infinite number of unstable periodic trajectories of different periods. Various control schemes for stabilizing particular orbits in chaotic circuits have been successfully demonstrated [14].

References

[1] F. M. Callier and C. A. Desoer, *Linear System Theory*, New York: Springer-Verlag, 1991.

[2] L. O. Chua, C. A. Desoer, and E. S. Kuh, *Linear and Nonlinear Circuits*, New York: McGraw-Hill, 1987.

[3] L. O. Chua and M. Hasler, Eds., *Special Issue on Chaos in Nonlinear Electronic Circuits*, Part A: Tutorials and Reviews, *IEEE Trans. Circuits Syst. I, Fundament. Theory Applicat.*, vol. 40, Oct. 1993; Part B: Bifurcation and Chaos, IEEE *Trans. Circuits Syst. I, Fundament. Theory Applicat.*, vol. 40, Nov. 1993; Part C: Applicat., *IEEE Trans. Circuits Syst. II, Analog and Digital Signal Process.*, vol. 40, Oct. 1993.

[4] L. O. Chua, M. Itoh, L. Kočarev, and K. Eckert, "Chaos synchronization in Chua's circuit," *J. Circuits Syst. Comput.*, vol. 3, no. 1, pp. 93–108, Mar. 1993.

[5] L. O. Chua, C. W. Wu, A. Huang, and G.-Q. Zhong, "A universal circuit for studying and generating chaos — Part I: Routes to chaos," *IEEE Trans. Circuits Syst. I, Fundamental Theory Applicat.*, vol. 40, pp. 732–744, Oct. 1933.

[6] J. A. Glazier and A. Libchaber, "Quasi-periodically and dynamical systems: An experimentalist's view," *IEEE Trans. Circuits Syst.*, vol. 35, pp. 790–809, July 1988.

[7] J. Guckenheimer and P. Holmes, *Nonlinear Oscillations, Dynamical Systems, and Bifurcations of Vector Fields*, New York: Springer-Verlag, 1983.

[8] M. W. Hirsch, "The dynamical systems approach to differential equations," *Bull. Amer. Math. Soc.*, vol. 11, no. 1, pp. 1–64, July 1984.

[9] M. Itoh, H. Murakami, K. S. Halle, and L. O. Chua, "Transmission of signals by chaos synchronization," IEICE Tech. Rep., CAS93-39, NLP93-27, pp. 89–96, 1993.

[10] M. P. Kennedy, "Three steps to chaos — Part I: Evolution," *IEEE Trans. Circuits and Systems I, Fundament. Theory Applicat.*, vol. 40, pp. 640–656, Oct. 1993; "Three steps to chaos — Part II: A Chua's circuit primer," *IEEE Trans. Circuits Syst. I, Fundament. Theory Applicat.*, vol. 40, pp. 657–674, Oct. 1993.

[11] M. P. Kennedy and L. O. Chua, "Van der Pol and chaos," *IEEE Trans. Circuits Syst.*, vol. CAS-33, pp. 974–980, Oct. 1986.

[12] M. P. Kennedy, K. R. Krieg, and L. O. Chua, "The Devil's staircase: The electrical engineer's fractal," *IEEE Trans. Circuits Syst.*, vol. 36, pp. 1133–1139, 1989.
[13] W. Lauterborn and U. Parlitz, "Methods of chaos physics and their application to acoustics," *J. Acoust. Soc. Amer.*, vol. 84, no. 6, pp. 1975–1993, Dec. 1988.
[14] M. Ogarzalek, "Taming chaos — Part I: Synchronization," *IEEE Trans. Circuits Syst. I, Fundament. Theory Applicat.*, vol. 40, pp. 693–699, Oct. 1993; "Taming chaos — Part II: Control," *IEEE Trans. Circuits Syst. I, Fundament. Theory Applicat.*, vol. 40, pp. 700–706, Oct. 1993.
[15] T. S. Parker and L. O. Chua, *Practical Numerical Algorithms for Chaotic Systems*, New York: Springer-Verlag, 1989.
[16] L. M. Pecora and T. Carroll, "Driving systems with chaotic signals," *Phys. Rev.*, vol. 44, no. 4, pp. 2374–2383, Aug. 15, 1991.
[17] W. H. Press, B. P. Flannery, S. A. Teukolsky, and W. T. Vetterling, *Numerical Recipes in C*, Cambridge: Cambridge Univ., 1988.
[18] J. M. T. Thompson and H. B. Stewart, *Nonlinear Dynamics and Chaos*, New York: Wiley, 1986.
[19] M. Vidyasagar, *Nonlinear Systems Analysis*, Englewood Cliffs, NJ: Prentice-Hall, 1978.
[20] C. W. Wu and N. F. Rul'kov, "Studying chaos via 1-D maps — A tutorial," *IEEE Trans. Circuits Syst. I, Fundament. Theory Applicat.*, vol. 40, pp. 707–721, Oct. 1993.
[21] L. O. Chua, M. Komuro, and T. Matsumoto, "The Double Scroll Family, Parts I and II," *IEEE Trans. Circuits Syst.*, vol. 33, pp. 1073–1118, Nov. 1986.

Further Information

Current Research in Chaotic Circuits. The August 1987 issue of the *Proceedings of the IEEE* is devoted to "Chaotic Systems." The *IEEE Transactions on Circuits and Systems*, July 1988, focuses on "Chaos and Bifurcations of Circuits and Systems."

A three-part special issue of the *IEEE Transactions on Circuits and Systems* on "Chaos in Electronic Circuits" appeared in October (parts I and II) and November 1993 (part I). This Special Issue contains 42 papers on various aspects of bifurcations and chaos.

Two Special Issues (March and June 1993) of the *Journal of Circuits, Systems and Computers* are devoted to "Chua's Circuit: A Paradigm for Chaos." These works, along with several additional papers, a pictorial guide to forty-five attractors in Chua's oscillator, and the ABC simulator, have been compiled into a book of the same name — *Chua's Circuit: A Paradigm for Chaos,* R. N. Madan, Ed. Singapore: World Scientific, Singapore 1993.

Developments in the field of bifurcations and chaos, with particular emphasis on the applied sciences and engineering, are reported in *International Journal of Bifurcation and Chaos,* which is published quarterly by World Scientific, Singapore 9128.

Research in chaos in electronic circuits appears regularly in the *IEEE Transactions on Circuits and Systems.*

Simulation of Chaotic Circuits. A variety of general-purpose and custom software tools has been developed for studying bifurcations and chaos in nonlinear circuits systems.

ABC (Adventures in Bifurcations and Chaos) is a graphical simulator of Chua's oscillator, which runs on IBM-compatible PCs. ABC contains a database of component values for all known attractors in Chua's oscillator, initial conditions, and parameter sets corresponding to homoclinic and heteroclinic trajectories, and bifurcation sequences for the period-doubling, intermittency, and quasiperiodic routes to chaos. The program and database are available from

Dr. Michael Peter Kennedy
Department of Electronic and Electrical Engineering
University College Dublin
Dublin 4, Ireland

E-mail: mpk@midir.ucd.ie

In "Learning about Chaotic Circuits with SPICE," *IEEE Transactions on Education*, vol. 36, pp. 28–35, Jan. 1993, David Hamill describes how to simulate a variety of smooth chaotic circuits using the general-purpose circuit simulator SPICE. A commercial variant of SPICE, called PSpice, is available from

MicroSim Corporation
20 Fairbanks
Irvine, CA 92718

Telephone: (714) 770-3022

The free student evaluation version of this program is sufficiently powerful for studying simple chaotic circuits. PSpice runs on both workstations and PCs.

INSITE is a software toolkit that was developed at the University of California, Berkeley, for studying continuous-time and discrete-time dynamical systems. The suite of nine programs calculates and displays trajectories, power spectra, and state delay maps, draws vector fields of two-dimensional systems, reconstructs attractors from time series, calculates Poincaré sections, dimension, and Lyapunov exponents. The package runs on DEC, HP, IBM, and Sun workstations under UNIX, and on IBM-compatible PCs under DOS. For additional information, contact

INSITE Software
P.O. Box 9662
Berkeley, CA 94709

Telephone: (510) 530-9259

9

Transmission Lines

T. K. Ishii
Marquette University, Wisconsin

9.1 Generic Relations

Equivalent Circuit

The equivalent circuit of a generic transmission line in monochromatic single frequency operation is shown in Fig. 9.1 [1–3], where $\dot{Z}$ is a series impedance per unit length of the transmission line (Ω/m) and $\dot{Y}$ is a shunt admittance per unit length of the transmission line (S/m). For a uniform, nonlinear transmission line, either $\dot{Z}$ or $\dot{Y}$ or both $\dot{Z}$ and $\dot{Y}$ are functions of the transmission line voltage and current, but both $\dot{Z}$ and $\dot{Y}$ are not functions of location on the transmission line. For a nonuniform, linear transmission line, both $\dot{Z}$ and $\dot{Y}$ are functions of location, but not functions of voltage and current on the transmission line. For a nonuniform and nonlinear transmission line, both $\dot{Z}$ and $\dot{Y}$ or $\dot{Z}$ or $\dot{Y}$ are functions of the voltage, the current, and the location on the transmission line.

Transmission Line Equations

Ohm's law of a transmission line, which is the amount of voltage drop on a transmission line per unit distance of voltage transmission is expressed as

$$\frac{d\dot{V}}{dz} = -\dot{I}\dot{Z} \quad \left(\text{V/m}\right) \tag{9.1}$$

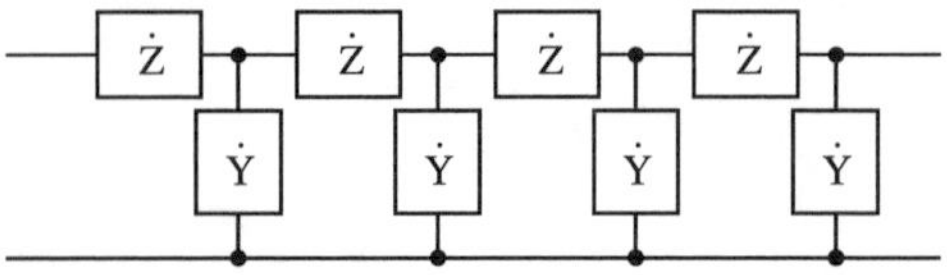

FIGURE 9.1 Equivalent circuit of a generic transmission line.

where $\dot{V}$ is the transmission line voltage (Volt), $\dot{I}$ is the transmission line current (Ampere), Z is the series impedance per unit length of the transmission line (Ω), and z is a one-dimensional coordinate placed in parallel to the transmission line (meter). The equation of current decrease per unit length is

$$\frac{d\dot{I}}{dz} = -\dot{Y}\dot{V} \quad (\text{A/m}) \tag{9.2}$$

where $\dot{Y}$ is the shunt admittance per unit distance of the transmission line (S/m).

Combining (9.1) and (9.2), the Telegrapher's equation or Helmholtz's wave equation for the transmission line voltage is [1–3].

$$\frac{d^2\dot{V}}{dz^2} - \dot{Z}\dot{Y}\dot{V} = 0 \quad (\text{V/m})^2 \tag{9.3}$$

The Telegrapher's equation or Helmholtz's wave equation for the transmission line current is [1–3]

$$\frac{d^2\dot{I}}{dz^2} - \dot{Z}\dot{Y}\dot{I} = 0 \quad (\text{A/m}^2) \tag{9.4}$$

General Solutions and Propagation Constant

The general solution of the Telegrapher's equation for transmission line voltage is [1–3]

$$\dot{V} = \dot{V}_F \varepsilon^{-\dot{\gamma}z} + \dot{V}_R \varepsilon^{\dot{\gamma}z} \tag{9.5}$$

where $|\dot{V}_F|$ is the amplitude of the voltage waves propagating in +z-direction, $|\dot{V}_R|$ is the amplitude of the voltage waves propagating –z-direction, and $\dot{\gamma}$ is the propagation constant of the transmission line [1].

$$\dot{\gamma} = \pm\sqrt{\dot{Z}\dot{Y}} = \alpha + \dot{\gamma}\beta \quad (\text{m}^{-1}) \tag{9.6}$$

where the + sign is for forward propagation or propagation in +z-direction, and the – sign is for the backward propagation or propagation in –z-direction. α is the attenuation constant and it is the real part of propagation constant $\dot{\gamma}$:

$$\alpha = \Re\dot{\gamma} = \Re\sqrt{\dot{Z}\dot{Y}} \quad (\text{m}^{-1}) \tag{9.7}$$

In (9.6), β is the phase constant and it is the imaginary part of the propagation constant $\dot{\gamma}$:

$$\beta = \Im\dot{\gamma} = \Im\sqrt{\dot{Z}\dot{Y}} \quad (\text{m}^{-1}) \tag{9.8}$$

Characteristic Impedance

The characteristic impedance of the transmission line is [1–3]

$$Z_0 = \sqrt{\frac{\dot{Z}}{\dot{Y}}} \quad (\Omega) \tag{9.9}$$

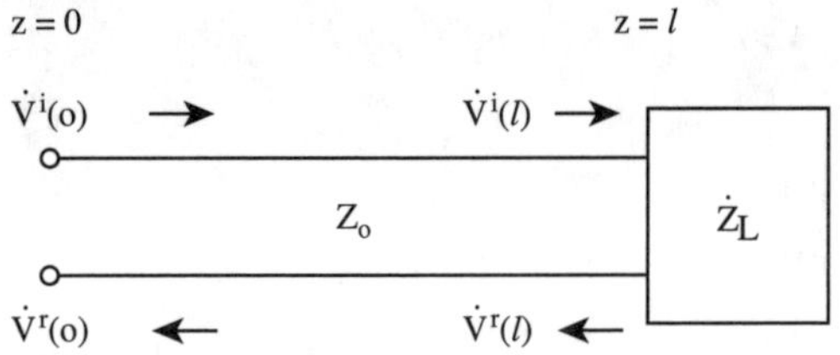

FIGURE 9.2 Incident wave $\dot{V}^i$ and reflected wave $\dot{V}^r$.

Wavelength

The wavelength of the transmission line voltage wave and transmission line current wave is

$$\lambda = \frac{2\pi}{\beta} = \frac{2\pi}{\Im\sqrt{\dot{Z}\dot{Y}}} \quad (\text{m}) \tag{9.10}$$

Phase Velocity

The phase velocity of voltage wave propagation and current wave propagation is

$$\upsilon_p = f\lambda = \frac{\omega}{\beta} \quad (\text{m/s}) \tag{9.11}$$

where f is the frequency of operation and the phase constant is

$$\beta = \frac{2\pi}{\lambda} \quad (\text{m}^{-1}) \tag{9.12}$$

Voltage Reflection Coefficient at the Load

If a transmission line of characteristic impedance Z_0 is terminated by a mismatched load impedance $\dot{Z}_L$, as shown in Fig. 9.2, a voltage wave reflection occurs at the load impedance $\dot{Z}_L$. The voltage reflection coefficient is [1–3]

$$\dot{\rho}(l) = \frac{\dot{V}^r(l)}{\dot{V}^i(l)} = \frac{\dot{Z}_L - Z_0}{\dot{Z}_L + Z_0} = \frac{\dot{\tilde{Z}}_L - 1}{\dot{\tilde{Z}}_L + 1} \tag{9.13}$$

where $\dot{\rho}(l)$ is the voltage reflection coefficient at $z = l$, and $\dot{\tilde{Z}} = \dot{Z}_L / Z_0$ is the normalized load impedance. $\dot{V}_i(l)$ is the incident voltage at the load at $z = l$. When Z_0 is a complex quantity $\dot{Z}_0$, then

$$\dot{\rho}(l) = \frac{\dot{Z}_L - Z_0^*}{\dot{Z}_L + Z_0^*} \tag{9.14}$$

where Z_0^* is a conjugate of $\dot{Z}_0$.

Voltage Reflection Coefficient at the Input

Input Fig. 9.2 $\dot{V}^i(l)$ is caused by $\dot{V}^i(0)$, which is the incident voltage at the input $z = 0$ of the transmission line. $\dot{V}^r(l)$ produces, $\dot{V}^r(0)$, which is the reflected voltage at the input $z = 0$. The voltage reflection coefficient at the input occurs, then, by omitting transmission line loss [1–3]

$$\dot{\rho} = \frac{\dot{V}^r(0)}{\dot{V}^i(0)} = \dot{\rho}(l)\varepsilon^{-2j\beta l} \tag{9.15}$$

Input Impedance

At the load $z = l$, from (9.13),

$$\dot{\tilde{Z}}_L = \frac{1+\dot{\rho}(l)}{1-\dot{\rho}(l)} \tag{9.16}$$

or

$$\dot{Z}_L = Z_0 \frac{1+\dot{\rho}(l)}{1-\dot{\rho}(l)} \tag{9.17}$$

At the input of the transmission line $z = 0$

$$\dot{\tilde{Z}}(0) = \frac{1+\dot{\rho}(0)}{1-\dot{\rho}(0)} \tag{9.18}$$

Using (9.15),

$$\dot{\tilde{Z}}(0) = \frac{1+\dot{\rho}(l)\varepsilon^{-2j\beta l}}{1-\dot{\rho}(l)\varepsilon^{-2j\beta l}} \tag{9.19}$$

Inserting (9.13), [1–3]

$$\dot{\tilde{Z}}(0) = \frac{\dot{\tilde{Z}} + j\tan\beta l}{1 + j\dot{\tilde{Z}}_L \tan\beta l} \tag{9.20}$$

or

$$\dot{\tilde{Z}}(0) = \frac{\dot{Z}_L + jZ_0 \tan\beta l}{Z_0 + j\dot{Z}_L \tan\beta l} \tag{9.21}$$

If the line is lossy and

$$\dot{\gamma} = \alpha + j\beta \tag{9.22}$$

then [1]

$$\dot{Z}(0) = \dot{Z}_0^* \frac{\dot{Z}_L + \dot{Z}_0^* \tanh\dot{\gamma} l}{\dot{Z}_0^* + \dot{Z}_L \tanh\dot{\gamma} l} \tag{9.23}$$

where $\dot{Z}_0^*$ is the complex conjugate of $\dot{Z}_0$.

Input Admittance

The input admittance at $z = 0$ is

$$\dot{Y}(0) = \dot{Y}_0^* \frac{\dot{Y}_L + \dot{Y}_0^* \tanh\dot{\gamma} l}{\dot{Y}_0^* + \dot{Y}_L \tanh\dot{\gamma} l} \tag{9.24}$$

where $\dot{Y}(0)$ is the input admittance of the transmission line, $\dot{Y}_0$ is the characteristic admittance of the transmission line, which is

$$\dot{Y}_0 = \frac{1}{\dot{Z}_0} \tag{9.25}$$

and $\dot{Y}_0^*$ is the conjugate of $\dot{Y}_0 \cdot \dot{Y}_L$ is the load admittance; i.e.,

$$\dot{Y}_L = \frac{1}{\dot{Z}_L} \tag{9.26}$$

When the line is lossless,

$$\dot{\gamma} = j\beta \tag{9.27}$$

then [1–3]

$$\dot{Z}(0) = \frac{\dot{Z}_L + jZ_0 \tan\beta l}{Z_0 + jZ_L \tan\beta l} \tag{9.28}$$

9.2 Two-Wire Lines

Geometric Structure

A structural diagram of the cross-sectional view of a commercial two-wire line is shown in Fig. 9.3. As observed in this figure, two parallel conductors, in most cases made of hard-drawn copper, are positioned by a plastic dielectric cover.

Transmission Line Parameters

In a two-wire line

$$\dot{Z} = R + jX = R + j\omega L \quad (\Omega/\text{m}) \tag{9.29}$$

where $\dot{Z}$ is the impedance per unit length of the two-wire line (Ω/m), R is the resistance per unit length (Ω/m), X is the reactance per unit length (Ω/m), $\omega = 2\pi f$ is the operating angular frequency (s^{-1}), and L is the inductance per unit length (H/m).

For a two-wire line made of hard-drawn copper [4]

$$R = 8.42 \frac{\sqrt{f}}{a} \quad (\mu\Omega/\text{m}) \tag{9.30}$$

where f is the operating frequency and a is the radius of the conductor [4]:

$$L = 0.4 \ln\frac{b}{a} \quad (\mu\text{H/m}) \tag{9.31}$$

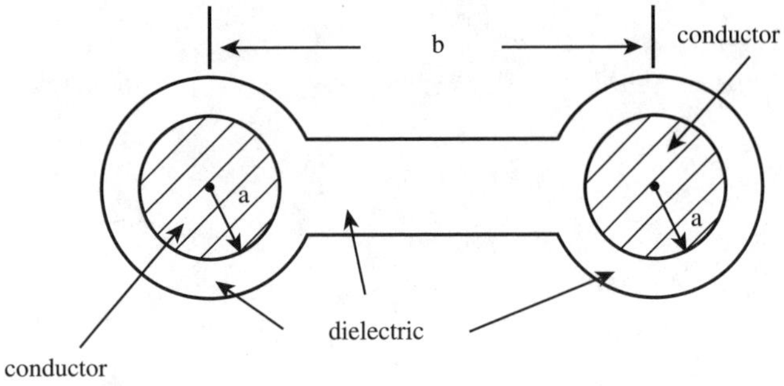

FIGURE 9.3 Cross-sectional view of a two-wire line.

where b is the wire separation or the center-to-center distance of the two-wire line as illustrated in Fig. 9.3.

$$\dot{Y} = G + jB = G + j\omega C \quad (\mathrm{S/m}) \tag{9.32}$$

where $\dot{Y}$ is a shunt admittance per unit length of the two-wire line (S/m), G is a shunt conductance per unit length of the two-wire line (S/m), B is a shunt susceptance per unit length (S/m), and C is a shunt capacitance per unit length (F/m)

$$G = \frac{3.14\sigma_d}{\cosh^{-1}\left(\frac{b}{2a}\right)} \quad (\mathrm{pS/m}) \tag{9.33}$$

where σ_d is the insulation conductivity of the plastic dielectric surrounding the two parallel conductors, and

$$C = \frac{27.8\varepsilon_r}{\cosh^{-1}\left(\frac{b}{2a}\right)} \quad (\mathrm{pF/m}) \tag{9.34}$$

where ε_r is the relative permittivity of the plastic insulating material [5].

If R and G are negligibly small, the characteristic impedance is [6]

$$Z_0 = 277 \log_{10} \frac{b}{a} \quad (\Omega) \tag{9.35}$$

The attenuation constant of a generic two-wire line is, including both R and G [1],

$$\alpha = \frac{R}{2Z_0} + \frac{GZ_0}{2} \quad \left(\mathrm{m}^{-1}\right) \tag{9.36}$$

and the phase constants is [1]

$$\beta = \left\{ \frac{(\omega LB - RG) + \sqrt{(RG - \omega LB)^2 + (\omega LG + BR)^2}}{2} \right\}^{\frac{1}{2}} \quad \left(\mathrm{m}^{-1}\right) \tag{9.37}$$

Wavelength and Phase Velocity

The wavelength on a lossless two-wire line ($R = 0$, $G = 0$) is

$$\lambda_0 = \frac{\omega}{\beta_0} = \frac{1}{\sqrt{LC}} \quad (\mathrm{m}) \tag{9.38}$$

where β_0 is the phase constant of the lossless two-wire line

$$\beta_0 = \omega\sqrt{LC} \quad \left(\mathrm{m}^{-1}\right) \tag{9.39}$$

The wavelength on a lossy two-wire line ($R \neq 0, G \neq 0$) is

$$\lambda = \frac{2\pi}{\beta} = \left\{ \frac{8\pi^2}{(\omega LB - RG) + \sqrt{(RG - \omega LB)^2 + (\omega LG + BR)^2}} \right\}^{\frac{1}{2}} \quad (\mathrm{m}) \tag{9.40}$$

The phase velocity of transverse electromagnetic (TEM) waves on a lossless two-wire line is

$$v_0 = f\lambda_0 = \frac{\omega}{\beta_0} = \frac{1}{\sqrt{LC}} \quad (\text{m/s}) \tag{9.41}$$

The phase velocity of TEM waves on a lossy two-wire line is

$$v = f\lambda = \frac{\omega}{\beta} = \left\{ \frac{2\omega^2}{(\omega LB - RG) + \sqrt{(RG - \omega LB)^2 + (\omega LG + BR)^2}} \right\}^{\frac{1}{2}} \quad (\text{m/s}) \tag{9.42}$$

9.3 Coaxial Lines

Geometric Structure

A generic configuration of a coaxial line is shown in Fig. 9.4. The center and outer conductors are coaxially situated and separated by a coaxial insulator. Generally, coaxial lines are operated in the TEM mode, in which both the electric and magnetic fields are perpendicular to the direction of propagation. The propagating electric fields are in the radial direction and propagating magnetic fields are circumferential to the cylindrical surfaces.

In Fig. 9.4, a is the radius of the center conductor, b is the inner radius of the outer conductor, and c is the outer radius of the outer conductor.

Transmission Line Parameters

The series resistance per unit length of the line for copper is [7],

$$R = 4.16\sqrt{f}\left(\frac{1}{a} + \frac{1}{b}\right) \quad (\mu\Omega/\text{m}) \tag{9.43}$$

The series inductance per unit length is

$$L = 0.2 \ln \frac{b}{a} \quad (\mu\text{H/m}) \tag{9.44}$$

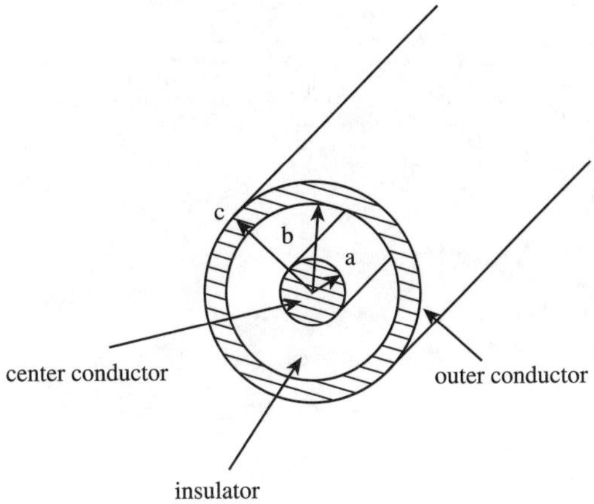

FIGURE 9.4 Generic configuration of a coaxial line.

The shunt conductance per unit length is

$$G = \frac{6.28\sigma_i}{\ln(b/a)} \quad (\mathrm{S/m}) \tag{9.45}$$

where σ_i is the conductivity of the *insulator* between the conductors. The shunt capacitance per unit length is

$$C = \frac{55.5\varepsilon_r}{\ln(b/a)} \quad (\mathrm{pF/m}) \tag{9.46}$$

where ε_r is the relative permittivity of the insulator. When the loss of the line is small, the characteristic impedance of the coaxial line is

$$Z_0 = \frac{138}{\sqrt{\varepsilon_r}} \log_{10}\frac{b}{a} \quad (\Omega) \tag{9.47}$$

when the line is lossy [1]

$$\dot{Z}_0 = \sqrt{\frac{\dot{Z}}{\dot{Y}}} = \sqrt{\frac{R + j\omega L}{G + j\omega C}} = R_0 + jX_0 \tag{9.48}$$

$$R_0 = \frac{\left\{RG + \omega^2 LC + \sqrt{\left(RG + \omega^2 LC\right)^2 + \left(\omega LG - \omega RC\right)^2}\right\}^{\frac{1}{2}}}{\sqrt{G^2 + \omega^2 C^2}} \tag{9.49}$$

$$X_0 = \frac{1}{2R_0} \cdot \frac{\omega LG - \omega CR}{G^2 + \omega^2 C^2} \tag{9.50}$$

The propagation constant of the coaxial line is

$$\dot{\gamma} = \alpha + j\beta \tag{9.51}$$

The attenuation constant is [1]

$$\alpha = \frac{\omega LG + \omega CR}{2\beta} \tag{9.52}$$

where the phase constant is

$$\beta = \left\{\frac{\left(\omega^2 LC - RG\right) + \sqrt{\left(RG - \omega^2 LC\right)^2 + \left(\omega LG + \omega RC\right)^2}}{2}\right\}^{\frac{1}{2}} \tag{9.53}$$

Wavelength and Phase Velocity

The phase velocity on the coaxial line is

$$\upsilon_p = \frac{\omega}{\beta} \tag{9.54}$$

The wavelength on the line is

$$\lambda_1 = \frac{2\pi}{\beta} \tag{9.55}$$

9.4 Waveguides

Rectangular Waveguides

Geometric Structure

A rectangular waveguide is a hollow conducting pipe of rectangular cross section as depicted in Fig. 9.5. Electromagnetic microwaves are launched inside the waveguide through a coupling antenna at the transmission site. The launched waves are received at the receiving end of the waveguide by a coupling antenna. In this case, a rectangular coordinate system is set up on the rectangular waveguide (Fig. 9.5). The *z*-axis is parallel to the axis of the waveguide and is set coinciding with the lower left corner of the waveguide. The wider dimension of the cross section of the waveguide *a* and the narrower dimension of the cross section of the waveguide is *b*, as shown in the figure.

Modes of Operation

The waveguide can be operated in either *H*- or *E*-modes, depending on the excitation configuration. An *H*-mode is a propagation mode in which the magnetic field, *H*, has a *z*-component, *Hz*, as referred to in Fig. 9.5. In this mode, the electric field, *E*, is perpendicular to the direction of propagation, which is the +*z*-direction. Therefore, an *H*-mode is also called a transverse electric (TE) mode. An E-mode is a propagation mode in which the electric field, *E*, has a *z*-component, E_z, as referred to in Fig. 9.5. In this mode, the magnetic field, *H*, is perpendicular to the direction of propagation, which is the +z-direction. Therefore, an *E*-mode is also called a transverse magnetic (TM) mode.

Solving Maxwell's equations for *H*-modes [1],

$$\dot{H}_z = \dot{H}_0 \cos\frac{m\pi x}{a} \cos\frac{n\pi y}{b} \varepsilon^{-\dot{\gamma}z + j\omega t} \tag{9.56}$$

where $|\dot{H}_0|$ is the amplitude of $\dot{H}_z$. Waveguide loss was neglected. Constants *m* and *n* are integral numbers 0, 1, 2, 3, ... and are called the mode number, and $\dot{\gamma}$ is the propagation constant. Both *m* and *n* cannot equal 0 simultaneously. Solving Maxwell's equations for *E*-modes [1],

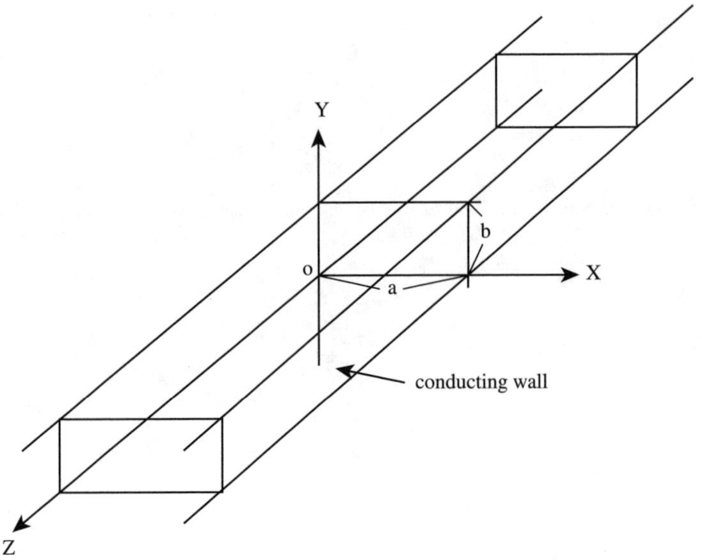

FIGURE 9.5 A rectangular waveguide and rectangular coordinate system.

$$\dot{E}_z = \dot{E}_0 \sin\frac{m\pi x}{a}\sin\frac{n\pi y}{b}\varepsilon^{-\dot{\gamma}z+j\omega t} \tag{9.57}$$

where $|\dot{E}_0|$ is the amplitude of $\dot{E}_z$. Neither m nor n can equal 0. The waveguide loss was neglected.

An H-mode is expressed as the H_{mn}-mode or TE_{mn}-mode. An E-mode is expressed as the E_{mn}-mode or TM_{mn}-mode.

Waveguide Parameters

Propagation constant $\dot{\gamma}$ of a rectangular waveguide made of a good conductor [1]:

$$\dot{\gamma} = j\beta_0\sqrt{1-\left(\frac{\lambda}{\lambda_c}\right)^2} \quad \left(\mathrm{m}^{-1}\right) \tag{9.58}$$

where β_0 is the phase constant of free space, which is

$$\beta_0 = \frac{2\pi}{\lambda} \quad \left(\mathrm{m}^{-1}\right) \tag{9.59}$$

Here, λ is the wavelength in free space, and λ_c is the cutoff wavelength of the waveguide. Electromagnetic waves with $\lambda > \lambda_c$ cannot propagate inside the waveguide. It is given for both E_{mn}-mode and H_{mn}-mode operation by [1]:

$$\lambda_c = \frac{2}{\sqrt{\left(\frac{m}{a}\right)^2+\left(\frac{n}{b}\right)^2}} \quad (\mathrm{m}) \tag{9.60}$$

This means that if the waveguide is made of a good conductor, the attenuation constant

$$\alpha \approx 0 \quad \left(m^{-1}\right) \tag{9.61}$$

and the phase constant is

$$\beta_g = \beta_0\sqrt{1-\left(\frac{\lambda}{\lambda_c}\right)^2} \tag{9.62}$$

The wavelength in the waveguide, i.e., waveguide wavelength λ_g, is longer than the free-space wavelength λ:

$$\lambda_g = \frac{\lambda}{\sqrt{1-\left(\frac{\lambda}{\lambda_c}\right)^2}} \tag{9.63}$$

Then, the speed of propagation v_p is $> c = f\lambda$.

$$v_p = f\lambda_g = \frac{f\lambda}{\sqrt{1-\left(\frac{\lambda}{\lambda c}\right)^2}} = \frac{c}{\sqrt{1-\left(\frac{\lambda}{\lambda c}\right)^2}} \tag{9.64}$$

For an H_{mn}-mode, the wave impedance is

$$\eta_H = \frac{-\dot{E}_y}{\dot{H}_x} = \frac{\dot{E}_x}{\dot{H}_y} = \frac{\sqrt{\dfrac{\mu_0}{\varepsilon_0}}}{\sqrt{1-\left(\dfrac{\lambda}{\lambda_c}\right)^2}} \quad (\Omega) \tag{9.65}$$

For an E_{mn}-mode the wave impedance is

$$\eta_E = \frac{-\dot{E}_y}{\dot{H}_x} = \frac{\dot{E}_x}{\dot{H}_y} = \sqrt{1-\left(\frac{\lambda}{\lambda_c}\right)^2} \Bigg/ \sqrt{\frac{\mu_0}{\varepsilon_0}} \quad (\Omega) \tag{9.66}$$

Circular Waveguides

Geometric Structure

A circular waveguide is a hollow conducting pipe of circular cross section, as depicted in Fig. 9.6. Electromagnetic microwaves are launched inside the waveguide through a coupling antenna at the transmission site. The launched waves are received at the receiving end of the waveguide by a coupling antenna. In this case, a circular coordinate system (r, ϕ, z) is set up in the circular waveguide, as depicted in Fig. 9.6. The z-axis is coincident with the axis of the cylindrical waveguide. The inside radius of the circular waveguide is a.

Modes of Operation

The circular waveguide can be operated in either H- or E-modes, depending on the excitation configuration. An H-mode is a propagation mode in which the magnetic field, H, has a z-component, H_z, as referred to in Fig. 9.6. In this mode the electric field, E, is perpendicular to the direction of propagation, which is the $+z$-direction. Therefore, an H-mode is also called a TE mode. In the mode $E_z = 0$. An E-mode is a propagation mode in which the electric field, E, has a z-component, E_z, as referred to in Fig. 9. 6. In this mode the magnetic field, H, is perpendicular to the direction of propagation, which is the $+z$-direction. Therefore, an E-mode is also called a TM mode.

Solving Maxwell's equations [1],

$$\dot{H}_z = \dot{H}_0 J_n\left(k'_{cm} r\right) \cos n\phi\, \varepsilon^{-\dot{\gamma}z + j\omega t} \tag{9.67}$$

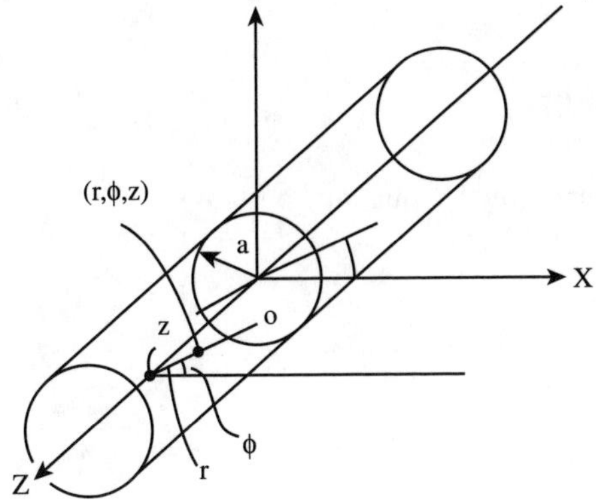

FIGURE 9.6 A circular waveguide and cylindrical coordinate system.

Here, $|\dot{H}_0|$ is the amplitude of $\dot{H}_z$, n and m are integral numbers 0, 1, 2, 3, … and are called the mode number, $J_n(k'_{cm} r)$ is the Bessel function of nth order, with the argument that $k'_{cm} r$, k'_{cm} is the mth root of $J'_n(k_{cm} a) = 0$, which is

$$k'_{cm} = \frac{u'_{nm}}{a} \tag{9.68}$$

where u'_{nm} is the mth root of the derivative of the Bessel function of order n, i.e., $J'_n(x) = 0$, where x is a generic real argument. The propagation constant is $\dot{\gamma}$.

Solving Maxwell's equations for E-modes,

$$\dot{E}_z = \dot{E}_0 J_n(k_{cm} r)\cos n\phi \varepsilon^{-\dot{\gamma}z + j\omega t} \tag{9.69}$$

k_{cm} is an mth root of $J_n(k_c a) = 0$, which is

$$k_{cm} = \frac{u_{nm}}{a} \tag{9.70}$$

where u_{nm} is the mth root of the Bessel function of order n, i.e., $J_n(x) = 0$, where x is a generic real argument.

An H-mode in a circular waveguide is expressed as the H_{nm}-mode or the TM_{nm}-mode. An E-mode is expressed as the E_{nm}-mode or the TM_{nm}-mode.

Waveguide Parameters

The propagation constant $\dot{\gamma}$ of a circular waveguide made of a good conductor is [1]

$$\dot{\gamma} = j\beta_0 \sqrt{1 - \left(\frac{\lambda}{\lambda_c}\right)} \quad (\mathrm{m}^{-1}) \tag{9.71}$$

where β_0 is the phase constant of free space, which is

$$\beta_0 = \frac{2\pi}{\lambda} \quad (\mathrm{m}^{-1}) \tag{9.72}$$

Here, λ is the wavelength in free space, and λ_c is the cutoff wavelength of the waveguide. Electromagnetic waves with $\lambda > \lambda_c$ cannot propagate inside the waveguide. It is given for an H_{nm}-mode [1]

$$\lambda_{CH} = \frac{2\pi a}{u'_{nm}} \quad (\mathrm{m}) \tag{9.73}$$

For E_{nm}-mode operation,

$$\lambda_{CE} = \frac{2\pi a}{u_{nm}} \quad (\mathrm{m}) \tag{9.74}$$

This means that if the waveguide is made of a good conductor, the attenuation constant is

$$\alpha \approx 0 \quad (\mathrm{m}^{-1}) \tag{9.75}$$

and the phase constant is

$$\beta_g = \beta_0 \sqrt{1 - \left(\frac{\lambda}{\lambda_c}\right)^2} \quad (\mathrm{m}^{-1}) \tag{9.76}$$

The waveguide wavelength is

$$\lambda_g = \frac{\lambda}{\sqrt{1-\left(\frac{\lambda}{\lambda_c}\right)^2}} \quad \left(\mathrm{m}^{-1}\right) \tag{9.77}$$

The speed of propagation (phase velocity) is

$$v_p = f\lambda_g = \frac{c}{\sqrt{1-\left(\frac{\lambda}{\lambda c}\right)^2}} \quad (\mathrm{m/s}) \tag{9.78}$$

For an H_{nm}-mode, the wave impedence is

$$\begin{aligned} \eta_H &= \frac{\dot{E}_r}{\dot{H}_\phi} = \frac{-\dot{E}\phi}{\dot{H}_r} = \frac{j\omega\mu_0}{\dot{\gamma}} \\ &= \frac{\omega\mu_0}{\beta_0\sqrt{1-\left(\frac{\lambda}{\lambda_c}\right)^2}} \quad (\Omega) \end{aligned} \tag{9.79}$$

For an E_{nm}-mode, the wave impedance is

$$\begin{aligned} \eta_E &= \frac{\dot{E}_r}{\dot{H}_\phi} = \frac{-\dot{E}_\phi}{\dot{H}_r} = \frac{\dot{\gamma}}{j\omega\varepsilon_0} \\ &= \frac{\beta_0\sqrt{1-\left(\frac{\lambda}{\lambda_c}\right)^2}}{\omega\varepsilon_0} \quad (\Omega) \end{aligned} \tag{9.80}$$

9.5 Microstrip Lines

Geometric Structure

Figure 9.7 presents a general geometric structure of a microstrip line. A conducting strip of width, w, and thickness, t, is laid on an insulating substrate of thickness, H, and permittivity, $\varepsilon = \varepsilon_0\varepsilon_r$. The dielectric substrate has a groundplate underneath, as presented in Fig. 9.7.

Transmission Line Parameters

The characteristic impedance, Z_0, of a microstrip line, as shown in Fig. 9.7, is given by [8–10]

$$Z_0 = \frac{42.4}{\sqrt{\varepsilon_r+1}} \ln \left\{1+\left(\frac{4H}{w'}\right)\left[\left(\frac{14+8/\varepsilon_r}{11}\right)\left(\frac{4H}{W'}\right)+\sqrt{\left(\frac{14+8\varepsilon_r}{11}\right)^2\left(\frac{4H}{w'}\right)+\frac{1+1/\varepsilon_r}{2}\pi^2}\right]\right\} \tag{9.81}$$

where w' is an effective width of the microstrip, which is given by

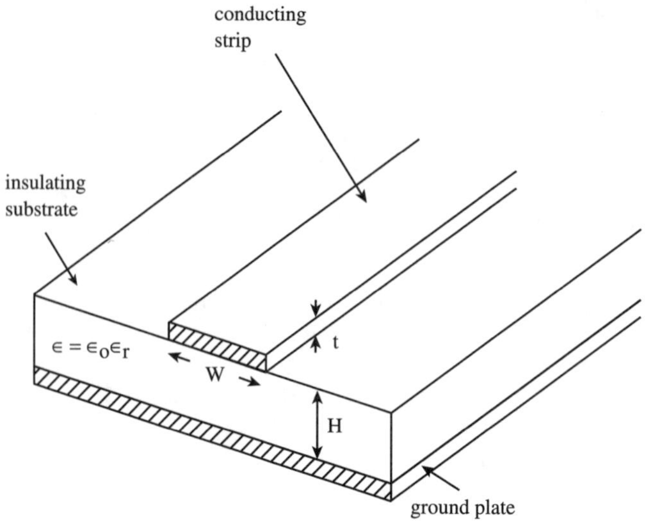

FIGURE 9.7 Geometric structure of a microstrip line.

$$w' = w + \frac{1 + 1/\varepsilon_r}{2} \cdot \frac{t}{\pi} \ln \frac{10.87}{\sqrt{\left(\frac{t}{H}\right)^2 + \left(\frac{1/\pi}{w/t + 1.10}\right)^2}} \tag{9.82}$$

The attenuation constant of a microstrip line is

$$\alpha = \frac{p + p'}{2} \frac{Z_{01}}{Z_0} \frac{2\pi}{\lambda_0} \quad (\text{Np/m}) \tag{9.83}$$

where

$$Z_{01} \equiv 30 \ \ln \left\{ 1 + \frac{1}{2}\left(\frac{8H}{w'}\right) \left[\frac{8H}{w'} + \sqrt{\left(\frac{8H}{w'}\right)^2 + \pi^2} \right] \right\} \tag{9.84}$$

and

$$p \equiv 1 - \frac{Z_{01}}{Z_{0\delta}} \tag{9.85}$$

where

$$Z_{0\delta} \equiv 30 \ \ln \left\{ 1 + \frac{1}{2}\left(\frac{8(H+\delta)}{w'}\right) \left[\frac{8H+\delta)}{w'} + \sqrt{\left(\frac{8H+\delta}{w'}\right)^2 + \pi^2} \right] \right\} \tag{9.86}$$

and

$$\delta = \frac{1}{\sqrt{\pi f \mu \sigma}} \tag{9.87}$$

is the skin depth of the conducting strip of conductivity σ

$$p' \equiv \frac{P_k}{1 - \frac{1/q - 1}{\varepsilon_r}} \tag{9.88}$$

where

$$\varepsilon_r = \varepsilon_r' - j\varepsilon_r'' \quad (9.89)$$

and

$$P_K = \sin\left[\tan^{-1}\frac{\varepsilon_r''}{\varepsilon_r'}\right] \quad (9.90)$$

and

$$q = \frac{1}{\varepsilon_r - 1}\left[\left(\frac{Z_{01}}{Z_0}\right)^2 - 1\right] \quad (9.91)$$

Wavelength and Phase Velocity

The transmission line wavelength is

$$\lambda_1 = \frac{Z_0}{Z_{01}}\lambda_0 \quad (9.92)$$

The phase constant of the microstrip line is then

$$\beta = \frac{2\pi}{\lambda_0}\frac{Z_{01}}{Z_0} \quad (9.93)$$

The phase velocity of electromagnetic waves on the microstrip line is

$$v_p = 3 \times 10^8 \frac{Z_0}{Z_{01}} \quad (9.94)$$

9.6 Coplanar Waveguide

Geometric Structure

A cross-sectional view of a coplanar waveguide (CPW) is given in Fig. 9.8. The waveguide consists of a narrow, central conducting strip of width s(m) and very wide conducting plates on both sides of the central conducting strip, with gap widths *w*. These conductors are developed on a surface of dielectric substrate of thickness *d*, as presented in the figure. The electromagnetic waves propagate in the gap between the outer conducting plates and the center conducting strip.

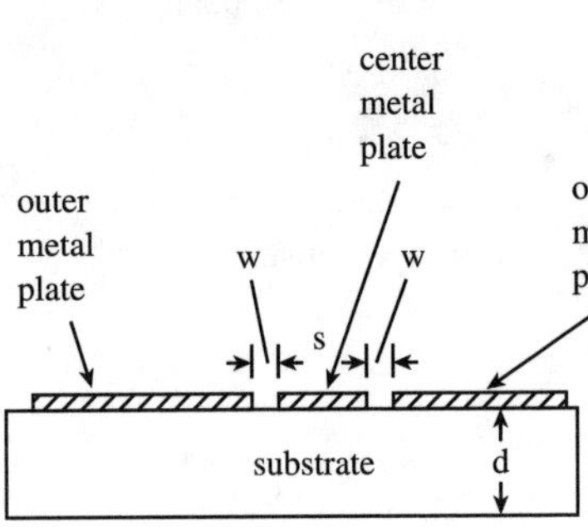

FIGURE 9.8 Cross-sectional view of a coplanar waveguide.

Transmission Line Parameters

The attenuation constant of a coplanar waveguide is given by [11]

$$\alpha = \left(\frac{\pi}{2}\right)^5 \cdot 2 \cdot \frac{1 - \dfrac{\varepsilon_{\text{eff}(f)}}{\varepsilon_r}}{\sqrt{\dfrac{\varepsilon_{\text{eff}(f)}}{\varepsilon_r}}} \cdot \frac{(s+2w)^2 \varepsilon_r^{3/2}}{c^3 K'(k) K(k)} \quad (\text{Np/m}) \tag{9.95}$$

where the effective dielectric constant is given by

$$\sqrt{\varepsilon_{\text{eff}}(f)} = \sqrt{\varepsilon_q} + \frac{\sqrt{\varepsilon_r} - \sqrt{\varepsilon_q}}{1 + a\left(\dfrac{f}{f_{\text{TE}}}\right)^{-b}} \tag{9.96}$$

and

$$\varepsilon_q = \frac{\varepsilon_r + 1}{2} \tag{9.97}$$

ε_r is the relative permittivity of the substrate material,

$$f_{\text{TE}} = \frac{c}{4d\sqrt{\varepsilon_r - 1}} \tag{9.98}$$

is the TE mode cutoff frequency,

$$k \equiv s/(s+2w) \tag{9.99}$$

$K(k)$ is the complete ellipitic integral of the first kind of the argument k, and c is the speed of light in vacuum, which is 3×10^8 m/s. The parameter a is [11]

$$a \approx \log^{-1}\left[u \log\frac{s}{w} + v\right] \tag{9.100}$$

$$u \approx 0.54 - 0.64q + 0.015q^2 \tag{9.101}$$

$$v \approx 0.43 - 0.86q + 0.54q^2 \tag{9.102}$$

$$q \approx \log\frac{s}{d} \tag{9.103}$$

The parameter b is an experimentally determined constant

$$b \approx 1.8 \tag{9.104}$$

$$K'(k) = K\left(\sqrt{1-k^2}\right) \tag{9.105}$$

The phase constant of the coplanar waveguide is

$$\beta(f) = 2\pi \frac{f}{c} \sqrt{\varepsilon_{\text{eff}}(f)} \quad (\text{rad/m}) \tag{9.106}$$

The characteristic impedence of the coplanar waveguide is [11]

$$Z_0 = \frac{120\pi}{\sqrt{\varepsilon_{\text{eff}}(f)}} \frac{K'(k)}{4K(k)} \quad (\Omega) \tag{9.107}$$

Wavelength and Phase Velocity

The wavelength of electromagnetic waves propagating on a coplanar waveguide is obtained from (9.106):

$$\lambda_l = \frac{2\pi}{\beta(f)} = \frac{c}{f} \cdot \frac{1}{\sqrt{\varepsilon_{\text{eff}}(f)}} = \frac{\lambda}{\sqrt{\varepsilon_{\text{eff}}(f)}} \quad (\text{m}) \tag{9.108}$$

The phase velocity of the waves on the coplanar waveguide is, then,

$$v_p = f\lambda_1 = \frac{c}{\sqrt{\varepsilon_{\text{eff}}(f)}} \quad (\text{m/s}) \tag{9.109}$$

References

[1] T. K. Ishii, *Microwave Engineering*, San Diego, CA: Harcount, Brace, Jovanovich, 1989.
[2] J. R. Wait, *Electromagnetic Wave Theory*, New York: Harper & Row, 1985.
[3] V. F. Fusco, *Microwave Circuits*, Englewood Cliffs, NJ: Prentice Hall, 1987.
[4] L. A. Ware and H. R. Reed, *Communications Circuits*. New York: John Wiley & Sons, 1949.
[5] E. A. Guillemin, *Communications Networks*, New York: John Wiley & Sons, 1935.
[6] F. E. Terman, *Radio Engineering*, New York: McGraw-Hill, 1941.
[7] H. J. Reich, P. F. Ordung, H. L. Krauss, and J. G. Skalnik, *Microwave Theory and Techniques*, Princeton, NJ: D. Van Nostrand, 1953.
[8] H. A. Wheeler, "Transmission-line properties of parallel strips separated by a dielectric sheet," *IEEE Trans. MTT*, vol. MTT-13, pp. 172–185, Mar. 1965.
[9] H. A. Wheeler, "Transmission-line properties of parallel strips by a conformal-mapping approximation," *IEEE Trans. MTT*, vol. MTT-12, pp. 280–289, May 1964.
[10] H. A. Wheeler, "Transmission-line properties of a strip on a dielectric sheet on a plane," *IEEE Trans. MTT*, vol. MTT-25, pp. 631–647, Aug. 1977.
[11] M. Y. Frankel, S. Gupta, J. A. Valdmanis, and G. A. Mourou, "Terahertz attenuation and dispersion characteristics of coplanar transmission lines." *IEEE Trans. MTT*, vol. 39, no. 6, pp. 910–916, June 1991.

10

Multiconductor Transmission Lines

Daniël De Zutter
Gent University, Belgium

Luc Martens
Gent University, Belgium

10.1 Introduction: Frequency vs. Time Domain Analysis

Multiconductor transmission lines (MTL), or multiconductor buses as they are also often called, are found in almost every electrical packaging technology and on every technology level from digital chips, over MMICs (monolithic microwave integrated circuits) to MCMs (multichip modules), boards, and backplanes. Multiconductor transmission lines are electrical conducting structures with a constant cross-section (the x, y-plane) that propagates signals in the direction perpendicular to that cross-section (the z-axis) (see also Fig. 10.1). Being restricted to a constant cross-section we are in fact dealing with the so-called uniform MTL. The more general case using a nonuniform cross-section is much more difficult to handle and constitutes a fully three-dimensional problem.

It is not the purpose of this chapter to give a detailed account of the physical properties and the use of the different types of MTL. The literature on this subject is abundant and any particular reference is bound to be both subjective and totally incomplete. Hence, we put forward only [1], [2], and [3] as references here, as they contain a wealth of information and additional references.

In the frequency domain, i.e., for harmonic signals, solution of Maxwell's sourceless equations yields a number of (evanescent and propagating) modes characterized by modal propagation factors $\exp(\pm j\beta z)$ and by a modal field distribution, which depends only upon the (x, y)-coordinates of the cross-section. In the presence of losses and for evanescent modes β can take complex values, and $j\beta$ is then replaced by $\gamma = \alpha + j\beta$. In the propagation direction, the modal field amplitudes essentially behave as voltage and current along a transmission line. This immediately suggests that MTL should be represented on the circuit level by a set of coupled circuit transmission lines. The relationship between the typical circuit quantities, such as voltages, currents, coupling impedances, and signal velocities, on the one hand, and the original field quantities (modal fields and modal propagation factors) is not straightforward [4]. In general, the circuit quantities will be frequency dependent. The frequency domain circuit model parameters can be used as the starting point for time domain analysis of networks, including multiconductor lines. This is, again, a vast research topic with important technical applications.

Section 10.2 describes the circuit modeling in the frequency domain of uniform MTL based on the Telegrapher's equations. The meaning of the voltages and currents in these equations is explained both at lower frequencies in which the quasi-transverse electromagnetic (TEM) approach is valid as well as

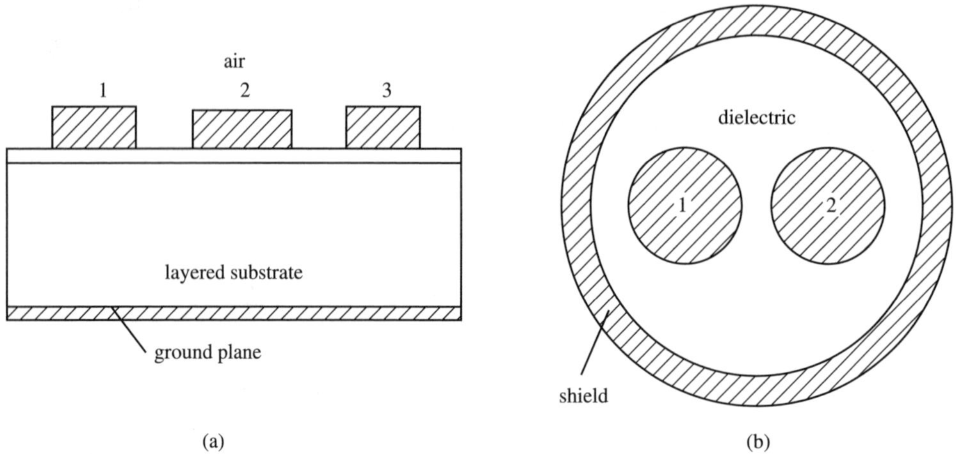

FIGURE 10.1 Two examples of cross sections of multiconductor lines.

in the so-called full-wave regime valid for any frequency. The notions TEM, quasi-TEM, and full-wave are elucidated. We introduce the capacitance, inductance, resistance, and conductance matrices together with the characteristic impedance matrix of the coupled transmission line model. Finally, for some simple MTL configurations analytical formulas are presented expressing the previous quantities and the propagation factors as a function of the geometric and electrical parameters of these configurations.

It would be a formidable task to give a comprehensive overview of all the methods that are actually used for the time domain analysis of MTL. In the remaining part of this paragraph a very short overview (both for uniform and nonuniform structures) is presented along with some references. In the case of linear loads and drivers, frequency domain methods in combination with (fast) Fourier transform techniques are certainly most effective [5–7]. In the presence of nonlinear loads and drivers other approaches must be used. Simulations based on harmonic balance techniques [8, 9] are, again, mainly frequency domain methods. All signals are approximated by a finite sum of harmonics and the nonlinear loads and drivers are taken into account by converting their time domain behavior to the frequency domain. Kirchhoff laws are then imposed for each harmonic in an iterative way. Harmonic balance techniques are not very well suited for transient analysis or in the presence of strong nonlinearities, but are excellent for mixers, amplifiers, filters, etc. Many recent efforts were directed toward the development of time domain simulation methods (for both uniform and nonuniform interconnection structures) based on advanced convolution-type approaches. It is, of course, impossible to picture all the ramifications in this research field. We refer the reader to a recent special issue of *IEEE Circuits and Systems Transactions* [10], to the "Simulation Techniques for Passive Devices and Structures" section of a special issue of *IEEE Microwave Theory and Techniques Transactions* [11], and to a 1994 special issue of the *Analog Integrated Circuits and Signal Processing Journal* [12] and to the wealth of references therein.

Both frequency and time domain experimental characterization techniques for uniform and nonuniform multiconductor structures can be found in Chapter 11.

10.2 Telegrapher's Equations for Uniform Multiconductor Transmission Lines

Generalities

Figures 10.1(a) and (b) show the cross-sections of two general coupled lossy MTLs consisting of $N+1$ conductors. These conductors either can be perfectly conducting or exhibit finite conductivity. Their cross-section remains constant along the propagation or longitudinal direction z. The $(N+1)$th conductor is taken as reference conductor. In many practical cases, this will be the ground plane at the top or bottom

[(Fig. 10.1(a)] of the layered dielectric in which the conductors are embedded or is the shielding surrounding the other conductors [(Fig. 10.1(b)]. We restrict the analysis to the frequency domain, i.e., all field components and all voltages and currents have a common time dependence, exp($j\omega t$), which is suppressed in the sequel. The generalized Telegrapher's equations governing the circuit representation of the MTL of Fig. 10.1 in terms of a set of C-coupled circuit transmission lines is given by [4]:

$$\frac{d\mathbf{V}}{dz} + \mathbf{ZI} = 0$$
$$\frac{d\mathbf{I}}{dz} + \mathbf{YV} = 0 \tag{10.1}$$

V and **I** are column vectors, the C elements of which are the voltages and currents of the circuit model; **Z** and **Y** are the $C \times C$ impedance and admittance matrices. Equation (10.1) is a good circuit description of the wave phenomena along a MTL if only the fundamental modes of the corresponding field problem are of importance. In that case $C = N$ ($C = 3$ in Fig. 10.1(a) and $C = 2$ in Fig. 10.1(b)) if a ground plane is present, and $C = N - 1$ in the absence of a ground plane. For the relationship between the actual electromagnetic field description in terms of modes and the circuit model (10.1), we refer the reader to [4] and [13]. The general solution to (10.1) is given by

$$\mathbf{V}(z) = 2\left(\mathbf{I}_m^T\right)^{-1} e^{-j\beta z}\mathbf{K}^+ + 2\left(\mathbf{I}_m^T\right)^{-1} e^{j\beta z}\mathbf{K}^-$$
$$\mathbf{I}(z) = \mathbf{I}_m e^{-j\beta z}\mathbf{K}^+ - \mathbf{I}_m e^{j\beta z}\mathbf{K}^- \tag{10.2}$$

$\mathbf{K}^+$ and $\mathbf{K}^-$ are column vectors with C elements. β is a diagonal $C \times C$ matrix with the propagation factors β_f ($f = 1, 2, \ldots, C$) of the C fundamental eigenmodes as diagonal elements. This matrix reduces to a single propagation factor for a single transmission line (see Eq. 9.6). For the calculation of the fields and of the propagation constants many different methods can be found in the literature [14]. Solution (10.2) of the differential equations (10.1) is the extension of the corresponding equations (9.5) for a single transmission line to the coupled line case. It also consists of waves respectively traveling in positive and negative z-directions. The $\mathbf{I}_m$ is a $C \times C$ matrix, the columns of which are the current eigenvectors of the circuit model. The following relationships hold:

$$\mathbf{Z} = j\omega\mathbf{L} + \mathbf{R} = 2j\left(\mathbf{I}_m^T\right)^{-1}\beta\left(\mathbf{I}_m\right)^{-1}$$
$$\mathbf{Y} = j\omega\mathbf{C} + \mathbf{G} = \frac{j}{2}\mathbf{I}_m\boldsymbol{\beta}\mathbf{I}_m^T \tag{10.3}$$

L, R, C and **G** are the $C \times C$ (frequency dependent) inductance, resistance, capacitance, and conductance matrices. The $C \times C$ characteristic impedance matrix of the transmission line models is given by $\mathbf{Z}_{\text{char}} = 2(\mathbf{I}_m^T)^{-1}(\mathbf{I}_m)^{-1}$. The matrix $\mathbf{Z}_{\text{char}}$ replaces the simple characteristic impedance number of (9.9). In general the mapping of the wave phenomenon onto the circuit model [(10.1) to (10.3)] depends on the choice of $\mathbf{I}_m$. We refer the reader to the detailed discussions in [4]. For MTLs the most adopted definition for the elements of $\mathbf{I}_m$ is [15]

$$I_{m,jf} = \oint_{c_j} \mathbf{H}_{\text{tr},f} \cdot \mathbf{dl} \quad j, f = 1, 2, \ldots, C \tag{10.4}$$

where c_j is the circumference of conductor j and where $\mathrm{H}_{\text{tr},f}$ is the transversal component of the magnetic field of eigenmode f. This means than $\mathrm{I}_{m,jf}$ is the total current through conductor j due to eigenmode f. This definition is used in the power-current impedance definition for microstrip and stripline problems

[9]. For slotline circuits, a formulation that parallels the one given above must be used, but in this case it makes much more sense to introduce the voltage eigenvectors and to define them as line integrals of the electric field (see Appendix B of [16]).

As **Z** and **Y** in (10.3) are frequency dependent, the time domain equivalent of (10.1) involves convolution integrals between **Z** and **I** on the one hand and **Y** and **V** on the other hand. The propagation factors β_f are also frequency dependent, hence the signal propagation will show dispersion, i.e., the signal waveform becomes distorted while propagating.

Low-Frequency or Quasi-Transverse Electromagnetic Description

In the previous section, the reader was given a very general picture of the MTL problem, valid for any frequency. This analysis is the so-called full-wave analysis. The present section is restricted to the low-frequency or quasi-static regime. Here, the cross-section of the MTL is small with respect to the relevant wavelengths, the longitudinal field components can be neglected and the transversal field components can be found from the solution of an electrostatic or magnetostatic problem in the cross-section of the MTL. A detailed discussion of the theoretical background can be found in [17]. In the quasi-TEM limit and in the absence of losses **R** and **G** are zero and **L** and **C** become frequency independent and take their classical meaning. Both skin-effect losses and small dielectric losses can be accounted for by a perturbation approach [18]. In that case a frequency dependent **R** and **G** must be reintroduced. If **R** and **G** are zero and **L** and **C** are frequency independent, the following Telegrapher's equations hold:

$$\begin{aligned} \frac{\partial \mathbf{v}}{\partial z} &= -\mathbf{L} \cdot \frac{\partial \mathbf{i}}{\partial t} \\ \frac{\partial \mathbf{i}}{\partial z} &= -\mathbf{C} \cdot \frac{\partial \mathbf{v}}{\partial t} \end{aligned} \tag{10.5}$$

Equation (10.5) is the time domain counterpart of (10.1). We have replaced the capital letters for voltages and currents with lower case letters to distinguish between time and frequency domain. **L** and **C** are related to the total charge Q_i per unit length carried by each conductor and to the total magnetic flux F_i between each conductor and the reference conductor:

$$\begin{aligned} \mathbf{Q} &= \mathbf{C} \cdot \mathbf{V} \\ \mathbf{F} &= \mathbf{L} \cdot \mathbf{I} \end{aligned} \tag{10.6}$$

where **Q** and **F** are $C \times 1$ column vectors with elements Q_i and F_i, respectively. For a piecewise homogeneous medium, one can prove that the inductance matrix **L** can be derived from an equivalent so-called vacuum capacitance matrix $\mathbf{C}_v$ with $\mathbf{L} = \mathbf{C}_v^{-1}$ and where $\mathbf{C}_v$ is calculated in the same way as **C**, but with the piecewise constant ε everywhere replaced by the corresponding value of $1/\mu$. For nonmagnetic materials, this operation corresponds with taking away all dielectrics and working with vacuum, thus explaining the name of the matrix $\mathbf{C}_v$. Other properties of **C** and **L** are:

C and **L** are symmetric, i.e., $C_{ij} = C_{ji}$ and $L_{ij} = L_{ji}$
C is real, $C_{ii} > 0$ and $C_{ij} < 0$ $(i \neq j)$
l is real, $L_{ii} > 0$ and $L_{ij} > 0$

The propagation factors β_f which form the elements of β in (10.2) are now given by the eigenvalues of $(\mathbf{LC})^{1/2}$ or equivalently of $(\mathbf{CL})^{1/2}$. The current eigenvectors which form the columns of $\mathbf{I}_m$ are now solutions of the following eigenproblem (where ω is the circular frequency):

$$\omega^2(\mathbf{CL})\mathbf{I} = (\beta_f)^2\mathbf{I} \tag{10.7}$$

The corresponding eigenvoltages are solutions of:

$$\omega^2(\mathbf{LC})\mathbf{V} = (\beta_f)^2\mathbf{V} \tag{10.8}$$

Hence, corresponding voltage and current eigenmodes propagate with the same propagation factors as **L**, **C**, **V**, and **I** and are frequency independent, β_f is proportional with ω and can be rewritten as $\beta_f = \omega\beta'_f$, proving that the propagation is nondispersive with velocity $v_f = 1/\beta'_f$. Remember that the subindex f takes the values 1, 2,…, C, i.e., for a three-conductor problem above a ground plane [N = C = 3, see Fig. 10.1(a)], three distinct propagation factors and corresponding eigenmode profiles exist for currents and voltages. Note, however, that for the same β_f the eigenvector for the currents differs from the eigenvector of the voltages.

We conclude this section by remarking that, for MTL embedded in a homogeneous medium (such as the simple stripline or the coaxial cable with homogeneous filling), $(LC) = \varepsilon\,\mu\mathbf{1}$, where **1** is the unity matrix. Thus, the eigenmodes are purely TEM, i.e., electric and magnetic fields have only transversal components and the longitudinal ones are exactly zero. All propagation factors β_f take the same value $[c/(\varepsilon_r\mu_r)^{1/2}]$, where c is the velocity of light in vacuum. Note, however, that even for identical β_f different eigenmodes will be found.

Numerical calculation of **L** and **C** can be performed by many different numerical methods (see the reference section of [18]), and for sufficiently simple configurations, analytical formulas are available. For a line with one conductor and a ground plane ($N = C = 1$) the characteristic impedance $Z_0 = (L/C)^{1/2}$ - and the signal velocity v_p is $v_p = (LC)^{-1/2}$.

Analytical Expressions for Some Simple Multiconductor Transmission Line Configurations

Symmetric Stripline

Sections 9.3 through 9.6 presented a number of MTL consisting of a single conductor and a ground plate ($N = C = 1$). Here, we add another important practical example, the symmetric stripline configuration of Fig. 10.2. We restrict ourselves to the lossless case. A perfectly conducting strip of width w and thickness t is symmetrically placed between two perfectly conducting ground planes with spacing b. The insulating substrate has a permittivity $\varepsilon = \varepsilon_0\varepsilon_r$, and is nonmagnetic. This case has a single fundamental mode. The characteristic impedance Z_0 is given by [19]:

$$Z_0\sqrt{\varepsilon_r} = 30\ln\left\{1 + \frac{4}{\pi}\frac{b-t}{W'}\left[\frac{8}{\pi}\frac{b-t}{W'} + \sqrt{\left(\frac{8}{\pi}\frac{b-t}{W'}\right)^2 + 6.27}\right]\right\} \tag{10.9}$$

with

$$\frac{W'}{b-t} = \frac{W}{b-t} + \frac{\Delta W}{b-t} \tag{10.10}$$

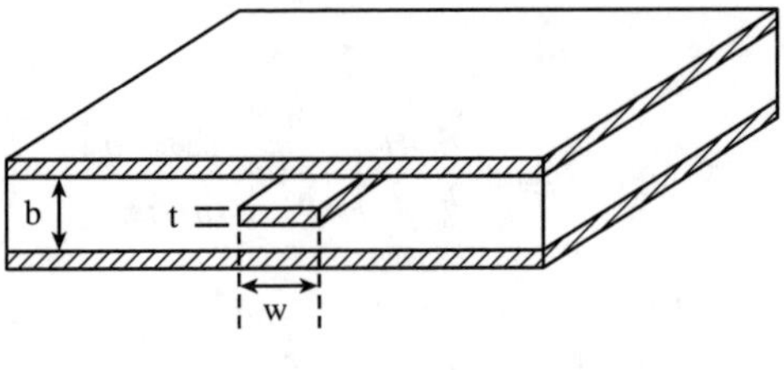

FIGURE 10.2 Strip line configuration.

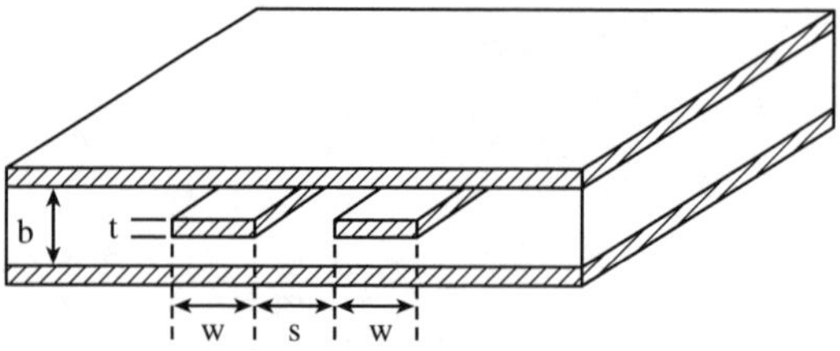

FIGURE 10.3 Coupled, symmetric strip line configuration.

where

$$\frac{\Delta W}{b-t} = \frac{x}{\pi(1-x)}\left\{1-\frac{1}{2}\ln\left[\left(\frac{x}{2-x}\right)^2+\left(\frac{0.0796x}{W/b+1.1x}\right)^m\right]\right\}$$

$$m = 2\left[1+\frac{2}{3}\frac{x}{1-x}\right]^{-1}, \quad x = t/b$$

For $W'/(b-t) < 10$ (10.9) is 0.5% accurate. The signal velocity v_p is given by $c/(\varepsilon_r)^{1/2}$, where c is the velocity of light in vacuum. The corresponding L and C are given by $L = Z_0/v_p$ and $C = 1/(Z_0 v_p)$.

Coupled Striplines

The configuration is depicted in Fig. 10.3. It consists of the symmetric combination of the structure of Fig. 10.2. There are now two fundamental TEM modes ($N = C = 2$). The even mode (index e) corresponds to the situation in which both central conductors are placed at the same voltage (speaking in low-frequency terms, of course). The odd mode (index o) corresponds to the situation where the central conductors are placed at opposite voltages. The impedances of the modes (respectively, $Z_{0,e}$ and $Z_{0,o}$) are given by:

$$Z_{0e,o} = \frac{30\pi(b-t)}{\sqrt{\varepsilon_r}\left(W+\frac{bC_f}{2\pi}A_{e,o}\right)} \tag{10.11}$$

with

$$A_e = 1+\frac{\ln(1+\tanh\theta)}{\ln 2}$$

$$A_o = 1+\frac{\ln(1+\coth\theta)}{\ln 2}$$

$$\theta = \frac{\pi S}{2b}$$

$$C_f(t/b) = 2\ln\left(\frac{2b-t}{b-t}\right)-\frac{t}{b}\ln\left[\frac{t(2b-t)}{(b-t)^2}\right]$$

The signal velocity is the same for both modes ($c/(\varepsilon_r)^{1/2}$), and the L and C of both modes can be found by replacing Z_0 in the "Symmetric Strip Line" section by $Z_{0,e}$ and $Z_{0,o}$, respectively.

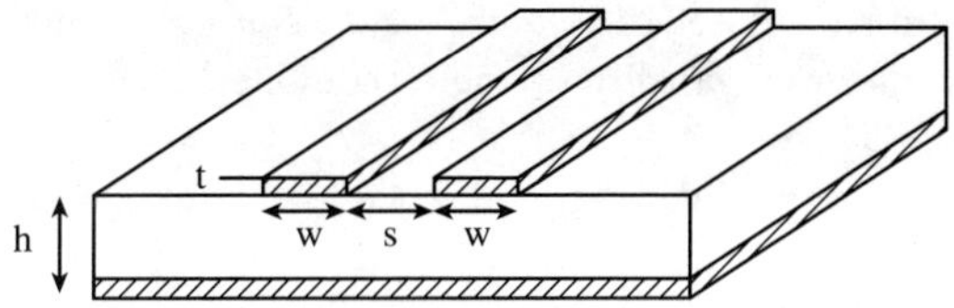

FIGURE 10.4 Coupled, symmetric microstrip configuration.

Coupled Microstrip Lines

The configuration is depicted in Fig. 10.4. It consists of the symmetric combination of the structure of Fig. 9.7. Again, we have two fundamental modes, but the modes are hybrid, i.e., not purely TEM. Much work has been done on this configuration. The formulas proposed in the literature are quite lengthy; and we refer the reader to [20] and [21]. Reference [20] gives a very good overview together with some simple approximations, and [21] gives the most accurate formulas, taking into account the frequency dependence. It is important to remark here that the two impedances, $Z_{0,e}$ and $Z_{0,o}$, can be found. They depend upon frequency. Both modes now have a different velocity. The data found in literature are typically expressed in terms of the effective dielectric constant. The modal field lines are both found in the air above the substrate and in the substrate itself. Hence, the field experiences an effective dielectric constant that is smaller than the dielectric constant of the substrate. The effective dielectric constant for the even mode ($\varepsilon_{r,e}$) will be higher than for the odd mode ($\varepsilon_{r,o}$), and are frequency dependent. The corresponding modal velocities are given by $c/(\varepsilon_{r,e})^{1/2}$ and $c/(\varepsilon_{r,o})^{1/2}$.

Two-Wire Line

See Section 9.2.

References

[1] T. Itoh, Ed., *Numerical Techniques for Microwave and Millimeter-Wave Passive Structures*, New York: John Wiley & Sons, 1989.

[2] J. A. Kong, Ed., *Progress in Electromagnetics Research*, Volumes 1–5, New York: Elsevier, 1989–1991.

[3] C. F. Coombs, Ed., *Printed Circuits Handbook*, 3rd ed., New York: McGraw-Hill, 1988.

[4] N. Faché, F. Olyslager, and D. De Zutter, *Electromagnetic and Circuit Modelling of Multiconductor Transmission Lines*, Oxford Engineering Series 35, Oxford: Clarendon Press, 1993.

[5] T. R. Arabi, T. K. Sarkar, and A. R. Djordjevic, "Time and frequency domain characterization of multiconductor transmission lines," *Electromagnetics*, vol. 9, no. 1, pp. 85–112, 1989.

[6] J. R. Griffith and M. S. Nakhla, "Time domain analysis of lossy coupled transmission lines," *IEEE Trans. MTT*, vol. 38, no. 10, pp. 1480–1487, Oct. 1990.

[7] B. J. Cooke, J. L. Prince, and A. C. Cangellaris, "S-parameter analysis of multiconductor, integrated circuit interconnect systems," *IEEE Trans. Computer-Aided Design*, vol. CAD-11, no. 3, pp. 353–360, March 1992.

[8] V. Rizzoli et al., "State of the art and present trends in nonlinear microwave CAD techniques," *IEEE Trans. MTT*, vol. 36, no. 2, pp. 343–363, Feb. 1988.

[9] R. Gilmore, "Nonlinear circuit design using the modified harmonic balance algorithm," *IEEE Trans. MTT*, vol. 34, no. 12, pp. 1294–1307, Dec. 1986.

[10] *IEEE Trans. Circuits Syst. Transactions, I: Fundamental Theory and Applications, Special Issue on Simulation, Modelling and Electrical Design of High-Speed and High-Density Interconnects*, vol. 39, no. 11, Nov. 1992.

[11] *IEEE Trans. MTT, Special Issue on Process-Oriented Microwave CAD and Modeling* vol. 40, no. 7, July 1992.

[12] *Analog Integrated Circuits Signal Process., Special Issue on High-Speed Interconnects*, vol. 5, no. 1, pp. 1–107, Jan. 1994.

[13] F. Olyslager, D. De Zutter, and A. T. de Hoop, "New reciprocal circuit model for lossy waveguide structures based on the orthogonality of the eigenmodes," *IEEE Trans. MTT,* vol. 42, no. 12, pp. 2261–2269, Dec. 1994.

[14] F. Olyslager and D. De Zutter, "Rigorous boundary integral equation solution for general isotropic and uniaxial anisotropic dielectric waveguides in multilayered media including losses, gain and leakage," *IEEE Trans. MTT,* vol. 41, no. 8, pp. 1385–1392, Aug. 1993.

[15] R. H. Jansen and M. Kirschning, "Arguments and an accurate model for the power-current formulation of microstrip characteristic impedance," *Arch. Elek. Übertragung,* vol. 37, no. 3/4, pp. 108–112, 1983.

[16] T. Dhaene and D. De Zutter, "CAD-oriented general circuit description of uniform coupled lossy dispersive waveguide structures," *IEEE Trans. MTT, Special Issue on Process-Oriented Microwave CAD and Modeling*, vol. 40, no. 7, pp. 1445–1554, July 1992.

[17] I. V. Lindell, "On the quasi-TEM modes in inhomogeneous multiconductor transmission lines," *IEEE Trans. MTT,* vol. 29, no. 8, pp. 812–817, 1981.

[18] F. Olyslager, N. Faché, and D. De Zutter, "New fast and accurate line parameter calculation of general multiconductor transmission lines in multilayered media," *IEEE Trans, MTT,* vol. MTT-39, no. 6, pp. 901–909, June 1991.

[19] H. A. Wheeler, "Transmission line properties of a stripline between parallel planes," *IEEE Trans. MTT,* vol. 26, pp. 866–876, Nov. 1978.

[20] T. Edwards, *Foundations for Microstrip Circuit Design,* 2nd ed., Chichester, U.K.: John Wiley & Sons, 1992.

[21] M. Kirschning and R. H. Jansen, "Accurate wide-range design equations for the frequency-dependent characteristics of parallel coupled microstrip lines," *IEEE Trans. MTT,* vol. 32, no. 1, pp. 83–90, Jan. 1984.

11

Time and Frequency Domain Responses

Luc Martens
Gent University, Belgium

Daniël De Zutter
Gent University, Belgium

11.1 Time Domain Reflectometry

Principles

Time domain reflectometry is used to characterize interconnections in the time domain. The setup essentially consists of a time domain step generator and a digital sampling oscilloscope (Figure 11.1) [1]. The generator produces a positive-going step signal with a well-defined rise time. The step is applied to the device under test. The reflected and the transmitted signals are shown on the oscilloscope. Measuring the reflected signal is called time domain reflectometry (TDR); the transmitted signal is measured using the time domain transmission (TDT) option.

The characteristic impedance levels and delay through an interconnection structure can be derived from the TDR measurements. The TDT measurement gives information about the losses (decrease of magnitude) and degradation of the rise time (filtering of high-frequency components). The TDR/TDT measurements also are used to extract an equivalent circuit consisting of transmission lines and lumped elements.

The fundamentals of TDR are discussed in detail in [2] and [3]. Reference [4] describes the applications of TDR in various environments including PCB/backplane, wafer/hybrids, IC packages, connectors, and cables.

One-Port Time Domain Reflectometry

Figure 11.2 demonstrates that the device under test is a simple resistor with impedance Z_L. In this case, a mismatch with respect to the reference or system impedance Z_0 exists. A reflected voltage wave will appear on the oscilloscope display algebraically added to the incident wave. The amplitude of the reflected voltage wave, E_r, is determined by the reflection coefficient of the load impedance, Z_L, with respect to the system impedance Z_0:

$$E_r = \rho E_i = \frac{Z_L - Z_0}{Z_L + Z_0} E_i \tag{11.1}$$

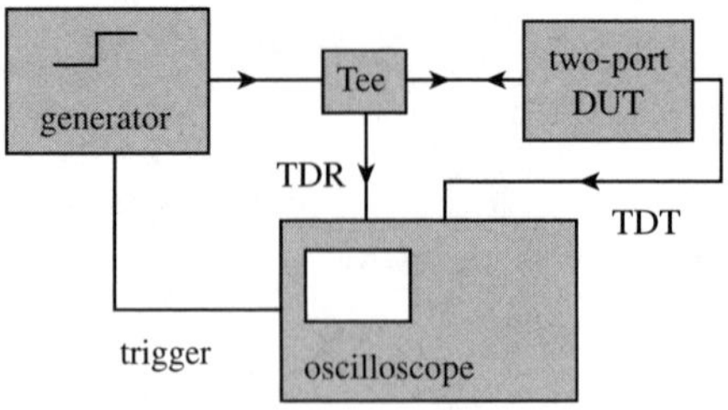

FIGURE 11.1 Setup of TDR/TDT measurements. DUT = device under test.

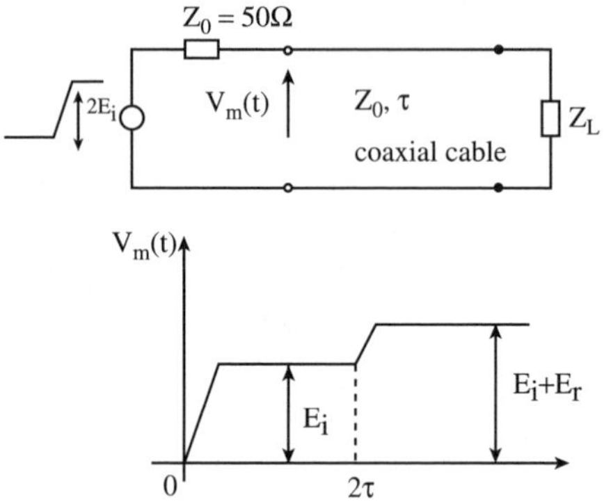

FIGURE 11.2 TDR measurement of an impedance Z_L.

Figure 11.2 also depicts the time domain picture shown on the oscilloscope for a load, the impedance Z_L, that is larger than Z_0. From the measurement of the magnitude E_r of the reflected voltage wave, the load impedance Z_L can be derived.

Time Domain Reflectometry Pictures for Typical Loads

The most simple loads to be measured are the open circuit and the short circuit. For ideal open-circuit and short-circuit loads the reflection coefficient is, respectively, 1 and –1. This means that the measured voltage doubles in the first case and goes to zero in the second case, when the reflected voltage wave arrives at the oscilloscope (Figure 11.3).

For any other real load impedance, the reflection coefficient lies between –1 and 1. If the real load impedance is larger than the reference impedance, the reflected voltage wave is a positive-going step signal. In this case the amplitude of the voltage is increased when the reflected wave is added to the input step (Figure 11.4). The reverse happens when the load impedance is lower than the reference impedance.

For complex load impedances, the step response is more complicated. For example, in the case of a series connection of a resistance and an inductance or a parallel connection of a resistance and a capacitance, a first-order step response is obtained. From the two pictures in Figure 11.5, we learn that a series inductance gives a positive dip, while the capacitance produces a negative dip.

Time Domain Reflectometric Characterization of an Interconnection Structure

One of the advantages of TDR is its ability to determine impedance levels and delays through an interconnection structure with multiple discontinuities. An example is shown in Figure 11.6 for a microstrip line connected to the measurement cable. We assume a perfect junction of the two transmission lines. The line is terminated in a load with impedance Z_L. Observe that two mismatches produce reflections that can be analyzed separately. The mismatch at the junction of the two transmission lines generates a reflected wave $E_{r1} = \rho_1 E_i$. Similarly, the mismatch at the load creates a reflection due to its reflection coefficient ρ_2. Both reflection coefficients are defined as:

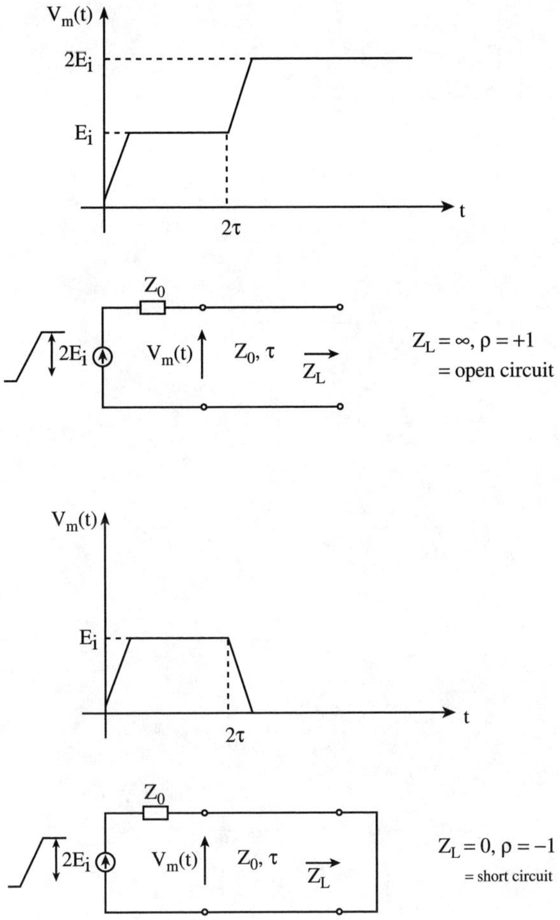

FIGURE 11.3 TDR pictures of an open- and a short-circuit termination.

$$\rho_1 = \frac{Z_0' - Z_0}{Z_0' + Z_0}$$
$$\rho_2 = \frac{Z_L - Z_0'}{Z_L + Z_0'} \tag{11.2}$$

After a time τ, the reflection at the junction of the transmission lines occurs. The voltage wave associated with this reflection adds to the oscilloscope's picture at the time instant 2τ. The voltage wave that propagates further in the microstrip line is $(1 + \rho_1)\, E_i$ and is incident on Z_L. The reflection at Z_L occurs at the time $\tau + \tau'$ and is given by:

$$E_{rL} = \rho_2 (1 + \rho_1) E_i \tag{11.3}$$

After a time $\tau + 2\tau'$, a second reflection is generated at the junction. The reflection is now determined by the reflection coefficient $\rho_1' = -\rho_1$. The voltage wave E_{r2} that is transmitted through the junction and propagates in the direction of the generator adds to the time domain picture at time instant $2\tau + 2\tau'$ and is given by:

$$E_{r2} = (1 + \rho_1) E_{rL} = (1 - \rho_1) \rho_2 (1 + \rho_1) E_i = (1 - \rho_1^2) \rho_2 E_i \tag{11.4}$$

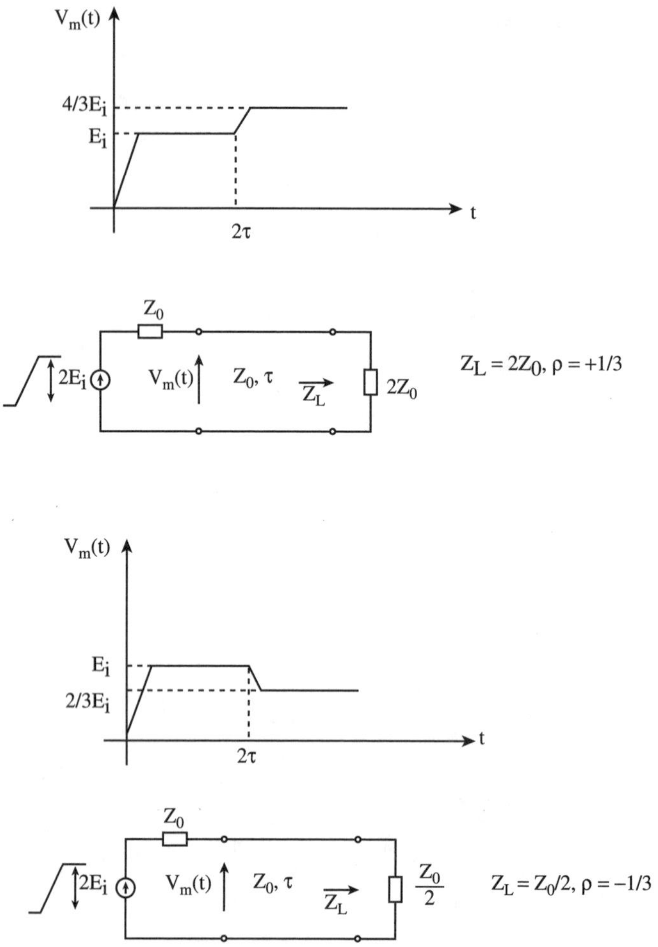

FIGURE 11.4 TDR pictures of real impedance terminations ($Z_L = 2Z_0$ and $Z_L = Z_0/2$).

If ρ_1 is small in comparison to 1, then

$$E_{r2} \approx \rho_2 E_i \tag{11.5}$$

which means that ρ_2 can be determined from the measurement of E_{r2}.

In this example, the measurement cable was perfectly matched to the generator impedance so that no reflection occurred at the generator side, which simplifies the time domain picture. In the case of an interconnection with many important discontinuities (high reflection coefficient), multiple reflections can prevent a straightforward interpretation of the oscilloscope's display.

11.2 Frequency Domain Network Analysis

Introduction

A distributed circuit also can by analyzed in the frequency domain. At low frequencies the circuits are characterized by their *Z*- or *Y*-parameters. At high frequencies, circuits are better characterized by *S*-parameters. We focus only on *S*-parameter characterization.

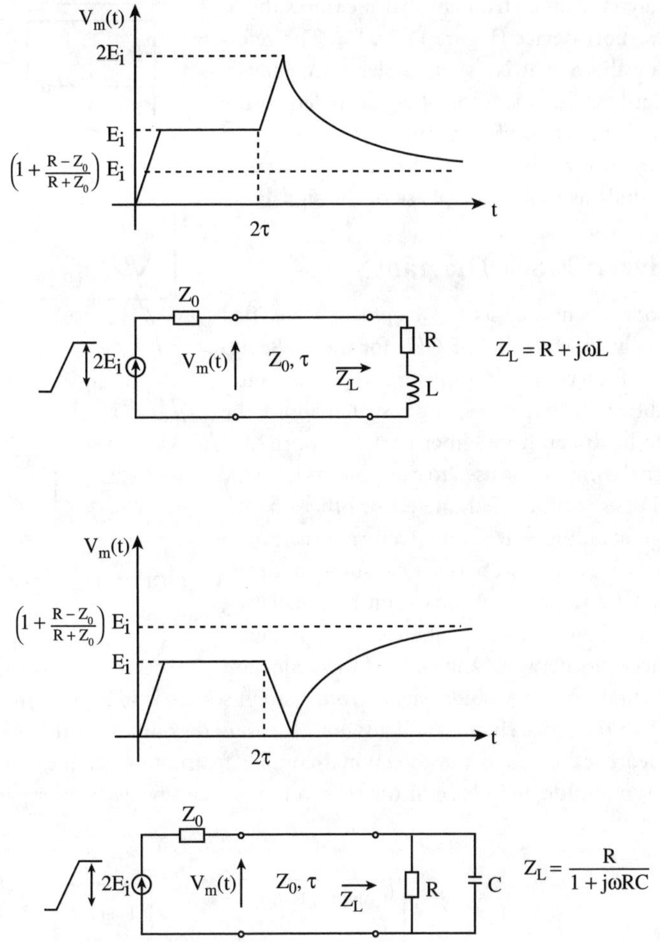

FIGURE 11.5 TDR pictures of two complex impedance terminations.

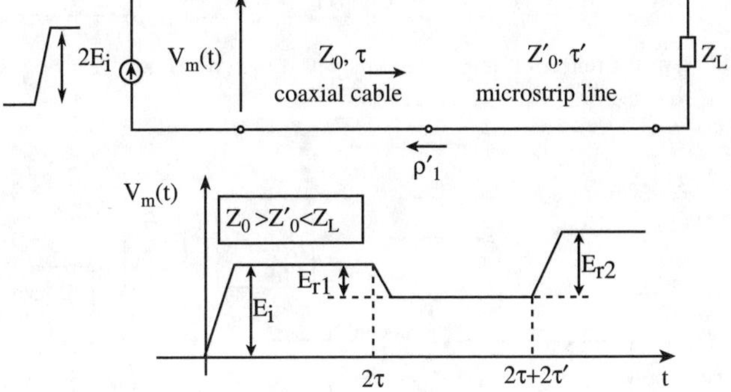

FIGURE 11.6 TDR measurement of a microstrip line terminated in an impedance Z_L.

The network analyzer is an instrument that measures the *S*-parameters of a two-port device (Figure 11.7). [5]. The reference impedance for the *S*-matrix is the system impedance, which is the standard 50 Ω. A sinusoidal signal is fed to the device under test and the reflected and transmitted signals are measured. If the network analyzer is of the vectorial type, it measures the magnitude as well as the phase of the signals.

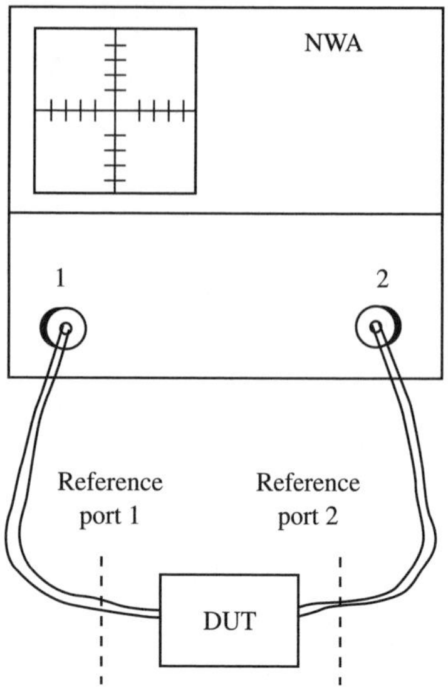

FIGURE 11.7 Network analyzer (NWA) setup.

Network Analyzer: Block Diagram

In operation, the source is usually set to sweep over a specified bandwidth (maximally 45 MHz to 26.5 GHz for the HP8510B network analyzer). A four-port reflectometer samples the incident, reflected, and transmitted waves, and a switch allows the network analyzer to be driven from either port 1 or port 2.

A powerful internal computer is used to calculate and display the magnitude and phase of the *S*-parameters or other quantities such as voltage standing wave radio, return loss, group delay, impedance, etc. A useful feature is the capability of determining the time domain response of the circuit by calculating the inverse Fourier transform of the frequency domain data.

Figure 11.8 depicts the network analyzer in the reflection measurement configuration. A sinusoidal signal from a swept source is split by a directional coupler into a signal that is fed to a reference channel and a signal that is used as input for the device under test. The reflected signal is fed back through the directional coupler to the test channel. In the complex ratio measuring unit the amplitude and phase of the test and reference signals are compared.

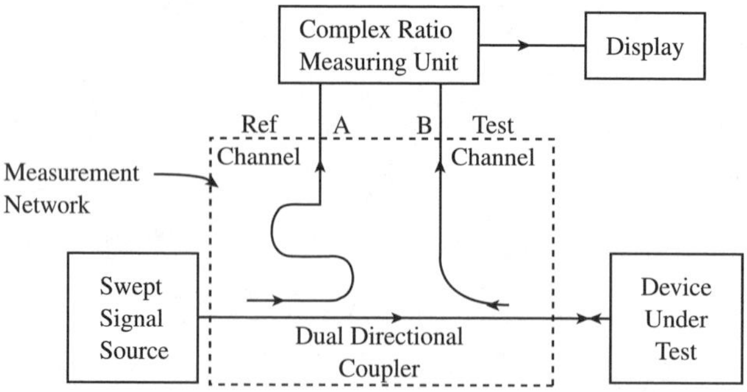

FIGURE 11.8 Block diagram for reflection measurements with the network analyzer.

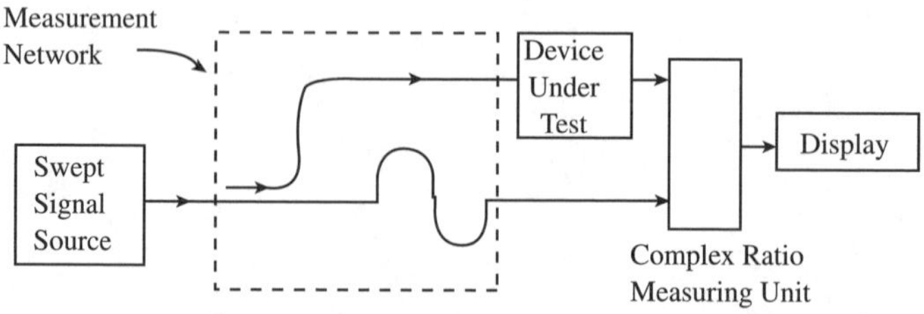

FIGURE 11.9 Block diagram for transmission measurements with the network analyzer.

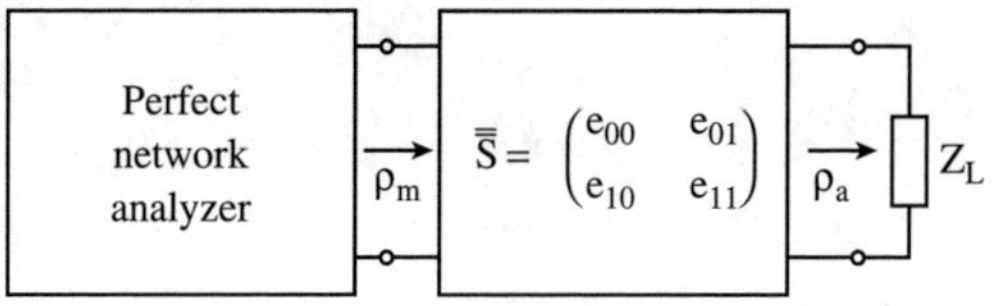

FIGURE 11.10 Error correction network for calibration of the S-parameter measurements.

The principle of the transmission measurement is the same as for the reflection configuration (Figure 11.9). In this setup the signal transmitted through the device under test is fed to the test channel. The other part of the signal of the swept source is transmitted through the directional coupler and fed to the reference channel. Again, amplitude and phase of the two signals are compared.

An important issue concerning the measurement of S-parameters is the calibration of the results. In order to perform a good calibration, the nature of the measurement errors must be well understood. These items are the subject of the following section.

Measurement Errors and Calibration

Measurement Errors

The network analyzer measurement errors [6] can be classified in two categories:

1. *Random Errors* — These are nonrepeatable measurement variations that occur due to noise, environmental changes, and other physical changes in the test setup. These are any errors that the system itself may not be able to measure.
2. *Systematic Errors* — These are repeatable errors. They include the mismatch and leakage terms in the test setup, the isolation characteristic between the reference and test signal paths, and the system frequency response.

In most S-parameters measurements, the systematic errors are those that produce the most significant measurement uncertainty. Because each of these errors produces a predictable effect upon the measured data, their effects can be removed to obtain a corrected value for the test device response. The procedure of removing these systematic errors is called calibration. We illustrate here the calibration for one-port measurements. The extension to two-ports is mathematically more involved, but the procedure is fundamentally the same.

One-Port Calibration: Error Correction Network

In the configuration for one-port measurements the real network analyzer can be considered as a connection of a "perfect" network analyzer and a two-port whose parameters are determined by the systematic errors. ρ_m is the measured reflection coefficient, while ρ_a is the actual one (Figure 11.10). The three independent coefficients of the error correction network [7] are:

1. *Directivity* e_{00} — The vector sum of all leakage signals appearing at the network analyzer test input due to the inability of the signal separation device to absolutely separate incident and reflected waves, as well as residual effects of cables and adapters between the signal separation device and the measurement plane.
2. *Source impedance mismatch* e_{11} — The vector sum of the signals appearing at the network analyzer test input due to the inability of the source to maintain absolute constant power at the test device input as well as cable and adaptor mismatches and losses outside the source leveling loop.
3. *Tracking or frequency response* $e_{10}e_{01}$ — The vector sum of all test setup variations in the magnitude and phase frequency response between the reference and test signal paths.

The relations between ρ_m and ρ_a are given by:

$$\rho_m = e_{00} + \frac{e_{01}e_{10}\rho_a}{1 - e_{11}\rho_a} \tag{11.6}$$

$$\rho_a = \frac{\rho_m - e_{00}}{e_{11}(\rho_m - e_{00}) + e_{10}e_{01}} \tag{11.7}$$

The error coefficients are determined by measuring standard loads. In the case of the SOL calibration, a short-circuit, open-circuit, and matched load are measured [5]. The standards are characterized over a specified frequency range (e.g., from 45 MHz to 26.5 GHz for the network analyzer HP8510B). The best-specified standards are coaxial. The quality of the coaxial standards is determined by the precision of the mechanical construction. The short circuit and the load are in most cases nearly perfect ($\rho_a^1 = -1$ and $\rho_a^3 = 0$, respectively). The open circuit behaves as a frequency-dependent capacitance:

$$\rho_a^2 = e^{-j2\text{arctg}(\omega C Z_0)} \quad \text{with } C = C_0 + f^2 C_2 \tag{11.8}$$

Z_0 is the characteristic impedance of the coaxial part of the standard. C_0 and C_2 are specified by the constructor of the hardware. Once the reflection coefficients of the standards are specified and the standards are measured, a set of linear equations in the error coefficients is derived:

$$a\rho_a^i + b - c\rho_a^i\rho_m^i = \rho_m^i \quad i = 1, 2, 3 \tag{11.9}$$

where a = $e_{01}e_{10} - e_{00}e_{11}$, $b = e_{00}$, and $c = -e_{11}$. This set is easily solved in the error coefficients which are then used together with the measurement of ρ_m to determine ρ_a.

Other calibration methods (TRL, LRM) are described in [8] and [9]. Analogous calibration techniques also were developed for time domain measurements (TDR/TDT) [10]. Applications of S-parameter measurements on circuits on PCBs are described in [11].

Error Checking and Verification of the Calibration

Checking errors in the calibration is done by applying the calibration to a measured standard. The specified reflection response of the standard must be found. In Figure 11.11, the uncalibrated measurement of the open-circuit standard is shown. After calibration, we obtain the specified response (capacitance model) of the standard (Figure 11.11). The calibration can be further verified on a standard transmission line.

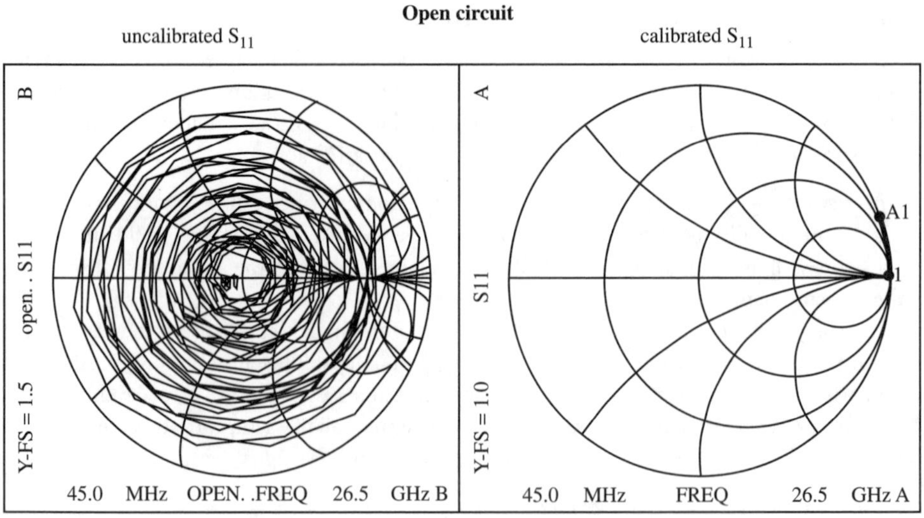

FIGURE 11.11 Smith chart representation of the uncalibrated and calibrated S_{11} data of an open circuit (measurements in the range of 45 MHz to 26.5 GHz).

References

[1] A. M. Nicolson, C. L. Benett, Jr., D. Lammensdorf, and L. Susman, "Applications of time domain metrology to the automation of broad-band microwave measurements," *IEEE Trans. MTT,* vol. MTT-20, no. 1, pp. 3–9, Jan. 1972.

[2] Agilent Technologies, "TDR fundamentals for use with HP54120T digitizing oscilloscope and TDR," HP-Appl. Note 62, Apr. 1988.

[3] Agilent Technologies, "Improving time domain network analysis measurements," HP-Appl. Note 62-1, Apr. 1988.

[4] Agilent Technologies, "Advanced TDR techniques," HP-Appl. Note 62-3, May 1990.

[5] R. A. Hackborn, "An automatic network analyzer system," *Microwave J.*, vol. pp. 45–52, May 1968.

[6] B. Donecker, "Accuracy predictions for a new generation network analyzer," *Microwave J.*, vol. 27, pp. 127–141, June 1984.

[7] J. Williams, "Accuracy enhancement fundamentals for network analyzers," *Microwave J.*, vol. 32, pp. 99–114, Mar. 1989.

[8] G. F. Engen and C. A. Hoer, "Thru-reflect-line: an improved technique for calibrating the dual six-port automatic network analyzer," *IEEE Trans. MTT,* vol. MTT-27, pp. 987–993, Dec. 1979.

[9] D. Wiliams, R. Marks, and K. R. Phillips, "Translate LRL and LRM calibrations," *Microwaves RF,* vol. 30, pp. 78–84, Feb. 1991.

[10] T. Dhaene, L. Martens, and D. De Zutter, "Calibration and normalization of time domain network analyzer measurements," *IEEE Trans. MTT,* vol. MTT-42, pp. 580–589, Apr. 1994.

[11] P. Degraeuwe, L. Martens, and D. De Zutter, "Measurement set-up for high-frequency characterization of planar contact devices," in Proc. of 39th Automatic RF Techniques (ARFTG) Meeting, Albuquerque, NM, pp. 19–25, June 1992.

12

Distributed RC Networks

Vladimír Székely

Budapest University of Technology and Economics, Hungary

In everyday practice, we often encounter RC networks that are inherently distributed (DRC lines). All the resistors and the interconnection lines of an IC, the channel regions of the FET, and MOS transistors are DRC lines. The electrical behavior of such networks is discussed in the current section.

12.1 Uniform Distributed RC Lines

First, let us consider the so-called lossless RC lines. The structure in Fig. 12.1 has only serial resistance and parallel capacitance. Both the resistance and the capacitance are distributed along the x axis. r and c denote the resistance and the capacitance pro unit length, respectively. In this section, these values are considered constants, which means that the RC line is *uniform*.

Consider a Δx length section of the structure. Using the notations of Fig. 12.1, the following relations can be given between the currents and voltages:

$$\Delta v = -ir\Delta x \qquad \Delta i = -\frac{\partial v}{\partial t}c\Delta x \tag{12.1}$$

By taking the limit as $\Delta x \to 0$, we obtain a pair of differential equations

$$\frac{\partial v}{\partial x} = -ri \qquad \frac{\partial i}{\partial x} = -c\frac{\partial v}{\partial t} \tag{12.2}$$

which describe the behavior of the lossless RC line. Substituting the first equation of (12.2) after derivation into the second equation of (12.2) results in

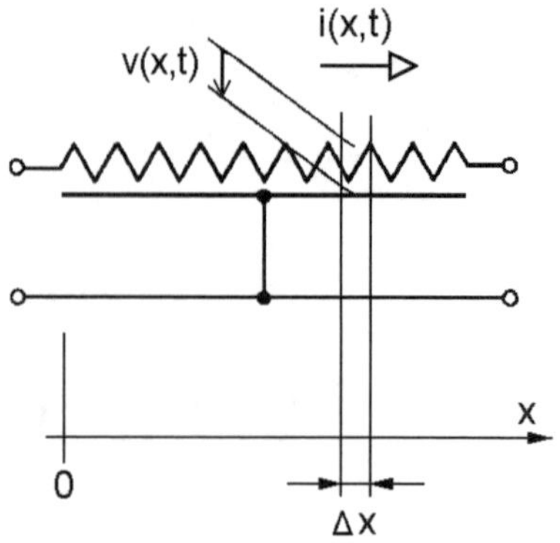

FIGURE 12.1 Lossless DRC line.

$$\frac{\partial v}{\partial t} = \frac{1}{rc}\frac{\partial^2 v}{\partial x^2} \tag{12.3}$$

This shows that the time-dependent voltage along the line $v(x, t)$ is determined by a homogeneous, constant-coefficient, second-order partial differential equation, with the first time derivative on one side and the second spatial derivative on the other side. This kind of equation is called a *diffusion equation*, as diffusion processes are described using such equations.

Solution in the Time Domain

The differential Eq. (12.3) is fulfilled by the simple, spatially Gauss-like function

$$v(x,t) = \frac{Q/c}{2\sqrt{\pi t/rc}}\exp\left(-\frac{x^2}{4t/rc}\right) \tag{12.4}$$

This can be proved easily by substituting the function into (12.3).

Solution (12.4) describes the physical situation in which the uniform RC structure is of infinite length in both directions, and a short, impulse-like Q charge injection is applied at $x = 0$ in the $t = 0$ instant. This means a Dirac-δ excitation at $x = 0$ both in charge and voltage. As time elapses, the charge spreads equally in both directions. The extension of the charge and voltage wave is ever increasing, but their amplitude is decreasing [see Fig. 12.2(a)].

The same result is represented in a different way in Fig. 12.2(b). The time dependence of the voltage is shown at the $x = H$, $x = 2H$, etc., spatial positions. As we move away from the $x = 0$ point, the maximum of the impulse appears increasingly later in time, and the originally sharp impulse is increasingly extended. This means that the RC line delays the input pulse and at the same time strongly spreads it. The RC line is *dispersive.*

Superposition

Equation (12.3) is homogeneous and linear. This means that any sum of the solutions is again a solution. Based on this fact, the problem of Fig. 12.3(a) can be solved. At $t = 0$, the voltage distribution of the RC line is given by an arbitrary $U(x)$ function. For $t > 0$, the distribution can be calculated by dividing the $U(x)$ function into $\Delta\xi \to 0$ elementary slices. These may be considered to be individual Dirac-δ excitations and the responses given to them can be summarized by integration:

$$v(x,t) = \frac{1}{2\sqrt{\pi t/rc}}\int_{-\infty}^{\infty} U(\xi)\exp\left(-\frac{(x-\xi)^2}{4t/rc}\right)d\xi \tag{12.5}$$

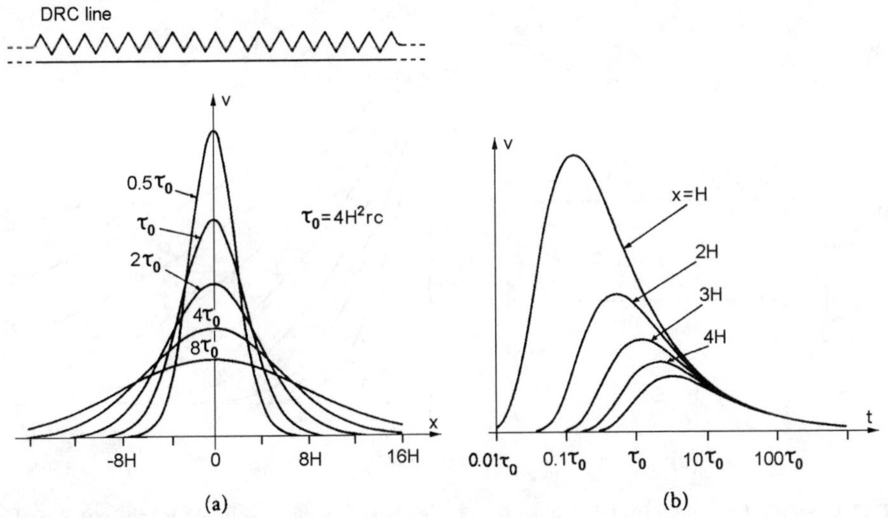

FIGURE 12.2 Effect of an impulse-like charge injection at $x = 0$. (a) Voltage distribution in subsequent time instants. (b) Voltage transients in different distances from the injection point.

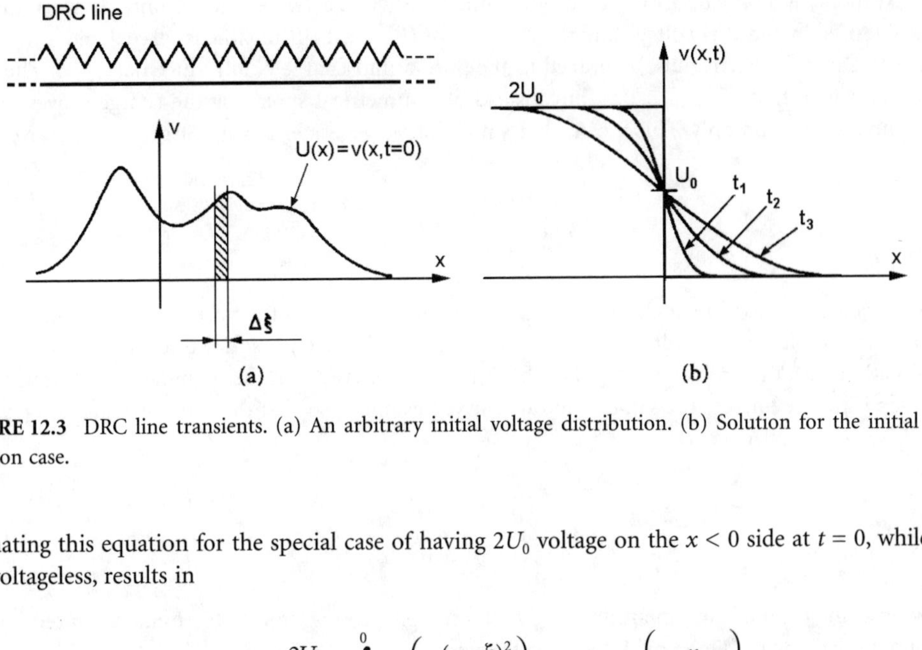

FIGURE 12.3 DRC line transients. (a) An arbitrary initial voltage distribution. (b) Solution for the initial step-function case.

Evaluating this equation for the special case of having $2U_0$ voltage on the $x < 0$ side at $t = 0$, while $x > 0$ is voltageless, results in

$$v(x,t) = \frac{2U_0}{2\sqrt{\pi t/rc}} \int_{-\infty}^{0} \exp\left(-\frac{(x-\xi)^2}{4t/rc}\right) d\xi = U_0 \operatorname{erfc}\left(\frac{x}{2\sqrt{t/rc}}\right) \tag{12.6}$$

where the integral of the Gauss function is notated by erfc(x), the complementary error function.[1] The originally abrupt voltage step is getting increasingly less steep with time [Fig. 12.3(b)]. In the middle at $x = 0$, the voltage remains U_0.

[1] $\operatorname{erfc}(x) = 1 - \frac{2}{\sqrt{\pi}}\int_0^x \exp(-y^2)\, dy = \frac{2}{\sqrt{\pi}}\int_x^\infty \exp(-y^2)\, dy$

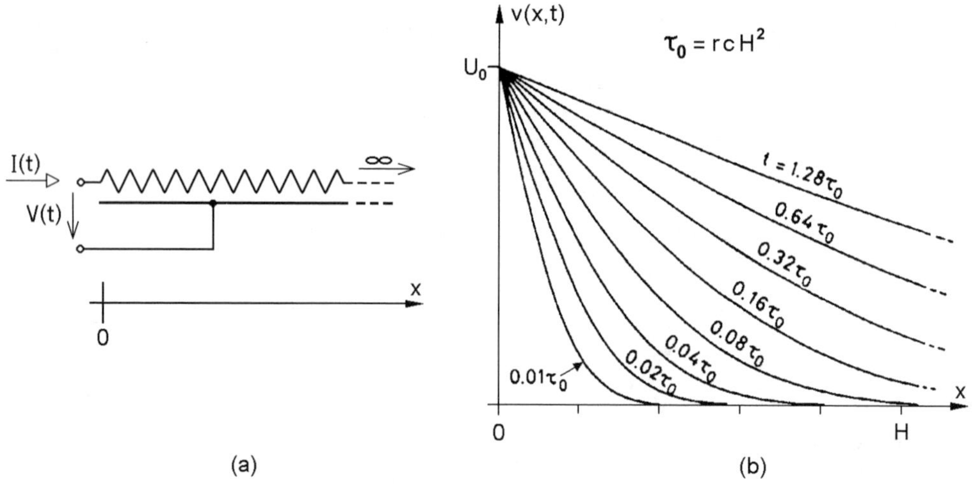

FIGURE 12.4 Semi-infinite uniform DRC line. (a) Notation. (b) Normalized solution for the initially relaxed line.

Semi-Infinite Line

Our next model is a bit closer to practice; the uniform RC line extends to $x \geq 0$ only. At $x = 0$ the port is characterized by the $V(t)$ voltage and the $I(t)$ current (Fig. 12.4). If the line is relaxed and a $I(t) = \delta(t)$ current (a Dirac-δ current pulse) is forced to the port, a unit charge is introduced at $x = 0$. The result will be similar to that of Fig. 12.2(a), but instead of symmetrical spreading the charge moves towards the positive x direction only. This means that a unit charge generates a twice-larger voltage wave

$$v(x,t) = \frac{1/c}{\sqrt{\pi t/rc}} \exp\left(-\frac{x^2}{4t/rc}\right) \tag{12.7}$$

Let us consider the case in which step function excitation is given to the port of Fig. 12.4. At $t < 0$, the port and the whole line are voltage free; at $t \geq 0$ a constant U_0 voltage is forced to the port. Comparing this situation with the problem of Fig. 12.3(b), it can be observed that the boundary conditions for the $x > 0$ semi-infinite line are the same as in our current example, so that the solution must to be similar as well [see Fig. 12.4(b)]:

$$v(x,t) = U_0 \operatorname{erfc}\left(\frac{x}{2\sqrt{t/rc}}\right) \tag{12.8}$$

Applying at the $t = 0$ instant an arbitrary $W(t)$ forced voltage excitation to the initially relaxed line, the response is given by the Duhamel integral as follows:

$$w(x,t) = \int_0^t \frac{dW(\tau)}{d\tau} \operatorname{erfc}\left(\frac{x}{2\sqrt{(t-\tau)/rc}}\right) d\tau \tag{12.9}$$

Finite DRC Line

Let the DRC line of length L of Fig. 12.5(a) be closed at $x = L$ with a short circuit. Let the line at $t < 0$ be relaxed and assume a $W(t > 0)$ voltage excitation at the $x = 0$ port. Using the $w(x, t)$ voltage response of the semi-infinite line (12.9), the response function for the short-terminated line of length L

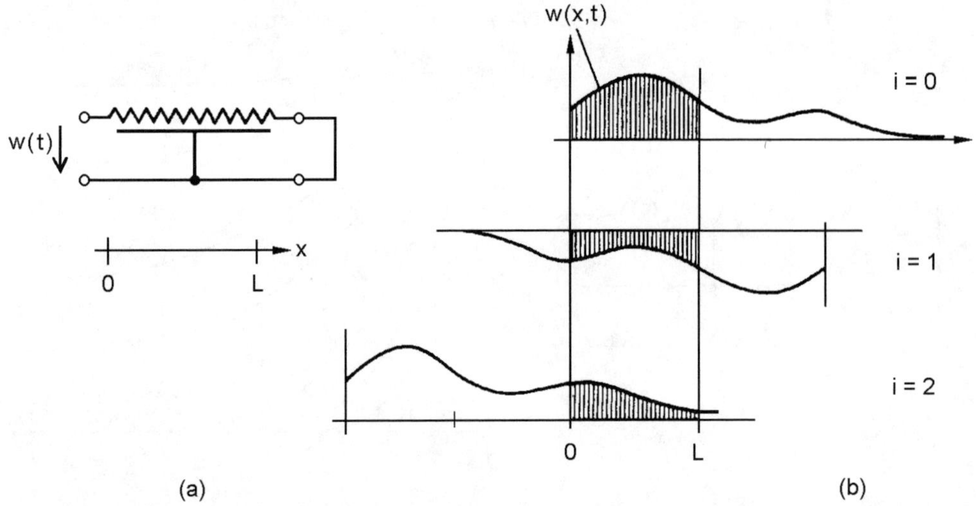

FIGURE 12.5 Finite-length uniform DRC line. (a) DRC line with short-circuit at $x = L$. (b) Visualization of the mirroring procedure.

$$\begin{aligned} v(x,t) &= \sum_{i=0}^{\infty} (-1)^i \cdot w\left(2iL + (-1)^i x, t\right) \\ &= \sum_{k=0}^{\infty} \left(w(2kL + x, t) - w(2kL + 2L - x, t)\right) \end{aligned} \tag{12.10}$$

This result is illustrated in Fig. 12.5(b). The $v(x, t)$ function is given as the sum of the shifted, negated, and mirrored replicas of the $w(x, t)$ function, so that it is a valid solution as well. The $x = L$ boundary condition is the short circuit $v(x = L, t) = 0$. The $i = 0$ and $i = 1$ functions are the same size with different signs at $x = L$, so they cancel each other. The same is true for $i = 2$ and 3, and so on. The $x = 0$ boundary condition $v = W(t)$ is fulfilled by the $i = 0$ function, while the further functions cancel each other in pairs (the $i = 1$ and 2, etc.).

The result can be interpreted as the response of the semi-infinite line being mirrored with negative sign on the short termination. In the case of Fig. 12.5(a), the termination on both the $x = 0$ and $x = L$ ends is assured by zero impedance short circuit; the resultant voltage function comes from the successive back and forth mirroring between these two "mirrors".

It is easy to understand that a termination with an open circuit results in mirroring without sign change. (At this termination the current equals to zero so that the dv/dx derivative equals to zero as well. This requirement is always fulfilled in the mirroring point summarizing the continuous incident function with its mirrored version.) According to this, the voltage on the open-terminated line of Fig. 12.6(a) is

$$v(x,t) = \sum_{k=0}^{\infty} (-1)^k \left(w(2kL + x, t) + w(2kL + 2L - x, t)\right) \tag{12.11}$$

which in the case of step function excitation with U_0 amplitude is

$$v(x,t) = U_0 \sum_{k=0}^{\infty} (-1)^k \left[\operatorname{erfc}\left(\frac{2kL + x}{2\sqrt{t/rc}}\right) + \operatorname{erfc}\left(\frac{2kL + 2L - x}{2\sqrt{t/rc}}\right)\right] \tag{12.12}$$

This function is given in Fig. 12.6(b) for some time instants.

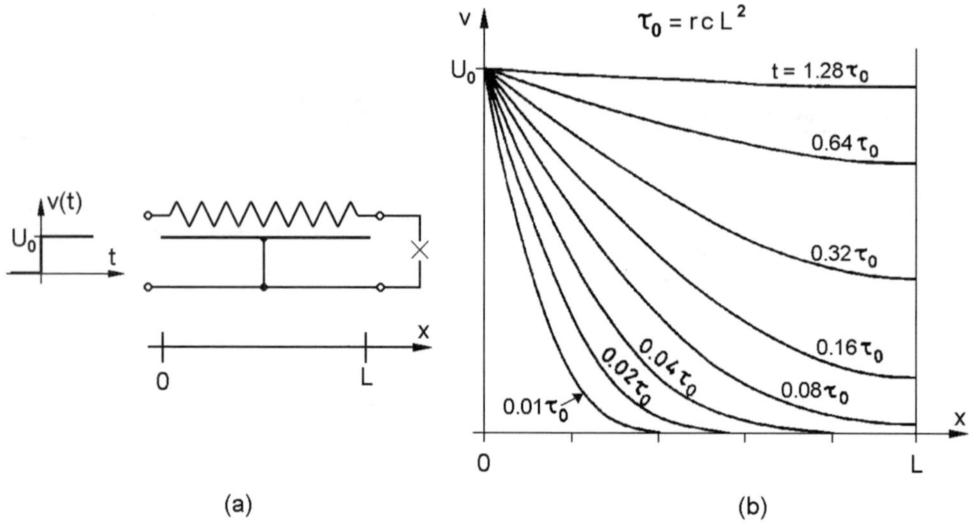

FIGURE 12.6 Finite-length uniform DRC line. (a) Open-circuit at the far end. (b) Normalized solution for the initially relaxed line.

Solution in the Frequency Domain

To find the solution of the differential Eq. (12.3), the following trial function can be used

$$v(x,t) = v \cdot \exp(j\omega t) \cdot \exp(\gamma x) \tag{12.13}$$

Substituting this function into (12.3) results in the following so-called *dispersion equation*:

$$\gamma = \sqrt{j\omega rc} = (1+j)\frac{1}{\sqrt{2}}\sqrt{\omega rc} \tag{12.14}$$

This means that a wave-like solution exists as well. However, it is strongly collapsing because the real and imaginary parts of γ are always equal, which means that the attenuation on a path of λ wavelength is exp $(-2\pi) \cong 1/535$.

The lossless DRC line can be considered to be a special telegraph line having neither serial inductance nor shunt conductance. The telegraph line theory can be conveniently used at the calculation of uniform DRC networks. The γ propagation constant and the $\mathbf{Z}_0$ characteristic impedance for the present case are:

$$\gamma = \sqrt{src} \qquad \mathbf{Z}_0 = \sqrt{\frac{r}{sc}} \tag{12.15}$$

With these the two-port impedance parameters and chain parameters of an RC line of length L can be given as follows:

$$\mathbf{Z}_{ij} = \mathbf{Z}_0 \begin{bmatrix} \operatorname{cth}\gamma L & 1/\operatorname{sh}\gamma L \\ 1/\operatorname{sh}\gamma L & \operatorname{cth}\gamma L \end{bmatrix} \qquad \begin{bmatrix} A & B \\ C & D \end{bmatrix} = \begin{bmatrix} \operatorname{ch}\gamma L & \mathbf{Z}_0 \operatorname{sh}\gamma L \\ \frac{1}{\mathbf{Z}_0}\operatorname{sh}\gamma L & \operatorname{ch}\gamma L \end{bmatrix} \tag{12.16}$$

If one of the ports is terminated by the impedance of $\mathbf{Z}_t$, the following impedance can be "seen" on the opposite port

$$\mathbf{Z}_{in} = \mathbf{Z}_0 \frac{\mathbf{Z}_t \operatorname{ch} \gamma L + \mathbf{Z}_0 \operatorname{sh} \gamma L}{\mathbf{Z}_t \operatorname{sh} \gamma L + \mathbf{Z}_0 \operatorname{ch} \gamma L} \tag{12.17}$$

Uniform, Lossy DRC Lines

In some cases, the DRC structure also has shunt conductance, which means that it is lossy. The value of this conductance for the unit length is notated by g. In such a case, without giving the details of the calculation, the $v(x, t)$ line voltage can be determined by the solution of the equation

$$\frac{\partial v}{\partial t} = \frac{1}{rc}\frac{\partial^2 v}{\partial x^2} - \frac{g}{c}v \tag{12.18}$$

The following forms of the characteristic impedance and the propagation constant can be used now in the frequency domain

$$\gamma = \sqrt{r(g + sc)} \qquad \mathbf{Z}_0 = \sqrt{\frac{r}{g + sc}} \tag{12.19}$$

It is an interesting fact that the charge carrier motion in the base region of homogeneously doped bipolar transistors can be described by formally similar equations, so that the intrinsic transients of the bipolar transistors can be exactly modeled by lossy DRC two-ports [5].

Example 12.1 Wiring Delays: Neither the series resistance nor the stray capacitances of the interconnection leads of integrated circuits are negligible. As an example, in the case of a polysilicon line of 1 μm width $r \cong 50$ kΩ/mm, $c \cong 0.04$ pF/mm. This means that these wires should be considered to be DRC lines. The input logical levels appear on their output with a finite delay. Let us determine the delay of a wire of length L. From (12.12),

$$v(L,t) = U_0 \sum_{k=0}^{\infty} 2(-1)^k \operatorname{erfc}\big((2k+1)\vartheta\big)$$

where

$$\vartheta = \frac{L}{2\sqrt{t/rc}}$$

The value of the summation over k will reach 0.9 at $\vartheta = 0.5$, so that the voltage at the end of the line will reach the 90% of U_0 after a time delay of

$$t_{delay} \cong rcL^2 \tag{12.20}$$

Note that the time delay increases with the square of the length of the wire. In the case of $L = 1$ mm the time delay of this polysilicon wire is already 2 ns, which is more than the time delay of a CMOS logical gate. For lengthy wires (> 0.2 ÷ 0.5 mm), metal wiring must be applied with its inherently small resistivity.

Example 12.2 Parasitic Effects of IC Resistors: In an IC amplifier stage the transistor is loaded with, e.g., $R = 10$ kΩ [Fig. 12.7(a)]. This resistor has been fabricated by the base diffusion, the sheet resistance is 200 Ω, and the parasitic capacitance is 125 pF/mm². The width of the resistor is 4 μm. Let us determine the impedance of the resistor in the 1 to 1000 MHz range.

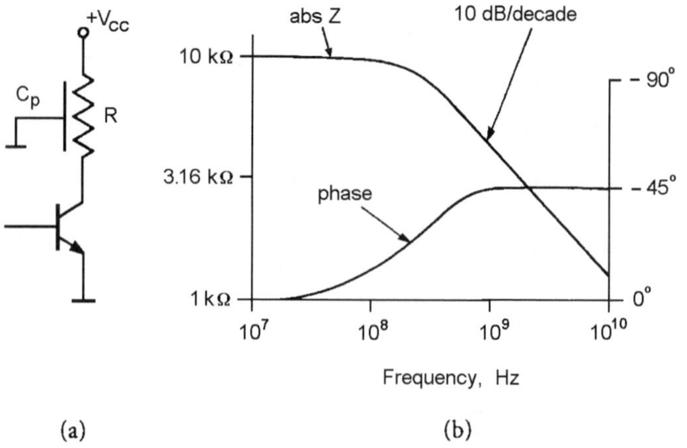

FIGURE 12.7 A simple IC amplifier stage. (a) The load resistor is in fact a DRC line. (b) The amplitude and phase plot of the load.

The resistance can be realized in 4 μm × 200 μm size. The total parasitic capacitance is $C_p = 0.1$ pF. The impedance of the transistor-side port can be calculated according to (12.17), considering that the opposite port is short-terminated, as

$$\mathbf{Z}_{port} = \mathbf{Z}_0 \operatorname{th} \gamma L = \sqrt{\frac{r}{\mathbf{s}c}} \operatorname{th}\left(\sqrt{\mathbf{s}rc}\, L\right) = \sqrt{\frac{R}{\mathbf{s}C_p}} \operatorname{th} \sqrt{\mathbf{s}RC_p} \tag{12.21}$$

Using the $\mathbf{s} = j\omega$ substitution, along with the actual data, the amplitude and phase functions of Fig. 12.7(b) can be obtained for the impedance. At 10 MHz, the phase shift caused by the parasitic capacitance is negligible, but at 100 MHz it is already considerable.

It is important to recognize that in the case of half size linewidths the size of the resistor will be only 2 × 100 μm, which results in one-fourth of the previous value in C_p. This means that the capacitance becomes disturbing only at four times larger frequencies.

Note in Fig. 12.7(b) that the amplitude function shows a 10 dB/decade decay and the phase keeps to 45°, as if the load would be characterized by a "half pole". This 10 dB/decade frequency dependence often can be experienced at DRC lines.

12.2 Nonuniform Distributed RC Lines

In some cases, the capacitance and/or the resistance of the DRC line shows a spatial dependency. This happens if the width of the lead strip is modulated in order to reach some special effects (Fig 12.8(a), *tapered* RC line). In case of a biased IC resistance, the capacitance changes along the length of the structure as well because of the voltage dependency of the junction capacitance. These structures are referred to as *nonuniform* DRC lines.

Let the spatially dependent resistance and capacitance pro unit length be notated by $r(x)$ and $c(x)$, respectively. The following equations can be given for the structure:

$$\frac{\partial v}{\partial x} = -r(x)i \qquad \frac{\partial i}{\partial x} = -c(x)\frac{\partial v}{\partial t} \tag{12.22}$$

With these, the following differential equation can be written:

$$\frac{\partial v}{\partial t} = \frac{1}{c(x)}\frac{\partial}{\partial x}\left(\frac{1}{r(x)}\frac{\partial v}{\partial x}\right) \tag{12.23}$$

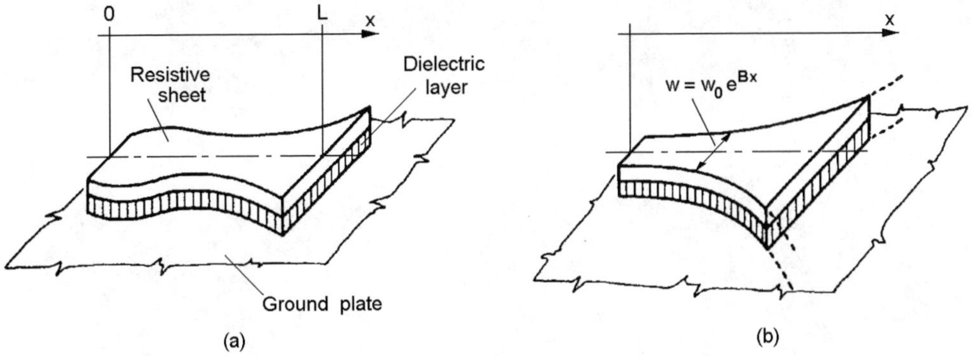

FIGURE 12.8 Nonuniform distributed RC lines. (a) Tapered line. (b) Exponentially tapered line.

We can obtain a more convenient form if we consider as an independent variable (instead of the x spatial co-ordinate) the total ρ resistance related to a given reference point (e.g., to the $x = 0$ point), as follows:

$$\rho(x) = \int_0^x r(\xi)\,d\xi \qquad r(x) = \frac{\partial \rho}{\partial x} \tag{12.24}$$

The variable defined this way can be considered as a kind of arc-length parameter. It has been introduced by [2]. With this new variable

$$\frac{\partial v}{\partial t} = \frac{1}{K(\rho)}\frac{\partial^2 v}{\partial \rho^2} \tag{12.25}$$

where

$$K(\rho) = K(\rho(x)) = \frac{c(x)}{r(x)} \tag{12.26}$$

The $K(\rho)$ function describes well the spatial parameter changes of the RC line; that is, the structure of the line. Therefore, the $K(\rho)$ function is called the **structure function.** Those DRC structures for which the $K(\rho)$ functions are the same are considered to be electrically equivalent.

Reference [2] uses the $\sigma(\rho)$ integral or *cumulative* version of the structure function:

$$\sigma(\rho) = \int_0^\rho K(\rho)\,d\rho = \int_0^{x(\rho)} \frac{c(x)}{r(x)}\frac{d\rho}{dx}\,dx = \int_0^{x(\rho)} c(x)\,dx \tag{12.27}$$

This is the total capacitance related to the $x = 0$ point. This means that the cumulative srtucture function is the total capacitance versus total resistance map of the structure. An example of such a map is plotted in Fig. 12.9.

The differential Eq. (12.25) is homogeneous and linear; therefore, superposition can be used. Because this equation is of variable coefficient type, however, analytic solution can be expected only rarely. Such a case is that of the $K = K_0/\rho^4$ structure function for which

$$v(\rho,t) = const\,\frac{1}{t^{3/2}}\exp\left(-\frac{K_0}{4\rho^2 t}\right) \tag{12.28}$$

Another form of (12.25) is also known. To obtain this form, we should turn to the **s** domain with

$$\mathbf{s}v = \frac{1}{K(\rho)}\frac{\partial^2 v}{\partial \rho^2} \tag{12.29}$$

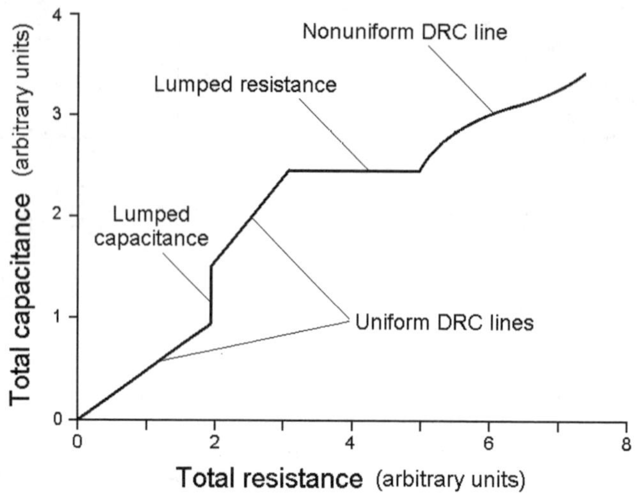

FIGURE 12.9 Cumulative structure function.

Let us introduce the following new variable:

$$\mathbf{Z}(\rho) = \frac{v(\mathbf{s},\rho)}{i(\mathbf{s},\rho)} = \frac{v}{-\frac{1}{r}\frac{\partial v}{\partial x}} = -\frac{v}{\frac{\partial v}{\partial \rho}} \tag{12.30}$$

This variable is in fact the impedance of the line at the location of ρ. After rearrangements, the

$$\frac{d\mathbf{Z}}{d\rho} = 1 + \mathbf{s}K(\rho)\mathbf{Z}^2 \tag{12.31}$$

equation can be obtained. This is called the Riccati differential equation. In the case of a known $K(\rho)$ structure function, the one-port impedance of the nonuniform line can be determined from it by integration. In some cases, even the analytic solution is known. Such a case is the exponentially tapered line of Fig. 12.8(b), for which

$$r(x) = \frac{R_s}{w_0}\exp(-Bx) \qquad c(x) = C_p w_0 \exp(Bx) \qquad K(\rho) = \frac{R_s C_p}{B^2}\frac{1}{\rho^2} \tag{12.32}$$

where R_s is the sheet resistance of the structure, C_p is the capacitance per unit area, and ρ is related to the point in the infinity. If the port in the infinity is shorted, the impedance of the location of ρ is

$$\mathbf{Z}(\mathbf{s}) = \frac{\sqrt{1 + 4\mathbf{s}R_s C_p / B^2} - 1}{2\mathbf{s}R_s C_p / B^2}\rho \tag{12.33}$$

In other cases, numerical integration of (12.31) leads to the solution.

Approximation with Concatenated Uniform Sections

The following model can be used for approximate calculation of nonuniform structures. We split the structure function into sections [see Fig. 12.10(a)] and use stepwise approximation. Inside the sections, K = constant, so that they are uniform sections. Concatenating them according to Fig. 12.10(b) an approximate model is obtained.

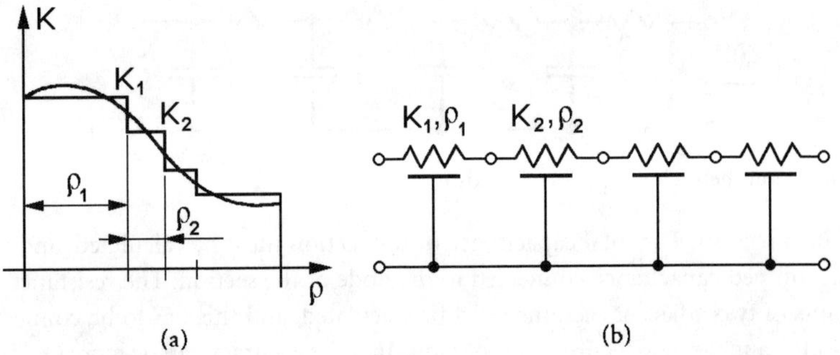

FIGURE 12.10 Approximation with concatenated uniform sections. (a) Stepwise approximation of the structure function. (b) Approximate model.

In the frequency domain, the overall parameters of the resultant two-port can be easily calculated. The chain parameter matrices of the concatenated sections have to be multiplied in the appropriate order. The time domain behavior can be calculated by inverse Laplace transformation.

Asymptotic Approximation for Large s

The chain parameters of a nonuniform DRC line can be approximately written as

$$\begin{bmatrix} A & B \\ C & D \end{bmatrix} \cong \frac{1}{2}\exp\left(\Delta\sqrt{\mathbf{s}}\right)\begin{bmatrix} \lambda & \mu/\sqrt{\mathbf{s}} \\ \sqrt{\mathbf{s}}/\mu & 1/\lambda \end{bmatrix} \tag{12.34}$$

where

$$\Delta = \int_0^{R_0} \sqrt{K(\rho)}\, d\rho \qquad R_0 = \rho(L) \qquad \lambda = \left(\frac{K(R_0)}{K(0)}\right)^{1/4} \qquad \mu = \left(K(R_0)K(0)\right)^{-1/4}$$

This approximation is valid for large **s** values and for sufficiently smooth function $K(\rho)$ [2].

Lumped Element Approximation

Distributed RC networks can be approximated by lumped element RC networks as well. The case of a lossless line is depicted in Fig. 12.11(a). The element values can be determined in either of the following two ways.

1. In the case of a known structure spatial discretization can be used. The nonuniform line must be split into sections of width h [see Fig. 12.11(b)]. A node of the network is associated with the

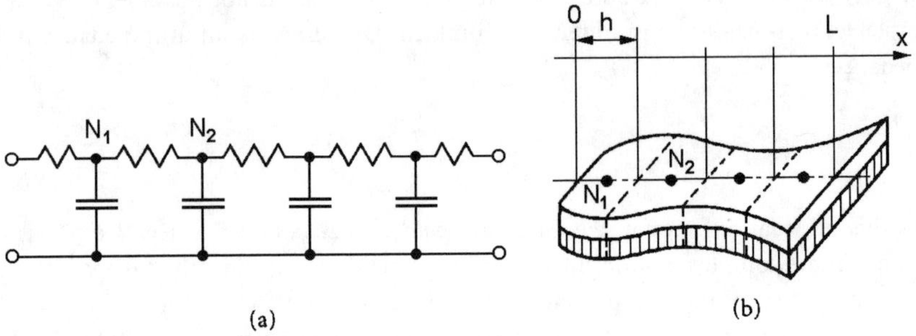

FIGURE 12.11 Lumped element approximation. (a) Network model. (b) The line split into sections.

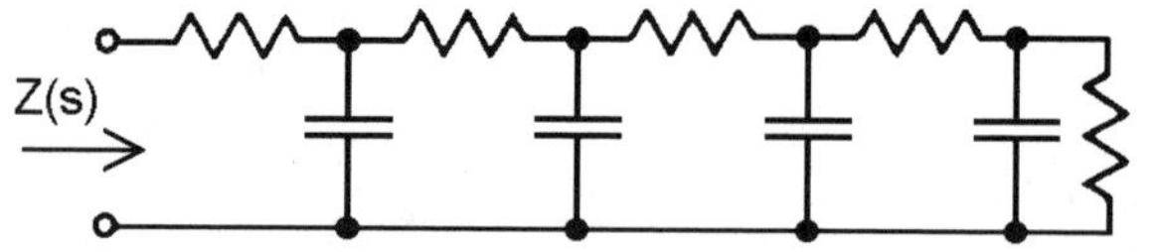

FIGURE 12.12 Cauer equivalent circuit (ladder structure).

middle of each section. The total capacitance of the section must be calculated, and this gives the value of the lumped capacitance connected to the node of the section. The resistance between the middle points of two adjacent sections must be calculated, and this has to be connected between the nodes of the appropriate sections. It is obvious that the accuracy can be increased by decreasing h. The price is the increasing number of lumped elements. With $h \to 0$, we obtain the exact model.

2. When we know the impedance function, we can build the model using the pole-zero pattern of the network. For example, let us investigate a uniform RC line of finite length L, short-circuited at the far end. The corresponding impedance expression, according to (12.21), is

$$Z(s) = \frac{1}{\sqrt{sK_0}} \operatorname{th} R_0 \sqrt{sK_0} \tag{12.35}$$

where $K_0 = c/r$, $R_0 = r \cdot L$. This function has poles and zeroes on the negative real axis in an infinite number. The zero and pole frequencies are

$$\sigma_{zi} = (2i)^2 \frac{\pi^2}{4} \frac{1}{R_0^2 K_0} \qquad \sigma_{pi} = (2i+1)^2 \frac{\pi^2}{4} \frac{1}{R_0^2 K_0} \tag{12.36}$$

where $i = 1, 2, \ldots, \infty$. Neglecting all the poles and zeroes situated well above the frequency range of interest and eliminating successively the remainder poles and zeroes from the (12.37) impedance function the element values of the ladder network in Fig. 12.12 (Cauer equivalent) can be obtained from

$$Z(s) = R_0 \frac{\prod_{i=1}^{z} (1 + s/\sigma_{zi})}{\prod_{i=1}^{p} (1 + s/\sigma_{pi})} \tag{12.37}$$

12.3 Infinite-Length RC Lines

It was demonstrated earlier in the chapter that the DRC network can be described with the help of the pole-zero set of its impedance, as in the case of lumped element circuits. However, the number of these poles and zeroes is infinite. The *infinite-length* DRC lines generally do not have this property. For this network category the exact description by discrete poles and zeroes is not possible.

For example, let us consider an infinitely long uniform DRC line. Its input impedance is the characteristic impedance:

$$Z(s) = \sqrt{\frac{r}{sc}} \tag{12.38}$$

Evidently, this impedance function *does not have poles and zeroes* on the negative σ axis. This is the general case for a more complex, nonuniform distributed network if the length of the structure is infinite. The characteristic feature of these impedance functions is that $\sqrt{j\omega}$ factors appear in them. This is why in the logarithmic amplitude vs. frequency diagram (Bode plot) regions with 10 dB/decade slope appear, as pointed out in [1].

This section provides a generalization of the pole and zero notions and the time-constant representation in order to make them suitable to describe infinitely long distributed one-ports as well.

Before developing new ideas let us summarize the normal, well-known descriptions of a lumped element RC one-port. The port impedance of such a circuit is described by a rational function with real coefficients, as

$$\mathbf{Z}(\mathbf{s}) = R_0 \frac{\left(1+\mathbf{s}/\sigma_{z1}\right)\left(1+\mathbf{s}/\sigma_{z2}\right)\ldots\left(1+\mathbf{s}/\sigma_{zn-1}\right)}{\left(1+\mathbf{s}/\sigma_{p1}\right)\left(1+\mathbf{s}/\sigma_{p2}\right)\ldots\left(1+\mathbf{s}/\sigma_{pn}\right)} \tag{12.39}$$

where R_0 is the overall resistance, σ_p are the poles, and σ_z are the zeroes (as absolute values). The pole and zero values, together with the overall resistance value, hold all the information about the one-port impedance. Thus, an unambiguous representation of this impedance is given by a set of pole and zero values, and an overall resistance value. This will be called the *pole-zero representation.*

Expression (12.39) can be rearranged as

$$\mathbf{Z}(\mathbf{s}) = \sum_{i=1}^{n} \frac{R_i}{1+\mathbf{s}/\sigma_{pi}} = \sum_{i=1}^{n} \frac{R_i}{1+\mathbf{s}\tau_i} \tag{12.40}$$

where

$$\tau_i = 1/\sigma_{pi} \tag{12.41}$$

which corresponds directly to the $v(t)$ voltage response for a step-function current excitation:

$$v(t) = \sum_{i=1}^{n} R_i\left(1-\exp\left(-t/\tau_i\right)\right) \tag{12.42}$$

In this case, the impedance is described in terms of the τ_i time-constants of its response and of the R_i magnitudes related to it. This will be called the *time-constant representation.*

Generalization of the Time Constant Representation[2]

A lumped element one-port can be represented by a finite number of τ time-constants and R magnitudes. A graphic representation of this is demonstrated in Fig. 12.13. Each line of this plot represents a time constant, and the height of the line is proportional to the magnitude. This figure can be regarded as

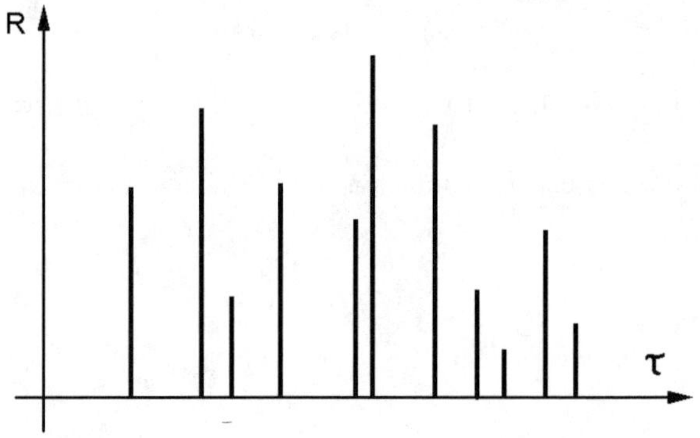

FIGURE 12.13 A lumped element one-port can be represented with a discrete set of time constants. (From [4], reprinted with permission, © 1991 IEEE.)

[2]Portions reprinted with permission from [4]. © 1991 IEEE.

some kind of a spectrum, the spectrum of the time constants that appeared in the step-function response of the network.

The port-impedance of a lumped element network has discrete "spectrum lines" in finite number. An infinite distributed network has no discrete lines, but it can be described with the help of a continuous time constant spectrum. The physical meaning of this idea is that in a general response any time constant can occur in some amount, some density, so that a density spectrum may suitably represent it.

We define the spectrum function by first introducing a new, logarithmic variable for the time and the time constants:

$$z = \ln t \qquad \zeta = \ln \tau \tag{12.43}$$

Let us consider a DRC one-port, the response of which contains numerous exponentials having different time constants and magnitudes. The **time constant density** is defined as

$$R(\zeta) = \lim_{\Delta\zeta \to 0} \frac{\text{sum of magnitudes between } \zeta \text{ and } \zeta + \Delta\zeta}{\Delta\zeta} \tag{12.44}$$

From this definition directly follows the fact that the step-function response can be composed from the time-constant density:

$$v(t) = \int_{-\infty}^{\infty} R(\zeta)\left[1 - \exp(-t/\exp(\zeta))\right] d\zeta \tag{12.45}$$

This integral is obviously the generalization of the summation in (12.42). If the $R(\zeta)$ density function consists of discrete lines (Dirac-δ pulses), (12.42) is given back.

Using the logarithmic time variable in the integral of (12.45)

$$v(z) = \int_{-\infty}^{\infty} R(\zeta)\left[1 - \exp(-\exp(z - \zeta))\right] d\zeta \tag{12.46}$$

a convolution-type differential equation is obtained. Differentiating both sides with respect to z, we obtain

$$\frac{d}{dz} v(z) = R(z) \otimes W(z) \tag{12.47}$$

where

$$W(z) = \exp(z - \exp(z)) \tag{12.48}$$

is a fixed weighting function with shape depicted in Fig. 12.14, and $\otimes$ is the symbol of the convolution operation [3].

It can be proved that the area under the function $W(z)$ is equal to unity

$$\int_{-\infty}^{\infty} W(z)\, dz = 1 \tag{12.49}$$

This means that

$$\int_{-\infty}^{\infty} R(z)\, dz = v(t \to \infty) = R_0 \tag{12.50}$$

where R_0 is the zero-frequency value of the impedance. In other words, the finite step-function response guarantees that the time-constant density has finite integral.

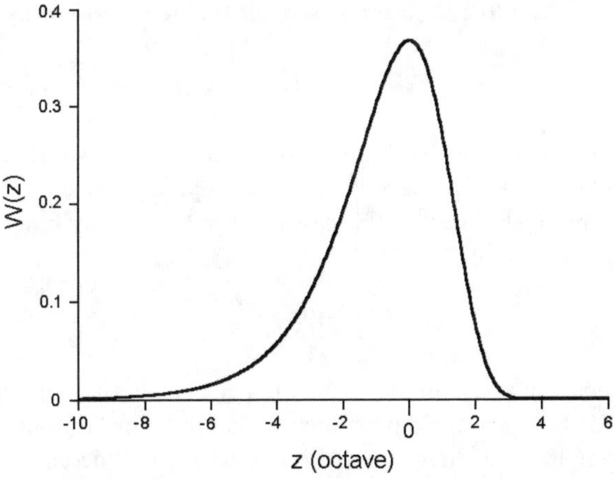

FIGURE 12.14 The shape of the $W(z)$ function. (From [4], reprinted with permission, © 1991 IEEE.)

Generalization of the Pole-Zero Representation

The task is now to substitute the pole-zero pattern of the lumped network with a continuous (eventually excepting some discrete points) function to describe the general distributed parameter network.

As emphasized above, the Bode plot of a distributed parameter network frequently shows regions with a 10 dB/decade slope. Figure 12.15 presents such an amplitude diagram. Using poles and zeroes, we can only approximate this behavior. If we place to point ω_1 a pole, the Bode plot turns to the decay of 20 dB/decade, which is too steep. If a zero is placed, the diagram returns to the zero-slope. However, if we *alternate* poles and zeroes in a manner that the *mean value* of the slope should give the prescribed one, then any slope can be approximated. (For the previously-mentioned case, if the zeroes are situated exactly midway between the adjacent poles, then the mean slope is 10 dB/decade.) The suitability of the approximation depends on the density of poles and zeroes and can be improved by increasing the density. In this case, the network-specific information is not carried by the number of poles and zeroes (their number tends to infinity), but by the *relative* position of the zeroes between the adjacent poles.

An alternative interpretation is also possible. The pair of a neighboring pole and zero constitutes a *dipole*. The "intensity" of that dipole depends on the distance between the pole and the zero. If they coincide and cancel each other, then the intensity is equal to zero. If the zero is situated at the maximal distance from the pole (i.e., it is at the next pole), the intensity reaches its maximal value. We choose this to be the unity.

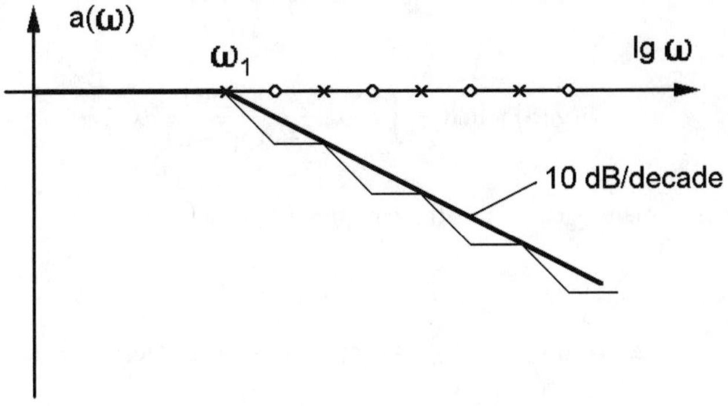

FIGURE 12.15 The 10 dB/decade decay of a DRC line amplitude plot can be approximated with an alternative sequence of poles and zeroes. (From [4], reprinted with permission, © 1991 IEEE.)

For later convenience, we turn to a logarithmic variable on the negative σ-axis:

$$\Sigma = \ln(-\sigma) \tag{12.51}$$

Let us investigate a ΔΣ interval of the logarithmic Σ-axis bounded by two adjacent poles. The distance between the left-hand pole and the inner zero is δΣ. Now, suppose that the density of the poles tends to infinity; i.e., ΔΣ becomes infinitely small. In this case the **dipole intensity function** is

$$I_d(\Sigma) = \lim_{\Delta\Sigma \to 0} \frac{\delta\Sigma}{\Delta\Sigma} \tag{12.52}$$

Considering that the poles and zeros of an RC port-impedance alternate, it follows that $0 \le I_d \le 1$. For an infinite, distributed RC two-pole the dipole intensity generally has regions in which the I_d value is between 0 and 1. For example, if the Bode plot shows a slope of 10 dB/decade, the value of I_d equals 0.5. This occurs in the case of an infinite, uniform RC line. For discrete circuits, the I_d function has only two possible values: 0 or 1.

Relations among $R(\zeta)$, $I_d(\Sigma)$ and the Impedance Function

Obviously, we needed one-to-one relations between the time constant density or the dipole intensity representation and the impedance expression of the one-port. [For the lumped element case (12.39) and (12.40) give these correspondences].

Rather simple relations exist among the $R(\zeta)$, $I_d(\Sigma)$, and the $\mathbf{Z}(\mathbf{s})$ impedance function (see below). An interesting feature of these relations is a striking mathematical symmetry: the same expression couples the time constant density to the impedance and the dipole intensity to the logarithmic impedance. The detailed proofs of the relations presented here are given in [4].

If the $\mathbf{Z}(\mathbf{s})$ complex impedance function is known, $R(\zeta)$ or $I_d(\Sigma)$ can be calculated as[3]

$$R(\zeta) = \frac{1}{\pi} \operatorname{Im} \mathbf{Z}(\mathbf{s} = -\exp(-\zeta)) \tag{12.53}$$

$$I_d(\Sigma) = \frac{1}{\pi} \operatorname{Im}(\ln \mathbf{Z}(\mathbf{s} = -\exp(\Sigma))) \tag{12.54}$$

If the $R(\zeta)$ or $I_d(\Sigma)$ function is known

$$\mathbf{Z}(\mathbf{S}) = R_0 - \int_{-\infty}^{\infty} R(-x) \frac{\exp(\mathbf{S} - x)}{1 + \exp(\mathbf{S} - x)} dx \tag{12.55}$$

$$\ln \mathbf{Z}(\mathbf{S}) = \ln R_0 - \int_{-\infty}^{\infty} I_d(x) \frac{\exp(\mathbf{S} - x)}{1 + \exp(\mathbf{S} - x)} dx \tag{12.56}$$

where **S** is the complex-valued logarithm of the complex frequency:

$$\mathbf{S} = \ln \mathbf{s} \tag{12.57}$$

Using the integral Eqs. (12.55) and (12.56), however, we must keep at least one from the two conditions:

[3] For (12.53) $\mathbf{Z}(\mathbf{s} \to \infty) = 0$ is supposed.

1. **s** is not located on the negative real axis.
2. If **s** is located on the negative real axis then, at this point, and in a $\varepsilon \to 0$ neighborhood, $R(\zeta)$ or $I_d(\Sigma)$ must be equal to 0.

Note that (12.53) to (12.56) are closely related to the Cauchy integral formula of the complex function theory. Substituting (12.53) into (12.55) and exploiting some inherent properties of the RC impedance functions after some mathematics, the Cauchy integral results. The same is true for (12.54) and (12.56).

An important feature of the transformations of (12.53) and (12.55) is that they are linear. This means that the $\mathbf{Z}(\mathbf{s}) \leftrightarrow R(\zeta)$ transformation and the summation are interchangeable.

Practical Calculation of the $R(\zeta)$ Function[4]

Equation (12.53) suggests that only the $j\omega$ imaginary frequency has to be replaced by the $\mathbf{s} = -\exp(-z)$ complex frequency, and then the imaginary part of the calculated complex response multiplied by $1/\pi$ provides the time-constant spectrum.

However, the procedure is not as simple as that because of the use of (12.53). This equation requires a great amount of caution. As the equation shows, the imaginary part of the **Z** impedance has to be calculated along the negative real axis of the complex plane. Along this axis, singularities usually lie: the *poles* of the network equation of lumped circuits or some *singular lines* in the case of distributed systems. These singularities can prevent the use of (12.53) for the calculation of the time-constant spectrum.

We can overcome these difficulties by adapting an approximate solution. In order to walk around the "dangerous" area, we have to avoid following the negative real axis. A line that is appropriately close to this axis might be used instead [9], like:

$$\mathbf{s} = -(\cos\delta + j\sin\delta)\exp(-z) \tag{12.58}$$

Obviously, the δ angle has to be very small, not more than 2 to 5°. Even if this angle is small, an error is introduced into the calculation. It can be proven that the calculated $R_C(z)$ time-constant spectrum can be expressed with the exact one by the following convolution equation:

$$R_C(z) = \frac{\pi - \delta}{\pi} R(z) \otimes e_r(z) \tag{12.59}$$

where

$$e_r(z) = \frac{1}{\pi - \delta} \frac{\sin\delta \exp(-z)}{1 - 2\cos\delta \exp(-z) + \exp(-2z))} \tag{12.60}$$

This function is a narrow pulse of unity area. The error of the calculation is represented by this function. Diminishing δ the $e_r(z)$ function becomes narrower and narrower. Thus, any accuracy requirement can be fulfilled by choosing an appropriately small δ angle. The half-value width, which is a measure of the resolution, is given by

$$\Delta_e = 2\ln\left(2 - \cos\delta + \sqrt{(2-\cos\delta)^2 - 1}\right) \cong 2\delta \tag{12.61}$$

If, for example, $\delta = 2°$, then the resolution is 0.1 octave, which means that two poles can be distinguished if the ratio between their frequencies is greater than 1.072.

Obviously, the calculated result has to be corrected with the factor of $\pi/(\pi - \delta)$.

Example 12.3 A tight analogy exists between electrical conductance and heat flow. Heat-conducting media, which can be characterized with distributed heat resistance and distributed heat capacitance, behave similarly to the electrical DRC networks. The analogous quantities are as follows:

---

[4]Portions reprinted, with permission, from [9]. © 2000 IEEE.

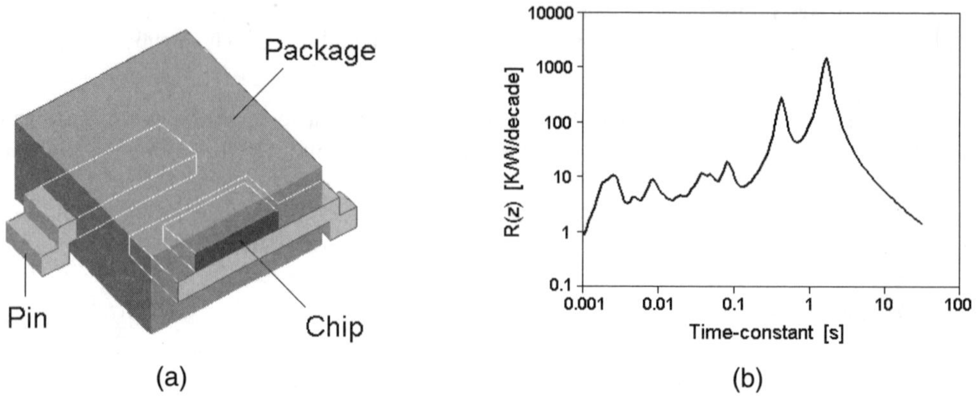

FIGURE 12.16 A transistor package and its time-constant spectrum. (From [9], reprinted with permission. © 2000 IEEE.)

Voltage	→	Temperature
Current	→	Power flow
Resistance	→	Thermal resistance
Capacitance	→	Heat capacitance

In the simplest model law, 1-V voltage corresponds to 1°C, 1 A current to 1 W power, etc., but different mapping can be applied as well.

The described analogy means that the tool set that is used to treat DRC networks can be applied to the calculation of heat-flow problems as well. This fact provides a direct way to calculate time-constant spectra in thermal field solver programs. These are thermal simulation tools suitable to solve the model equations in the **s**-domain. By using the substitution of (12.58), some of these programs calculate directly the thermal time-constant spectrum of different structures [9]. As an example, a transistor package, presented in Fig. 12.16(a), was simulated. The time-constant spectrum calculated by the field solver is plotted in Fig. 12.16(b). It is clearly visible that besides the two dominant time-constants a large number of further time-constants appear in the spectrum.

12.4 Inverse Problem for Distributed RC Circuits

Equation (12.47) offers a direct way to determine the time-constant density from the (measured or calculated) response function, which means that it is a method for the identification of RC one-ports. Using (12.47) for a measured time domain response function, the time-constant density of the one-port impedance can be determined. By using this method, equivalent circuits can be constructed easily. This possibility is of considerable practical importance [8]; however, only approximate results can be obtained because the calculation leads to the inverse operation of the convolution. This operation can be done only approximately.

A possibility exists for identification in the frequency domain as well. Introducing the $\Omega = \ln\omega$ notation for the frequency axis, convolution-type equations can be found [8] between the $R(\zeta)$ time-constant density and the $\mathbf{Z}(\omega)$ complex impedance:

$$-\frac{d}{d\Omega}\mathrm{Re}\,\mathbf{Z}(\Omega) = R(\zeta = -\Omega) \otimes W_R(\Omega) \tag{12.62}$$

$$-\mathrm{Im}\,\mathbf{Z}(\Omega) = R(\zeta = -\Omega) \otimes W_I(\Omega) \tag{12.63}$$

where the weight-functions are

$$W_R(\Omega) = \frac{2\cdot\exp(2\Omega)}{\left(1+\exp(2\Omega)\right)^2} \tag{12.64}$$

$$W_I(\Omega) = \frac{\exp(\Omega)}{1+\exp(2\Omega)} \tag{12.65}$$

Moreover, a direct convolution relation exists between the Bode diagram of the impedance and the dipole intensity. Considering the Bode amplitude and phase diagrams, i.e., by using $\ln(\mathbf{Z}(\Omega)) = \ln \text{abs}(\mathbf{Z}(\Omega)) + j\cdot\text{arcus}(\mathbf{Z}(\Omega))$, we obtain

$$-\frac{d}{d\Omega}\ln \text{abs}(\mathbf{Z}(\Omega)) = I_d(\Omega) \otimes W_R(\Omega) \tag{12.66}$$

$$-\text{arcus}(\mathbf{Z}(\Omega)) = I_d(\Omega) \otimes W_I(\Omega) \tag{12.67}$$

These equations may also be used for identification.

Network Identification by Deconvolution

All the relations between the $R(\zeta)$ time-constant spectrum and the different network responses are of convolution type (12.47), (12.62), and (12.63). Knowing some kind of network responses the inverse operation of the convolution: the *deconvolution* leads to the $R(\zeta)$ function. The same is true for the relations (12.66), (12.67) of the $I_d(\Sigma)$ dipole intensity. This means that the problem of identification of DRC networks is reduced to a deconvolution step. This method, called NID (network identification by deconvolution), is discussed in detail in [8], together with the appropriate deconvolution methods.

An important fact is that if we know the time response or only the real or the imaginary part of the frequency response, the network can be completely identified. (Noise effects produce practical limits — see later.)

Example 12.4 For the sake of simplicity, a **lumped circuit problem** will be discussed first. The investigated RC network is given in Fig. 12.17(a). We have calculated the frequency response of the $\mathbf{Z}(s)$ port-impedance of this circuit by using a standard circuit-simulator program, with a 40 point/frequency-decade resolution. Both the real and the imaginary parts of this impedance are plotted in Fig. 12.17(b).

In order to apply (12.62), the derivative of the real part was calculated numerically. The result is shown in Fig. 12.18(a). In the next step, this function was deconvolved by the $W_R(\Omega)$ function. The result is plotted in Fig. 12.18(b). This function is the approximate time-constant density of the network. We expect that this function depicts the pole-pattern of the circuit. This is in fact obtained: the four peaks of the function are lying at $f = 497.7$ Hz, 1585 Hz, 4908 Hz, and 15850 Hz. These values correspond to the time constants of 320 μs, 100.4 μs, 32.43 μs, and 10.04 μs, respectively. The ratios of the peak areas are about 1:2:1:2. These data agree well with the actual parameters of the circuit in Fig. 12.17(a).[5]

Notice that the noise corrupting the $\mathbf{Z}(\omega)$ function considerably affects the result of the identification. In order to reach 1 octave resolution of $R(\zeta)$ along the frequency axis, about 68 dB noise separation is needed in $\mathbf{Z}(\omega)$. Detailed discussion of the noise effects on the identification can be found in [8].

Example 12.5 As a second example, let us discuss the **thermal identification of a semiconductor package + heat sink structure**. The analogy between electrical current and heat flow introduced in Example 12.3 will be applied again.

Between a semiconductor chip and its ambient a complex distributed thermal structure exists consisting of many elements. The main parts of it are the chip itself, the soldering, the package, mounting to the heat sink, the heat sink itself, and the ambience. This is obviously a distributed thermal RC network, the input-port of which is the top surface of the chip and the far end is the ambience ("the world").

---

[5]Example reprinted with permission from [8]. © 1998 IEEE.

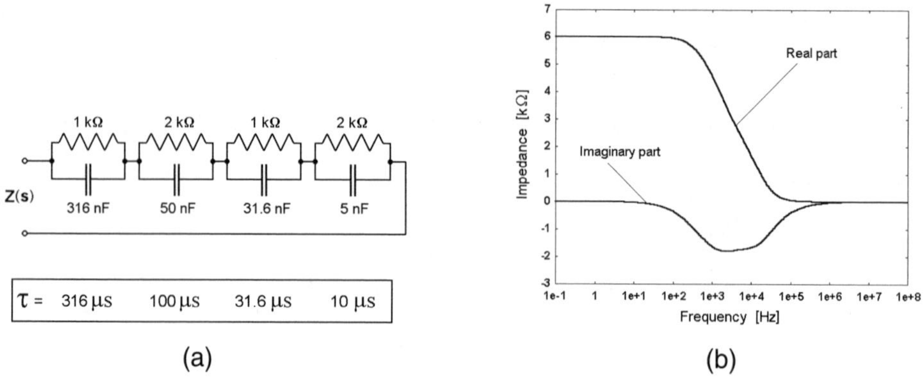

FIGURE 12.17 RC ladder and the frequency response of the **Z**(ω) impedance. (From [8], reprinted with permission. © 1998 IEEE.)

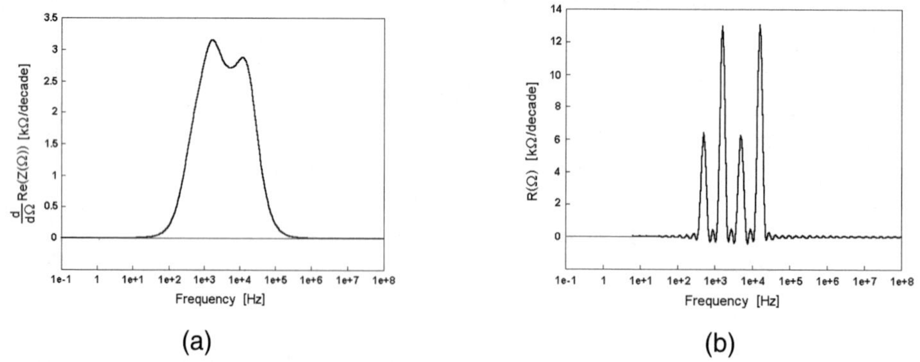

FIGURE 12.18 Identification steps. (a) Derivative of Real(**Z**). (b) The identified time-constant spectrum. (From [8], reprinted with permission. © 1998 IEEE.)

Thus, the structure can be considered practically infinite. This means that we have to examine a nonuniform infinite-length DRC network.

Investigations in the time domain require recording the thermal step-response of the system. A thermal test chip can be used for this purpose, containing appropriate heating elements that can assure step-function power excitation. The temperature rise is measured by the forward voltage change of a pn junction integrated into the test chip as well. This is the thermal response function. Such a thermal response is plotted in Fig. 12.19(a). The time range of the measurement is strikingly wide: 9 decades, from 10 μs to some thousand s. This is indispensable since the thermal time constants of the heat-flow structure vary over a wide range.

According to (12.47), after numerical derivation of the step-response and by a consecutive deconvolution, the time constant density function $R(z)$ can be obtained [see Fig. 12.19(b)]. Because of the quantization noise and measuring error the deconvolution operation can be done only approximately with a 1 ÷ 1.5 octave resolution. A suitable algorithm is discussed in [7]. Figure 12.19(b) illustrates that, in the 100 μs to 10 s interval, time constants spread over a relative wide range. This refers to the distributed structure of the chip and the package. At $\tau \approx 1000$ s, a relatively sharp, distinct time constant appears. This can be identified as originating from the heat capacitance of the whole heat sink and the heat sink-ambience thermal resistance.

Splitting the resultant time constant spectrum into $\Delta\tau$ time slots, each of these slots can be approximated by a Dirac-δ spectrum line proportional in height to the appropriate slot area. These give the data of a lumped element approximation according to (12.40). Now, the equivalent circuit of the heat-flow structure can be generated either in Foster or in Cauer normal form.

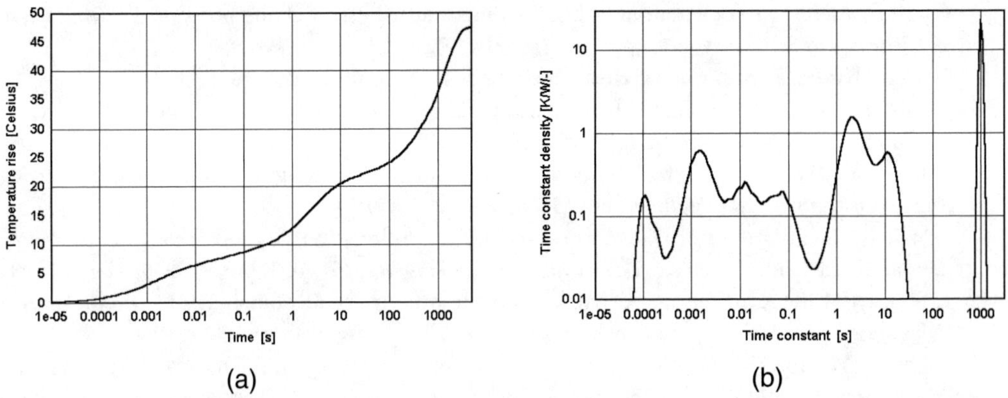

FIGURE 12.19 Thermal identification of a package + heat sink structure. (a) Thermal response between 10 μs and 4000 s. (b) The $R(z)$ time-constant density function.

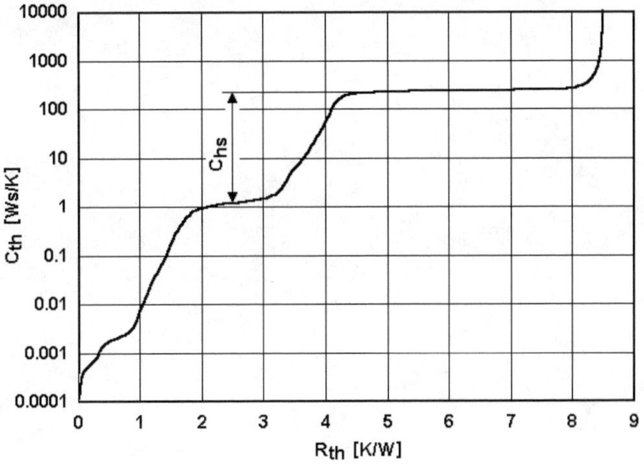

FIGURE 12.20 Cumulative structure function of the package + heat sink structure identified from the response function.

Using the Cauer-approximation of the DRC line we can calculate the approximate $K(\rho)$ and $\sigma(\rho)$ structure functions. From these functions, the heat-conducting cross section areas, the heat flow path length, etc. can be derived. This means that geometric and physical data of the heat-flow structure can be extracted and checked with the help of an electrical measurement. The structure function calculated from the measurement results of Fig. 12.19 is plotted in Fig. 12.20. It is easy to read out, e.g., the C_{hs} heat capacitance of the heat sink. For more details, see [6],[10].

References

[1] M. S. Ghausi and J. J. Kelly, *Introduction to Distributed Parameter Networks*, New York: Holt, Rinehart and Winston, 1968.

[2] E. N. Protonotarios and O. Wing, "Theory of nonuniform RC lines, Part I," *IEEE Transactions on Circuit Theory*, vol. 14, pp. 2-12, Mar. 1967

[3] D. G. Gardner, J. C. Gardner, G. Laush and W. W. Meinke: "Method for the analysis of multi-component exponential decay curves", *Journal of Chemical Physics*, vol. 31, no. 4, pp. 978-986, Oct. 1959

[4] V. Székely, "On the representation of infinite-length distributed RC one-ports", *IEEE Transactions on Circuits and Systems*, vol. 38, pp. 711-719, July 1991.
[5] R L. Pritchard, *Electrical Characteristics of Transistors,* New York: McGraw-Hill, 1967
[6] V. Székely and Tran Van Bien, "Fine structure of heat flow path in semiconductor devices: a measurement and identification method," *Solid-State Electronics*, vol. 31, pp. 1363-1368, Sept. 1988.
[7] T. J. Kennett, W. V. Prestwich, A. Robertson, "Bayesian deconvolution. I: convergent properties", *Nuclear Instruments and Methods*, no. 151, pp. 285-292, 1978.
[8] V. Székely, "Identification of RC Networks by Deconvolution: Chances and Limits", *IEEE Transactions on Circuits and Systems-I. Theory and applications*, vol. 45, no. 3, pp. 244-258, March 1998.
[9] V. Székely, M. Rencz, "Thermal dynamics and the time constant domain", *IEEE Transactions on Components and Packaging Technologies,* vol. 23, no. 3, pp. 587-594, Sept. 2000.
[10] V. Székely, "Restoration of physical structures: an approach based on the theory of RC networks", Proceedings of the *ECCTD'99, European Conference on Circuit Theory and Design*, 29 Aug.–2. Sept. 1999, Stresa, Italy, pp. 1131-1134.

13

Synthesis of Distributed Circuits

T. K. Ishii
Marquette University, Wisconsin

13.1 Generic Relations

The starting procedure for the synthesis of distributed circuits is the same as for the conventional synthesis of lumped parameter circuits. If a one-port network is to be synthesized, then a desired driving-point immittance $H(s)$ must be defined first, where

$$s = \sigma + j\omega \tag{13.1}$$

is the complex frequency, σ is the damping coefficient of the operating signal, and ω is the operating angular frequency. If a two-port network is to be synthesized, then a desired transmittance $T(s)$ must be defined first.

According to conventional principles of network synthesis [1], for the one-port network, $H(s)$ is represented by

$$H(s) = \frac{P(s)}{Q(s)} = \frac{a_n s^n + a_{n-1} s^{n-1} + \cdots + a_1 s + a_0}{b_m s^m + b_{m-1} s^{m-1} + \cdots + b_1 s + b_0} \tag{13.2}$$

where a_n and b_m are constants determined by the network parameters, $Q(s)$ is a driving function, and $P(s)$ is the response function. For a two-port network

$$T(s) = \frac{P(s)}{Q(s)} = \frac{a_n s^n + a_{n-1} s^{n-1} + \cdots + a_1 s + a_0}{b_m s^m + b_{m-1} s^{m-1} + \cdots + b_1 s + b_0} \tag{13.3}$$

Both $H(s)$ and $T(s)$ should be examined for realizability [1] before proceeding.

If the summation of even-order terms of $P(s)$ is $M_1(s)$ and the summation of odd-order terms of $P(s)$ is $N_1(s)$, then

$$P(s) = M_1(s) + N_1(s) \tag{13.4}$$

Similarly,

$$Q(s) = M_2(s) + N_2(s) \tag{13.5}$$

For a one-port network the driving-point impedance is synthesized by [1]

$$Z(s) = \frac{N_1(s)}{M_2(s)} \tag{13.6}$$

or

$$Z(s) = \frac{M_1(s)}{N_2(s)} \tag{13.7}$$

For a two-port network [1], if $P(s)$ is even, the transadmittance is

$$y_{21} = \frac{P(s)/N_2(s)}{1 + [M_2(s)/N_2(s)]} \tag{13.8}$$

the open-circuit transfer admittance is

$$y_{21} = \frac{P(s)}{N_2(s)} \tag{13.9}$$

and the open-circuit output admittance is

$$y_{22} = \frac{M_2(s)}{N_2(s)} \tag{13.10}$$

If $P(s)$ is odd,

$$y_{21} = \frac{P(s)/N_2(s)}{1 + [M_2(s)/N_2(s)]} \tag{13.11}$$

$$y_{21} = \frac{P(s)}{M_2(s)} \tag{13.12}$$

and

$$y_{22} = \frac{N_2(s)}{M_2(s)} \tag{13.13}$$

In both cases,

$$y_{11} = \frac{y_{21}(s)}{n} \tag{13.14}$$

where n is the current-ratio transfer function from port 1 to port 2. From these y- or z-parameters, the required values for the network components, i.e., L, C, and R, can be determined [1].

In high-frequency circuits the L, C, and R may be synthesized using distributed circuit components. The synthesis of distributed components in microstrip line and circuits is the emphasis of this chapter.

13.2 Synthesis of a Capacitance

If the required capacitive impedance is $-jXc\ \Omega$, the capacitance is

$$C = \frac{1}{\omega X_C} \tag{13.15}$$

where

$$X_C > 0 \tag{13.16}$$

and ω is the operating angular frequency. In a distributed circuit, the capacitance C is often synthesized using a short section of a short-circuited transmission line of negligibly small transmission line loss. If the characteristic impedance of such a transmission line is Z_0, the operating transmission line wavelength is λ_l, and the length of the transmission line is l in meters, then [2]

$$jX_c = jZ_0 \tan\frac{2\pi l}{\lambda_l} \tag{13.17}$$

where

$$\frac{\lambda_l}{4} < l < \frac{\lambda_l}{2}$$

or, more generally,

$$\frac{n\lambda_l}{2} + \frac{\lambda_l}{4} < l < \frac{n\lambda_l}{2} + \frac{\lambda_l}{2} \tag{13.18}$$

and n is an integer. Detailed information on Z_0 and λ_l is given in Chapter 9. Combining (13.15 and (13.17),

$$C = \frac{1}{\omega Z_0 \tan\dfrac{2\pi l}{\lambda_l}} \tag{13.19}$$

In practical synthesis, the transmission line length must be determined. The design equation is

$$l = \frac{\lambda_l}{2\pi}\tan^{-1}\left(\frac{1}{\omega C Z_0}\right) \tag{13.20}$$

Thus, the capacitance C can be synthesized using a section of a short-circuited transmission line.

If an open-circuited transmission line is used instead, then

$$l = \frac{\lambda_l}{2}\left[\frac{1}{\pi}\tan^{-1}\frac{1}{\omega C Z_0} + \frac{1}{2}\right] \tag{13.21}$$

Equation (13.19) is valid provided that (13.20) is used for l or,

$$l = \frac{\lambda_l}{2}\tan^{-1}\ \omega C Z_0 \tag{13.22}$$

where

$$\frac{n\lambda_l}{2} < l < \frac{\lambda_l}{4} + \frac{n\lambda_l}{2} \tag{13.23}$$

If an open-circuited transmission line section is used instead of the short-circuited transmission line section, just add $\lambda_l/4$ to the line length.

13.3 Synthesis of an Inductance

If the required inductive impedance is $+jX_L\ \Omega$, the inductance is

$$L = \frac{X_L}{\omega} \tag{13.24}$$

where

$$X_L > 0 \tag{13.25}$$

and ω is the operating angular frequency. In a distributed circuit the inductance L is often synthesized using a short section of a short-circuited transmission line of negligibly small transmission line loss. If the characteristic impedance of such a transmission line is Z_0, the operating transmission line wavelength is λ_l, and the length of transmission line is l meters, then [2]

$$jX_L = jZ_0 \tan\frac{2\pi l}{\lambda_l} \tag{13.26}$$

where

$$0 < l < \frac{\lambda_l}{4} \tag{13.27}$$

or, more generally,

$$\frac{n\lambda_l}{2} < l < \frac{n\lambda_l}{2} + \frac{\lambda_l}{4} \tag{13.28}$$

and n is an integer. Detailed information on Z_0 and λ_l is given in Chapter 9.

Combining (13.24) and (13.26)

$$L = \frac{Z_0}{\omega}\tan 2\pi\frac{l}{\lambda_l} \tag{13.29}$$

In practical synthesis, the transmission line length must be designed. The design equation is

$$l = \frac{\lambda_l}{2\pi} \tan^{-1} \frac{\omega L}{Z_0} \tag{13.30}$$

Thus, the inductance L can be synthesized using a section of a short-circuited transmission line.

If an open-circuited transmission line is used instead, then

$$l = \frac{\lambda_l}{2} \left(\frac{1}{\pi} \tan^{-1} \frac{\omega L}{Z_0} + \frac{1}{2} \right) \tag{13.31}$$

Equation (13.29) is valid for the open-circuited transmission line provided that (13.31) is used for l or

$$l = \frac{\lambda_l}{2\pi} \tan^{-1} \frac{Z_0}{\omega L} \tag{13.32}$$

where

$$\frac{n\lambda_l}{2} + \frac{\lambda_l}{4} < l < \frac{\lambda_l}{2} + \frac{n\lambda_l}{2} \tag{13.33}$$

If an open-circuited transmission line is used instead of the short-circuited transmission line section, just add $\lambda_l/4$ to the short-circuited transmission line design.

13.4 Synthesis of a Resistance

A distributed circuit resistance can be synthesized using a lossy transmission line of length l with the line end short- or open-circuited. When the line end is short-circuited, the input impedance of the line of length l and characteristic impedance $\dot{Z}_0$ is [2]

$$\dot{Z}_i = \dot{Z}_0 \tanh \dot{\gamma} l \tag{13.34}$$

where the propagation constant is

$$\dot{\gamma} = \alpha + j\beta \tag{13.35}$$

α is the attenuation constant, and β is the phase constant of the line. Assuming Z_0 is real, the input impedance becomes a pure resistance, i.e., the reactance becomes zero, when

$$\beta l = \frac{2\pi l}{\lambda_l} = \frac{1}{2}\pi, \frac{3}{2}\pi, \frac{5}{2}\pi, \ldots \tag{13.36}$$

When

$$\beta l = (2n+1)\frac{\pi}{2} \tag{13.37}$$

and where

$$n = 0, 1, 2, 3, \ldots \tag{13.38}$$

the short-circuited transmission line is at antiresonance. When

$$\beta l = 2n\frac{\pi}{2} \tag{13.39}$$

and where

$$n = 1, 2, 3, \ldots \tag{13.40}$$

the transmission line is at resonance.

When the transmission line is at antiresonance, the input impedance is [2]

$$\dot{Z}_i = \dot{Z}_0 \frac{1+\varepsilon^{-\alpha(2n+1)\frac{\lambda_l}{2}}}{1-\varepsilon^{-\alpha(2n+1)\frac{\lambda_l}{2}}} \tag{13.41}$$

If

$$\dot{Z}_0 \approx Z_0 \tag{13.42}$$

the input resistance is

$$R_i \approx Z_0 \frac{1+\varepsilon^{-\alpha(2n+1)\frac{\lambda_l}{2}}}{1-\varepsilon^{-\alpha(2n+1)\frac{\lambda_l}{2}}} \tag{13.43}$$

where

$$l = (2n+1)\frac{\lambda_l}{4} \tag{13.44}$$

When the transmission line section is at resonance, the input resistance is

$$R_i \approx Z_0 \frac{1-\varepsilon^{-\alpha n\lambda_l}}{1+\varepsilon^{-\alpha n\lambda_l}} \tag{13.45}$$

if (13.41) holds, where

$$l = n\frac{\lambda_l}{2} \tag{13.46}$$

From transmission line theory, α and β can be determined from the transmission line model parameters as [2]

$$\alpha = \frac{\omega LG + BR}{2\beta} \tag{13.47}$$

$$\beta = \left\{ \frac{\omega LB - RG + \sqrt{(RG - \omega LB)^2 + (\omega LG + BR)^2}}{2} \right\}^{\frac{1}{2}} \tag{13.48}$$

where L is the series inductance per meter, G is the shunt conductance per meter, B is the shunt susceptance per meter, and R is the series resistance per meter of the tranmission line section.

The characteristic impedance of a lossy line is [2]

$$\dot{Z}_0 = R_0 + jX_0 \tag{13.49}$$

$$R_0 = \frac{\left\{RG + \omega LB + \sqrt{(RG - \omega LB)^2 + (\omega LG - BR)^2}\right\}^{\frac{1}{2}}}{\sqrt{2}\sqrt{G^2 + B^2}} \tag{13.50}$$

$$X_0 = \frac{1}{2R_0} \cdot \frac{\omega LG - BR}{G^2 + B^2} \tag{13.51}$$

or simply from (13.34)

$$l = \frac{1}{\dot{\gamma}} \tanh^{-1} \frac{R_i}{\dot{Z}_0} \tag{13.52}$$

to determine l for desired R_i for synthesis.

In practical circuit synthesis, convenient, commercially available, surface-mountable chip resistors are often used. Integrated circuit resistors are monolithically developed. Therefore, the technique described here is seldom used.

13.5 Synthesis of Transformers

An impedance $\dot{Z}_1$ can be transformed into another impedance $\dot{Z}_2$ using a lossless or low-loss transmission line of length l and characteristic impedance Z_0 [2]:

$$\dot{Z}_2 = Z_0 \frac{\dot{Z}_1 + jZ_o \tan \frac{2\pi l}{\lambda_l}}{Z_0 + j\dot{Z}_1 \tan \frac{2\pi l}{\lambda_l}} \tag{13.53}$$

where λ_l is the wavelength on the transmission line. Solving (13.53) for l,

$$l = \frac{\lambda_l}{2\pi} \tan^{-1} \frac{Z_0(\dot{Z}_2 - \dot{Z}_1)}{j(Z_0^2 - \dot{Z}_1 \dot{Z}_2)} \tag{13.54}$$

In (13.53), if

$$l = \frac{\lambda_l}{4} \tag{13.55}$$

then

$$\dot{Z}_2 = Z_0^2 / \dot{Z}_1 \tag{13.56}$$

This is a quarter wavelength transmission line transformer. A low-impedance $\dot{Z}_1$ is transformed into a new high-impedance $\dot{Z}_2$, or vice versa. A capacitive $\dot{Z}_1$ is transformed into an inductive $\dot{Z}_2$, or vice versa. However, Z_0 is usually real, which restricts the available transformed impedances.

An admittance $\dot{Y}_1$ can be transformed into another admittance $\dot{Y}_2$ using a lossless or low-loss transmission line of length l and characteristic admittance $\dot{Y}_0$ [2]

$$\dot{Y}_2 = Y_0 \frac{\dot{Y}_1 + jY_0 \tan \frac{2\pi l}{\lambda_l}}{Y_0 + j\dot{Y}_1 \tan \frac{2\pi l}{\lambda_l}} \tag{13.57}$$

Solving (13.56) for l,

$$l = \frac{\lambda_l}{2\pi} \tan^{-1} \frac{Y_0\left(\dot{Y}_2 - \dot{Y}_1\right)}{j\left(Y_0^2 - \dot{Y}_1\dot{Y}_2\right)} \tag{13.58}$$

In (13.57), if

$$l = \frac{\lambda_1}{4} \tag{13.59}$$

then

$$\dot{Y}_2 = Y_0^2 / \dot{Y}_1 \tag{13.60}$$

This is a one quarter-wavelength transmission line transformer. A low-admittance $\dot{Y}_1$ is transformed into a new high-admittance $\dot{Y}_2$, or vice versa. A capacitive $\dot{Y}_1$ is transformed into an inductive $\dot{Y}_2$, or vice versa. However, Y_0 is usually real, which restricts the available transformed admittances.

The transforming method (13.53) and (13.57) cannot transform $\dot{Z}_1$ to every possible value of $\dot{Z}_2$, especially when the value of Z_0 is given. In practice, this is often the case. If this is the case and $\dot{Z}_1$ is complex, use (13.53) first to transform $\dot{Z}_1$ to $\dot{Z}_2'$, where

$$\dot{Z}_2' = R_2' + jX_2' \tag{13.61}$$

and

$$R_2' = R_2 \tag{13.62}$$

where

$$\dot{Z}_2 = R_2 + jX_2 \tag{13.63}$$

Then, synthesize $X_2 - X_2'$ and add it in series at the input of the transmission line, as depicted in Figure 13.1. Then,

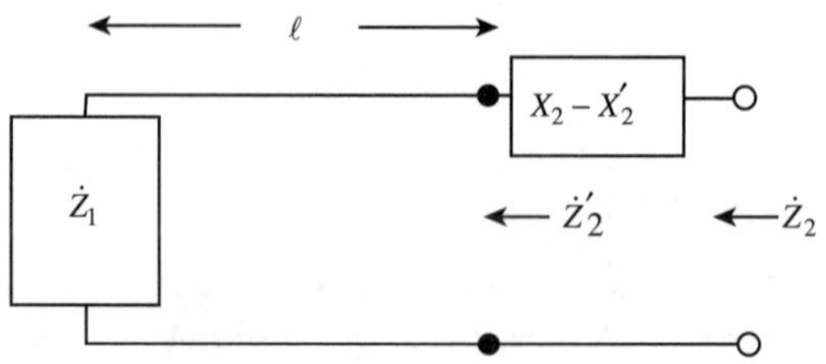

FIGURE 13.1 Synthesis of $\dot{Z}_1$ to any $\dot{Z}_2$ transformer.

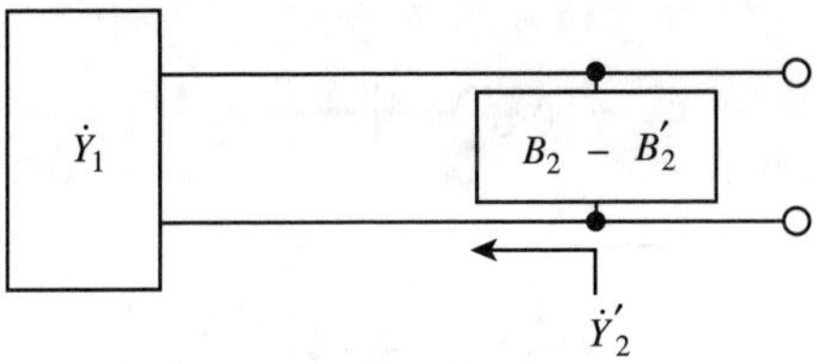

FIGURE 13.2 Synthesis of any $\dot{Y}_2$, from $\dot{Y}_1$ transformation.

$$\begin{aligned}\dot{Z}_2 &= \dot{Z}_2' + j(X_2 - X_2') \\ &= R_2' + jX_2' + jX_2 - jX_2' \\ &= R_2' + jX_2 \\ &= R_2 + jX_2\end{aligned} \tag{13.64}$$

Thus, $\dot{Z}_1$ can be transformed to any $\dot{Z}_2$ desired.

In a similar manner, to transform a shunt admittance $\dot{Y}_1$ to any other shunt admittance $\dot{Y}_2$, first transform $\dot{Y}_1$ to $\dot{Y}_2'$ so that

$$\dot{Y}_2' = G_2' + jB_2' \tag{13.65}$$

and

$$G_2' = G_2 \tag{13.66}$$

where

$$\dot{Y}_2 = G_2 + jB_2 \tag{13.67}$$

Then, synthesize $B_2 - B_2'$ and add it in shunt at the input of the transmission line as depicted in Figure 13.2. Then,

$$\begin{aligned}\dot{Y}_2 &= \dot{Y}_2' + j(B_2 - B_2') \\ &= G_2' + jB_2' + jB_2 - jB_2' \\ &= G_2' + jB_2 = G_2 + jB_2\end{aligned} \tag{13.68}$$

Thus, $\dot{Y}_1$ can be transformed to any $\dot{Y}_2$ desired.

13.6 Synthesis Examples

Distributed circuits can be synthesized using a number of transmission line sections. The transmission line sections include waveguides, coaxial lines, two-wire lines, and microstrip lines and coplanar waveguides. In this section, distributed circuit synthesis examples of distributed circuits using microstrip lines are presented for convenience. Similar techniques can be utilized for other types of transmission lines.

Series *L-C* Circuit

A series *L-C* circuit as illustrated in Figure 13.3 is considered to be a phase delay line in distributed line technology. The series impedance of an *L-C* circuit is $j(\omega L - 1/\omega C)$. If the load is a resistance of R_L Ω, then the phase delay in the output voltage is

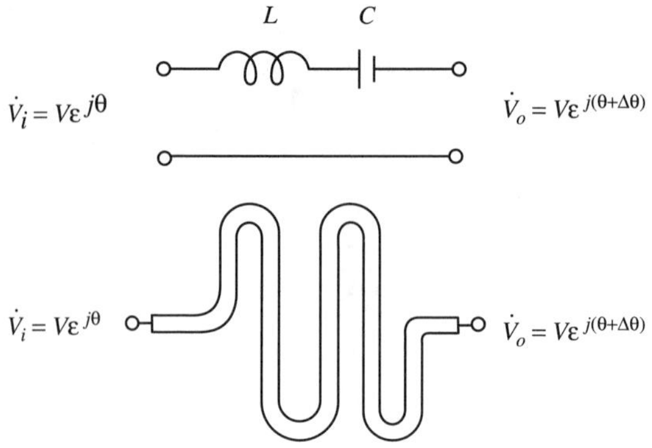

FIGURE 13.3 Synthesis of a series *L-C* circuit using a microstrip meander line.

$$\Delta\theta = \tan^{-1}\frac{\omega L - \dfrac{1}{\omega C}}{R_L} \tag{13.69}$$

The phase "delay" should be interpreted algebraically. Depending on the size of ωL and ωC, $\Delta\theta$ can be either + or –. If it is delay, the sign of $\Delta\theta$ is – and if it is advance, the sign of $\Delta\theta$ must be + in Figure 13.3.

If the phase constant of the microstrip line is

$$\beta = \frac{2\pi}{\lambda_i} \tag{13.70}$$

where λ_1 is the wavelength on the line, then the synthesizing equation is

$$\Delta\theta = \beta l \tag{13.71}$$

or

$$l = \frac{\Delta\theta}{\beta} = \frac{\lambda_l}{2\pi}\tan^{-1}\frac{\omega L - \dfrac{1}{\omega C}}{R_L} \tag{13.72}$$

Usually, the delay line takes the form of a meander line in microstrip line, as illustrated in Figure 13.3.

Parallel *L-C* Circuit

A parallel *L-C* circuit (see Figure 13.4) can be synthesized using distributed circuit components. In this figure, the open-circuited microstrip line length l_1 represents inductive shunt admittance and the synthesis equation is given by (13.31). In Figure 13.4 another open-circuited microstrip line length l_2 represents capacitive shunt admittance. The synthesis equation for this part is given by (13.21).

Series *L-C-R* Circuit

If the load resistance connected to the output circuit of a series *L-C-R* circuit (see Figure 13.5) is R_L, then the output voltage $\dot{V}_0$ of this network across R_L due to the input voltage $\dot{V}_i$ is

$$V_0 = \frac{R_L\dot{V}_i}{R_L + R + j\left(\omega L - \dfrac{1}{\omega C}\right)} = \frac{R_L\dot{V}_i}{Z\varepsilon^{j\phi}} \tag{13.73}$$

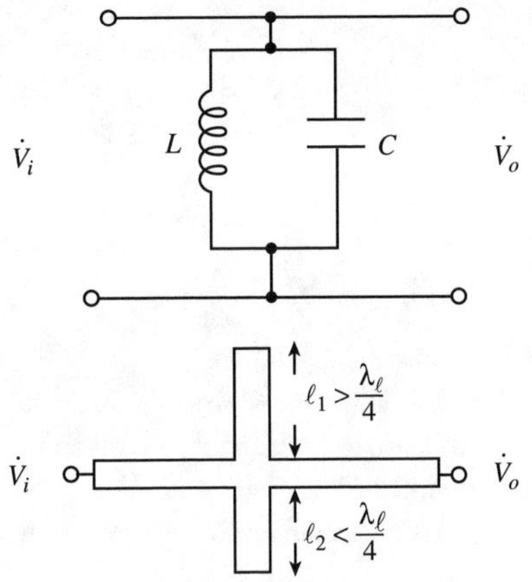

FIGURE 13.4 Synthesis of a parallel *L*-*C* circuit using a microstrip line circuit.

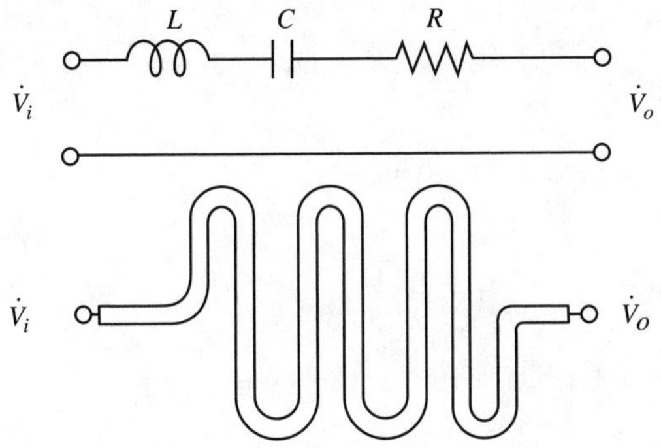

FIGURE 13.5 Synthesis of a series *L*-*C*-*R* circuit.

where

$$Z = \sqrt{(R_L + R)^2 + \left(\omega L - \frac{1}{\omega C}\right)^2} \tag{13.74}$$

$$\phi = \tan^{-1} \frac{\omega L - \dfrac{1}{\omega C}}{R_L + R} \tag{13.75}$$

Thus,

$$\dot{V}_0 = \frac{R_L}{Z} \dot{V}_i \varepsilon^{-j\phi} \tag{13.76}$$

In a distributed circuit, such as the microstrip line in Figure 13.5,

$$V_0 = \dot{V}_i \varepsilon^{-\alpha l} \varepsilon^{-j\beta l} \tag{13.77}$$

where α is the attenuation constant, β is the phase constant, and l is the total length of the microstrip line.

Then, comparing (13.76) with (13.75),

$$\varepsilon^{-\alpha l} = \frac{R_L}{Z} \tag{13.78}$$

or

$$\alpha l = \ln \frac{Z}{R_L} \tag{13.79}$$

and

$$\beta l = \phi \tag{13.80}$$

Parallel *L-C-R* Circuit

A parallel *L-C-R* circuit can be synthesized using the microstrip line (see Figure 13.6). In this figure, the microstrip of length l_1, characteristic impedance $\dot{Z}_{01}$, attenuation constant α_1, and phase constant β_1 represents a parallel circuit of inductance L and $2R$. Another microstrip line of length l_2, characteristic impedance $\dot{Z}_{02}$, attenuation constant α_2, and phase constant β_2 represents a parallel circuit of capacitance C and $2R$.

The input admittance of the open-circuited lossy microstrip line is

$$\dot{Y}_i = \frac{1}{\dot{Z}_{01}} \tanh(\alpha_1 + j\beta_1)l_1 \tag{13.81}$$

In a parallel L-$2R$ circuit, the parallel admittance is

$$Y_i = \frac{1}{2R} + \frac{1}{j\omega L} \tag{13.82}$$

The synthesis equation for the inductive microstrip line is

$$(\alpha_1 + j\beta_1)l_1 = \tanh^{-1} \dot{Z}_{01}\left(\frac{1}{2R} - j\frac{1}{\omega L}\right) \tag{13.83}$$

Similarly, for the capacitive microstrip line,

$$(\alpha_2 + j\beta_2)l_2 = \tanh^{-1} \dot{Z}_{02}\left(\frac{1}{2R} + j\omega C\right) \tag{13.84}$$

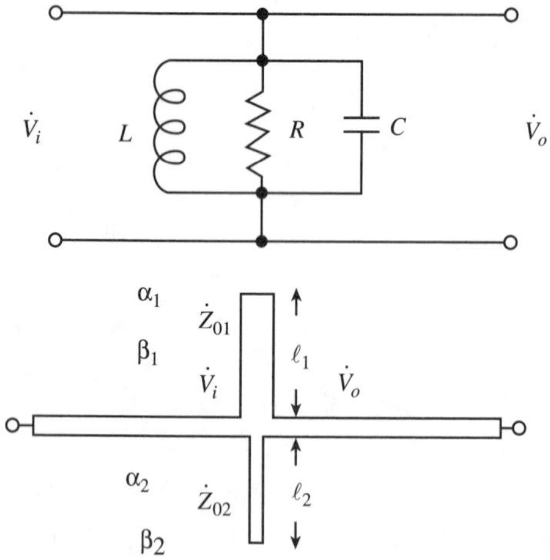

FIGURE 13.6 Synthesis of a parallel *L-C-R* circuit.

To synthesize this distributed circuit, (13.82) and (13.83) must be solved. These are transcendental equations of an excess number of unknowns to be determined by trial and error. Therefore, a digital computer is needed. In many cases, lossless lines are used for L and C and a shunt chip resistor is used for R to avoid complications.

Low-Pass Filters

If the load resistance of a low-pass filter, as presented in Figure 13.7 is R_L, then the admittance across R_L is

$$\dot{Y}_L = \frac{1}{R_L} + j\omega C \tag{13.85}$$

If this is synthesized using the microstrip line shown in Figure 13.7, then the normalized admittance of the microstrip line is

$$\tilde{\dot{Y}}_L \equiv \frac{\dot{Y}_L}{Y_0} = \frac{\frac{1}{R_L} + j\omega C}{Y_0} \tag{13.86}$$

where Y_0 is the characteristic admittance of the micropstrip line. $\tilde{\dot{Y}}_L$ should be plotted on a Smith chart as presented in Figure 13.7, and its angle of reflection coefficient ϕ_L must be noted. The input admittance of this circuit is

$$\dot{Y}_i = \cfrac{1}{j\omega L + \cfrac{R_L \cfrac{1}{j\omega C}}{R_L + \cfrac{1}{j\omega C}}} = \frac{R_L + j\left\{\omega C R_C^2\left(1-\omega^2 LC\right) - \omega L\right\}}{R_L^2\left(1-\omega^2 LC\right)^2 + (\omega L)^2} \tag{13.87}$$

This admittance is normalized as

$$\tilde{\dot{Y}}_i = \dot{Y}_i / Y_0 \tag{13.88}$$

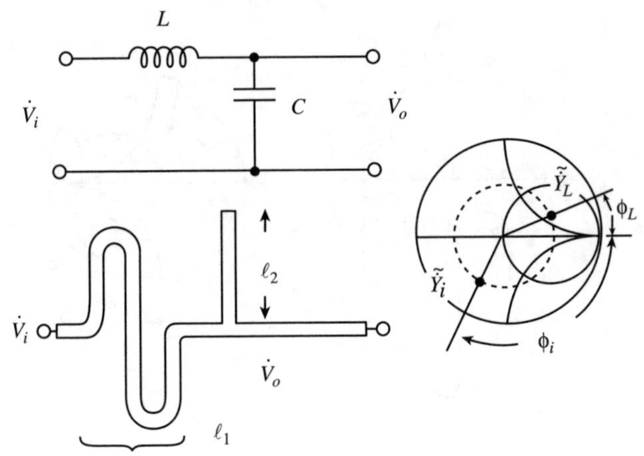

FIGURE 13.7 Synthesis of a low-pass filter.

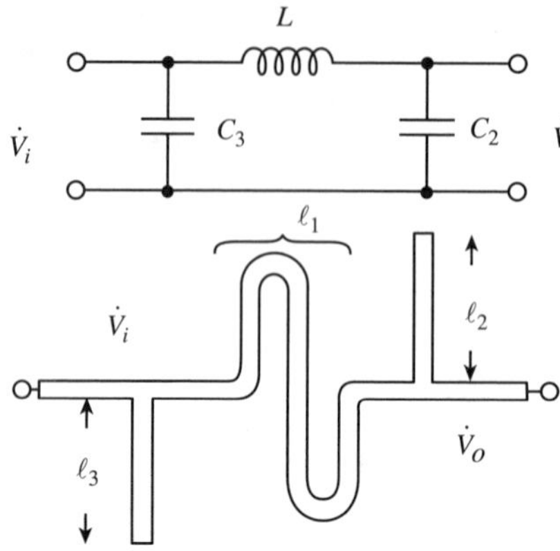

FIGURE 13.8 Synthesis of a π-network low-pass filter.

and it must be plotted on the Smith chart (Figure 13.7). The phase angle of the reflection coefficient ϕ_i at this point must be noted. Therefore, the synthesis equation is

$$\beta_1 l_1 = \phi_L + \phi_i \tag{13.89}$$

In Figure 13.7, the meander line of length l_1, representing the inductance L, and a straight microstrip line of length l_2, representing the capacitance C, are joined to synthesize a microwave low-pass filter. The synthesis of the shunt capacitance C in Figure 13.7 is accomplished using an open-circuited microstrip line of the phase constant β_2. The length of the microstrip line l_2 is determined from (13.21). If the low-pass filter takes on a π-shape, the microstrip version of synthesis will be similar to Figure 13.8. The microstrip line stub l_3 can be synthesized in the same way as the capacitive stub l_2.

High-Pass Filters

If a high-pass filter consists of a capacitance C and an inductance L, as shown in Figure 13.9, then normalized admittance across the load resistance R_L is

$$\tilde{\dot{Y}}_L = \dot{Y}_L / Y_0 = \left(\frac{1}{R_L} - j\frac{1}{\omega L} \right) \Big/ Y_0 \tag{13.90}$$

where Y_0 is the characteristic admittance of the microstrip line to be used for synthesis. This must be plotted on the Smith chart, as depicted in Figure 13.9, and the angle of reflection coefficient ϕ_L should be noted.

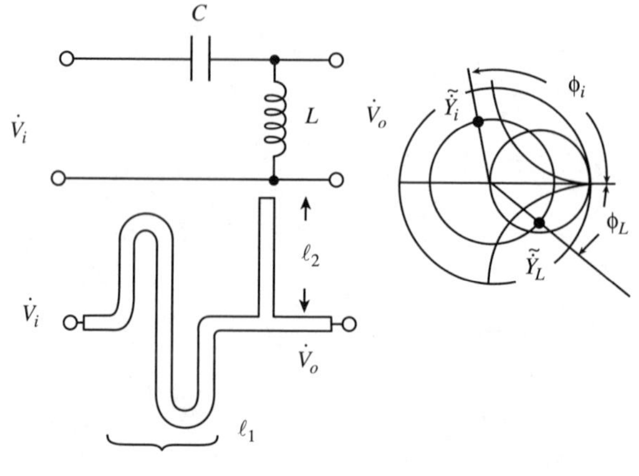

FIGURE 13.9 Synthesis of a high-pass filter.

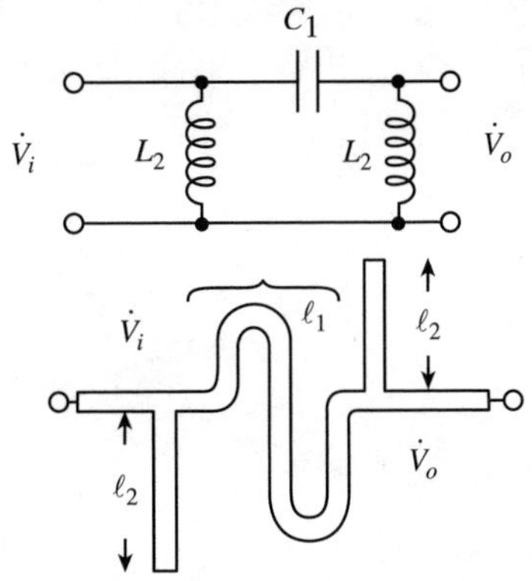

FIGURE 13.10 Synthesis of a π-network high-pass filter.

Normalized admittance at the input is

$$\tilde{\dot{Y}}_i \equiv \dot{Y}_i / Y_0 = \frac{1}{\dfrac{1}{j\omega C} + \dfrac{(j\omega L)R_L}{j\omega L + R_L}} \bigg/ Y_0$$
$$= \frac{(\omega^2 LC)^2 R_L + j\omega\{R_L^2 C(1-\omega^2 LC) + \omega^2 L^2 C\}}{R_L^2(1-\omega^2 LC)^2 + (\omega L)^2} \bigg/ Y_0 \tag{13.91}$$

Plot this on the Smith chart (Figure 13.9), and the angle of the voltage reflection coefficient ϕ_i should be noted.

The phase delay of the distributed microstrip line is

$$\dot{v}_0 = \dot{v}_i \varepsilon^{-j\beta l_1} \tag{13.92}$$

where β is a phase constant of the microstrip line and

$$\beta = 2\pi/\lambda_l \tag{13.93}$$

where λ_l is the wavelength on the microstrip line.

In (13.91), l_1 is the length of microstrip line representing the C-section of the low-pass filter. Then, the synthesis equation is

$$\beta l_1 = 360^\circ - (\phi_i + \phi_L) \tag{13.94}$$

The shunt inductance L in Figure 13.9 is synthesized using (13.31). Thus, a high-pass filter can be synthesized using a distributed microstrip line section. If the low-pass filter is in a π-network form, then microstrip line configuration will be at the designed frequency (see Figure 13.10).

Bandpass Filters

A bandpass filter may be a series connection of an inductor L and a capacitor of capacitance C, as shown in Figure 13.11. This figure is identical to Figure 13.3. Therefore, the microstrip line synthesis equation is (13.72).

If the bandpass filter is in the shape of a π-network (Figure 13.12(a)), then the series L_1C_1, represented by the meander microstrip line of length l_5, can be designed using (13.72). Both of the shunt L_2 and C_2

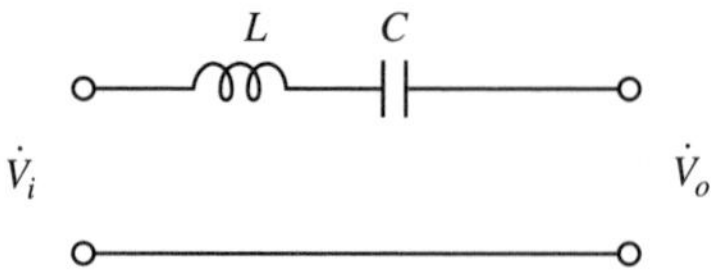

FIGURE 13.11 Configuration of a series L-C bandpass filter.

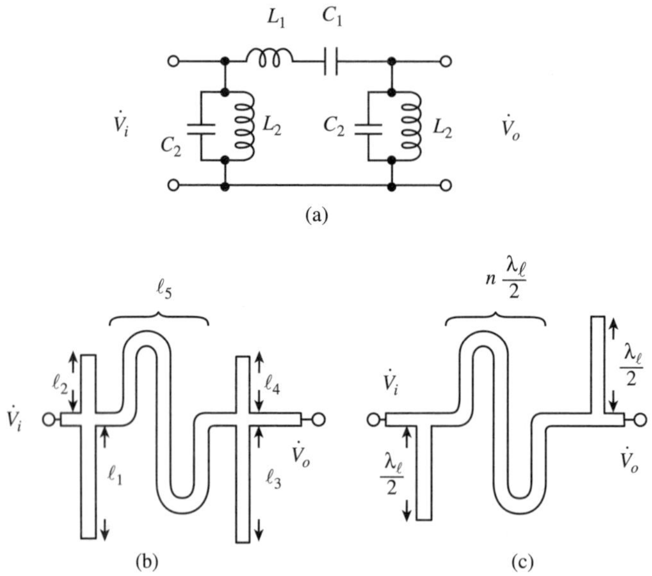

FIGURE 13.12 Synthesis of a distributed microstrip line π-network bandpass filter.

arms are identical to Figure 13.4. Therefore, this part is synthesized by utilizing the technique in the section, "Parallel L-C Circuit", which results in the configuration presented in Figure 13.12(b). An alternative distributed version of Figure 13.12(a) is given in Figure 13.12(c). In this figure, the technique in λ_l is the microstrip line wavelength at the center of the desired passband. The frequency bandwidth is determined by the quality factor of the resonating element.

Bandstop Filters

Various types of bandstop filters are available. A schematic diagram of a series type bandstop filter is shown in Fig. 13.13. It is basically a resonant circuit inserted in series with the transmission line. The center frequency of the bandstop is the resonant frequency of the resonant circuit. The loaded quality factor of the resonator determines the bandstop width.

In the distributed microstrip line the resonator is a section of a half-wavelength microstrip line resonator of either low impedance [Figure 13.13(a)] or high impedance [Figure 13.13(b)], or a dielectric resonator placed in proximity to the microstrip line, [see Figure 13.13(c)] or a ring resonator placed in proximity to the microstrip line [Figure 13.13(d)]. The resonant frequency of the resonators is the center frequency of the bandstop. The bandwidth is determined by the loaded Q and coupling of the resonator.

A bandstop filter can also be synthesized in microstrip line if the filter is a parallel type (Figure 13.14). Simply attaching a one quarter-wavelength, open-circuited stub will create a bandstop filter. The center of the bandstop has a wavelength of λ_l (Figure 13.14).

A bandstop filter may take a π-network form, as illustrated in Figure 13.15, by the combination of the $\lambda_l/2$ resonator and $\lambda_l/4$ open-circuited stubs. The frequency bandwidth of the filter depends on the quality factor of the resonating elements.

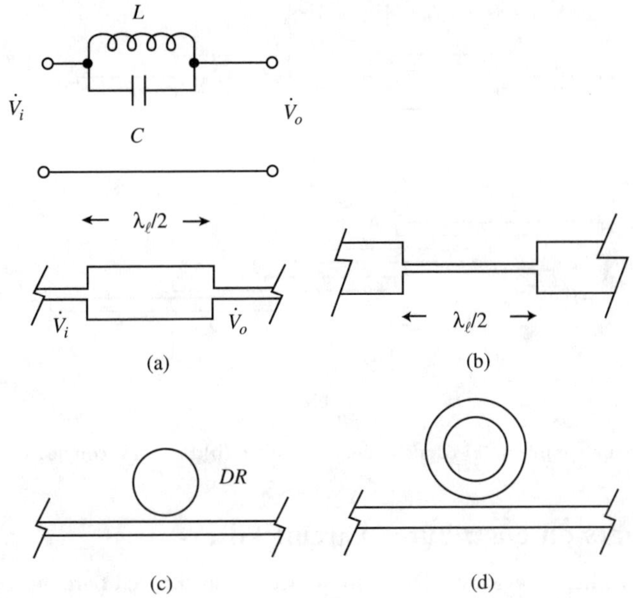

FIGURE 13.13 Synthesis of distributed series type bandstop filters.

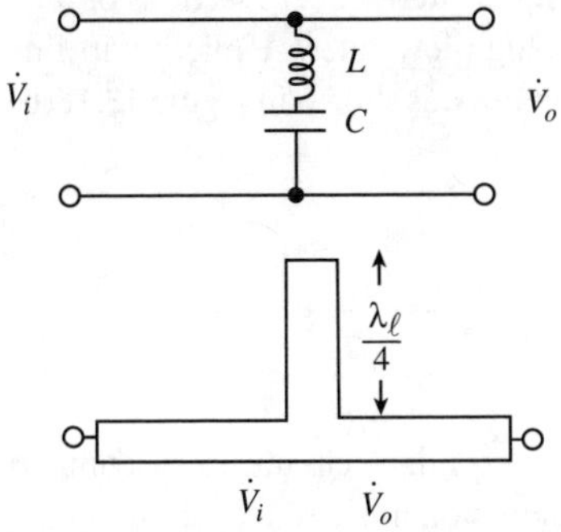

FIGURE 13.14 Synthesis of distributed parallel type bandstop filters.

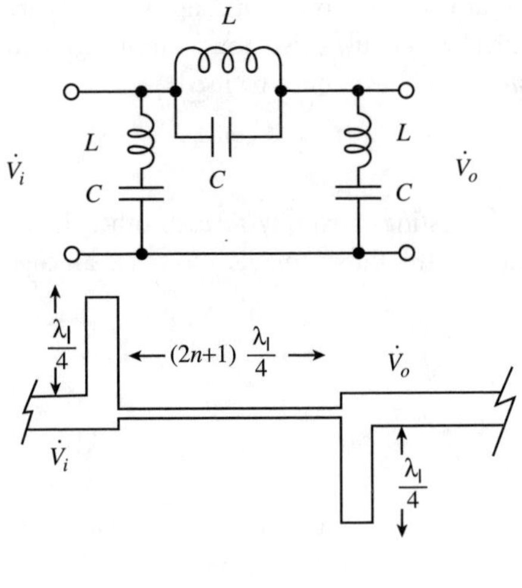

FIGURE 13.15 Synthesis of distributed π-network bandstop filters.

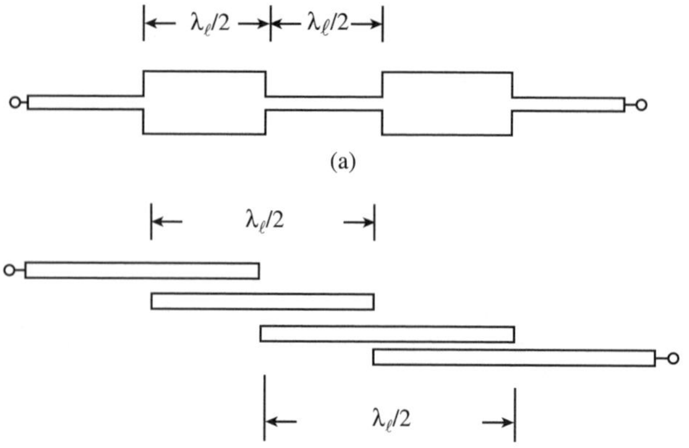

FIGURE 13.16 Microstrip line filters. (a) Bandstop high-low filter. (b) Bandpass coupled resonator filter.

Further Comments on Distributed Circuit Filters

If the unit sections of filters previously described are treated as lumped parameter filters, they can be cascaded into a number of stages to obtain desired filter characteristics, as exemplified by Chebyshev and Butterworth filters. The center frequencies of the stages can be staggered to obtain the desired frequency bandwidth [5].

A radio frequency choke is a special case of a bandstop filter. Cascading of alternate $\lambda_l/2$ sections of high and low impedance [Figure 13.16(a)] is common. This is called the high-low filter. A microstrip line bandpass filter can be made using a number of $\lambda_l/2$ microstrip line resonators, as shown in Figure 13.16(b) [5].

13.7 Synthesis of Couplers

Generic Relations

A coupler is a circuit with coupling between a primary circuit and a secondary circuit. In distributed circuit technology, it often takes the form of coupling between two transmission line circuits. The coupling may be done by simply placing the secondary transmission line in proximity to the primary transmission line, as presented in Figure 13.17(a) and (c), or both the primary and the secondary transmission lines are physically connected, as shown in Figure 13.17(b). These examples of distributed coupler circuits are shown in microstrip line forms. No lumped parameter equivalent circuit exists for these distributed coupler circuits. Therefore, the previously described synthesis techniques cannot be used.

Proximity Couplers

Coupling of two transmission lines is accomplished simply by existing proximity to each other. If the transmission lines are microstrip lines [Figure 13.17(a) or Figure 13.18], the coupler is also called an edge coupler.

The coupling factor (CF) of a coupler is defined as [3]

$$\mathrm{CF} = \frac{P_4}{P_1} \tag{13.95}$$

where P_1 is the microwave power fed to port 1 and P_4 is the microwave power output at port 4. If the power coupling factor per meter of the line edge is $k(x)$ in the coupling region of length l, then [3]

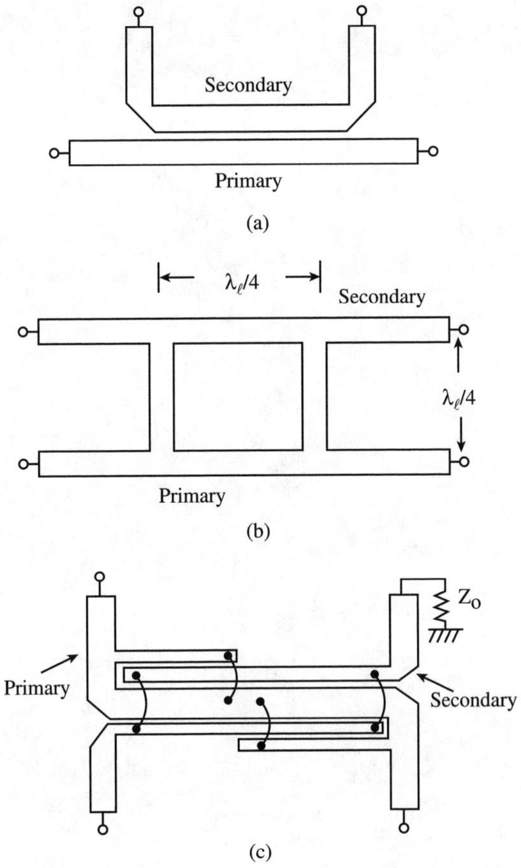

FIGURE 13.17 Distributed circuit couplers. (a) Proximity coupler. (b) Quarter-wavelength coupler. (c) Lange coupler.

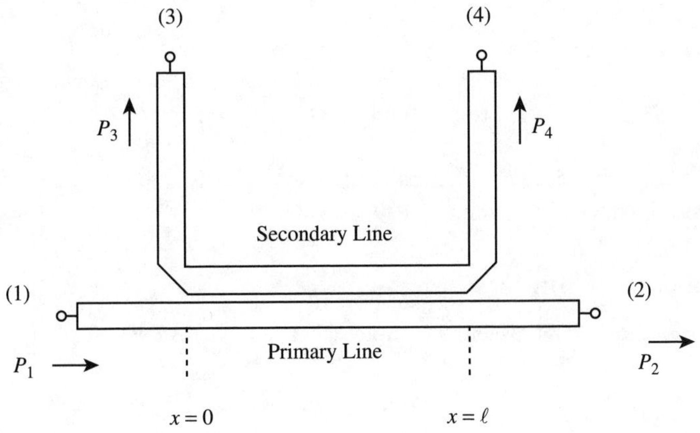

FIGURE 13.18 Microstrip line proximity coupler.

$$\mathrm{CF} = \left| \int_0^l k(x) \varepsilon^{-j\beta x} dx \right| \tag{13.96}$$

This can be expressed in decibels as

$$\mathrm{CF(dB)} = 10 \log \frac{P_4}{P_1} = 10 \log \left| \int_0^l k(x) \varepsilon^{-j\beta x} dx \right| \tag{13.97}$$

The directivity of a coupler is defined as [3]

$$\mathrm{dir} = \frac{P_3}{P_4} \tag{13.98}$$

$$\mathrm{dir} = \frac{1}{\mathrm{CF}} \left| \int_0^l k(x) \varepsilon^{-j2\beta x} dx \right| \tag{13.99}$$

The decibel expression of directivity, then, is

$$\begin{aligned} \mathrm{dir(dB)} &= 10 \log_{10} \frac{P_3}{P_4} \\ &= 10 \log_{10} \frac{1}{\mathrm{CF}} \left| \int_0^l k(x) \varepsilon^{-j2\beta x} dx \right| \end{aligned} \tag{13.100}$$

Generally, $P_3 \neq P_4$; therefore, the coupling has a directional property. This is the reason that this type of coupler is termed the directional coupler.

Insertion loss (IL) of a coupler is defined as

$$\mathrm{IL} = \frac{P_2}{P_1} = \frac{P_1 - (P_3 + P_4)}{P_1} \tag{13.101}$$

Inserting (13.95) and (13.98) into (13.101),

$$\mathrm{IL} = 1 - \mathrm{CF}(1 + \mathrm{dir}) \tag{13.102}$$

or

$$\mathrm{IL(dB)} = 10 \log\{1 - \mathrm{CF}(1 + \mathrm{dir})\} \tag{13.103}$$

A complete analytical synthesis of microstrip line couplers is not available. However, computer software based on semiempirical equations is commercially available [5].

Quarter-Wavelength Couplers

The coupling factor of a quarter-wavelength coupler (Figure 13.19) is defined as [3]

$$\mathrm{CF} = \frac{P_4}{P_1} \tag{13.104}$$

The amount of P_4 or coupling factor can be controlled by the width of microstrip lines connecting the secondary line to the primary line.

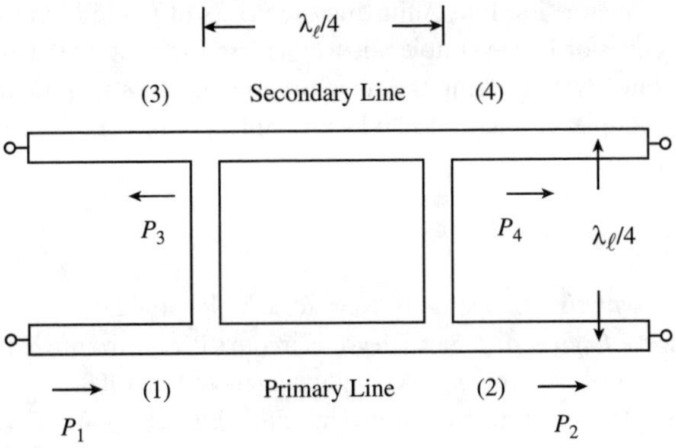

FIGURE 13.19 Quarter-wavelength coupler.

$$\mathrm{CF}[\mathrm{dB}] = 10\log_{10} P_4/P_1 \tag{13.105}$$

The directivity is

$$\mathrm{dir} = \frac{P_3}{P_4} \tag{13.106}$$

or

$$\mathrm{dir}[\mathrm{dB}] = 10\log\frac{P_3}{P_4} \tag{13.107}$$

At the precise design operating frequency, P_3 should be zero. The insertion loss for this coupler is

$$\mathrm{IL} = \frac{P_1 - (P_3 + P_4)}{P_1} = 1 - \mathrm{CF}(1 + \mathrm{dir}) \tag{13.108}$$

Lange Couplers

A schematic diagram of a generic Lange coupler [4] is given in Figure 13.20. In this figure, the primary transmission line is from port (1) to port (2). The secondary transmission line is from port (3) to port (4). In Lange couplers, both the primary and the secondary lines are coupled by edge coupling or proximity coupling. It should be noted that the transmission lines are not physically connected to each other by bonding wires. The bond wires on both ends form connections of the secondary transmission line from port (3) to port (4). The bond wires in the middle of the structure are the electrical potential equalizing

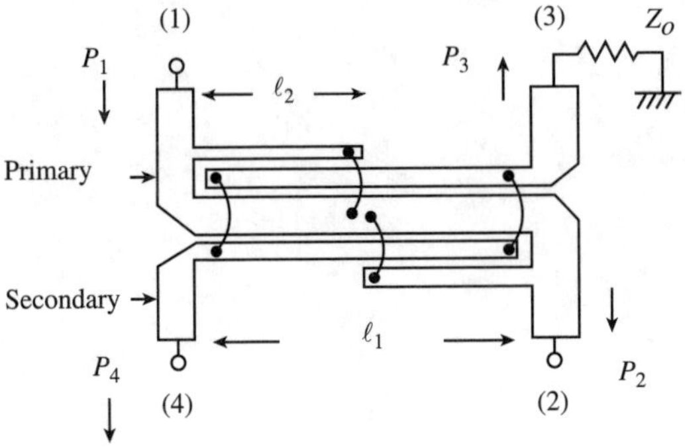

FIGURE 13.20 Lange coupler.

bonding in the primary transmission line. Adjusting the values of l_1 and l_2 creates a variety of coupling factors. One of the objectives of Lange couplers is to increase the coupling factor between the primary lines and the secondary line. By comparing the structure to that of a straight edge-coupled coupler (see Figure 13.18), the Lange coupler in Figure 13.20 has more length of coupling edge. Therefore, a higher coupling factor is obtainable [5].

References

[1] F. F. Kuo, *Network Analysis and Synthesis*, New York: Wiley, 1962.

[2] T. K. Ishii, *Microwave Engineering*, San Diego: Harcourt Brace Jovanovich, 1989.

[3] F. J. Tischer, *Mikrowellen-Messtechnik*, Berlin: Springer-Verlag, 1958.

[4] J. Lange, "Interdigitated stripline quadrature hybrid," *IEEE Trans. MTT*, vol. 17, no. 11, pp. 1150, 1151, Dec. 1969.

[5] V. F. Fusco, *Microwave Circuit*, Englewood Cliffs, NJ: Prentice Hall, 1987.

Index

A

B

C

D

E

F

G

H

I

J

K

L

M

N

T

U